TELEPEN
84 8740619 2
DILEWYD O STOC
WITHDRAWN FROM STOCK
CPC UCW
LLYFRGELL
LIBRARY
ABERYSTWYTH

AF598683

THE JACOBSON RADICAL OF GROUP ALGEBRAS

NORTH-HOLLAND MATHEMATICS STUDIES 135
Notas de Matemática (115)

Editor: Leopoldo Nachbin

Centro Brasileiro de Pesquisas Físicas
Rio de Janeiro
and University of Rochester

NORTH-HOLLAND – AMSTERDAM • NEW YORK • OXFORD • TOKYO

THE JACOBSON RADICAL OF GROUP ALGEBRAS

Gregory KARPILOVSKY

Department of Mathematics
University of the Witwatersrand
Johannesburg, South Africa

1987

NORTH-HOLLAND – AMSTERDAM • NEW YORK • OXFORD • TOKYO

ISBN: 0 444 70190 7

Publishers:

ELSEVIER SCIENCE PUBLISHERS B.V.
P.O. BOX 1991
1000 BZ AMSTERDAM
THE NETHERLANDS

Sole distributors for the U.S.A. and Canada:

ELSEVIER SCIENCE PUBLISHING COMPANY, INC.
52 VANDERBILT AVENUE
NEW YORK, N.Y. 10017
U.S.A.

Library of Congress Cataloging-in-Publication Data

Karpilovsky, Gregory, 1940-
The Jacobson radical of group algebras.

(North-Holland mathematics studies ; 135) (Notas de matemática ; 115)
Bibliography: p.
Includes index.
1. Group algebras. 2. Jacobson radical. 3. Modules (Algebra) I. Title. II. Series. III. Series: Notas de matemátics (Rio de Janeiro, Brazil) ; no. 115.
QA1.N86 no. 115 [QA171] 510 s [512'.24] 86-32918
ISBN 0-444-70190-7 (U.S.)

PRINTED IN THE NETHERLANDS

To the memory of my father

ILIA KARPILOVSKY

(1915-1942)

Preface

Let G be a finite group and let F be a field. It is well known that linear representations of G over F can be interpreted as modules over the group algebra FG. Thus the investigation of ring-theoretic structure of the Jacobson radical $J(FG)$ of FG is of fundamental importance. During the last two decades the subject has been pursued by a number of researchers and many interesting results have been obtained. The present book is intended to give a systematic account of this work but the treatment is by no means intended to be exhaustive. The author has tried to be fairly complete in what he considers as the main body of the theory, and the reader should get a considerable amount of knowledge of central ideas, the basic results, and the fundamental methods. We have tried to avoid making the discussion too technical. With this view in mind, maximum generality has not been achieved in those places where this would entail a loss of clarity or a lot of technicalities.

The present monograph is written on the assumption that the reader has had the equivalent of a standard first-year graduate algebra course. Thus we assume a familiarity with basic ring-theoretic and group-theoretic concepts and an understanding of elementary properties of modules, tensor products, and fields. For the convenience of the reader, a chapter on algebraic preliminaries is included. This chapter provides a survey of topics needed later in the book. There is a fairly large bibliography of works which are either directly relevant to the text or offer supplementary material of interest.

A word about notation. As is customary, Theorem 4.3.2 denotes the second result of Section 3 of Chapter 4; however, for simplicity, all references to this result within Chapter 4 itself are designated as Theorem 3.2.

A systematic description of the material is supplied by the introductions to individual chapters and therefore will not be repeated here.

I would like to express my gratitude to my wife for the encouragement she has given me in the preparation of this book. My thanks to P.Brockhaus, E.Formanek, S.Koshitani, B.Külshammer, K.Motose, Y.Ninomiya, Y.Tshushima and D.A.R.Wallace for sending me reprints and preprints of their work, and also to Lucy Rich for her excellent typing.

Gregory Karpilovsky

Contents

1
Ring-theoretic background

This chapter reviews some basic ring-theoretic facts and techniques as well as fixes conventions and notation for the rest of the book. Because we presuppose a familiarity with various elementary ring-theoretic terms, only a brief description of them is presented. Many readers may wish to glance briefly at the contents of this chapter, referring back to the relevant sections when they are needed later. Occasionally, results are stated somewhat more generally than necessary for later use.

1. NOTATION AND TERMINOLOGY.

All rings in this book are *associative* with $1 \neq 0$ and subrings of a ring R are assumed to have the same identity element as R. Each ring homomorphism will be assumed to preserve identity elements.

Let R be a ring. An element $x \in R$ is *nilpotent* if $x^n = 0$ for some positive integer n. An ideal J in R is *nil* if every element of J is nilpotent, while J is *nilpotent* if there is a positive integer n such that $J^n = 0$, where J^n is the product of J with itself n times. Two idempotents u,v of R are *orthogonal*, if $uv = vu = 0$. A nonzero idempotent is *primitive* if it cannot be written as a sum of two nonzero orthogonal idempotents. We denote by $Z(R)$ the centre of R. An idempotent e is called *centrally primitive* if e is a primitive idempotent of $Z(R)$. The ring of all $n \times n$-matrices over R is denoted by $M_n(R)$.

Let R be a ring. From this we construct a new ring R^o, called the *opposite ring* of R. Both the underlying set and the additive structure of R^o are just those of R. But the multiplication, denoted by $*$, is given by $x*y = yx$, $x,y \in R$. It is easy to verify that R^o is a ring with these operations. Clearly, $Z(R) = Z(R^o)$ and R is commutative if and only if $R = R^o$.

Let A be a ring, R a commutative ring, and $\lambda : R \to Z(A)$ a ring homomorphism. The resulting system (A,R,λ) is called an *R-algebra*. It will be convenient to suppress the λ and speak of A as an R-algebra or as an algebra over R. Thus A is an R-algebra (with respect to some λ) if and only if there is an ideal I of R with R/I isomorphic to a subring of $Z(R)$. Setting $ra = \lambda(r)a, r \in R, a \in A$, it follows that A is an R-module such that

$$r(xy) = x(ry) = (rx)y \qquad \text{for all } r \in R, x,y \in A$$

Conversely, if A is an R-module for which the above equalities hold, then the map $R \to Z(A), r \mapsto r\cdot 1$ is a ring homomorphism and thus A is an R-algebra. By a *homomorphism of R-algebras* we understand a ring homomorphism which is also a homomorphism of R-modules.

Let $(R_i), i \in I$ be a family of rings and let R be the direct product set $\prod_{i\in I} R_i$. We can define addition and multiplication on R by the rules

$$(x_i)+(y_i) = (x_i+y_i) \quad , \quad (x_i)(y_i) = (x_iy_i) \qquad (x_i, y_i \in R_i)$$

It is straightforward to verify that R is a ring; we shall refer to R as the *direct product of the family* $(R_i), i \in I$.

By an R-module we always understand a unital R-module; the symbols ${}_RM$ or M_R will be used to underline the fact that M is a left or a right R-module, respectively. We usually consider *left* R-modules, and, in this case, speak simply about an *R-module*.

Let V and W be R-modules. We denote by $\operatorname{Hom}_R(V,W)$ the additive abelian group of all R-homomorphisms $V \to W$. The same notation will be used in case V and W are right R-modules. In case $V = W$ we shall write $\operatorname{End}_R(V)$ instead of $\operatorname{Hom}_R(V,V)$. Note that $\operatorname{End}_R(V)$ is a ring under the multiplication

$$(fg)(v) = f(g(v)) \qquad (f,g \in \operatorname{End}_R(V), v \in V)$$

A nonzero R-module V is *irreducible* if 0 and V are the only submodules of V. The module V is *reducible* if $V \neq 0$ and V is not irreducible.

Let $(V_i), i \in I$, be a family of modules over the same ring R, and let V

be the direct product set $\prod_{i \in I} V_i$. Then under the operations

$$(x_i)+(y_i) = (x_i+y_i) \quad , \quad r(x_i) = (rx_i) \qquad (x_i, y_i \in V_i, r \in R)$$

V becomes an R-module. We shall refer to V as the *direct product* of the family $(V_i), i \in I$. The submodule of V consisting of all (x_i) with finitely many $x_i \neq 0$ is called the *external direct sum* (or simply *direct sum*) of the family of modules (V_i) and is denoted by $\bigoplus_{i \in I} V_i$. Assume that each V_i is a submodule of an R-module W. We say that W is an *(internal) direct sum* of the family (V_i) if each $w \in W$ can be written uniquely in the form $w = \sum_{i \in I} v_i$ with $v_i \in V_i$ for all $i \in I$ and with finitely many $v_i \neq 0$. Since external and internal direct sums of $(V_i), i \in I$, are isomorphic, we shall use the same symbol $\bigoplus_{i \in I} V_i$ to denote both of them. The precise situation will always be clear from the context. A nonzero R-module V is said to be *indecomposable* if for any two submodules X and Y of V, the equality $V = X \oplus Y$ implies $X = 0$ or $Y = 0$.

A partially ordered set in which any two elements have a supremum and an infimum is called a *lattice*. For any R-module V, the set of all submodules of V forms a lattice under the partial ordering by inclusion, with intersection as infimum and sum as supremum. For any three submodules V_1, V_2, V_3 of V the following general relation is called the *modular law* :

if $V_1 \subseteq V_3$, then $V_1 + (V_2 \cap V_3) = (V_1 + V_2) \cap V_3$

Let $\mathrm{Lat}(V)$ denote the lattice of all submodules of V. A map $f : \mathrm{Lat}(V) \rightarrow \mathrm{Lat}(W)$ is said to be a *lattice homomorphism* if $f(V_1+V_2) = f(V_1) + f(V_2)$ and $f(V_1 \cap V_2) = f(V_1) \cap f(V_2)$ for all $V_1, V_2 \in \mathrm{Lat}(V)$. In case f is also a bijection, we say that f is a lattice isomorphism. Note that an order-preserving bijection with an order-preserving inverse between $\mathrm{Lat}(V)$ and $\mathrm{Lat}(W)$ is always a lattice isomorphism.

A partially ordered set is said to be *inductive* if it is nonempty and every chain in it has an upper bound. According to Zorn's lemma, every partially ordered set which is inductive has a maximal element. Let V be an R-module. A submodule W of V is called *maximal* if W is a proper submodule (i.e. $W \neq V$) and W is not contained in any proper submodule of V.

An R-module V is said to *have an R-basis* $(v_i), i \in I$, if there exist elements $v_i \in V$ such that each $v \in V$ can be written as a finite sum $v = \sum r_i v_i$ with uniquely determined coefficients $r_i \in R$. Such modules V are called *free R-modules*.

Unless explicitely stated otherwise, all groups are assumed to be *multiplicative*. The multiplicative group of a field F is denoted by F^*. If X is a subset of G, $<X>$ will denote the subgroup of G generated by X, while $C_G(X)$ and $N_G(X)$ is the *centralizer* and *normalizer* of X in G, respectively. As usual, $Z(G)$ denotes the centre of G.

Let H be a subgroup of a group G. A subset of G containing just one element from each left coset xH is called a *left transversal* of H in G, and right transversals are defined correspondingly.

Let g be an element of a finite group G and let p be a prime. We say that g is a *p-element* if g has order a power of p; we say that g is a p'-element (or is *p-regular*) if its order is prime to p. Each $g \in G$ can be written in a unique way $g = g_p g_{p'}$, where g_p is a p-element, $g_{p'}$ is a p'-element, and g_p and $g_{p'}$ commute. The elements g_p and $g_{p'}$ are called the p and *p'-parts* of g, respectively. The symbols $C\ell(G)$ and $\mathrm{Syl}_p(G)$ denote respectively the sets of conjugacy classes and Sylow p-subgroups of G. A conjugacy class C of G is called *p-regular* (respectively, *p-singular*) if the order of the elements of C is not divisible by p (respectively, divisible by p).

Let X be an arbitrary set. An *action* of G on X is a homomorphism θ of G into the permutation group of X. When the action is understood, we shall suppress the symbol θ and instead write $x \mapsto {}^g x$ for $\theta(g), g \in G, x \in X$. The *stabilizer* $G(x)$ of $x \in X$ is defined by

$$G(x) = \{g \in G \mid {}^g x = x\}$$

and the *orbit* ${}^G x$ of $x \in X$ is defined by

$${}^G x = \{{}^g x \mid g \in G\}$$

We say that G acts *transitively* if ${}^{G}x = X$ for some (and hence all) $x \in X$. One readily verifies that $G(x)$ is a subgroup of G and that $|{}^{G}x| = |G : G(x)|$. Throughout the book, S_n and A_n denote symmetric and alternating groups of degree n, respectively, while $\mathbb{Z}_n$ is a cyclic group of order n.

2. ARTINIAN AND NOETHERIAN MODULES

An R-module V is said to be *noetherian* (respectively, *artinian*) if every ascending (respectively, descending) chain of submodules of V stops. We say that the ring R is *noetherian* (respectively, *artinian*) if the regular R-module ${}_{R}R$ is noetherian (respectively, artinian). The module V is called *finitely generated* if $V = \sum_{i=1}^{n} Rv_i$ for suitable $v_1, \ldots, v_n \in V$. We say that V is *finitely cogenerated* if for every family $(V_i), i \in I$ of submodules of V with $\bigcap_{i\in I} V_i = 0$, there exists a finite subset J of I such that $\bigcap_{j\in J} V_j = 0$.

2.1. PROPOSITION. For a module V the following statements are equivalent

(i) V is artinian

(ii) Every nonempty set of submodules of V has a minimal element

(iii) Every factor-module of V is finitely cogenerated.

Proof. (i) $\Rightarrow$ (ii): Let X be a nonempty subset of submodules of V and assume that X does not have a minimal element. Then for each $W \in X$ the set $\{W' \in X | W' \subset W\}$ is nonempty. Thus, by the axiom of choice, there is a function $W \mapsto W'$ with $W \supset W'$ for each $W \in X$. Fixing $W \in X$, we therefore obtain an infinite descending chain $W \supset W' \supset W'' \supset \ldots$ of submodules of V.

(ii) $\Rightarrow$ (iii): We need only show that if W is a submodule of V and $(V_i), i \in I$ is a family of submodules of V with $\bigcap_{i\in I} V_i = W$, then $W = \bigcap_{j\in J} V_j$ for some finite subset J of I. To this end, put

$$Y = \{\bigcap_{k\in K} V_k | K \subseteq I \text{ is finite}\}$$

Then, by hypothesis, Y has a minimal element, say $\bigcap_{j\in J} V_j$ with $J \subseteq I$ and J finite. Clearly, $W = \bigcap_{j\in J} V_j$.

(iii) $\Rightarrow$ (i): Assume that V has a descending chain

$$V_1 \supseteq V_2 \supseteq \ldots \supseteq V_n \supseteq \ldots$$

of submodules and put $W = \bigcap_{n \in \mathbb{N}} V_n$. By hypothesis, V/W is finitely cogenerated so there exists n with $W = V_n$. Thus $V_{n+i} = V_n$ for $i = 1,2,\ldots,$ as required. ■

2.2. COROLLARY. Let V be a nonzero artinian module. Then V has an irreducible submodule. ■

2.3. PROPOSITION. For a module V the following statements are equivalent:

(i) V is noetherian

(ii) Every nonempty set of submodules of V has a maximal element

(iii) Every submodule of V is finitely generated.

Proof. The proof of this proposition is dual to that of Proposition 2.1 and therefore will be omitted. ■

2.4. COROLLARY. Let V be a nonzero noetherian module. Then V has a maximal submodule. ■

Note that Corollary 2.4 is also a consequence of the following observation.

2.5. PROPOSITION. Let V be a finitely generated R-module. Then every proper submodule of V is contained in a maximal submodule. In particular, V has a maximal submodule if $V \neq 0$.

Proof. Let V be generated by $x_1,\ldots,x_r$ and let W be a proper submodule of V. Consider the set X of proper submodules of V which contain W. This is inductive: it contains W and if (W_λ) is a chain in X, then $\bigcup_\lambda W_\lambda$ is a submodule. If $V = \bigcup_\lambda W_\lambda$, then each $x_i \in W_{\lambda_i}$ for some λ_i. Let W_μ be the largest of the modules $W_{\lambda_1},\ldots,W_{\lambda_r}$. Then W_μ contains $x_1,\ldots,x_r$ and so $W_\mu = V$, a contradiction. Thus X is inductive and, by Zorn's lemma, it has a maximal element. This is a maximal submodule of V containing W, as we wished to show. ■

In particular, this result can be applied to the ring R itself. As left R-module, R is generated by the single element 1, and the submodules are left ideals, hence we obtain the

2.6. COROLLARY. Let R be an arbitrary ring. Then any proper left ideal of R

is contained in a maximal left ideal of R. In particular, R has a maximal left ideal. ■

2.7. PROPOSITION. Let

$$0 \longrightarrow U \longrightarrow V \longrightarrow W \longrightarrow 0$$

be an exact sequence of R-modules. Then V is artinian (noetherian) if and only if both U and W are artinian (noetherian).

Proof. Assume that V is artinian. Then since U is isomorphic to a submodule of V, U is artinian by the definition. Because every factor-module of W is isomorphic to a factor-module of V, W is artinian, again by the definition.

Conversely, suppose that both U and W are artinian. To prove that V is artinian, we may assume that $U \subseteq V$ and $W = V/U$. Let

$$V_1 \supseteq V_2 \supseteq \ldots \supseteq V_n \supseteq \ldots$$

be a descending chain of submodules of V. Because V/U is artinian, there is an integer m such that

$$V_m + U = V_{m+i} + U \qquad (i = 1,2,\ldots)$$

Since U is artinian, there is an integer $n \geqslant m$ such that

$$V_n \cap U = V_{n+i} \cap U \qquad (i = 1,2,\ldots)$$

Taking into account modularity and that $V_n \supseteq V_{n+i}$, we have for each $i = 1,2,\ldots$

$$\begin{aligned} V_n &= V_n \cap (V_n + U) = V_n \cap (V_{n+i} + U) = V_{n+i} + (V_n \cap U) \\ &= V_{n+i} + (V_{n+i} \cap U) = V_{n+i} \end{aligned}$$

Thus V is artinian. The proof of the noetherian case is dual. ■

2.8. COROLLARY. Let $V = V_1 \oplus \ldots \oplus V_n$. Then V is artinian (noetherian) if and only if each $V_i, i \in \{1,2,\ldots,n\}$, is artinian (noetherian). ■

2.9. PROPOSITION. Assume that the nonzero module V is either artinian or noetherian. Then V is the direct sum

$$V = V_1 \oplus \ldots \oplus V_n$$

of a finite set of indecomposable submodules.

Proof. For each nonzero module V that does not have a finite indecomposable decomposition choose a proper decomposition

$$V = V' \oplus X'$$

where V' has no finite indecomposable decomposition. Assume that V is nonzero and is not a finite direct sum of indecomposable modules. Then

$$V = V' \oplus X' \ , \ V' = V'' \oplus X'', \ldots,$$

is a sequence of proper decompositions. Therefore there exist infinite chains

$$X' \subset X' \oplus X'' \subset \ldots \quad \text{and} \quad V \supset V' \supset V'' \supset \ldots$$

proving that V is neither artinian nor noetherian. So the proposition is true. ■

Let V be a nonzero module. A finite chain of $n+1$ submodules of V

$$V = V_0 \supset V_1 \supset \ldots \supset V_n = 0$$

is called a *composition series of length* n for V provided that each V_{i-1}/V_i is irreducible $(i = 1,2,\ldots,n)$, i.e. provided each term in the chain is maximal in its predecessor. The irreducible modules V_0/V_1, $V_1/V_2, \ldots, V_{n-1}/V_n$ are called the *composition factors* of the series.

2.10. PROPOSITION. A nonzero module V has a composition series if and only if V is both artinian and noetherian.

Proof. Assume that V is both artinian and noetherian. Since any factor-module of V is artinian, it follows from Corollary 2.2. that there is an ascending chain

$$0 \subset V_1 \subset V_2 \subset \ldots$$

of submodules of V where each term in the chain is maximal in its successor.

Since V is noetherian, the chain terminates at V after a finitely many terms. Thus V has a composition series.

Conversely, suppose that V has a composition series. We shall use induction on n, where n is the minimum length of all such series. If $n = 1$, then V is irreducible and there is nothing to prove. Otherwise, if

$$V = V_0 \supset V_1 \supset \ldots \supset V_n = 0$$

is a composition series of minimal length for V, then V/V_1 is irreducible and V_1 has a composition series of length $n-1$. The desired conclusion is therefore a consequence of Proposition 2.7. ■

Assume that

$$V = V_0 \supset V_1 \supset \ldots \supset V_n = 0$$

and

$$V = W_0 \supset W_1 \supset \ldots \supset W_m = 0$$

are two composition series of V. Then these two series are called *equivalent* in case $n = m$ and there is a permutation σ of $\{1,2,\ldots,n\}$ such that

$$V_i/V_{i+1} \cong W_{\sigma(i)}/W_{\sigma(i)+1} \qquad (i = 1,2,\ldots,n)$$

2.11. PROPOSITION. (The Jordan-Hölder Theorem). If a module V has a composition series, then every pair of composition series for V are equivalent.

Proof. We use induction on $c(V)$, where $c(V)$ is the minimum length of composition series for V. The case $c(V) = 1$ being trivial, we may assume that $c(V) = n > 1$ and that any module with a composition series of smaller length has all of its composition series equivalent. Fix a composition series

(1) $\quad V = V_0 \supset V_1 \supset \ldots \supset V_n = 0$

of minimal length and choose any other composition series for

(2) $\quad V = W_0 \supset W_1 \supset \ldots \supset W_m = 0$

If $V_1 = W_1$, then by the induction hypothesis and in view of $c(V_1) \leqslant n-1$, (1) and (2) are equivalent. Hence we may assume that $V_1 \neq W_1$ in which case $V_1 + W_1 = V$ since V_1 is a maximal submodule of V. It follows that

(3) $V/V_1 \cong W_1/(V_1 \cap W_1)$ and $V/W_1 \cong V_1/(V_1 \cap W_1)$

whence $V_1 \cap W_1$ is maximal in both W_1 and V_1. By Proposition 2.10, $V_1 \cap W_1$ has a composition series

$$V_1 \cap W_1 = X_o \supset X_1 \supset \dots \supset X_k = 0$$

and therefore

$$V_1 \supset X_o \supset \dots \supset X_k = 0 \quad \text{and} \quad W_1 \supset X_o \supset \dots \supset X_k = 0$$

are composition series for V_1 and W_1. Because $c(V_1) < n$, every two composition series for V_1 are equivalent, so the two series

$$V = V_o \supset V_1 \supset V_2 \supset \dots \supset V_n = 0$$

and

$$V = V_o \supset V_1 \supset X_o \supset \dots \supset X_k = 0$$

are equivalent. It follows that $k < n - 1$ and $c(W_1) < n$, proving that every two composition series for W_1 are equivalent. The conclusion is that the two series

$$V = W_o \supset W_1 \supset \dots \supset W_m = 0$$

and

$$V = W_o \supset W_1 \supset X_o \supset \dots \supset X_k = 0$$

are equivalent. Since, by (3), $V/V_1 \cong W_1/X_o$ and $V/W_1 \cong V_1/X_o$, the series (1) and (2) are also equivalent, as required. ■

A module V that is both artinian and noetherian is said to be of *finite length*. In view of the Jordan-Hölder Theorem, for such a module V we can define its *(composition) length* $c(V)$ unambiguously by

$$c(V) = \begin{cases} 0 & \text{if } V = 0 \\ n & \text{if } V \text{ has a composition series of length } n. \end{cases}$$

2.12. PROPOSITION. Let ${}_RR = V_1 \oplus \dots \oplus V_n$ with $V_i \neq 0$ for $i = 1,2,\dots,n$, and write

$$1 = e_1 + \dots + e_n \qquad (e_i \in V_i)$$

Then $\{e_1,\dots,e_n\}$ is a set of pairwise orthogonal idempotents in R and $V_i = Re_i, 1 \leqslant i \leqslant n$. Conversely, if $\{e_1,\dots,e_n\}$ is a set of pairwise orthogonal idempotents in R, then

$$R(\sum_{i=1}^{n} e_i) = \bigoplus_{i=1}^{n} Re_i$$

Proof. Given $r \in R$, we have $r = r\cdot 1 = 1\cdot r = re_1 + \dots + re_n$. Thus in particular

$$e_i = e_ie_1 + \dots + e_ie_n$$

proving that $e_ie_j = \delta_{ij}e_i$ for all i,j. Moreover $Re_i \subseteq V_i$ and ${}_RR = Re_1 \oplus \dots \oplus Re_n$ which implies that $Re_i = V_i$ for all i.

Conversely, assume that $\{e_1,\dots,e_n\}$ is a set of pairwise orthogonal idempotents and put $e = \sum_{i=1}^{n} e_i$. Then $e^2 = e$ and $ee_i = e_ie = e_i$ for all i. Thus $Re = \sum_{i=1}^{n} Re_i$. If $\sum r_ie_i = 0$ then multiplication on the right by e_j implies that $r_je_j = 0$ for all j. Hence $Re = \bigoplus_{i=1}^{n} Re_i$, as required. ∎

2.13. COROLLARY. Assume that R is artinian or noetherian. Then there exists a set $\{e_1,\dots,e_n\}$ of pairwise orthogonal primitive idempotents of R with $1 = \sum_{i=1}^{n} e_i$. Furthermore

$${}_RR = Re_1 \oplus \dots \oplus Re_n$$

where each Re_i is an indecomposable module.

Proof. Direct consequence of Propositions 2.12 and 2.9. ∎

We shall refer to the R-modules $Re_1,\dots,Re_n$ in Corollary 2.13 as being the *principal indecomposable* R-modules.

3. COMPLETELY REDUCIBLE MODULES.

An R-module V is said to be *completely reducible* if every submodule of V is a direct summand of V.

3.1. PROPOSITION. Let V be completely reducible.

(i) Every homomorphic image of V is isomorphic to a submodule of V and every submodule of V is isomorphic to a homomorphic image of V.

(ii) Every submodule and every homomorphic image of V is completely reducible.

(iii) If $V \neq 0$ then V contains an irreducible submodule.

Proof. (i) If W is a submodule of V, then $V = W \oplus W'$ for some submodule W' of V. Then $W \cong V/W'$ and $V/W \cong W'$ as required.

(ii) Let W be a submodule of V and let $U = V/W$. By (i), it suffices to verify that U is completely reducible. So assume that $U_1 = V_1/W$ is a submodule of U. Choose V_2 with $V = V_1 \oplus V_2$. Then $U = U_1 \oplus (V_2 + W)/W$ and thus U is completely reducible.

(iii) Let v be a nonzero element of V. By Zorn's lemma, there exists a submodule W of V maximal with respect to the property that $v \notin W$. Write $V = W \oplus W'$ for some submodule W' of V. If W' is not irreducible, then $W' = W_1 \oplus W_2$ for some nonzero submodules W_1, W_2 of V. Because

$$(W \oplus W_1) \cap (W \oplus W_2) = W$$

it follows that $v \notin W \oplus W_i$ for either $i = 1$ or $i = 2$, contrary to the maximality of W. Hence W' is irreducible. ∎

The following alternative description of completely reducible modules is often useful.

3.2. PROPOSITION. For a nonzero module V the following statements are equivalent:

(i) V is completely reducible

(ii) V is the direct sum of irreducible submodules

(iii) V is the sum of irreducible submodules.

Proof. (i) $\Rightarrow$ (ii): Consider the collection of sets of irreducible submodules of V whose sum is direct. By Proposition 3.1(ii), it is nonempty and, by Zorn's lemma, there is a maximal element, say $\{V_i\}$, in this collection. Let $W = \oplus V_i$ and let $V = W \oplus W'$. If $W' \neq 0$ then by Proposition 3.1 (ii),(iii), W' contains an irreducible submodule V'. Thus $W + V' = V' \oplus (\oplus V_i)$, contrary to the maximality of $\{V_i\}$. Hence $W' = 0$ and $V = W$, as asserted.

(ii) $\Rightarrow$ (iii): Obvious

(iii) $\Rightarrow$ (i): Let W be a submodule of V. Owing to Zorn's lemma, there exists a submodule W' of V maximal with the property that $W \cap W' = 0$. Hence $W + W' = W \oplus W'$. We are therefore left to verify that $W \oplus W' = V$. Assume by way of contradiction that $W \oplus W' \neq V$ and choose $v \in V$ with $v \notin W \oplus W'$. By hypothesis, $v = v_1 + \ldots + v_n$ where $v_i \in V_i$ and V_i is an irreducible submodule of V for $i = 1,2,\ldots,n$. Hence $v_j \notin W \oplus W'$ for some j and so $V_j \cap (W \oplus W') \neq V_j$. Since V_j is irreducible, we conclude that $V_j \cap (W \oplus W') = 0$. Thus

$$W \oplus W' + V_j = W \oplus W' \oplus V_j$$

and so $W \cap (W' \oplus V_j) = 0$, contrary to the maximality of W'. ∎

3.3. COROLLARY. Let $V \neq 0$ be completely reducible. The following are equivalent

(i) V is artinian

(ii) V is noetherian

(iii) V is the direct sum of a finite number of irreducible submodules.

Proof. Direct consequence of Corollary 2.8 and Proposition 3.2. ∎

3.4. PROPOSITION. If ${}_RR$ is completely reducible, then every R-module is completely reducible.

Proof. Let V be an R-module and let $v \in V$. Then the map from ${}_RR$ to Rv which sends r to rv is a homomorphism. Thus Rv is completely reducible by Proposition 3.1(ii). Since $V \subseteq \sum_{v \in V} Rv$, the result follows by virtue of

Proposition 3.2. ■

As a preparation for the proof of the next result, we record the following observation.

3.5. LEMMA. Let $V = \oplus V_i$ where each V_i is irreducible, and let W be an irreducible submodule of V. Then $W \subseteq \oplus V_j$ where j ranges over those i for which $W \cong V_i$.

Proof. Choose nonzero $w \in W$ and write $w = \Sigma v_i$ with $v_i \in V_i$. If $v_i \neq 0$ then the map sending rw into rv_i for $r \in R$ is a nonzero homomorphism from $W = Rw$ to $V_i = Rv_i$. Since W, V_i are irreducible, $W \cong V_i$ as required. ■

A completely reducible R-module V is said to be *homogeneous* if it can be written as a sum of irreducible isomorphic modules. The sum of all irreducible submodules of V which are isomorphic to a given irreducible submodule of V is called a *homogeneous component* of V.

3.6. PROPOSITION. Let $V = \bigoplus_{i \in I} V_i$, where each V_i is irreducible and let $J \subseteq I$ be such that $V_j, j \in J$ are the representatives of all isomorphism classes of V_i, $i \in I$. For each $j \in J$, denote by X_j the sum of all V_i with $V_i \cong V_j$. Then

(i) The $X_j, j \in J$ are all homogeneous components of V

(ii) $V = \bigoplus_{j \in J} X_j$

Proof. (i) Apply Lemma 3.5.

(ii) Direct consequence of the definition of X_j. ■

Let V be any R-module. Then a submodule W of V is said to be *fully invariant* in V if it admits all R-endomorphisms of V.

3.7. LEMMA. Let $V \neq 0$ be a completely reducible R-module. A submodule W of V is fully invariant if and only if W is a sum of certain homogeneous components of V.

Proof. We keep the notation of Proposition 3.6. If $f \in \mathrm{End}_R(V)$ and $V_i \cong V_j$, then either $f(V_i) = 0$ or $f(V_i) \cong V_i \cong V_j$ since V_i is irreducible Thus $f(X_j) \subseteq X_j$, proving that each X_j, and hence the sum of certain X_j, is fully invariant. Conversely, assume that W is fully invariant. To prove that W is a sum of certain homogeneous components of V, we need only show: if P is an irreducible submodule of W and Q is an irreducible submodule of V such that $Q \cong P$, then $Q \subseteq W$. If $Q = P$, this is clear; otherwise $P \cap Q = 0$, so $P + Q = P \oplus Q$ and therefore $V = P \oplus Q \oplus U$ for some submodule U of V. Let $f : P \to Q$ be the given isomorphism and define an endomorphism ψ of V by

$$\psi(x+y+z) = f(x) + f^{-1}(y) + z \qquad (x \in P, y \in Q, z \in U)$$

Then $W \supseteq \psi(W) \supseteq \psi(P) = f(P) = Q$ as required. ■

3.8. COROLLARY. Let $V \neq 0$ be a completely reducible R-module and express V as the sum of its homogeneous components: $V = \bigoplus_{j \in J} X_j$. Then

$$\operatorname{End}_R(V) \cong \prod_{j \in J} \operatorname{End}_R(X_j)$$

Proof. Any family (f_j) of endomorphisms of X_j defines an endormorphism f of V and it is clear that the correspondence $(f_j) \longmapsto f$ is an injective homomorphism

$$\prod_{j \in J} \operatorname{End}(X_j) \longrightarrow \operatorname{End}_R(V)$$

Because each X_j is fully invariant (Lemma 3.7), any endomorphism f of V maps each X_j into itself. Setting $f_j = f|X_j, j \in J$, it follows that $(f_j) \mapsto f$ as required. ■

4. DIRECT DECOMPOSITION OF RINGS.

Let $(R_i), i \in I$ be a family of rings and let R be their direct product. Then the natural projections $\pi_i : R \to R_i$ are ring homomorphisms, while the injections $\lambda_i : R_i \to R$ preserve addition and multiplication, but not the identity element, and so are not ring homomorphisms. Of course the latter implies that the images of the R_i are not subrings, although they are ideals in R.

4.1. PROPOSITION. Let $R_1, R_2, \ldots, R_n$ be arbitrary rings

(i) If $R = R_1 \times \ldots \times R_n$, then there exists a decomposition of 1:

$$1 = e_1 + \dots + e_n$$

as a sum of pairwise orthogonal central idempotents such that

$$R = Re_1 \oplus \dots \oplus Re_n \quad \text{and} \quad Re_i \cong R_i \qquad (1 \leqslant i \leqslant n)$$

(ii) If $e_1,\dots,e_n$ are pairwise orthogonal central idempotents of a ring R with $1 = \sum_{i=1}^{n} e_i$, then $R = Re_1 \oplus \dots \oplus Re_n$ and $R \cong Re_1 \times Re_2 \times \dots \times Re_n$.

(iii) If $I_1,\dots,I_n$ are ideals in a ring R such that $R = I_1 \oplus \dots \oplus I_n$ and $1 = \sum_{i=1}^{n} e_i$ with $e_i \in I_i$, then $e_1,\dots,e_n$ are pairwise orthogonal central idempotents of R such that $I_i = Re_i, 1 \leqslant i \leqslant n$. In particular, $R \cong I_1 \times \dots \times I_n$.

Proof. (i) Let $\lambda_i : R_i \longrightarrow R$ be the natural injection, and let e_i be the image of the identity element of $R_i, 1 \leqslant i \leqslant n$. Then $e_1,\dots,e_n$ are pairwise orthogonal central idempotents of R with $1 = \sum_{i=1}^{n} e_i$. Hence $R = Re_1 \oplus \dots \oplus Re_n$. Since $Im\lambda_i = Re_i$ and e_i is the identity element of the ring Re_i, the required assertion follows.

(ii) The first statement is obvious. The map $R \longrightarrow Re_1 \times \dots \times Re_n$, $r \longmapsto (re_1,\dots,re_n)$ is obviously an injective homomorphism of rings. Since for any $r_1,\dots,r_n$ in R, $r_1e_1 + \dots + r_ne_n$ is mapped to $(r_1e_1,\dots,r_ne_n)$, the map is an isomorphism.

(iii) If $i \neq j$, then $e_ie_j \in I_i \cap I_j = 0$ and so $e_j = 1 \cdot e_j = \sum_{i=1}^{n} e_ie_j = e_j^2$. Hence, by Proposition 2.12, $R = Re_1 \oplus \dots \oplus Re_n = e_1R \oplus \dots \oplus e_nR$ and therefore $I_i = Re_i = e_iR, 1 \leqslant i \leqslant n$. It follows that $xe_i = e_ix = x$ for all $x \in I_i$, $1 \leqslant i \leqslant n$. Since $I_iI_j \subseteq I_i \cap I_j = 0$ for $i \neq j$, we deduce that e_i is a central idempotent, $1 \leqslant i \leqslant n$. The last statement is a consequence of (ii). ■

A ring R is called *indecomposable* if for any ideals X and Y of $R, R = X \oplus Y$ implies $X = 0$ or $Y = 0$.

4.2. COROLLARY. The following conditions are equivalent.

(i) R is indecomposable

(ii) $Z(R)$ is indecomposable

(iii) R is not isomorphic to a direct product of $n \geqslant 2$ rings

(iv) 0 and 1 are the only central idempotents of R

Proof. Direct consequence of Proposition 4.1. ∎

4.3. PROPOSITION. Assume that R is either artinian or noetherian. Then there exist ideals $I_1, I_2, \ldots, I_n$ of R such that $R = I_1 \oplus \ldots \oplus I_n$ and each I_i is an indecomposable ring.

Proof. Assume that R is noetherian. We may choose a direct decomposition $R = I_1 \oplus I_1'$ in which I_1' is (proper) maximal, in which case I_1 will be indecomposable. Next write $I_1' = I_2 \oplus I_2'$, where $I_2 \neq 0$ and I_2' is maximal. Continuing in this way, we obtain a sequence of decompositions

$$R = I_1 \oplus \ldots \oplus I_n \oplus I_n'$$

Because the associated sequence of ideals $I_1 \oplus \ldots \oplus I_n$ $(n = 1,2,\ldots)$ must terminate, we deduce that $I_n' = 0$ at some stage. In case R is artinian, we choose I_n minimal at each stage, then I_n' is a descending chain which must again stop. So in either case we obtain the desired decomposition. ∎

4.4. COROLLARY. Any artinian or noetherian ring is isomorphic to a direct product of finitely many indecomposable rings.

Proof. Apply Propositions 4.1 and 4.3. ∎

4.5. COROLLARY. Let $R_1, R_2, \ldots, R_n$ be arbitrary rings and let $R = R_1 \times \ldots \times R_n$. Then R is artinian (noetherian) if and only if each R_i is artinian (noetherian)

Proof. Direct consequence of Proposition 4.1 and Corollary 2.8. ∎

5. MATRIX RINGS.

Let R be an arbitrary ring and let $S = M_n(R)$. We shall identify R with its image in S consisting of all scalar matrices $\mathrm{diag}(r,\ldots,r), r \in R$. For every ordered pair $(i,j), 1 \leqslant i,j \leqslant n$, let e_{ij} be the matrix with (i,j)-th entry 1 and 0 elsewhere. The elements e_{ij}, called the *matrix units*, satisfy the following properties

(i) $e_{ij}e_{ks} = 0$ if $j \neq k$ and $e_{ij}e_{ks} = e_{ij}$ if $j = k$

(ii) $1 = e_{11} + \ldots + e_{nn}$

(iii) The centralizer of $\{e_{ij}\}$ in S is R.

Observe also that $R \cong e_{11} S e_{11}$. The following simple observation shows that (i), (ii) and (iii) determine S up to isomorphism.

5.1. PROPOSITION. Let S be a ring that contains a set of elements $\{v_{ij} \mid 1 \leqslant i,j \leqslant n\}$ satisfying (i) and (ii) and let R be the centralizer of this set in S. Then the map

$$\begin{cases} M_n(R) \xrightarrow{\psi} S \\ (a_{ij}) \longmapsto \sum_{i,j} a_{ij} v_{ij} \end{cases}$$

is an isomorphism of rings and R-modules. Furthermore, $R \cong v_{11} S v_{11}$.

Proof. We know that $M_n(R)$ is a free R-module with the matrix units e_{ij} as a basis. Hence $e_{ij} \mapsto v_{ij}$ extends to a homomorphism of R-modules, which clearly coincides with ψ. In view of (i), (ii) and (iii), ψ is easily seen to be a ring homomorphism.

Assume that $\sum_{i,j} a_{ij} v_{ij} = 0$ and let $k,s \in \{1,2,\dots,n\}$. Then for all $t \in \{1,\dots,n\}$

$$0 = v_{tk} \left(\sum_{i,j} a_{ij} v_{ij}\right) v_{st} = a_{ks} v_{tt}$$

and so

$$0 = \sum_t a_{ks} v_{tt} = a_{ks} \left(\sum_t v_{tt}\right) = a_{ks}$$

Hence $(a_{ij}) = 0$ and thus ψ is injective.

Finally, fix $s \in S$ and for each i,j set

$$r_{ij} = \sum_k v_{ki} s v_{jk}$$

Then for all v_{rt} we have

$$v_{rt} r_{ij} = \sum_k v_{rt} v_{ki} s\, v_{jk} = v_{ri} s\, v_{jt}$$

and

$$r_{ij} v_{rt} = \sum_k v_{ki} s\, v_{jk}\, v_{rt} = v_{ri} s\, v_{jt}$$

Thus r_{ij} commutes with all v_{rt} and therefore $r_{ij} \in R$. Furthermore, by the foregoing

$$\psi((r_{ij})) = \sum_{i,j} r_{ij}v_{ij} = \sum_{i,j} v_{ii}s\, v_{jj}$$

$$= (\sum_i v_{ii})s(\sum_j v_{jj}) = s,$$

proving that ψ is surjective. Since $\psi(e_{11} M_n(R)e_{11}) = v_{11}Sv_{11}$ and since $e_{11}M_n(R)e_{11} \cong R$, the result follows. ∎

Given a (left) R-module V and a positive integer n, we write V^n for n-th direct power of V. The following result shows that matrix rings arise as endomorphism rings of direct powers of modules.

5.2. PROPOSITION. $\operatorname{End}_R(V^n) \cong M_n(\operatorname{End}_R(V))$

Proof. Given $i,j \in \{1,2,\ldots,n\}$, define $\varepsilon_{ij} \in \operatorname{End}_R(V^n)$ by

$$\varepsilon_{ij}(v_1,\ldots,v_n) = (0,\ldots,\underset{\leftarrow i \rightarrow}{v_j},0,\ldots,0)$$

Then $\varepsilon_{ij}\varepsilon_{ks} = \delta_{jk}\varepsilon_{is}$ and $\sum\varepsilon_{ii} = 1$. Let $\psi \in \operatorname{End}_R(V^n)$ and write $\psi(v_1,\ldots,v_n) = (\psi_1(v_1),\ldots,\psi_n(v_n))$. Then $\psi\varepsilon_{ij} = \varepsilon_{ij}\psi$ for all i,j if and only if $\psi_1 = \psi_2 = \ldots = \psi_n$. Hence the centralizer of $\{\varepsilon_{ij}\}$ is identifiable with $\operatorname{End}_R(V)$. Now apply Proposition 5.1. ∎

In the preceding discussion we treated V^n as an R-module. But we can also consider V^n as an $M_n(R)$-module in the following natural way. We visualize the elements of V^n as column vectors and, for each $A \in M_n(R)$ and $x \in V^n$, define Ax as the matrix multiplication. With this convention, we now prove

5.3. PROPOSITION. (i) If W is a submodule of V, then the map $W \mapsto W^n$ provides an isomorphism from the lattice of submodules of V onto the lattice of submodules of V^n

(ii) $\operatorname{End}_{M_n(R)}(V^n) \cong \operatorname{End}_R(V)$

(iii) The map $V \mapsto V^n$ induces a bijective correspondence between the isomorphism classes of R and $M_n(R)$-modules. The inverse of this correspondence is given by $W \mapsto e_{11}W$.

Proof. (i) The correspondence $W \overset{f}{\mapsto} W^n$ is clearly order-preserving. Therefore we need only show that it has an inverse which is also order-preserving. Consider the projection on the first factor $\pi_1 : V^n \longrightarrow V$ and, for any submodule

X of V^n, put $g(X) = \pi_1(X)$. Then the correspondence $X \mapsto g(X)$ is order preserving. It is evident that $(gf)(W) = W$. From the action of $M_n(R)$ on V^n we see that X can be written in the form $X = V_1^n$ for a submodule V_1 of V. Hence $(fg)(X) = f(V_1) = V_1^n = X$, proving the assertion.

(ii) Put $S = M_n(R)$ and fix $f \in \mathrm{End}_S(V^n)$. Then, by the nature of action of S on V^n, f has the same projections, say λ_f, on all factors. Conversely, any $\psi \in \mathrm{End}_R(V)$ determines an element of $\mathrm{End}_S(V^n)$ whose projections on all factors are equal to ψ. Hence the map $f \mapsto \lambda_f$ provides the desired isomorphism.

(iii) It suffices to prove that $e_{11}V^n \cong V$ and $(e_{11}W)^n \cong W$, where V and W are R and $M_n(R)$-modules, respectively. By the definition of e_{11}, $e_{11}V^n = V \times 0 \times \ldots \times 0 \cong V$. Note also that $W = e_{11}W \oplus \ldots \oplus e_{n1}W$ as R-modules. Since $e_{k1}w = 0$ if and only if $e_{11}w = 0, 1 \leqslant k \leqslant n, w \in W$, the map $e_{11}w_1 + \ldots + e_{n1}w_n \overset{f}{\longmapsto} (e_{11}w_1, \ldots, e_{11}w_n)^T$ is well defined and is at least an R-isomorphism of W onto $(e_{11}W)^n$. We are therefore left to verify that $f(e_{ij}e_{k1}w) = e_{ij}f(e_{k1}w)$ for all $w \in W$ and all $i,j,k \in \{1,2,\ldots,n\}$. The latter being a consequence of the action of $M_n(R)$ on $(e_{11}W)^n$, the result follows. ■

Given an ideal I of R, we write

$$M_n(I) = \{(a_{ij}) \in M_n(R) \mid a_{ij} \in I \text{ for all } i,j \in \{1,\ldots,n\}\}$$

5.4. PROPOSITION. (i) The map $I \longrightarrow M_n(I)$ is a bijection between the sets of ideals of R and $M_n(R)$.

(ii) $M_n(R)/M_n(I) \cong M_n(R/I)$

Proof. (i) It is clear that $M_n(I)$ is an ideal of $M_n(R)$ and that the given map is injective. Let J be an ideal of $M_n(R)$ and $I \subseteq R$ consist of all entries of elements in J. Then I is an ideal of R such that $J = M_n(I)$.

(ii) The natural map $R \longrightarrow R/I$ induces a surjective homomorphism $M_n(R) \longrightarrow M_n(R/I)$ whose kernel is $M_n(I)$. ■

For future use, we next record

5.5. LEMMA. Let V be an R-module and let e be an idempotent of R. Then

$$\operatorname{Hom}_R(Re,V) \cong eV \qquad \text{as additive groups}$$

Similarly, if V is right R-module, then $\operatorname{Hom}_R(eR,V) \cong Ve$

Proof. If $f \in \operatorname{Hom}_R(Re,V)$, then $ef(e) = f(e^2) = f(e) \in eV$. Hence the map $f \mapsto f(e)$ is a homomorphism from $\operatorname{Hom}_R(Re,V)$ to eV. Conversely, if $v \in eV$ then the mapping $g_v : xe \mapsto xv$ is an R-homomorphism from Re to V. Since the map $v \mapsto g_v$ is an inverse of $f \mapsto f(e)$, the first isomorphism follows. The second isomorphism is proved by a similar argument. ■

5.6. PROPOSITION. Let e be an idempotent of a ring R. Then

$$\operatorname{End}_R(Re) \cong (eRe)^{\circ} \quad \text{and} \quad \operatorname{End}_R(eR) \cong eRe \qquad \text{as rings}$$

In particular

$$\operatorname{End}_R({}_RR) \cong R^{\circ} \quad \text{and} \quad \operatorname{End}_R(R_R) \cong R$$

Proof. Applying Lemma 5.5 for $V = Re$, we see that the map $f \mapsto f(e)$ is an isomorphism of the additive group of $\operatorname{End}_R(Re)$ onto the additive group of eRe. Given $f,g \in \operatorname{End}_R(Re)$, write $f(e) = er_1e$ and $g(e) = er_2e$ for some $r_1,r_2 \in R$. Then

$$(fg)(e) = f(er_2e) = er_2er_1e = (er_2e)(er_1e) = g(e)f(e),$$

proving that $f \mapsto f(e)$ reverses the multiplication. Since e is the identity element of the ring eRe, the above map preserves identity elements. This proves the first isomorphism. The second isomorphism is proved by applying Lemma 5.5 for $V = eR$. ■

5.7. LEMMA. (Schur's lemma). Let V be an irreducible module. Then $\operatorname{End}_R(V)$ is a division ring.

Proof. Let $f : V \to V$ be a nonzero R-homomorphism. Since $f(V)$ is a nonzero submodule of V, $f(V) = V$. On the other hand, since $f \neq 0$, $\operatorname{Ker} f \neq V$ and thus $\operatorname{Ker} f = 0$. Hence f is an isomorphism, proving the assertion. ■

We are now ready to prove

5.8. PROPOSITION. Assume that ${}_RR$ is completely reducible. Then R is artinian and there exist primitive idempotents $e_1, e_2, \ldots, e_r$ of R and positive integers $n_1, n_2, \ldots, n_r$ such that

$$R \cong \prod_{i=1}^{r} M_{n_i}(e_iRe_i)$$

and each e_iRe_i is a division ring.

Proof. Since 1 lies in the sum of finitely many irreducible submodules of ${}_RR$, the same is true for ${}_RR$. Hence R is artinian, so by Corollary 2.13, there exists a set $\{e_1, \ldots, e_n\}$ of pairwise orthogonal primitive idempotents of R with ${}_RR = Re_1 \oplus \ldots \oplus Re_n$, where each Re_i is indecomposable and hence irreducible. We may assume that $Re_1, Re_2, \ldots, Re_r$ are all nonisomorphic among the $Re_i, 1 \leqslant i \leqslant n$. If X_i is a homogeneous component of ${}_RR$ corresponding to Re_i, then ${}_RR = X_1 \oplus \ldots \oplus X_r$ (Proposition 3.7(ii)) and $X_i \cong (Re_i)^{n_i}$ for some positive integer $n_i, 1 \leqslant i \leqslant r$. Applying Propositions 5.6, 5.2 and Corollary 3.9, we derive

$$R^{\circ} \cong \prod_{i=1}^{r} \operatorname{End}_R(X_i) \cong \prod_{i=1}^{r} M_{n_i}(\operatorname{End}_R(Re_i)) \cong \prod_{i=1}^{r} M_{n_i}[(e_iRe_i)^{\circ}] \cong \left(\prod_{i=1}^{r} M_{n_i}(e_iRe_i)\right)^{\circ}$$

and therefore $R \cong \prod_{i=1}^{r} M_{n_i}(e_iRe_i)$. The fact that e_iRe_i is a division ring being a consequence of Lemma 5.7 and Proposition 5.6, the result follows. ■

6. THE RADICAL AND SOCLE OF MODULES AND RINGS.

Let V be a module over an arbitrary ring R. Then the *radical* of V, written $J(V)$, is defined to be the intersection of all maximal submodules of V. In case V contains no maximal submodule we set $J(V) = V$. The *Jacobson radical* $J(R)$ of R is defined by

$$J(R) = J({}_RR)$$

Expressed otherwise, $J(R)$ is the intersection of all maximal left ideals of R. We say that R is *semisimple* if $J(R) = 0$.

The sum of irreducible submodules of V, written $\operatorname{Soc}V$, is called the *socle* of V. If V contains no irreducible submodules, then we set $\operatorname{Soc}V = 0$. The *socle* of R, written $\operatorname{Soc}R$, is defined by $\operatorname{Soc}R = \operatorname{Soc}({}_RR)$.

6.1. PROPOSITION. (i) If $V \neq 0$ is a finitely generated R-module, then $J(V) \neq V$.

(ii) If $V \neq 0$ is an artinian module, then $\mathrm{Soc}(V) \neq 0$

(iii) If V is completely reducible, then $J(V) = 0$

(iv) V is completely reducible if and only if $\mathrm{Soc}V = V$

(v) $\mathrm{Soc}V$ is the unique largest completely reducible submodule of V.

Proof. (i) By Proposition 2.5, V has a maximal submodule and so $J(V) \neq V$.
(ii) By Proposition 2.1, V has an irreducible submodule and thus $\mathrm{Soc}(V) \neq 0$.
(iii) We may clearly assume that $V \neq 0$, in which case V is the direct sum of irreducible submodules (Proposition 3.2). Hence, by the definition of $J(V)$, $J(V) = 0$.
(iv) Again we may assume that $V \neq 0$. Now apply Proposition 3.2.
(v) Direct consequence of Proposition 3.2. ■

6.2. PROPOSITION. If $V \neq 0$ is an R-module, then the following conditions are equivalent:

(i) V is a direct sum of finitely many irreducible modules

(ii) V is artinian and completely reducible

(iii) V is artinian and $J(V) = 0$

Proof. The implication (i) $\Rightarrow$ (ii) is a consequence of Proposition 3.2 and Corollary 2.8, while the implication (ii) $\Rightarrow$ (iii) follows from Proposition 6.1(iii). Assume that (iii) holds. By Proposition 2.1, V is finitely cogenerated. Since $J(V) = 0$, this implies that $\bigcap_{i=1}^{n} V_i = 0$ for some maximal submodules $V_1, \ldots, V_n$ of V. Hence V is isomorphic to a submodule of $\prod_{i=1}^{n} (V/V_i)$ and so is completely reducible, by Proposition 3.1(ii). Now apply Corollary 3.3. ■

6.3. PROPOSITION. Let $f : W \longrightarrow V$ be a homomorphism of R-modules. Then

(i) $f(J(W)) \subseteq J(V)$ with equality if f is surjective and $\mathrm{Ker}f \subseteq J(W)$

(ii) $f(\mathrm{Soc}W) \subseteq \mathrm{Soc}V$

Proof. (i) If V has no maximal submodules, then $J(V) = V$ and hence $f(J(W)) \subseteq J(V)$. Assume that M is a maximal submodule of V. Then the map

$$f^* : W \longrightarrow V/M$$

given by

$$f^*(w) = f(w) + M$$

is an R-homomorphism. Since $\mathrm{Ker}f^*$ is a maximal submodule of $W, J(W) \subseteq \mathrm{Ker}f^*$ and therefore $f(J(W)) \subseteq M$. Thus $f(J(W)) \subseteq J(V)$.

Assume that f is surjective and $\mathrm{Ker}f \subseteq J(W)$. If W has no maximal submodules, then so does V, in which case

$$f(J(W)) = f(W) = V = J(V)$$

Assume that the set $\{M_i | i \in I\}$ of all maximal submodules of W is nonempty. By hypothesis, $\mathrm{Ker}f \subseteq M_i$ for all $i \in I$. If X and Y are the lattices of submodules of W containing $\mathrm{Ker}f$ and submodules of V, respectively, the map $U \mapsto f(U), U \in X$, is an isomorphism of X onto Y. Thus $\{f(M_i) | i \in I\}$ is the set of all maximal submodules of V and

$$J(V) = \bigcap_{i\in I} f(M_i) = f(\bigcap_{i\in I} M_i) = f(J(W))$$

as required.

(ii) Direct consequence of Propositions 3.1(ii) and 6.1(v). ■

6.4. COROLLARY. Let V be an R-module and let W be a submodule of V.

(i) $J(W) \subseteq J(V)$

(ii) $J(V/W) \supseteq (J(V)+W)/W$

(iii) If $W \subseteq J(V)$, then $J(V/W) = J(V)/W$

(iv) $\mathrm{Soc}W = W \cap \mathrm{Soc}V$

Proof. (i) Direct consequence of Proposition 6.3(i) applied for the case where f is the inclusion map

(ii) and (iii) Apply Proposition 6.3(i) for the case where $f : V \longrightarrow V/W$ is the natural homomorphism

(iv) By Proposition 6.3(ii), applied for the case where f is the inclusion map, we have $\mathrm{Soc}W \subseteq \mathrm{Soc}V$ and hence $\mathrm{Soc}W \subseteq W \cap \mathrm{Soc}V$. The opposite inclusion being a consequence of Proposition 3.2(ii), and 6.1(v), the result follows. ■

6.5. COROLLARY. Let V be an R-module. Then $J(V/J(V)) = 0$ and $J(V) \subseteq W$ for any submodule W of V for which $J(V/W) = 0$.

Proof. The first equality is a consequence of Corollary 6.4(iii), while the second follows from Corollary 6.4(ii). ■

6.6. COROLLARY. Let V be an R-module and let W be a submodule of V.

(i) $W = J(V)$ if and only if $W \subseteq J(V)$ and $J(V/W) = 0$

(ii) $W = \text{Soc}V$ if and only if $W \supseteq \text{Soc}V$ and $\text{Soc}W = W$

Proof. (i) Direct consequence of Corollary 6.5

(ii) Direct consequence of Proposition 6.1(v). ■

6.7. PROPOSITION. Let $(V_i), i \in I$ be a family of R-modules. Then

(i) $J(\prod_{i\in I} V_i) \subseteq \prod_{i\in I} J(V_i)$

(ii) $J(\bigoplus_{i\in I} V_i) = \bigoplus_{i\in I} J(V_i)$

(iii) $\text{Soc}(\bigoplus_{i\in I} V_i) = \bigoplus_{i\in I} \text{Soc}V_i$

Proof. (i) We may clearly assume that each V_i has a maximal submodule M_i. Then the product $M_i \times \prod_{j\neq i} V_j$ is a maximal submodule in $V = \prod_{i\in I} V_i$. Hence $J(V)$ is contained in each product $J(V_i) \times \prod_{j\neq i} V_j$ and the intersection of these is $\prod_{i\in I} J(V_i)$, whence the assertion.

(ii) By the argument of (i), $M_i \oplus (\bigoplus_{j\neq i} V_j)$ is a maximal submodule in $S = \bigoplus_{i\in I} V_i$ so that $J(S) \subseteq \bigoplus_{i\in I} J(V_i)$. But, by Proposition 6.4(i), $J(V_i) \subseteq J(S)$ for each $i \in I$. Hence the opposite containment is also true and the assertion follows.

(iii) We may clearly assume that $V = \bigoplus_{i\in I} V_i$ has an irreducible submodule W. If $\pi_i : W \longrightarrow V_i$ is the projection map, then $W \subseteq \bigoplus_{i\in I} \pi_i(W)$. Now each $\pi_i(W)$ is either 0 or irreducible. Thus $\pi_i(W) \subseteq \text{Soc}V_i$ and therefore $\text{Soc}V \subseteq \bigoplus_{i\in I} \text{Soc}V_i$. The opposite containment being a consequence of Proposition 6.1(v), the result follows. ■

6.8. COROLLARY. For any R-module V, $J(V) = 0$ if and only if V is isomorphic to a submodule of the direct product of a family of irreducible R-modules.

Proof. Let $(V_i), i \in I$ be a family of irreducible R-modules. By Proposition 6.7(i), $J(\prod_{i\in I} V_i) = 0$. Hence the "if" part is a consequence of Corollary 6.4. Conversely, assume that $J(V) = 0$. If $V = 0$ then there is nothing to prove. Hence we may assume that $V \neq 0$, in which case there is a nonempty family $(W_k), k \in K$, of maximal submodules of V with $\bigcap_{k\in K} W_k = 0$. Thus V admits an injective homomorphism into $\prod_{k\in K}(V/W_k)$. Since each V/W_k is irreducible, the result follows. ■

A submodule W of V is called *essential* in V in case for every submodule X of V,

$$W \cap X = 0 \text{ implies } X = 0$$

Dually, a submodule W of V is called *superfluous* in V in case for every submodule X of V

$$W + X = V \text{ implies } X = V$$

6.9. PROPOSITION. Let V be an R-module.

(i) $J(V)$ is the sum of all superfluous submodules in V. Furthermore, if V is finitely generated, then $J(V)$ is the unique largest superfluous submodule of V.

(ii) Soc$V$ is the intersection of all essential submodules in V. Furthermore, if V is artinian, then Soc$V$ is the unique smallest essential submodule of V.

Proof. (i) Let W be a superfluous submodule in V. To prove that $W \subseteq J(V)$, we may assume that V has a maximal submodule K. If $W \nsubseteq K$, then $K + W = V$ and hence $K = V$, a contradiction. Thus every superfluous submodule of V is contained in $J(V)$. On the other hand, let $x \in V$. If W is a submodule of V with $Rx + W = V$, then either $W = V$ or there is a maximal submodule K of V with $W \subseteq K$ and $x \notin K$. If $x \in J(V)$, then the latter cannot occur. Thus $x \in J(V)$ implies that Rx is a superfluous submodule of V, proving the first statement.

Now assume that V is finitely generated. By the foregoing, we need only verify that $J(V)$ is a superfluous submodule of V. So assume that W is a submodule of V with $V = W + J(V)$. Then, by Corollary 6.4(ii), $J(V/W) = V/W$. But V/W is finitely generated, so by Proposition 6.1(i) $V/W = 0$. Thus $W = V$ as required.

(ii) Let M be the intersection of all essential submodules in V. To prove that $\operatorname{Soc}V \subseteq M$, we may assume that V has an irreducible submodule W. If L is an essential submodule of V, then $L \cap W \neq 0$ and so $W \subseteq L$. Thus $\operatorname{Soc}V \subseteq M$ and therefore we need only verify that M is completely reducible. So fix a submodule N of M. If N' is a submodule of V maximal with respect to $N \cap N' = 0$, then $N + N' = N \oplus N'$ is an essential submodule of V. But then $N \subseteq M \subseteq N \oplus N'$, and by modularity

$$M = M \cap (N \oplus N') = N \oplus (M \cap N')$$

This N is a direct summand of M and therefore $\operatorname{Soc}V = M$.

Now assume that V is artinian. We need only verify that $\operatorname{Soc}V$ is an essential submodule of V. Assume that $W \cap \operatorname{Soc}V = 0$ for some nonzero submodule W of V. Since V is artinian, so is W and hence, by Proposition 2.1, W has an irreducible submodule X. But then $X \oplus \operatorname{Soc}V$ is a completely reducible module strictly containing $\operatorname{Soc}V$, contrary to Proposition 6.1(v). Hence $\operatorname{Soc}V$ is essential in V, as required. ■

6.10. PROPOSITION. Let V be an R-module generated by $v_1,\ldots,v_n$, and let $v \in V$. Then $v \in J(V)$ if and only if for all $r_i \in R$, $1 \leqslant i \leqslant n$, the elements $v_i + r_i v$, $1 \leqslant i \leqslant n$, generate V.

Proof. Let $v \in J(V)$ and let W be the submodule of V generated by the elements $v_i + r_i v$, $1 \leqslant i \leqslant n$. It is clear that $W + J(V) = V$ so that $W = V$, by Proposition 6.9(i).

Conversely, assume that $v \notin J(V)$. Then there is a maximal submodule W of V such that $v \notin W$. Hence we must have $Rv + W = V$. It follows that for each i there exists $r_i \in R$ such that $v_i \in -r_i v + W$, whence $v_i + r_i v \in W$ and thus the elements $v_i + r_i v$, $1 \leqslant i \leqslant n$, do not generate V. ■

We now turn our attention to the radical and socle of rings. Let V be an R-module. The *annihilator* of V, written $\operatorname{ann}(V)$, is defined by

$$\operatorname{ann}(V) = \{r \in R \mid rV = 0\}$$

It is clear that $\operatorname{ann}(V)$ is an ideal and that V may be viewed as an $R/\operatorname{ann}(V)$-

module. We say that V is *faithful* if $\operatorname{ann}(V) = 0$.

An ideal I of R is called *primitive* if the ring R/I has a faithful irreducible module. Clearly I is primitive if and only if I is the annihilator of an irreducible R-module.

6.11. LEMMA. Let V be an R-module. Then V is irreducible if and only if $V \cong R/X$ for some maximal left ideal X of R.

Proof. Clearly R/X is irreducible. Suppose that V is irreducible and let v be a nonzero element of V. Then Rv is a nonzero submodule of V and so $V = Rv$. The map $R \to V$, $r \mapsto rv$ is a surjective R-homomorphism. Hence $R/X \cong V$ for some left ideal X of R. Since V is irreducible, X is maximal. ■

6.12. LEMMA. Every maximal left ideal contains a primitive ideal and every primitive ideal is the intersection of the maximal left ideals containing it.

Proof. Let X be a maximal left ideal of R. By Lemma 6.11, R/X is an irreducible R-module. Hence the annihilator of R/X is a primitive ideal contained in X.

Let I be a primitive ideal in R and let V be an irreducible R-module whose annihilator is I. Given a nonzero v in V, put $M_v = \{r \in R \mid rv = 0\}$. Clearly M_v is a left ideal in R. Since V is irreducible, $V = Rv \cong R/M_v$ and so M_v is a maximal left ideal in R. Since I is the intersection of all M_v with $0 \neq v \in V$, the result follows. ■

6.13. PROPOSITION. Let R be a ring.

(i) $J(R)$ is the intersection of the annihilators of irreducible R-modules. In particular, $J(R)$ is an ideal of R.

(ii) $J(R)$ is the annihilator of a completely reducible R-module. Furthermore, $J(R) \subseteq I$ if I is the annihilator of a completely reducible R-module.

(iii) $\operatorname{Soc} R$ is an ideal of R.

Proof. (i) Direct consequence of Lemmas 6.11 and 6.12.

(ii) If $V = \bigoplus_{i \in I} V_i$, then $\operatorname{ann}(V) = \bigcap_{i \in I} \operatorname{ann}(V_i)$. Setting $(V_i), i \in I$ to be the set of representatives of the isomorphism classes of irreducible R-modules, we see that $J(R) = \mathrm{ann}(V)$. The second statement is a consequence of the fact that any completely reducible R-module is either 0 or a direct sum of a family of irreducible submodules.

(iii) For a given $r \in R$, the right multiplication by r is an endomorphism of $_R R$. Hence, by Proposition 3.1(ii), $(\mathrm{Soc}R)r$ is completely reducible. Thus $(\mathrm{Soc}R)r \subseteq \mathrm{Soc}R$, by Proposition 6.1(v). ■

6.14. PROPOSITION. If V is an R-module, then $J(R)V \subseteq J(V)$

Proof. If V has no maximal submodules, then $J(V) = V$ and there is nothing to prove. Let M be a maximal submodule of V. Then V/M is an irreducible R-module. Since $J(R)$ annihilates V/M, we see that $J(R)V \subseteq M$. Thus $J(R)V \subseteq J(V)$. ■

6.15. PROPOSITION. (Nakayama's lemma). Let V be a finitely generated R-module. If W is a submodule of V with $W + J(R)V = V$, then $W = V$.

Proof. By Proposition 6.14, $W + J(V) = V$. Since $J(V)$ is a superfluous submodule of V (Proposition 6.9(i)), the result follows. ■

6.16. PROPOSITION. Let $f : R_1 \longrightarrow R_2$ be a surjective homomorphism of rings. Then $f(J(R_1)) \subseteq J(R_2)$ with equality if $\mathrm{Ker}f \subseteq J(R_1)$.

Proof. Let $\{M_i | i \in I\}$ be the set of all maximal left ideals of R_1 containing $\mathrm{Ker}f$. Then $\{f(M_i) | i \in I\}$ is the set of all maximal left ideals of R_2 and

$$f(J(R_1)) \subseteq f(\bigcap_{i\in I} M_i) = \bigcap_{i\in I} f(M_i) = J(R_2)$$

If $\mathrm{Ker}f \subseteq J(R_1)$, then $\{M_i | i \in I\}$ is the set of all maximal left ideals of R_1. Hence $J(R_1) = \bigcap_{i\in I} M_i$ and, by the above, $f(J(R_1)) = J(R_2)$. ■

6.17. COROLLARY. Let I be an ideal of a ring R.

(i) $J(R/I) \supseteq (J(R) + I)/I$

(ii) If $I \subseteq J(R)$, then $J(R/I) = J(R)/I$. In particular, $J(R/J(R)) = 0$.

(iii) If $J(R/I) = 0$, then $J(R) \subseteq I$

(iv) $I = J(R)$ if and only if $I \subseteq J(R)$ and $J(R/I) = 0$.

Proof. (i) and (ii). Apply Proposition 6.16 for the case where $f : R \longrightarrow R/I$

is the natural homomorphism.

(iii) Direct consequence of (i)

(iv) Direct consequence of (ii) and (iii). ■

An element x of R is called a *left* (respectively, *right*) *unit* if there exists $y \in R$ such that $yx = 1$ (respectively, $xy = 1$). By a *unit* we mean an element x of R which is both a left and a right unit. Expressed otherwise, x is a unit if there exists $y \in R$, denoted by x^{-1}, such that $xy = yx = 1$.

6.18. PROPOSITION. Let x be an element of a ring R. Then $x \in J(R)$ if and only if for all $r \in R$, $1 - rx$ is a left unit. In particular, $J(R)$ contains no nonzero idempotents.

Proof. The element $1 - rx$ is a left unit if and only if $R(1-rx) = R$. So the first statement follows by virtue of Proposition 6.10. Assume that e is an idempotent in $J(R)$. By the above $x(1-e) = 1$ for some $x \in R$. Multiplying on the right by e yields $e = 0$. ■

6.19. PROPOSITION. Let R be a ring. Then $J(R)$ is the unique largest ideal I of R such that $1 - rx$ is a unit for all $r \in R, x \in I$.

Proof. By Proposition 6.18, it suffices to verify that $1 - x$ is a unit for all $x \in J(R)$. Since $1 - x$ is a left unit, $y(1-x) = 1$ for some $y \in R$. Then $z = 1 - y = -yx$ is in $J(R)$ and hence $1 = y'(1-z) = y'y$ for some $y' \in R$. Hence y is a unit and therefore $1 - x = y^{-1}$ is also a unit. ■

6.20. COROLLARY. Let R be a ring.

(i) $J(R^{\circ}) = J(R)$

(ii) $J(R)$ is the intersection of all maximal right ideals of R.

Proof. (i) Direct consequence of Proposition 6.19.

(ii) Follows from (i). ■

6.21. COROLLARY. If I is a left or right nil ideal of R, then $I \subseteq J(R)$.

Proof. Assume that I is a left nil ideal of R and let $x \in I$. Then, for any $r \in R, rx \in I$ and therefore $(rx)^n = 0$ for some $n \geqslant 1$. Invoking the identity

$$(1-y)\left(\sum_{i=0}^{n-1} y^i\right) = \left(\sum_{i=0}^{n-1} y^i\right)(1-y) = 1 - y^n$$

we infer that $1 - rx$ is a unit. Hence, by Proposition 6.18, $I \subseteq J(R)$. If I is a right nil ideal, then the same argument applied to R^o yields the result. ■

As a partial converse we have

6.22. PROPOSITION. If R is artinian, then $J(R)$ is nilpotent.

Proof. Put $J = J(R)$ and consider the chain $J \supseteq J^2 \supseteq \dots$. By hypothesis, $J^k = J^{k+1} = \dots$ for some $k \geqslant 1$. Setting $I = J^k$ we have $I^2 = I$. Assume by way of contradiction that $I \neq 0$. There exist left ideals X of R with $IX \neq 0$, for example $X = R$. Let M be a minimal element in the set of all such X. Then $I(JM) = (IJ)M = IM \neq 0$ and, since $JM \subseteq M$ we have $JM = M$. By Proposition 6.15, we are left to verify that M is a finitely generated R-module. By hypothesis, there exists $x \in M$ such that $Ix \neq 0$ and so $I(Rx) \neq 0$. As $Rx \subseteq M$ we must have $Rx = M$, as required. ■

6.23. PROPOSITION. Let I be a left ideal of R. Then $I \subseteq J(R)$ if and only if $IV \neq V$ for any nonzero finitely generated module V.

Proof. Let $V \neq 0$ be a finitely generated R-module. Then $J(V) \neq V$ by Proposition 6.1(i). Hence, by Proposition 6.14, if $I \subseteq J(R)$, then $IV \subseteq J(R)V \subseteq J(V) \neq V$. Conversely, if $IV \neq V$ for any nonzero finitely generated module V, then $IV = 0$ for any irreducible R-module V. Thus $I \subseteq J(R)$ by Proposition 6.13(i). ■

6.24. PROPOSITION. Let $(R_i), i \in I$, be a family of rings. Then

$$J\left(\prod_{i\in I} R_i\right) = \prod_{i\in I} J(R_i)$$

Proof. An element $(r_i) \in \prod_{i\in I} R_i$ is a left unit if and only if r_i is a left unit of R_i for all $i \in I$. Now apply Proposition 6.18. ■

6.25. PROPOSITION. For any ring R and any positive integer n,

$$J(M_n(R)) = M_n(J(R))$$

Proof. Let V be an R-module and let V^n be an $M_n(R)$-module as in

Proposition 5.3. Then $\mathrm{ann}(V^n) = M_n(\mathrm{ann}(V))$, by the definition of V^n. Hence the desired conclusion follows by virtue of Propositions 5.3 and 6.13(i). ■

6.26. PROPOSITION. Let R be a ring. Then the following conditions are equivalent:

(i) $_RR$ is a direct sum of finitely many irreducible modules

(ii) $_RR$ is completely reducible

(iii) R is semisimple artinian

(iv) Every R-module is completely reducible

(v) R is a finite direct product of full matrix rings over division rings.

Proof. The equivalence of (i) - (iv) is a consequence of Propositions 5.8, 2.1,6.2 and 3.4. The implication (ii) ⇒ (v) follows from Proposition 5.8. The implication (v) ⇒ (iii) is a consequence of Proposition 6.25,6.24 and 5.3 together with Corollary 2.8. ■

A ring R is said to be *simple* if 0 and R are the only ideals of R.

6.27. PROPOSITION. Let R be a ring. Then the following conditions are equivalent.

(i) R is simple artinian

(ii) $_RR$ is a direct sum of finitely many isomorphic irreducible modules

(iii) R is isomorphic to a full matrix ring over a division ring

(iv) R is semisimple artinian and all irreducible R-modules are isomorphic.

Proof. (i) ⇒ (ii): Direct consequence of Propositions 6.26 and 5.8

(ii) ⇒ (iii): Follows from Propositions 5.2 and 5.6 and the Schur's lemma.

(iii) ⇒ (iv): Follows from Propositions 6.26 and 5.3.

(iv) ⇒ (i): Follows from Propositions 6.26,5.2 and 5.4. ■

As a partial converse of Proposition 6.14, we prove

6.28. PROPOSITION. Let V be an R-module. If $R/J(R)$ is artinian, then

$$J(V) = J(R)V.$$

Proof. By Proposition 6.14, $J(R)V \subseteq J(V)$ and hence, by Corollary

6.6(i), we need only show that $J(V/J(R)V) = 0$. Since $J(R/J(R)) = 0$ (Corollary 6.17(ii)), every $R/J(R)$-module is completely reducible (Proposition 6.26). In particular, the $R/J(R)$-module $V/J(R)V$ is completely reducible. Hence $V/J(R)V$ is a completely reducible R-module, and therefore $J(V/J(R)V) = 0$, by Proposition 6.1(iii). ■

6.29. PROPOSITION. Let $R/J(R)$ be an artinian ring. Then

$$\mathrm{Soc}R = \{r \in R \mid J(R)r = 0\}$$

Proof. Since $\mathrm{Soc}R$ is a completely reducible R-module, $J(R)x = 0$ for all $x \in \mathrm{Soc}R$. Setting $V = \{r \in R \mid J(R)r = 0\}$, it follows that $\mathrm{Soc}R \subseteq V$. Since V can be regarded an $R/J(R)$-module, V is a completely reducible $R/J(R)$-module (Proposition 6.2.6). Hence V is a completely reducible R-module and so, by Proposition 6.1(v), $V \subseteq \mathrm{Soc}R$. ■

6.30. PROPOSITION. Let R be a ring such that every irreducible R-module is isomorphic to a submodule of $_RR$. Then

$$J(R) = \{r \in R \mid r\mathrm{Soc}R = 0\}$$

Proof. The hypothesis on R ensures that $r\mathrm{Soc}R = 0$ if and only if $rV = 0$ for every irreducible R-module V. Now apply Proposition 6.13(i). ■

6.31. PROPOSITION. Let R be an artinian ring and let V be an R-module. Then the following conditions are equivalent:

(i) V is of finite length

(ii) V is artinian

(iii) V is noetherian

Proof. By definition, (i) ⇒ (ii) and (i) ⇒ (iii). Assume that V is either artinian or noetherian. By Proposition 6.22, $J(R)^n = 0$ for some $n \geqslant 1$. By Proposition 2.1, it suffices to show that the modules $V/J(R)V$, $J(R)V/J(R)^2V$, $\ldots, J(R)^{n-1}V/J(R)^nV$ have composition series. However these modules are annihilated by $J(R)$ and so can be regarded as $R/J(R)$-modules. Furthermore, by Proposition 6.26, each of these modules is completely reducible and, by hypothesis, each of them is either artinian or noetherian. The desired conclusion is therefore a consequence of Corollary 3.3. ■

6.32. COROLLARY. Let R be an artinian ring.

(i) R is noetherian

(ii) ${}_RR$ has a composition series

(iii) Every finitely generated R-module is of finite length.

Proof. (i) Apply Proposition 6.31 for $V = {}_RR$

(ii) Apply (i) and Proposition 2.10

(iii) Let V be a finitely generated R-module. Since V is a homomorphic image of a direct sum of finitely many copies of R, V is artinian. Now apply Proposition 6.31. ■

6.33. PROPOSITION. Let R be an arbitrary ring, let V be an R-module, and let S be a subring of R.

(i) If V has a composition series $V_1 \supset V_2 \supset \dots \supset V_{n+1} \supset 0$ of length n, then $J(R)^n V = 0$. In particular, $J(R)^n = 0$ if V is a faithful R-module.

(ii) If every irreducible R-module is completely reducible as an S-module, then $J(S) \subseteq J(R)$.

Proof. (i) Because $J(R)$ annihilates each factor V_i/V_{i-1}, $J(R)^n$ must annihilate V.

(ii) Let $x \in J(S)$ and let V be an irreducible R-module. Since V is completely reducible as an S-module, $xV = 0$. Hence $x \in J(R)$ and the required assertion follows. ■

The following consequence is a sharpened version of Proposition 6.22.

6.34. COROLLARY. Let R be an artinian ring and let n be the composition length of ${}_RR$. Then

$$J(R)^n = 0.$$

Proof. By Corollary 6.32, ${}_RR$ has a composition series. Now apply Proposition 6.33 for $V = {}_RR$ ■

6.35. PROPOSITION. Let $e \neq 0$ be an idempotent of a ring R. Then

$$J(eRe) = eJ(R)e$$

Proof. Let I be a primitive ideal in R and let V be a faithful irreducible R/I-module. Then eV is an eRe-module. If $eV = 0$ then $J(eRe) \subseteq eRe \subseteq I$. Assume that $eV \neq 0$. Let $0 \subset W = eW \subseteq eV$ where W is an eRe-module. Then $V = RW$ since V is irreducible. Hence $eV = eRe\, W \subseteq W$ and so $eV = W$. It follows that eV is an irreducible eRe-module whence $J(eRe)V = J(eRe)eV = 0$. Thus $J(eRe) \subseteq I$. Since I was an arbitrary primitive ideal in R, this yields $J(eRe) \subseteq J(R)$ and so $J(eRe) = eJ(eRe)e \subseteq eJ(R)e$.

Suppose that $a \in J(R)$. Then $eae \in J(R)$ and so, by Proposition 6.18,(1-b) $(1-eae) = 1$ for some $b \in R$, in which case $beae = eae+b$. Multiplying by e on both sides yields $eae+ebe = ebeae$. Hence $(e-ebe)(e-eae) = e$ so that $e-eae$ is a left unit. Since $eJ(R)e$ is an ideal of eRe, Proposition 6.18 implies that $eJ(R)e \subseteq J(eRe)$. ■

6.36. PROPOSITION. For any idempotent e of R, $J(Re) = J(R)e$.

Proof. Since ${}_RR = Re \oplus R(1-e)$, we have $J(R) = J(Re) \oplus J(R(1-e))$ (Proposition 6.7(ii)) and $J(R) = J(R)e \oplus J(R)(1-e)$. But, by Proposition 6.14, $J(R)e \subseteq J(Re)$ and $J(R)(1-e) \subseteq J(R(1-e))$. Hence $J(Re) = J(R)e$, as required. ■

6.37. PROPOSITION. Let I be a nilpotent left ideal of a ring R, and let $x \in R$ be a nonnilpotent element such that $x^2-x \in I$. Then the left ideal Rx contains a nonzero idempotent y such that $y-x \in I$.

Proof. Let $I^k = 0$, and set $z_1 = x^2-x \in I$. If $z_1 = 0$, choose $y = x$, and we are done. If $z_1 \neq 0$, let $x_1 = x+z_1-2xz_1 \in Rx$. Then x, x_1 and z_1 commute with each other, and thus if x_1 is nilpotent, then so is $x = x_1-z_1+2xz_1$, a contradiction. Hence x_1 is a nonnilpotent element of Rx, and direct calculation shows that

$$x_1^2 - x_1 = 4z_1^3 - 3z_1^2$$

The element $z_2 = x_1^2 - x_1$ is nilpotent, contains z_1^2 as factor, and commutes with x_1. Continuing in this manner, we may construct a sequence $\{x_i\}$ of non-nilpotent elements of Rx, such that $z_1^{2^i}$ occurs as a factor in $x_i^2 - x_i$. If we choose i so that $2^i \geqslant k$, then we have $x_i^2 - x_i = 0$. Moreover, $x_i \neq 0$ since x_i is nonnilpotent. Thus $y = x_i$ is the desired idempotent. ∎

6.38. COROLLARY. Let I be a nonnilpotent left ideal in a (left) artinian ring R. Then I contains a nonzero idempotent.

Proof. Let X be a minimal member of the nonempty set of nonnilpotent left ideals contained in I. Then $I^2 \subseteq I$ and so $I^2 = I$. Now let Y be minimal in the set of left ideals contained in X such that $XY \neq 0$, and choose $a \in Y$ so that $Xa \neq 0$. Then $X{\cdot}Xa \neq 0$ and $Xa \subseteq Y$, whence $Y = Xa$. Thus $a = xa$ for some $x \in X$ and so $a = x^k a$ for all $k \geqslant 0$. It follows that Y contains the nonnilpotent element x. Setting

$$L = \{b \in X | ba = 0\}$$

we have $x^2 - x \in L$. Moreover, $Xa = Y \neq 0$ so $L \subset X$ and thus L is nilpotent. Invoking Proposition 6.37, we see that Rx contains a nonzero idempotent, whence so does I, since $I \supseteq X \supseteq Rx$. ∎

7. THE KRULL-SCHMIDT THEOREM

This section deals with the uniqueness of direct sum decompositions. The main result (Theorem 7.5) is Azumaya's generalization of a classical theorem of Krull and Schmidt (Corollary 7.6).

A ring R is said to be a *local ring* if $R/J(R)$ is a division ring. In what follows, we write $U(R)$ for the group of units of R.

7.1. LEMMA. The following conditions are equivalent:

(i) R is local

(ii) R has a unique maximal left ideal

(iii) $J(R) = R - U(R)$

(iv) The set of nonunits of R is a left ideal

Proof. (i) ⇒ (ii): Let I be a maximal left ideal of R. Then $I/J(R)$ is a proper ideal of $R/J(R)$ and hence $I = J(R)$.

(ii) ⇒ (iii): Let J be a unique maximal left ideal of R. Since $J(R)$ is the intersection of all maximal left ideals of R, we have $J(R) = J$, and so J is an ideal. The inclusion $J(R) \subseteq R-U(R)$ being obvious, assume that $x \in R - U(R)$. If $Rx \neq R$, then Rx lies in a maximal left ideal of R, whence $Rx \subseteq J(R)$, so $x \in J(R)$. On the other hand, if $Rx = R$ then $yx = 1$ for some $y \in R$. Clearly $y \notin J(R)$, otherwise $1 = yx \in J(R)$. Hence $Ry = R$, so $zy = 1$ for some $z \in R$ and hence $z = x$. Thus $x \in U(R)$, a contradiction.

(iii) ⇒ (iv) Obvious

(iv) ⇒ (i): Let $x \notin J(R)$ and let $I = R - U(R)$. If M is a maximal left ideal of R, then $M \subseteq I \neq R$, whence $M = I = J(R)$. Hence x is a unit and therefore $x + J(R)$ is a unit of $R/J(R)$. Thus $R/J(R)$ is a division ring, as required. ∎

7.2. COROLLARY. Let R be a local ring. Then the only idempotents of R are 0 and 1.

Proof. Let $e \neq 1$ be an idempotent of R. Then e is a nonunit and hence, by Lemma 7.1, $e \in J(R)$. Now apply Proposition 6.18. ∎

7.3. LEMMA. Let V be an R-module. Then V is indecomposable if and only if 0 and 1 are the only idempotents of $\mathrm{End}_R(V)$. In particular, by Corollary 7.2, if $\mathrm{End}_R(V)$ is local, then V is indecomposable.

Proof. Assume that $V = X \oplus Y$ is a proper decomposition. Then the projection $V \longrightarrow X$ is an idempotent of $\mathrm{End}_R(V)$ distinct from 0 and 1. Conversely, assume that $f \neq 0,1$ is an idempotent of $\mathrm{End}_R(V)$. Then $V = f(V) \oplus \mathrm{Ker} f$ is obviously a proper decomposition. ∎

7.4. LEMMA. Let R be a ring and let V be an R-module of finite length.

(i) If V is indecomposable, then every $f \in \mathrm{End}_R(V)$ is either a unit or nilpotent.

(ii) V is indecomposable if and only if $\mathrm{End}_R(V)$ is a local ring.

Proof. (i) We first show that an R-homomorphism $f : V \longrightarrow V$ is injective if and only if it is surjective. Assume that f is injective. Then there exists an integer n such that

$$V \supset f(V) \supset \dots \supset f^n(V) = f^{n+1}(V)$$

Therefore, for each $u \in V$, there exists $v \in V$ such that $f^n(u) = f^{n+1}(v)$. Since f is injective, $f(v) = u \in f(V)$ and so f is surjective.

Now assume that f is surjective. By hypothesis, there exists an integer n such that

$$0 \subset \mathrm{Ker}f \subset \dots \subset \mathrm{Ker}f^n = \mathrm{Ker}f^{n+1}$$

Since f (and hence f^n) is surjective, each $u \in \mathrm{Ker}f$ may be written in the form $u = f^n(v)$ for some $v \in V$. Because $f(u) = 0$, it follows that $v \in \mathrm{Ker}f^{n+1} = \mathrm{Ker}f^n$, and thus $u = 0$. This proves that f is injective.

It will next be shown that

$$V = f^n(V) \oplus \mathrm{Ker}f^n \qquad \text{for some } n \geqslant 1$$

Since V is indecomposable, the required assertion will follow. Choose an integer $n \geqslant 1$ such that $f^n(V) = f^{n+1}(V)$. Then the restriction of f to the submodule $f^n(V)$ is surjective, so by the above this restriction is also injective. This shows that $f^n(V) \cap \mathrm{Ker}f^n = 0$. Because $f^n(V) = f^{2n}(V)$, for each $u \in V$ there exists $v \in V$ such that $f^n(u - f^n(v)) = 0$. It follows that

$$u \in f^n(v) + \mathrm{Ker}f^n \subseteq f^n(V) + \mathrm{Ker}f^n$$

so $V = f^n(V) \oplus \mathrm{Ker}f^n$, as required.

(ii) Assume that V is indecomposable and put $E = \mathrm{End}_R(V)$. By Lemma 7.3, we need only verify that E is a local ring. Let f be a nonunit of E and let g be an arbitrary element of E. Then gf is a nonunit and hence is nilpotent, by virtue of (i). Thus $1-gf$ is a unit of E, so by Proposition 6.18, $f \in J(E)$ Invoking Lemma 7.1, we deduce that E is local. ■

We are now ready to prove

7.5. THEOREM. Let R be a ring and let V be an R-module. Suppose that

$$V \cong \bigoplus_{i=1}^{m} V_i \quad \text{and} \quad V \cong \bigoplus_{j=1}^{n} W_j$$

where for all i,j the rings $\text{End}_R(V_i)$ and $\text{End}_R(W_j)$ are local. Then $m = n$ and after a suitable rearrangement $V_i \cong W_i$ for all $i \in \{1,2,\ldots,m\}$.

Proof. We may harmlessly assume that $V = \bigoplus_{i=1}^{m} V_i = \bigoplus_{j=1}^{n} W_j$. Using induction on $\min\{m,n\}$, observe that if $m = 1$ or $n = 1$, the result is clear since by Lemma 7.3 V is indecomposable.

Denote by e_i, f_j the projections of V onto V_i, W_j, respectively. Then e_1 and $e_1 f_j e_1$ are all in $\text{End}_R(V_1)$. Because $e_1 = \sum_{j=1}^{n} e_1 f_j e_1$ and $\text{End}_R(V_1)$ is local, it follows from Lemma 7.1 that $e_1 f_j e_1$ is a unit in $\text{End}_R(V_1)$ for some j. Changing notation, if necessary, it may be assumed that $e_1 f_1 e_1$ is an automorphism of $V_1 = e_1(V)$. Thus $f_1(V_1) \subseteq W_1$ and the kernel of f_1 on V_1 is 0. We now prove that $V = f_1(V_1) \oplus (\bigoplus_{i=2}^{m} V_i)$.

If $x \in f_1(V_1) \cap \bigoplus_{i=2}^{m} V_i$, then $e_1(x) = 0$ and $x = (f_1 e_1)(y)$ for some $y \in V$. Thus $(e_1 f_1 e_1)(y) = e_1(x)$ and $e_1(y) = 0$ as $e_1 f_1 e_1$ is an automorphism on V_1. Hence $x = (f_1 e_1)(y) = 0$, proving that $f_1(V_1) \cap (\bigoplus_{i=2}^{m} V_i) = 0$. Assume that $v \in V$. Then $e_1(v) \in e_1(V) = (e_1 f_1 e_1)(V)$. Hence $e_1(v) = (e_1 f_1 e_1)(w)$ for some $w \in W$. It follows that $e_1(v - (f_1 e_1)(w)) = 0$ and $v - (f_1 e_1)(w) \in \bigoplus_{i=2}^{m} V_i$. Thus

$$v = (f_1 e_1)(w) + (v - (f_1 e_1)(w)) \in f_1(V) + \sum_{i=2}^{m} V_i$$

proving that $V = f_1(V_1) \oplus (\bigoplus_{i=2}^{m} V_i)$.

Because $f_1(V_1) \subseteq W_1$, this implies that $W_1 = f_1(V_1) \oplus \{W_1 \cap \bigoplus_{i=2}^{m} V_i\}$. By Lemma 7.3, W_1 is indecomposable. Since $f_1(V_1) \neq 0$, we see that $W_1 = f_1(V_1)$ is isomorphic to V_1 and $V = W_1 \oplus (\bigoplus_{i=2}^{m} V_i)$. The conclusion is that

$$\bigoplus_{i=2}^{m} V_i \cong V/W_1 \cong \bigoplus_{j=2}^{n} W_j$$

and hence, by induction, $m = n$ and after a suitable rearrangement $V_i \cong W_i$ for $i \in \{1,\ldots,m\}$. ■

7.6. COROLLARY. (Krull-Schmidt theorem). Let V be a nonzero R-module of finite length. Then V can be written as a direct sum $V = \bigoplus_{i=1}^{m} V_i$, where V_i are indecomposable submodules. Moreover, if $V = \bigoplus_{j=1}^{n} W_j$ is another decomposition of this kind, then $n = m$ and (after possible reordering the W_j) we have $V_i \cong W_i$ for each i.

Proof. Apply Proposition 2.9, Lemma 7.4(ii) and Theorem 7.5. ∎

8. PROJECTIVE, INJECTIVE AND FLAT MODULES.

Let R be a ring. A sequence of R-modules and R-homomorphisms

$$V_1 \xrightarrow{f_1} V_2 \xrightarrow{f_2} V_3 \longrightarrow \cdots \xrightarrow{f_{n-1}} V_n$$

is *exact* at V_i if $\mathrm{Ker} f_i = \mathrm{Im} f_{i-1}$. The sequence is *exact* if it is exact at each place, that is

$$\mathrm{Ker} f_i = \mathrm{Im} f_{i-1} \qquad 2 \leqslant i \leqslant n-1$$

A *short exact sequence* is an exact sequence of the form

$$(*) \qquad 0 \longrightarrow X \xrightarrow{\alpha} Y \xrightarrow{\beta} Z \longrightarrow 0$$

Exactness of (*) is equivalent to the following three conditions: α is injective $\mathrm{Ker}\beta = \mathrm{Im}\alpha$ and β is surjective. In this case β induces an isomorphism

$$Y/\alpha(X) \cong Z$$

The short exact sequence (*) is *split* if $\alpha(X)$ is a direct summand of Y. This is equivalent to either of the following:

(i) There exists a $\gamma \in \mathrm{Hom}_R(Y,X)$ such that $\gamma\alpha = 1_X$, where 1_X denotes the identity map on X.

(ii) There exists a $\delta \in \mathrm{Hom}_R(Z,Y)$ such that $\beta\delta = 1_Z$

The homorphisms γ and δ are called *splitting homomorphisms*.

A diagram of R-modules and R-homomorphisms

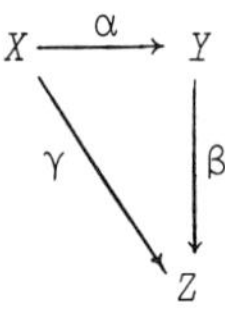

is *commutative* if $\gamma = \beta\alpha$. The same terminology applies to more complicated diagrams.

Let V be an R-module. Then V is called *projective*, if for any given diagram with exact bottom row

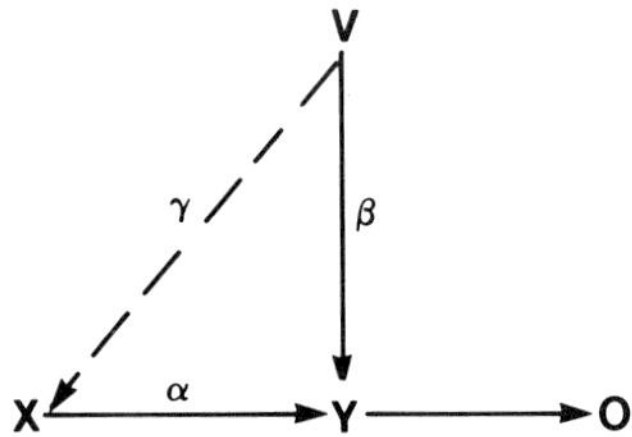

there exists an R-homomorphism γ such that $\beta = \alpha\gamma$.

8.1. PROPOSITION. The following conditions are equivalent

(i) An R-module V is projective

(ii) Any exact sequence $0 \longrightarrow X \xrightarrow{\alpha} Y \xrightarrow{\beta} V \longrightarrow 0$ of R-modules splits.

(iii) V is a direct summand of a free R-module

Proof. (i) ⇒ (ii): Because β is surjective and V is projective, we have the following commutative diagram:

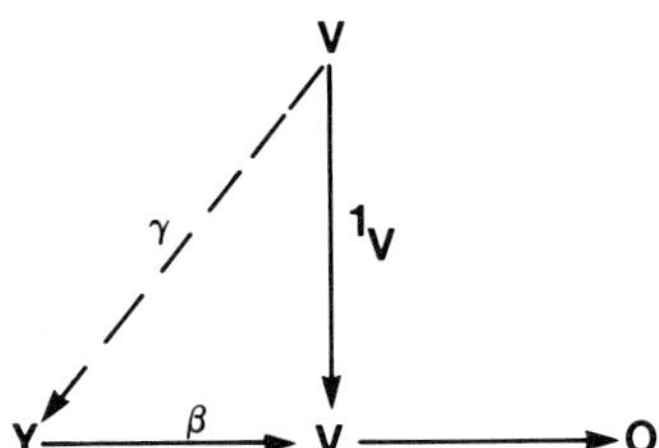

Hence γ is a splitting homomorphism and (ii) follows.

(ii) ⇒ (iii): Take as Y a suitable free R-module, and let $\gamma : V \to Y$ be a splitting homomorphism. Then $\gamma(V) \cong V$ is a direct summand of Y.

(iii) ⇒ (i): Let V be a direct summand of the free R-module F, and let $\{y_i\}$ be a free basis for F. Then we can let $\pi : F \longrightarrow V$ be the natural projection and $\varepsilon : V \longrightarrow F$ the natural injection so that $\pi\varepsilon = 1_V$. Now, suppose X and Y are R-modules and that $\beta : V \longrightarrow Y$ and $\alpha : X \longrightarrow Y$ are a homomorphism and a surjective homomorphism, respectively. We then have a natural map $\beta\pi$ from F to Y, and we can describe the situation pictorially as follows:

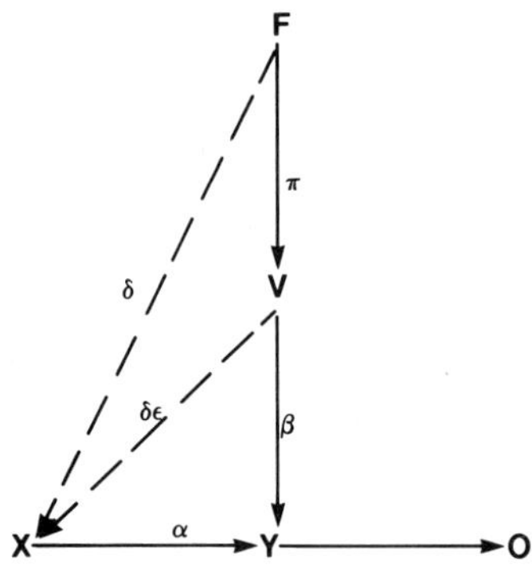

Now, for each generator $y_i \in F$, $\beta\pi(y_i) \in Y$ and because α is surjective there exists $x_i \in X$ with $\beta\pi(y_i) = \alpha(x_i)$. Thus if we define $\delta : F \longrightarrow X$ by $\delta(y_i) = x_i$, then the outer portion of the diagram is commutative, that is, $\alpha\delta = \beta\pi$. Finally, $\varepsilon : V \longrightarrow F$ so that $\delta\varepsilon : V \longrightarrow X$ and

$$\alpha(\delta\varepsilon) = (\alpha\delta)\varepsilon = (\beta\pi)\varepsilon$$
$$= \beta(\pi\varepsilon) = \beta$$

Thus the lower portion of the diagram is commutative, and by definition we conclude that V is projective. ∎

8.2. COROLLARY. Let V be a finitely generated R-module. Then V is projective if and only if V is a direct summand of a free R-module of a finite rank.

Proof. Let $v_1, v_2, \ldots, v_n$ be a generating set for V and let F be a free module freely generated by $x_1, x_2, \ldots, x_n$. Then the map $x_i \longmapsto v_i$ determines a surjective homomorphism $F \longrightarrow V$. Thus if V is projective, then V is a direct summand of F. The converse being true by virtue of Proposition 8.1, the result follows. ∎

If we reverse the direction of all the arrows in the definition of a projective module, we then obtain the definition of an injective module. Thus a given R-module V is *injective*, if for any given diagram with exact bottom row

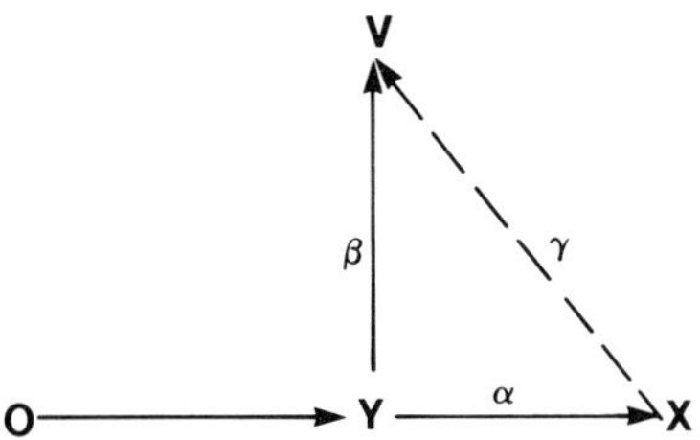

there exists an R-homomorphism γ such that $\beta = \gamma\alpha$.

8.3. PROPOSITION. The following conditions are equivalent:

(i) An R-module V is injective

(ii) Any exact sequence $0 \longrightarrow V \xrightarrow{\alpha} X \xrightarrow{\beta} Y \longrightarrow 0$ of R-modules splits.

Proof. (i) ⇒ (ii): Because α is injective, we have the following commutative daigram:

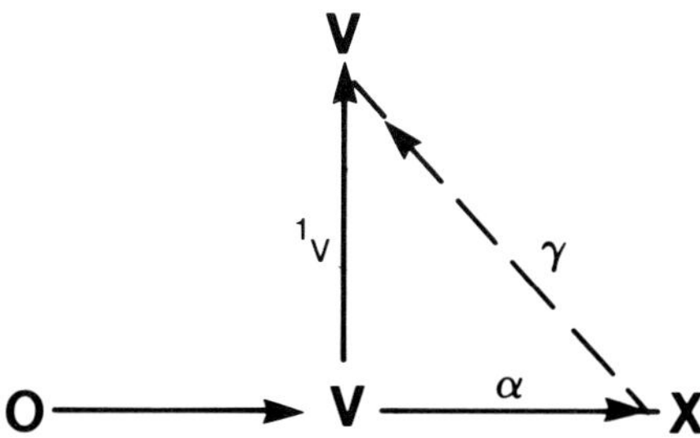

Hence γ is a splitting homomorphism and (ii) follows

(ii) ⇒ (i): Consider a commutative diagram with exact bottom row

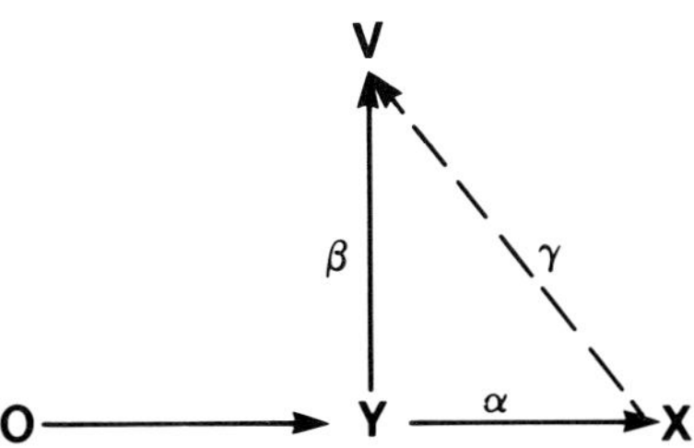

We may harmlessly assume that α is the inclusion map, in which case Y is a direct summand of X, by hypothesis. Write $X = Y \oplus Z$ and define $\gamma : X \longrightarrow V$ by $\gamma(y+z) = \beta(y), y \in V, z \in Z$. Then γ is an R-homomorphism and $\beta = \gamma\alpha$, as required. ■

8.4. PROPOSITION. Let $(V_i), i \in I$ be a family of R-modules. Then

(i) $\bigoplus_{i\in I} V_i$ is projective if and only if each V_i is projective

(ii) $\prod_{i\in I} V_i$ is injective if and only if each V_i is injective.

Proof. (i) If $V = \bigoplus_{i \in I} V_i$, then V is a direct summand of a free module if and only if so is each V_i. Hence, by Proposition 8.1, V is projective if and only if each V_i is projective.

(ii) Let $V = \prod_{i \in I} V_i$ and let $\pi_i : V \longrightarrow V_i$ and $\mu_i : V_i \longrightarrow V$ be the natural projections and injections, respectively. Assume that each V_i is injective and let $\alpha : V \to X$ be an injective R-homomorphism. Then, for each $i \in I$, there exists an R-homomorphism $\beta_i : X \longrightarrow V_i$ with $\pi_i = \beta_i\alpha$. Define $\beta : X \longrightarrow V$ by $\beta(x) = (\beta_i(x))$. Then β is an R-homomorphism with $\beta\alpha = 1_V$ and so, by Proposition 8.3, V is injective.

Conversely, assume that V is injective and let $\alpha : V_i \longrightarrow X$ be an injective R-homomorphism. Then there exists an R-homomorphism $\beta : X \to V$ with $\mu_i = \beta\alpha$. Define $\gamma : X \to V_i$ by $\gamma = \pi_i\beta$. Then $\gamma\alpha = \pi_i(\beta\alpha) = \pi_i\mu_i = 1_{V_i}$ and hence, by Proposition 8.3, V_i is injective. ■

Given a pair of rings R,S we say that V is an (R,S)-*bimodule* if V is a left R-module and a right S-module, with the actions of R and S on V commuting

$$(rv)s = r(vs) \qquad (r \in R, v \in V, s \in S)$$

Given a right R-module V and a left R-module W, the abelian group $V \otimes_R W$, called the *tensor product* of V and W, is defined as follows. Let F be a free $\mathbb{Z}$-module with $V \times W$ as a basis; then each element of F can be uniquely written in the form

$$\sum z_{ij}(v_i,w_j) \qquad (z_{ij} \in \mathbb{Z}, v_i \in V, w_j \in W)$$

with finitely many z_{ij} distinct from 0. Let T be the subgroup of F generated by all elements of the form

$$(v_1+v_2,w) - (v_1,w) - (v_2,w)$$

$$(v,w_1+w_2) - (v,w_1) - (v,w_2)$$

$$(v,rw) - (vr,w)$$

$v, v_i \in V$ and $w, w_i \in W$. Then $V \otimes_R W$ is defined as the factor group F/T. The image of (v,w) under the natural homomorphism $F \longrightarrow F/T$ is denoted by $v \otimes w$. Observe that the $\mathbb{Z}$-module $V \otimes_R W$ consists of all finite sums $\sum v_i \otimes w_i$ $(v_i \in V, w_i \in W)$.

Suppose that V is an (S,R)-bimodule. Then $V \otimes_R W$ can be regarded as a (left) S-module by putting

$$s(v \otimes w) = sv \otimes w \qquad (v \in V, w \in W, s \in S)$$

In particular, if V and W are modules over a commutative ring R, then $V \otimes_R W$ is an R-module. Let V be an (S,R)-bimodule, and let $W_1, \ldots, W_n$ be left R-modules. It is an immediate consequence of the definition that:

(i) $V \otimes_R R \cong V$ as S-modules

(ii) $V \otimes_R (W_1 \oplus \ldots \oplus W_n) \cong V \otimes_R W_1 \oplus \ldots \oplus V \otimes_R W_n$ as S-modules

Furthermore, if V and W are two free modules over a commutative ring R and (v_λ) (respectively, (w_μ)) is an R-basis of V (respectively, W), then $(v_\lambda \otimes w_\mu)$ is an R-basis of $V \otimes_R W$.

Let R be a ring, V, V' two right R-modules, W, W' two left R-modules, and $\phi : V \longrightarrow V'$, $\psi : W \longrightarrow W'$ two R-homomorphisms. Then the map

$$\phi \otimes \psi : V \otimes_R W \to V' \otimes_R W'$$

defined by

$$(\phi \otimes \psi)(v \otimes w) = \phi v) \otimes \psi(w)$$

is a homomorphism of additive groups. Furthermore, if V and V' are (S,R)-bimodules and ϕ is an S-homomorphism, then $\phi \otimes \psi$ is also an S-homomorphism.

Let R be a ring and let V be a right R-module. The module V is said to be *flat* if for every injective homomorphism $f : W' \longrightarrow W$ of left R-modules, the homomorphism

$$1_V \otimes f : V \otimes_R W' \longrightarrow V \otimes_R W$$

is injective. It is an easy consequence of the definition that, given a family $(V_i), i \in I$ of right R-modules, the module $\bigoplus_{i \in I} V_i$ is flat if and only if each V_i is flat. The following simple observation will be useful for our subsequent investigations.

8.5. PROPOSITION. Every projective module is flat.

Proof. Since projective modules are isomorphic to direct summands of free modules (Proposition 8.1), we need only show that the regular module R_R is flat. Let $f : W' \longrightarrow W$ be an injective homomorphism of left R-modules. The maps $\mu' : R \otimes_R W' \longrightarrow W', r \otimes w' \mapsto rw'$, and $\mu : R \otimes_R W \longrightarrow W, r \otimes w \mapsto rw$ are R-isomorphisms which render commutative the following diagram

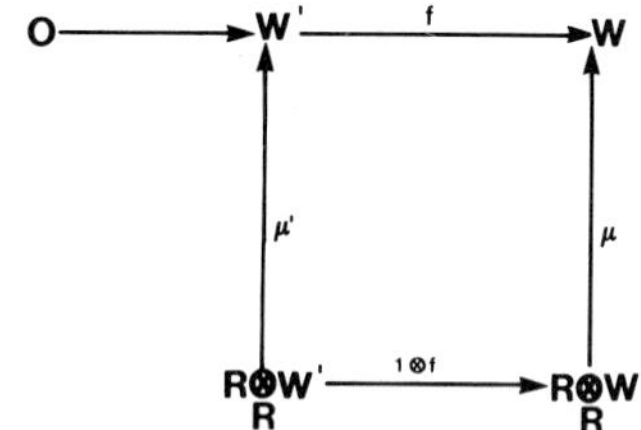

Thus $1 \otimes f$ is injective and R_R is flat. ■

9. PROJECTIVE COVERS.

In what follows all modules are assumed to be *finitely generated left modules*. Let R be a ring. We know that each R-module is a homomorphic image of a projective module. For some modules V an even stronger assertion is possible. A homomorphism $f : V \longrightarrow W$ of R-modules is said to be *essential* if for every proper submodule V' of V, $f(V') \neq f(V)$. We say that an R-module P is a *projective cover* of V in case P is projective and there is an *essential* epimorphism $P \longrightarrow V$.

9.1. LEMMA. (i) A homomorphism $f : V \longrightarrow W$ of R-modules is essential if and only if $\operatorname{Ker} f$ is superfluous.

(ii) If $V_1, V_2, \ldots, V_n$ are superfluous submodules of V, then so is $V_1 + \ldots + V_n$

(iii) If P_i is a projective cover of V_i, then $P_1 \oplus \ldots \oplus P_n$ is a projective cover of $V_1 \oplus \ldots \oplus V_n$.

(iv) A projective cover of an irreducible module is indecomposable.

Proof. (i) Assume that $f : V \longrightarrow W$ is an essential homomorphism. If W is a submodule of V such that $V = W + \operatorname{Ker} f$, then $f(V) = f(W)$ and so W cannot be a proper submodule of V. This proves that $V = W$ and hence that $\operatorname{Ker} f$ is a superfluous submodule of V. Conversely, assume that $\operatorname{Ker} f$ is superfluous. If V' is a submodule of V with $f(V) = f(V')$, then $V = V' + \operatorname{Ker} f$. Since $\operatorname{Ker} f$ is superfluous, we must have $V = V'$ and so f is essential.

(ii) Assume that W is a submodule of V such that $(V_1 + \dots + V_n) + W = V$. Then $V_1 + (V_2 + \dots + V_n + W) = V$ and, since V_1 is superfluous, we have $V_2 + \dots + V_n + W = V$. The desired assertion now follows by induction on n.

(iii) If $f_i : P_i \to V_i$ is an essential epimorphism, then $\oplus f_i : \oplus P_i \longrightarrow \oplus V_i$ is an epimorphism whose kernel $\overset{n}{\underset{i=1}{\oplus}} \operatorname{Ker} f_i$ is superfluous, by virtue of (ii). Hence, by (i), $\oplus f_i$ is essential, which means that $\oplus P_i$ is a projective cover of $\oplus V_i$.

(iv) Assume that $\psi : P \longrightarrow V$ is an essential epimorphism, where V is irreducible. If $P = P' \oplus P''$, for some proper submodules P' and P'', then $\psi(P') = \psi(P'') = 0$ since 0 is the only proper submodule of V. But then $V = \psi(P') + \psi(P'') = 0$, a contradiction. Hence, P is indecomposable, as required. ■

One of the consequences of the following lemma is that if a module does have a projective cover, then it is unique up to isomorphism.

9.2. LEMMA. Assume that an R-module V has a projective cover P, and let M be an R-module such that V is a homomorphic image of M.

(i) If M is projective, then P is isomorphic to a direct summand of M

(ii) If M has a projective cover L, then P is isomorphic to a direct summand of L

(iii) If M is another projective cover of V, then $M \cong P$.

Proof. (i) Fix an epimorphism $\phi : M \longrightarrow V$ and an essential epimorphism $\psi : P \longrightarrow V$. Since M is projective, there is a commutative diagram

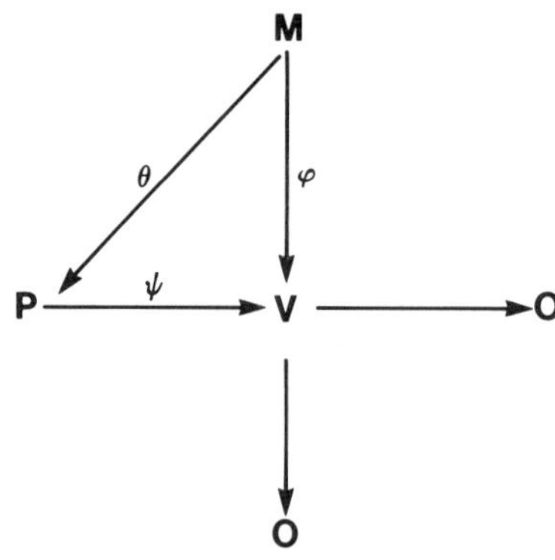

with exact row and column. Because ψ is an essential epimorphism and $\psi\theta$ is an epimorphism, θ is also an epimorphism. But P is projective, so there is a splitting homomorphism $f : P \longrightarrow M$, and hence $M = \text{Im} f \oplus \text{Ker}\,\theta$. Since f is injective, (i) is established.

(ii) Assume that L is a projective cover of M, and let $f : L \longrightarrow M$ be an epimorphism. Then $\phi f : L \longrightarrow V$ is an epimorphism, hence by (i) P is isomorphic to a direct summand of L.

(iii) Suppose that M is a projective cover of V, and in the diagram above choose ϕ to be essential. It will be shown that θ is also essential, from which will follow that $M = \text{Im} f \cong P$. Assume that M' is a proper submodule of M and let $P' = \theta(M')$. Then

$$\phi(M') = (\psi\theta)(M') = \psi(P') \neq V$$

and hence $P' \neq P$. Thus θ is essential, as we wished to show. ∎

We have accumulated all the information necessary to prove the following fundamental result. The statement of this result uses the fact that if R is artinian, then $V/J(R)V$ is completely reducible (Propositions 6.2 and 6.28).

From now on we shall write $P(V)$ for a projective cover of V.

9.3. THEOREM. Let R be an artinian ring and let V be an R-module.

(i) V has a projective cover

(ii) A projective R-module W is a projective cover of V if and only if

$$W/J(R)W \cong V/J(R)V$$

(iii) If V is projective and if $V/J(R)V = \bigoplus_{i=1}^{n} V_i$ is a decomposition as a direct sum of irreducible R-modules, then $V = \bigoplus_{i=1}^{n} P(V_i)$ is a decomposition as a direct sum of indecomposable R-modules. Furthermore, $V_i \cong V_j$ if and only if $P(V_i) \cong P(V_j)$.

(iv) If V and W are projective R-modules, then $V \cong W$ if and only if $V/J(R)V \cong W/J(R)W$

Proof. (i) Write V in the form L/S, where L is projective and S a submodule of L. Given a submodule W of S, let $f_W : L/W \longrightarrow L/S$ be the canonical homomorphism. Now choose W to be minimal in S such that f_W is essential; such a submodule exists, since f_S is essential and S is artinian. We claim that W is a direct summand of L; if sustained, it will follow that L/W is a projective module and hence that L/W is a projective cover of V.

Choose W' to be a submodule of L minimal among those whose projection $W' \xrightarrow{\phi} L/W$ is surjective, and let $p : L \longrightarrow L/W$ be the projection. Since L is projective, there is a commutative diagram

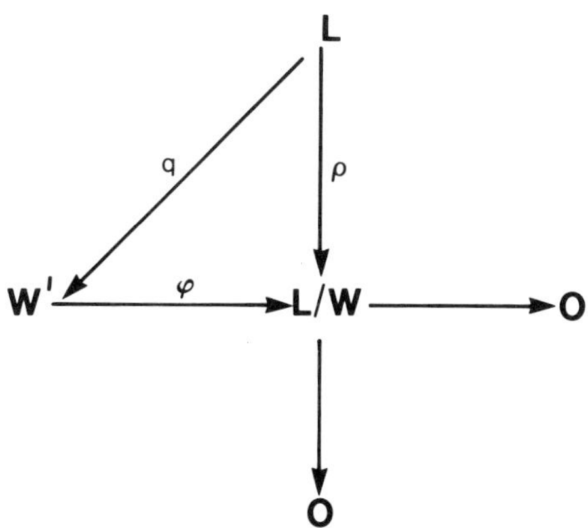

with exact row and column, and the minimality of W' ensures that $q(L) = W'$. Let L' denote the kernel of q. The projection $f_{L'}$ factors into

$$L/L' \longrightarrow L/W \longrightarrow L/S$$

and the two factors are essential. Because L' is contained in W, the minimality of W forces $W = L'$ (i.e. that ϕ is an isomorphism). It follows that

$L = W \oplus W'$, as required.

(ii) Let W be a projective module, let $f : W \longrightarrow V$ be an essential epimorphism and let $\pi : V \rightarrow V/J(R)V$ be the projection. Because π is essential, so is πf and thus $\mathrm{Ker}\pi f \subseteq J(R)W$ (Lemma 9.1(i) and Proposition 6.9(i)). On the other hand, $f(J(R)W) \subseteq J(R)V$ and therefore $J(R)W \subseteq \mathrm{Ker}\pi f$. Hence $\mathrm{Ker}\pi f = J(R)W$ and thus $W/J(R)W \cong V/J(R)V$. Conversely, assume that W is a projective R-module such that $W/J(R)W \cong V/J(R)V$. Let $f : V \longrightarrow V/J(R)V$ be the projection and $g : W \longrightarrow V/J(R)V$ be the homomorphism induced by the isomorphism $W/J(R)W \cong V/J(R)V$. Since f is an essential epimorphism and since W is projective, there exists an epimorphism $g' : W \longrightarrow V$ with $fg' = g$. Hence $\mathrm{Ker}g' \subseteq \mathrm{Ker}g = J(R)W$ is a superfluous submodule of W. Therefore, W is a projective cover of V, as asserted.

(iii) By Propositions 6.9(i) and 6.28, $J(R)V$ is a superfluous submodule of V. Since V is projective, it follows that $P(V/J(R)V) = V$. Invoking Lemma 9.1(iii) and (iv), we therefore infer that

$$V = P(V_1) \oplus \ldots \oplus P(V_n),$$

where each $P(V_i)$ is indecomposable.

Assume that $P(V_i) \cong P(V_j)$. Then, obviously, $P(V_i)/J(R)P(V_i) \cong P(V_j)/J(R)P(V_j)$ and, since V_i is irreducible, $V_i/J(R)V_i = V_i$. It follows from (ii) that $V_i \cong V_j$ and since $V_i \cong V_j$ obviously implies that $P(V_i) \cong P(V_j)$, the required assertion follows.

(iv) If $V \cong W$ then obviously $V/J(R)V \cong W/J(R)W$. Conversely, assume that V and W are projective R-modules such that $V/J(R)V \cong W/J(R)W$. By the Krull-Schmidt theorem, we may write

$$V/J(R)V = V_1 \oplus \ldots \oplus V_n \qquad W/J(R)W = W_1 \oplus \ldots \oplus W_n$$

where each V_i, W_i is irreducible and $V_i \cong W_i$, $1 \leqslant i \leqslant n$. Then, by (iii), we have

$$W = P(W_1) \oplus \ldots \oplus P(W_n) \cong P(V_1) \oplus \ldots \oplus P(V_n)$$

as required. ∎

Suppose that R is an artinian ring and let $n \geqslant 1$ be such that

$$R = U_1 \oplus \ldots \oplus U_n$$

where the U_i are principal indecomposable R-modules. Owing to the Krull-Schmidt Theorem, the U_i are uniquely determined up to isomorphism and the order in which they appear. By Proposition 2.12, the above decomposition determines a complete set $\{e_1,\ldots,e_n\}$ of primitive idempotents in R such that $U_i = Re_i$. Except when $J(R) = 0$ the principal indecomposable R-modules form only a small subclass of all indecomposable R-modules. However, it is the class which plays a very important role, as can be seen from the following:

9.4. PROPOSITION. Let R be an artinian ring

(i) The following conditions are equivalent:

(a) V is a projective cover of an irreducible R-module

(b) V is a projective indecomposable R-module

(c) V is a principal indecomposable R-module

(ii) If $Re_1,\ldots,Re_m$ are all nonisomorphic principal indecomposable R-modules, then $Re_1/J(R)e_1,\ldots,Re_m/J(R)e_m$ are all nonisomorphic irreducible R-modules.

Proof. (i) The implications (a) $\Rightarrow$ (b) and (c) $\Rightarrow$ (a) follow from Lemma 9.1 (iv) and Theorem 9.3(iii), respectively. Assume that (b) holds. Since V is projective, there exists $n \geqslant 1$ such that $R^n = V \oplus V'$, where R^n denotes a direct sum of n copies of R. Now the indecomposable components of R^n are principal indecomposable R-modules and V is indecomposable; hence (c) follows by virtue of the Krull-Schmidt theorem.

(ii) By (i), there exist irreducible R-modules $V_1,\ldots,V_m$ such that $Re_i = P(V_i)$, $1 \leqslant i \leqslant m$. By Theorem 9.3(iii), $V_1,\ldots,V_m$ are all nonisomorphic irreducible R-modules. By Propositions 6.9(i) and 6.36, $J(R)e_i$ is a superfluous submodule of Re_i and thus

$$P(Re_i/J(R)e_i) = Re_i = P(V_i) \qquad (1 \leqslant i \leqslant m)$$

Because $Re_i/J(R)e_i$ is completely reducible and Re_i is indecomposable, Lemma

9.1(iii) implies that $Re_i/J(R)e_i$ is in fact irreducible. Hence, by Theorem 9.3(iii), $V_i \cong Re_i/J(R)e_i$ and the result follows. ■

10. ALGEBRAS OVER FIELDS.

Throughout this section, A denotes a finite-dimensional algebra over a field F. All modules are assumed to be finitely generated. Let V be a vector space over F, let L be a field extension of F, and form the tensor product $V_L = L \otimes_F V$, which is then a vector space over L. If $\{v_1,\dots,v_n\}$ is a basis of V over F, then the elements $\{1\otimes v_1,\dots,1\otimes v_n\}$ form a basis of V_L over L. With the aid of the injective F-homomorphism.

$$\begin{cases} V \longrightarrow V_L \\ v \longmapsto 1 \otimes v \end{cases}$$

we shall often identify V with its image in V_L. With this identification, each element of V_L has a unique representation of the form $\sum_{i=1}^{n} \lambda_i v_i$ with $\lambda_i \in L$ and $\sum_{i=1}^{n} \lambda_i v_i \in V$ if and only if all $\lambda_i \in F$. Therefore, in passing from V to V_L we are extending the field of operators from F to L.

Suppose now that A is an F-algebra and let $A_L = L \otimes_F A$. Then A_L is an L-algebra and the map

$$\begin{cases} A \longrightarrow A_L \\ a \longmapsto 1 \otimes a \end{cases}$$

is an injective homomorphism of F-algebras. Again, identifying A with its image in A_L and denoting by $\{a_1,\dots,a_m\}$ an F-basis of A, we see that each element of A_L can be uniquely written in the form $\sum_{i=1}^{m} \lambda_i a_i$ with $\lambda_i \in L$; moreover, $\sum_{i=1}^{m} \lambda_i a_i \in A$ if and only if all $\lambda_i \in F$.

If V is an A-module, then the vector space V_L becomes an A_L-module under a module action

$$(\lambda_1 \otimes a)(\lambda_2 \otimes v) = \lambda_1\lambda_2 \otimes av \qquad (a \in A, v \in V, \lambda_1,\lambda_2 \in L)$$

By identifying V and A with their images in V_L and A_L, respectively, the action above can be written as

$$(\Sigma\lambda_i a_i)(\Sigma\mu_j v_j) = \sum_{i,j} \lambda_i\mu_j(a_i v_j) \qquad (\lambda_i,\mu_j \in L)$$

Thus the action of A_L on V_L is just the extension of the action of A on V by L-linearity.

Suppose now that B is an F-subspace of A. Then there is an injective L-homomorphism $\theta : L \otimes_F B \longrightarrow L \otimes_F A$ defined by $\theta(\ell \otimes b) = \ell \otimes b, \ell \in L, b \in B$. We write B_L for the image of B under θ. If B is a subring (ideal), then B_L is a subring (ideal) of A_L.

10.1 LEMMA. Let L be a field extension of F and let A be an algebra over F.

(i) If $A_1, A_2, \ldots, A_n$ are F-algebras, then

$$(\prod_{i=1}^{n} A_i)_L \cong \prod_{i=1}^{n} (A_i)_L \qquad \text{as } L\text{-algebras}$$

(ii) For all $n \geqslant 1$,

$$[M_n(A)]_L \cong M_n(A_L) \qquad \text{as } L\text{-algebras}$$

(iii) If I is an ideal of A, then

$$(A/I)_L \cong A_L/I_L \qquad \text{as } L\text{-algebras}$$

(iv) $J(A)_L \subseteq J(A_L)$

(v) If V, W are A-modules with $V \supseteq W$, then

$$(V/W)_L \cong V_L/W_L \qquad \text{as } A_L\text{-modules}$$

(vi) If V is an A-module, then $J(V)_L \subseteq J(V_L)$

Proof. (i) The map $(\prod_{i=1}^{n} A_i)_L \longrightarrow \prod_{i=1}^{n} (A_i)_L$, $\lambda \otimes (a_1, \ldots, a_n) \longmapsto (\lambda \otimes a_1, \ldots, \lambda \otimes a_n)$ is obviously an injective homomorphism of L-algebras. Since both of them are of the same L-dimension, the assertion follows.

(ii) The required isomorphism is given by $\lambda \otimes (a_{ij}) \longmapsto (\lambda \otimes a_{ij}), \lambda \in L, a_{ij} \in A$.

(iii) The map $\lambda \otimes a \longmapsto \lambda \otimes (a+I)$ is an L-algebra homomorphism of A_L onto

$(A/I)_L$ whose kernel contains I_L. The desired conclusion now follows from the fact that the algebras $(A/I)_L$ and A_L/I_L are of the same L-dimension.

(iv) By Proposition 6.22, $J(A)$ is nilpotent, hence so is $J(A)_L$. Now apply Corollary 6.21.

(v) The proof is similar to that of (iii)

(vi) Applying (iv), we have

$$J(V)_L = L \underset{F}{\otimes} J(A)V = (L \underset{F}{\otimes} J(A))(L \underset{F}{\otimes} V) = J(A)_L V_L \subseteq J(A_L)V_L = J(V_L),$$

as required. ■

10.2. LEMMA. Let V,W be A-modules and let L be a field extension of F. Then

$$\text{Hom}_{A_L}(V_L,W_L) \cong (\text{Hom}_A(V,W))_L \qquad \text{as } L\text{-spaces}$$

If $V = W$, the two sides of this are isomorphic L-algebras.

Proof. Given $f \in \text{Hom}_A(V,W)$, we have $1 \otimes f \in \text{Hom}_A(V_L,W_L)$. Hence the map

$$\begin{cases} (\text{Hom}_A(V,W))_L \to \text{Hom}_{A_L}(V_L,W_L) \\ \lambda \otimes f \mapsto \lambda(1 \otimes f) \end{cases}$$

is an injective L-homomorphism. It therefore suffices to verify that each element ψ in $\text{Hom}_{A_L}(V_L,W_L)$ is of the form $\psi = \sum_i \lambda_i(1 \otimes f_i)$ with $\lambda_i \in L, f_i \in \text{Hom}_A(V,W)$.

Let $\{\lambda_i | i \in I\}$ be an F-basis of L. Then each element x of V_L can be uniquely written in the form $\sum_{i\in I} \lambda_i \otimes v_i$ with $v_i \in V$ and with finitely many v_i distinct from 0. Define $f_i : V \longrightarrow W$ by

$$\psi(1 \otimes v) = \sum_{i\in I} \lambda_i \otimes f_i(v) \qquad (v \in V)$$

Then obviously each f_i is an A-homomorphism such that $\psi = \sum_i \lambda_i(1 \otimes f_i)$. ■

An A-module V is said to be *absolutely irreducible* if for any field extension L of F, V_L is an irreducible A_L-module.

10.3. PROPOSITION. Let V be an irreducible A-module. Then the following statements are equivalent:

(i) V is absolutely irreducible

(ii) If L is the algebraic closure of F, then V_L is an irreducible A_L-module.

(iii) $\mathrm{End}_A(V) = F$, i.e. each A-endomorphism of V is a left multiplication by an element of F.

Proof. (i) $\Rightarrow$ (ii) Obvious

(ii) $\Rightarrow$ (iii) We may harmlessly assume that V is faithful, in which case V_L is a faithful irreducible A_L-module. Then A_L is a simple algebra over an algebraically closed field L, so $A_L \cong M_n(L)$ for some $n \geqslant 1$ (Proposition 6.27). Since A is a simple algebra, we also have $A \cong M_k(D)$ for some division algebra D over F and some $k \geqslant 1$. Invoking Lemma 10.1(ii), we deduce that $M_n(L) \cong M_k(L \otimes_F D)$. By looking at the centers of both algebras, we have $L \cong L \otimes_F D$. Hence $D = F$ and, since $\mathrm{End}_A(V) \cong D^{\circ}$ (Propositions 5.6 and 5.8), we conclude that $\mathrm{End}_A(V) = F$.

(iii) $\Rightarrow$ (i): Again, we may assume that V is faithful in which case $A \cong M_n(F)$ and hence $A_L \cong M_n(L)$, for any field extension L of F. Identifying V with a minimal left ideal of $M_n(F)$, we see that V_L is a left ideal of $M_n(L)$ with $\dim_L V_L = n$. Hence V_L is irreducible and the result follows. ∎

Let A be an algebra over a field F. Then F is called a *splitting field* for A if every irreducible A-module is absolutely irreducible.

10.4. COROLLARY. F is a splitting field for A if and only if

$$A/J(A) \cong \prod_{i=1}^{r} M_{n_i}(F)$$

for some positive integers $n_1, n_2, \ldots, n_r$.

Proof. Let $V_1, \ldots, V_r$ be all nonisomorphic irreducible A-modules and let $D_i = \mathrm{End}_A(V_i)$, $1 \leqslant i \leqslant r$. By Propositions 5.6 and 5.8, $A/J(A) \cong \prod_{i=1}^{r} M_{n_i}(D_i^{\circ})$ for some positive integers $n_1, \ldots, n_r$. By Proposition 10.3, F is a splitting field for A if and only if $D_i = D_i^{\circ} \cong F$ for all $i \in \{1, \ldots, r\}$. So the corollary is true. ■

10.5. PROPOSITION. Let an ideal I of A possess a basis consisting of nilpotent elements. Then I is nilpotent.

Proof. We may harmlessly assume that F is algebraically closed. Furthermore, since $(I+J(A))/J(A)$ is an ideal of $A/J(A)$ having a basis consisting of nilpotent elements, we may assume that $J(A) = 0$. Thus $A = \prod_{i=1}^{r} M_{n_i}(F)$ for some positive integers $n_1,\dots,n_r$ and we denote by π_i the projection of I into $M_{n_i}(F)$, $1 \leqslant i \leqslant r$. It clearly suffices to verify that $\pi_i(I) = 0$ for all $i \in \{1,\dots,r\}$. Since $\pi_i(I)$ is an ideal of $M_{n_i}(F)$, either $\pi_i(I) = 0$ or $\pi_i(I) = M_{n_i}(F)$. In the latter case it will follow that $M_{n_i}(F)$ has a basis consisting of nilpotent elements. But then the trace of each matrix in $M_{n_i}(F)$ would be zero, which is impossible. Thus $\pi_i(I) = 0$ as required. ■

10.6. PROPOSITION. Let A be an F-algebra. Then $J(Z(A)) = Z(A) \cap J(A)$.

Proof. If $z \in J(Z(A))$, then z is nilpotent and hence $I = Az$ is a nilpotent ideal of A. Thus $z \in I \subseteq J(A)$ and therefore $J(Z(A)) \subseteq Z(A) \cap J(A)$. Conversely, if $z \in Z(A) \cap J(A)$, then z is nilpotent and hence $z \in J(Z(A))$. ■

10.7. PROPOSITION. If F is a splitting field for A, then F is a splitting field for $Z(A)$.

Proof. By Proposition 10.6, we have

$$Z(A)/J(Z(A)) \cong (Z(A)+J(A))/J(A) \subseteq Z(A/J(A))$$

Invoking Corollary 10.4, we deduce that $Z(A)/J(Z(A))$ is a subalgebra of a direct product of finitely many copies of F. But any such subalgebra must also be a direct product of finitely many copies of F. Hence the result follows by virtue of Corollary 10.4. ■

We now turn our attention to direct decomposition of algebras. First, however, we must develop our vocabulary.

By an *F-representation* of an F-algebra A (or simply a representation of A) we understand any homomorphism $A \longrightarrow \mathrm{End}_F(V)$ of F-algebras, where V is a (finite

-dimensional) vector space over F. If $n = \dim_F V$, then $\operatorname{End}_F(V)$ is identifiable with the F-algebra $M_n(F)$. Thus the given representation of A defines a homomorphism $A \longrightarrow M_n(F)$; we shall refer to any such homomorphism as a *matrix representation* of A.

If $f : A \longrightarrow \operatorname{End}_F(V)$ is a representation of A, define $xv = f(x)v$ for $x \in A, v \in V$. In this way V becomes an A-module, called the *underlying module* of f. Conversely, if V is an A-module, define $f : A \longrightarrow \operatorname{End}_F(V)$ by $f(x)v = xv$ for $x \in A, v \in V$. Then f is a representation of A; in case V is the regular A-module, we shall refer to f as the *regular representation* of A.

Two representations $f_i : A \longrightarrow \operatorname{End}_F(V_i)$, $i = 1,2$, are said to be *equivalent* if there exists an F-isomorphism $\psi : V_1 \longrightarrow V_2$ such that

$$f_2(x) = \psi f_1(x)\psi^{-1} \qquad \text{for all } x \in A$$

It is easily seen that two representations of A are equivalent if and only if the underlying modules are isomorphic. We say that a representation $f : A \longrightarrow \operatorname{End}_F(V)$ is *irreducible (completely reducible, indecomposable)* if the underlying module V is irreducible (completely reducible, indecomposable); the same terminology will be applied to matrix representations.

10.8. PROPOSITION. Let A be an F-algebra.

(i) There exists a direct decomposition

$$A = B_1 \oplus \ldots \oplus B_n$$

of A into indecomposable two-sided ideals $B_i \neq 0$ with $B_iB_j = 0$ for $i \neq j$

(ii) Write $1 = e_1 + \ldots + e_n$ with $e_i \in B_i$. Then the e_i are mutually orthogonal centrally primitive idempotents and $B_i = Ae_i = e_iA$.

(iii) $Z(A) = Z(A)e_1 \oplus \ldots \oplus Z(A)e_n$ is a direct decomposition of $Z(A)$ into indecomposable ideals and $Z(A)e_i = B_i \cap Z(A)$.

(iv) $J(Z(A)e_i) = J(Z(A))e_i$ and $Z(A)e_i/J(Z(A)e_i)$ is a finite field extension of F.

(v) Let $F_i = Z(A)e_i/J(Z(A)e_i)$, let $\theta_i : Z(A) \longrightarrow F_i$ be the natural homomorphism and let $f_i : X_i \longrightarrow M_{n_i}(F)$ be the regular representation of F_i. Define $\gamma_i : Z(A) \longrightarrow M_{n_i}(F)$ by $\gamma_i = f_i\theta_i$. Then

$$\mathrm{Ker}\gamma_i = \mathrm{Ker}\theta_i = Z(A)(1-e_i) + J(Z(A))e_i \qquad (1 \leqslant i \leqslant n)$$

and $\{\gamma_1,\ldots,\gamma_n\}$ is a complete set of nonequivalent irreducible matrix representation of $Z(A)$. Each γ_i satisfies $\gamma_i(e_i) = 1$ and $\gamma_i(e_j) = 0$ for $j \neq i$. Furthermore, if F is a splitting field for $Z(A)$ then each $n_i = 1$.

Proof. (i) and (ii). Apply Propositions 4.1(iii) and 4.3

(iii) The first statement and the inclusion $Z(A)e_i \subseteq B_i \cap Z(A)$ follows from (ii). Suppose that $z \in B_i \cap Z(A)$. Since e_i is the identity element of B_i, we have $z = ze_i \in Z(A)e_i$, proving that $B_i \cap Z(A) \subseteq Z(A)e_i$.

(iv) By Proposition 6.35, $J(Z(A)e_i) = J(Z(A))e_i$. Because each e_i is primitive, Proposition 9.4(ii) ensures that the $Z(A)$-module $Z(A)e_i/J(Z(A)e_i)$ is irreducible. The latter implies that the commutative algebra $Z(A)e_i/J(Z(A)e_i)$ is simple and hence is a field. That this field is a finite extension of F follows from the assumptions that A is finite-dimensional over F.

(v) It is obvious that $\mathrm{Ker}\gamma_i = \mathrm{Ker}\theta_i$. Assume that $z \in \mathrm{Ker}\theta_i$. Then $ze_i = xe_i$ for some $x \in J(Z(A))$, so

$$z = z(1-e_i) + xe_i \in Z(A)(1-e_i) + J(Z(A))e_i$$

and therefore $\mathrm{Ker}\theta_i \subseteq Z(A)(1-e_i) + J(Z(A))e_i$. The opposite inclusion being obvious, we conclude that $\mathrm{Ker}\theta_i = Z(A)(1-e_i) + J(Z(A))e_i$.

It follows from (iii) and (iv) that

$$Z(A)/J(Z(A)) = \bigoplus_{i=1}^{n} F_i$$

and that each F_i is a field. Furthermore, if F is a splitting field for $Z(A)$, then each F_i is isomorphic to F and hence $n_i = 1$. Since f_i is the only irreducible representation of F_i, the result follows. ■

We now introduce the important concept of blocks. Let

$$A = \bigoplus_{i=1}^{n} B_i = \bigoplus_{i=1}^{n} Ae_i$$

be the decomposition defined in Proposition 10.8. We shall refer to B_i as a *block* and to e_i as a *block idempotent* of A. We shall also write $B = B(e)$ to indicate that B is a block containing the block idempotent e. The representation $\gamma_i : Z(A) \longrightarrow M_{n_i}(F)$ described in Proposition 10.8 is uniquely determined up to equivalence by e_i. We call γ_i the *irreducible representation* of $Z(A)$ *associated with* e_i.

By a *central character* of A we understand any F-algebra homomorphism $Z(A) \longrightarrow F$. If F is a splitting field for $Z(A)$, then by Proposition 10.8, $\gamma_1,\ldots,\gamma_n$ are all central characters of A; we shall refer to γ_i as the *central character of* A *associated with* e_i (or B_i).

10.9. PROPOSITION. Let F be a splitting field for $Z(A)$ and let n be the number of blocks of A. Then

$$\dim_F J(Z(A)) = \dim_F Z(A) - n$$

Proof. As we have seen in the proof of Proposition 10.8(v), $Z(A)/J(Z(A))$ is isomorphic to a direct sum of n copies of F. Hence

$$\dim_F Z(A) - \dim_F J(Z(A)) = n$$

and the result follows. ■

10.10. PROPOSITION. Let $A = B(e_1) \oplus \ldots \oplus B(e_n)$ be the direct decomposition of A into sum of blocks.

(i) For every A-module V, $V = \bigoplus_{i=1}^{n} e_i V$ is a direct decomposition of V into A-modules. In particular, if V is indecomposable, then $V = e_i V$ for a unique $i \in \{1,2,\ldots,n\}$

(ii) For each left ideal I of A, we have

$$I = \bigoplus_{i=1}^{n} (I \cap B(e_i))$$

In particular, each indecomposable left ideal lies in exactly one of the $B(e_i)$.

(iii) If A_1 and A_2 are ideals of A and $A = A_1 \oplus A_2$, then A_1 and A_2 are direct sums of blocks. In particular, the blocks are the only indecomposable ideals of A.

(iv) Let I be an ideal of $Z(A)$ and let γ_i be the irreducible representation of $Z(A)$ associated with e_i. Then $e_i \in I$ if and only if $\gamma_i(I) \neq 0$.

Proof. (i) Because e_i is central, $e_i V$ is an A-module. Since the e_i are orthogonal, we also have

$$\sum_{i=1}^{n} e_i V = \bigoplus_{i=1}^{n} e_i V$$

Taking into account that

$$V = 1\cdot V = (\sum_{i=1}^{n} e_i)V \subseteq \sum_{i=1}^{n} e_i V \subseteq V$$

the assertion is proved.

(ii) By (i), $I = \bigoplus_{i=1}^{n} e_i A$ and so

$$e_i I \subseteq I \cap B(e_i) = I \cap e_i A \subseteq e_i I$$

Hence

$$I = \bigoplus_{i=1}^{n} (I \cap B(e_i))$$

(iii) By (ii),

$$A = A_1 \oplus A_2 = \bigoplus_{i=1}^{n} (A_1 \cap B(e_i)) \bigoplus_{i=1}^{n} (A_2 \cap B(e_i))$$

Hence for all i

$$B(e_i) = (A_1 \cap B(e_i)) \oplus (A_2 \cap B(e_i))$$

and since $B(e_i)$ is indecomposable we have $B(e_i) \subseteq A_1$ or $B(e_i) \subseteq A_2$. Thus

$$A_k = \oplus B(e_i) \qquad (k = 1,2)$$

$$B(e_i) \subseteq A_k$$

as asserted.

(iv) If $e_i \in I$ then $\gamma_i(e_i) = 1 \neq 0$ and hence $\gamma_i(I) \neq 0$. Conversely, assume that $\gamma_i(I) \neq 0$. Because $\gamma_i(I)$ is a nonzero ideal of the field $\gamma_i(Z(A))$, we have

$$\gamma_i(I) = \gamma_i(Z(A))$$

Thus $\gamma_i(e_i) \in \gamma_i(I)$. It therefore follows, from Proposition 10.8(v), that there are elements $x \in I, y \in J(Z(A))$ and $z \in Z(A)$ such that

$$x = e_i + z(1-e_i) + y$$

Multiplying both sides by e_i yields $e_i = xe_i - ye_i$. Hence, by raising both sides to the k-th power, we find

$$e_i = u + (-1)^k (ye_i)^k$$

with $u \in I$. Since ye_i is nilpotent, the result follows. ■

Let $e_1, e_2, \ldots, e_n$ be the block idempotents of A and let V be an indecomposable A-module. It follows from Proposition 10.10 that there exists $i \in \{1,2,\ldots,n\}$ such that

$$e_i V = V \quad \text{and} \quad e_j V = 0 \quad \text{for all} \quad j \neq i$$

In this case we say that V *lies in the block* $B(e_i)$. Observe that if V lies in the block $B(e_i)$, then

(a) All A-modules isomorphic to V lie in $B(e_i)$

(b) $v = 1 \cdot v = e_i v$ for all $v \in V$

The above gives a classification of all indecomposable (and in particular, all irreducible) A-modules into blocks. Our next aim is to tie together blocks and principal indecomposable modules. The following preliminary observations will clear our path.

10.11. LEMMA. Let e be an idempotent of A. Then

(i) For any ideal I of A, $Ae \cap I = Ie$

(ii) $Ae/J(A)e \cong \bar{A}\bar{e}$ where $\bar{e}$ denotes the image of e in $\bar{A} = A/J(A)$.

Proof. (i) Suppose that $x \in Ae \cap I$. Then $x = ae$ for some $a \in A$. Therefore

$$x = ae = ae^2 = xe \in Ie$$

proving that $Ae \cap I \subseteq Ie$. The opposite containment being trivial, (i) follows

(ii) Applying (i) for $I = J(A)$, we have

$$\begin{aligned} Ae/J(A)e &= Ae/(Ae \cap J(A)) \cong (Ae + J(A))/J(A) \\ &= (A/J(A))(e + J(A)) = \bar{A}\bar{e} \end{aligned}$$

as required. ■

Note that if Ae is a principal indecomposable A-module, then by Proposition 9.4(ii) and Lemma 10.11(ii), $\bar{A}\bar{e}$ is an irreducible A-module.

10.12. **LEMMA.** Let $A = \bigoplus_{i=1}^{n} Ae_i$ be a decomposition of A into principal indecomposable A-modules. Let $\bar{e}_i$ denote the image of e_i in $\bar{A} = A/J(A)$. Then for any A-module V,

$$\dim_F e_i V = \dim_F \operatorname{Hom}_A(Ae_i, V) = m \dim_F \operatorname{End}_A(\bar{A}\bar{e}_i);$$

where m is the multiplicity of the irreducible A-module $\bar{A}\bar{e}_i$ as a composition factor of V.

Proof. By Lemma 5.5, the map $\operatorname{Hom}_A(Ae_i, V) \longrightarrow e_i V$, $f \mapsto f(e)$ is an isomorphism of additive groups. Since this map is obviously F-linear, we have

$$\operatorname{Hom}_A(Ae_i, V) \cong e_i V \qquad \text{as } F\text{-spaces}$$

The equality $\dim_F \operatorname{Hom}_A(Ae_i, V) = m \dim_F \operatorname{End}(\bar{A}\bar{e})$ is trivial if V is irreducible. Therefore we may assume that V has a maximal nonzero submodule X. Because Ae_i is projective, the exact sequence $0 \to X \to V \to V/X \to 0$ induces an exact sequence

$$0 \longrightarrow \operatorname{Hom}_A(Ae_i, X) \longrightarrow \operatorname{Hom}_A(Ae_i, V) \longrightarrow \operatorname{Hom}_A(Ae_i, V/X) \longrightarrow 0$$

Thus we have

$$\dim_F \operatorname{Hom}_A(Ae_i, V) = \dim_F \operatorname{Hom}_A(Ae_i, X) + \dim_F \operatorname{Hom}_A(Ae_i, V/X)$$

If m_1 and m_2 are the multiplicities of $\bar{A}\bar{e}_i$ as a composition factor of X and V/X, respectively, then applying induction on $\dim V$, we have

$$\dim_F \operatorname{Hom}_A(Ae_i, X) = m_1 \dim_F \operatorname{End}_A(\bar{A}\bar{e}_i)$$

$$\dim_F \operatorname{Hom}_A(Ae_i, V/X) = m_2 \dim_F \operatorname{End}_A(\bar{A}\bar{e}_i)$$

Adding up these equalities gives

$$\begin{aligned}\dim_F \operatorname{Hom}_A(Ae_i, V) &= (m_1+m_2)\dim_F \operatorname{End}_A(\bar{A}\bar{e}_i)\\ &= m\dim_F \operatorname{End}_A(\bar{A}\bar{e}_i)\end{aligned}$$

as asserted. ∎

As an immediate consequence, we derive

10.13. COROLLARY. Let e be a primitive idempotent of A, let $\bar{e}$ be the image of e in $\bar{A} = A/J(A)$ and let V be an A-module. Then the following conditions are equivalent:

(i) $\bar{A}\bar{e}$ is a composition factor of V

(ii) $\operatorname{Hom}_A(Ae, V) \neq 0$

(iii) $eV \neq 0$. ∎

Let U and V be principal indecomposable A-modules. We say that U and V are *linked*, written $U \sim V$, if there exists a sequence $U_1 = U, \dots, U_n = V$ of principal indecomposable modules such that any two neighbouring left ideals in this sequence have a common composition factor. It is clear that $\sim$ is an equivalence relation on the set of all principal indecomposable A-modules. In what follows we denote by $X_1, \dots, X_m$ the equivalence classes of $\sim$.

10.14. PROPOSITION. Let $A = \bigoplus_{i=1}^{n} Ae_i$ be a decomposition of A into principal indecomposable modules, and for each $j \in \{1, \dots, m\}$ put

$$I_j = \bigoplus_{Ae_i \in X_j} Ae_i$$

(i) $I_1, \dots, I_m$ are all blocks of A

(ii) Two principal indecomposable A-modules are linked if and only if they belong to the same block.

Proof. (i) It is obvious that $A = \bigoplus_{j=1}^{m} I_j$. By Proposition 1.10(iii), it suffices to show that each I_j is a two-sided ideal contained in a block. Assume that $Ae_i \subseteq I_j$ and $Ae_k \subseteq I_\ell$ with $j \neq \ell$. Then Ae_i and Ae_k have no composition factor in common. In particular, Ae_k has no composition factor isomorphic to $\bar{A}\bar{e}_i$. Thus, by Corollary 10.13, $e_iAe_k = 0$. This proves that $I_j I_\ell = 0$ for $j \neq \ell$. Hence

$$I_jA = I_j(\bigoplus_{\ell=1}^{m} I_\ell) = I_j^2 \subseteq I_j$$

and so I_j is a two-sided ideal of A.

If Ae_i and Ae_j have the composition factor $\bar{A}\bar{e}_k$ in common, then by Corollary 10.13, $e_kAe_i \neq 0$ and $e_kAe_j \neq 0$. By Proposition 10.10(ii), there exist blocks B, B', B'' such that $Ae_i \subseteq B$, $Ae_j \subseteq B'$ and $Ae_k \subseteq B''$. It follows that $0 \neq e_kAe_i \subseteq B''B$ and hence $B = B''$. Similarly, $0 \neq e_kAe_j \subseteq B''B'$ and so $B' = B''$. Consequently $Ae_i + Ae_j \subseteq B$ and repeated application of this argument shows that I_j is contained in a certain block of A. This establishes the required assertion

(ii) Direct consequence of (i). ■

Assume that F is a splitting field for A and let V be an irreducible A-module. Then $\mathrm{End}_A(V) = F$ and each element $z \in Z(A)$ acts as an element of $\mathrm{End}_A(V)$ on V. Thus there is a function $\psi : Z(A) \longrightarrow F$ such that

$$zv = \psi(z)v \qquad \text{for all } z \in Z(A), v \in V$$

Clearly ψ is a central character of A. We shall refer to ψ as the *central character of A afforded by V.*

10.15. PROPOSITION. Suppose that F is a splitting field for A. Then two irreducible A-modules belong to the same block if and only if they afford the same central characters.

Proof. Assume that the irreducible A-modules U and V afford the central characters ρ and ψ, respectively. If U and V belong to a block $B = B(e)$

then $eu = u$ and $ev = v$ for all $u \in U, v \in V$. Hence $\rho(e) = \psi(e) = 1$, proving that $\rho = \psi$ by virtue of Proposition 10.8(v). Conversely, suppose that $\rho = \psi$, and let U belong to $B(e)$. Then $\rho(e) = \psi(e) = 1$ and hence $ev = \psi(e)v = v$ for all $v \in V$. This shows that V lies in B and the result follows. ■

Let $e_1, \dots, e_n$ be a complete set of primitive idempotents of A. Then

$$A = Ae_1 \oplus \dots \oplus Ae_n$$

is a decomposition of A as a direct sum of principal indecomposable modules. Let $\overline{e_i}$ denote the image of e_i in $\bar{A} = A/J(A)$. We know, from Lemma 10.11(ii), that

$$Ae_i/J\,A\,e_i \cong \bar{A}\,\overline{e_i}$$

Suppose that the numbering is so chosen that $Ae_1, \dots, Ae_m$ are all nonisomorphic indecomposable A-modules. Then, by Proposition 9.4, $\bar{A}\bar{e}_1, \dots, \bar{A}\bar{e}_m$ are all nonisomorphic irreducible A-modules. Given $i, j \in \{1, \dots, m\}$, let c_{ij} be the multiplicity of the irreducible A-module $\bar{A}\,\bar{e}_j$ as a composition factor of Ae_i. The nonnegative integers c_{ij} are called the *Cartan invariants* of A, and the $m \times m$ matrix $C = (c_{ij})$ is called the *Cartan matrix* of A. The Cartan matrix C is only determined up to a permutation of its rows and columns (depending on the numbering of $Ae_1, \dots, Ae_m$).

10.16. PROPOSITION. Let F be a splitting field for A, let

$$A = Ae_1 \oplus \dots \oplus Ae_n$$

be a decomposition of A into direct sum of principal indecomposable modules, and let $C = (c_{ij})$ be the Cartan matrix of A. Then

(i) $c_{ij} = \dim_F e_j A e_i$

(ii) The multiplicity of Ae_i as a direct summand of A in the sense of the Krull-Schmidt theorem is $\dim_F \bar{A}\,\bar{e}_i$

(iii) The multiplicity of $\bar{A}\,\bar{e}_i$ as a composition factor of A is $\dim_F e_i A$

Proof. (i) Since F is a splitting field for A, $\underset{A}{\mathrm{End}}(\bar{A}\bar{e}_i) = F$. Applying Lemma 10.12, we deduce that the multiplicity of $\bar{A}\bar{e}_j$ as a composition factor of Ae_i is equal to $\dim_F e_j Ae_i$, as required.

(ii) Since F is a splitting field for A, $\bar{A}$ is a direct sum of full matrix rings over F. Hence the multiplicity of $\bar{A}\bar{e}_i$ in $\bar{A}$ is equal to $\dim_F \bar{A}\bar{e}_i$. Owing to Proposition 9.4, Ae_i is determined, up to isomorphism, by $\bar{A}\bar{e}_i$ and therefore $\dim_F \bar{A}\bar{e}_i$ also equals the multiplicity of Ae_i as a direct summand of A.

(iii) Direct consequence of Lemma 10.12. ∎

As an immediate consequence, we derive

10.17. COROLLARY. Let F be a splitting field for A, let $V_1,\ldots,V_r$ be all nonisomorphic irreducible A-modules and let $n_i = \dim_F V_i$, $1 \leqslant i \leqslant r$. Then

$$A/J(A) \cong \bigoplus_{i=1}^{r} n_i V_i \quad \text{and} \quad A \cong \bigoplus_{i=1}^{r} n_i P(V_i)$$

In particular,

$$\dim_F J(A) = \dim_F A - \sum_{i=1}^{r} n_i^2 \; .$$

2 Group algebras and their modules

In this chapter we introduce group algebras and discuss a number of their properties to be applied in subsequent investigations. These include support of central idempotents, the notion of defect group of a block, the number of irreducible FG-modules and some formal properties of induced modules. Later chapters will treat various aspects of these topics in greater detail and depth. In selecting material we have held steadfast to a single-minded purpose: to present only those results deemed essential for the application to the study of the Jacobson radical of the group algebra FG.

1. GROUP ALGEBRAS

Let R be a commutative ring and let G be a group, possibly infinite. The group algebra RG is the free R-module on the elements of G, with multiplication induced by that in G. More explicitely, RG consists of all formal linear combinations $\Sigma x_g \cdot g, x_g \in R, g \in G$, with finitely many $x_g \neq 0$ subject to

(i) $\Sigma x_g \cdot g = \Sigma y_g \cdot g$ if and only if $x_g = y_g$ for all $g \in G$

(ii) $\Sigma x_g \cdot g + \Sigma y_g \cdot g = \Sigma (x_g + y_g) \cdot g$

(iii) $(\Sigma x_g \cdot g)(\Sigma y_h \cdot h) = \Sigma z_t \cdot t$ where $z_t = \sum_{gh=t} x_g y_h$

(iv) $r(\Sigma x_g \cdot g) = \Sigma (r x_g) \cdot g$ for all $r \in R$

It is straightforward to verify that these operations define RG as an associative R-algebra with $1 = 1_R \cdot 1_G$ where 1_R and 1_G are identity elements of R and G, respectively. With the aid of the injective homomorphisms

$$\begin{cases} R \longrightarrow RG \\ r \longmapsto r \cdot 1_G \end{cases} \qquad \begin{cases} G \longrightarrow RG \\ g \longmapsto 1_R \cdot g \end{cases}$$

we shall in the future identify R and G with their images in RG. With these identifications, the formal sums and products become ordinary sums and products

For this reason, from now on we drop the dot in $x_g \cdot g$. We shall also adopt the convention that $RG \cong RH$ means an isomorphism of R-algebras.

Let $x = \Sigma x_g g \in RG$. Then the *support* of x, Supp$x$, is defined by

$$\text{Supp}x = \{g \in G | x_g \neq 0\}$$

It is clear that Supp$x$ is a finite subset of G that is empty if and only if $x = 0$. We next exhibit some elementary properties of group algebras.

1.1. PROPOSITION. Let A be an R-algebra and let

$$\psi : G \longrightarrow U(A)$$

be a homomorphism of G into the unit group of A. Then the mapping $f : RG \longrightarrow A$ defined by

$$f(\Sigma x_g g) = \Sigma x_g \psi(g)$$

is a homomorphism of R-algebras. In particular, if ψ is injective and A is R-free with $\psi(G)$ as a basis, then $RG \cong A$.

Proof. Because RG is R-free with G as a basis, f is a homomorphism of R-modules. Let

$$x = \Sigma x_g g \quad \text{and} \quad y = \Sigma y_g g$$

be two elements of RG. Then

$$\begin{aligned} f(xy) &= f(\sum_{a,b \in G} x_a y_b ab) = \Sigma x_a y_b \psi(a)\psi(b) \\ &= \sum_{a \in G} x_a \psi(a) \sum_{b \in G} y_b \psi(b) \\ &= f(x)f(y) \end{aligned}$$

as asserted. ■

The *augmentation ideal* $I(R,G)$ of RG is defined to be the kernel of the homomorphism from RG to R induced by collapsing G to 1. In other words, $I(R,G)$ consists of all $x = \Sigma x_g g \in RG$ for which

$$\text{aug}(x) = \Sigma x_g = 0$$

We shall refer to aug(x) as the *augmentation* of x.

It follows from the equality

$$\Sigma x_g g = \Sigma x_g (g-1) + \Sigma x_g$$

that as an R-module, $I(R,G)$ is a free module with the elements $g-1$ $(1 \neq g \in G)$ as a basis. In the future we shall often suppress reference to R and simply denote the augmentation ideal of RG by $I(G)$.

If X is a subset of RG, we write $RG \cdot X$ and $X \cdot RG$ for the left and write ideals of RG, respectively, generated by X, i.e.

$$RG \cdot X = \sum_{x \in X} RGx \quad \text{and} \quad X \cdot RG = \sum_{x \in X} xRG$$

Of course, if $RGx = xRG$ for all $x \in X$, then $RG \cdot X = X \cdot RG$ is a two sided ideal of RG. For example, if N is a normal subgroup of G, then the equalities

$$(n-1)g = g(g^{-1}ng-1) \quad \text{and} \quad g(n-1) = (gng^{-1}-1)g \qquad (n \in N, g \in G)$$

show that $RG \cdot I(N) = I(N) \cdot RG$ is a two-sided ideal of RG. The signigicance of this ideal comes from the following fact.

1.2. PROPOSITION. Let $\psi : G \longrightarrow H$ be a surjective homomorphism of groups and let $N = \text{Ker}\psi$. Then the mapping $f : RG \longrightarrow RH$ which is the R-linear extension of ψ is a surjective homomorphism of R-algebras whose kernel is $RG \cdot I(N)$. In particular,

$$RG/RG \cdot I(N) \cong R(G/N)$$

Proof. That f is a surjective homomorphism of R-algebras is a consequence of Proposition 1.1. It is plain that $RG \cdot I(N) \subseteq \text{Ker}f$. Consequently, f induces a homomorphism $\bar{f} : RG/RG \cdot I(N) \longrightarrow RH$. The restriction of $\bar{f}$,

$$\lambda : [G + RG \cdot I(N)]/RG \cdot I(N) \longrightarrow H$$

is an isomorphism. Thanks to Proposition 1.1, λ^{-1} can be extended to a homomorphism $RH \longrightarrow RG/RG\cdot I(N)$ which is the inverse to $\bar{f}$. Thus $\text{Ker} f = RG\cdot I(N)$, as desired. ■

Let I be an ideal of RG and let

$$G \cap (1+I) = \{g \in G | g-1 \in I\}$$

Then $G \cap (1+I)$ is the multiplicative kernel of the natural map $G \longrightarrow RG/I$, and hence a normal subgroup of G. In view of this obsefvation, the next corollary arises from Proposition 1.2 by taking $I = RG\cdot I(N)$.

1.3. COROLLARY. Let N be a normal subgroup of G. Then

$$G \cap (1+RG\cdot I(N)) = N.$$

Let H be a subgroup of G. Because RH is a subring of RG, we can view RG as a left or right RH-module by way of ordinary multiplication.

1.4. PROPOSITION. Let H be a subgroup of G. If T is a right (left) transversal for H in G, then RG is a free left (right) RH-module with T as a basis.

Proof. Assume that T is a right transversal for H in G. Then, for any $t \in T, (RH)t$ is the R-linear span of the coset Ht. Accordingly, for any $t_1, t_2, \ldots, t_n \in T, (RH)t_1 + \ldots + (RH)t_n$ is the R-linear span of $\bigcup_{i=1}^{n} Ht_i$. As is apparent from the definition of RG, if X and Y are disjoint subsets of G, their R-linear spans meet at 0. Hence $RG = \bigoplus_{t \in T} (RH)t$, proving that RG is a free left RH-module with T as a basis. A similar argument proves the case where T is a left transversal. ■

1.5. PROPOSITION. Let S be a subring of a commutative ring R. Then for any group G, $R \otimes_S SG \cong RG$ as R-algebras.

Proof. Observe that $\{1 \otimes g | g \in G\}$ is an R-basis for $R \otimes_S SG$ and that the mapping

$$\begin{cases} G \longrightarrow R \otimes_S SG \\ g \mapsto 1 \otimes g \end{cases}$$

is an injective homomorphism. Now apply Proposition 1.1. ■

1.6. PROPOSITION. For any groups G and H,

$$R(G\times H) \cong RG \underset{R}{\otimes} RH$$

Proof. Again note that $\{x \otimes y \mid x,y \in G\}$ is an R-basis for $RG \underset{R}{\otimes} RH$ and that the mapping

$$\begin{cases} G\times H \longrightarrow RG \underset{R}{\otimes} RH \\ (x,y) \longmapsto x \otimes y \end{cases}$$

is an injective homomorphism. Now apply Proposition 1.1. ■

1.7. PROPOSITION. Let $\{C_i \mid i \in I\}$ be the set of all finite conjugacy classes of G and, for each $i \in I$, let $C_i^+ = \sum_{x\in C_i} x$. Then $Z(RG)$ is the R-linear span of all C_i^+.

Proof. Since for all $g \in G$, $g^{-1}C_i^+g = C_i^+$, we have $C_i^+ \subseteq Z(RG)$ for all $i \in I$. On the other hand, let $x = \sum x_g g \in Z(RG)$. Then, for all $t \in G, t^{-1}xt = x$ or, equivalently $x_{t^{-1}gt} = x_g$, for all $t \in G, g \in \text{Supp}x$. Hence x is an R-linear combination of those C_i^+ for which $C_i \subseteq \text{Supp}x$, as required. ■

The "if" part of the following result is a classical Maschke's theorem .

1.8. PROPOSITION. Let G be a finite group and let F be a field of characteristic $p \geqslant 0$. Then FG is semisimple if and only if p does not divide the order of G.

Proof. If $p > 0$ divides $|G|$, then $x = \sum_{g\in G} g$ satisfies $x^2 = |G|x = 0$ and $x \in Z(FG)$. Hence $FGx \subseteq J(FG) \neq 0$. Conversely, assume that $p \nmid |G|$, and let W be a submodule of an FG-module V. Write $V = W \oplus W'$ for a suitable F-subspace W' and let $\theta : V \longrightarrow W$ be the projection map. Define $\psi : V \longrightarrow V$ by

$$\psi(v) = |G|^{-1} \sum_{x\in G} x\theta x^{-1}v \qquad (v \in V)$$

Since for all $v \in V$ and $y \in G$,

$$\psi(yv) = |G|^{-1} \sum_{x\in G} x\theta x^{-1}yv = |G|^{-1} \sum_{z\in G} (yz)\theta(yz)^{-1}yv$$

$$= |G|^{-1} \sum_{g\in G} yz\theta z^{-1}v = y\psi(v),$$

ψ is an FG-homomorphism.

Suppose now that $v \in W$. Then, for any $x \in G, x^{-1}v \in W$, so $\theta(x^{-1}v) = x^{-1}v$. Accordingly, $x\theta x^{-1}v = v$ and $\psi(v) = v$. Setting $W'' = \mathrm{Ker}\psi$, it follows that W'' is an FG-submodule of V such that $W'' \cap W = 0$. Finally, let $v \in V$. Then, by the above, $v - \psi(v) \in W''$, so $v = \psi(v) + (v - \psi(v)) \in W + W'$. Thus $V = W \oplus W''$ and the result follows. ■

1.9. PROPOSITION. Let F be a field of characteristic $p > 0$, let N be a normal p'-subgroup of G and let $e = |N|^{-1} \sum_{x\in N} x$. Then e is a central idempotent of FG such that $FGe \cong F(G/N)$. In particular, $F(G/N)$ is a direct factor of FG.

Proof. It is easy to verify that e is a central idempotent. Let $f : FG \longrightarrow F(G/N)$ be the natural homomorphism. Since $FG = FGe \oplus FG(1-e)$ and $\mathrm{Ker}f = FG\cdot I(N)$, it suffices to show that $FG\cdot I(N) = FG(1-e)$.

Since $f(e) = 1$, we have $1-e \in \mathrm{Ker}f$ and therefore $FG(1-e) \subseteq FG\cdot I(N)$. Given $n \in N$, we also have $ne = e$ so that $(n-1)e = 0$ and $FG\cdot I(N)e = 0$. Thus $FG\cdot I(N) \subseteq FG(1-e)$ and the result follows. ■

We close this section by providing the following piece of information. Let R be a commutative ring and G a group. Given a nonzero R-module V, by a *representation* of G on V we understand a homomorphism

$$\rho : G \longrightarrow \mathrm{Aut}_R(V)$$

We say that ρ is *faithful* if $\mathrm{Ker}\rho = 1$. Owing to Proposition 1.1, to each representation ρ of G on V, there corresponds an RG-module structure on V given by

$$(\sum x_g g)v = \sum x_g \rho(g)v \qquad (v \in V,\ x_g \in R)$$

Conversely, any nonzero RG-module V determines a representation ρ of G on V given by $\rho(g)v = gv$ for all $g \in G$, $v \in V$. Hence there is a bijective correspondence between the class of all nonzero RG-modules and the class of all representations of G on R-modules. Moreover, if V is R-free of finite rank n, then $\mathrm{Aut}_R(V) \cong GL(n,R)$, the latter being the group of all $n \times n$ non-singular matrices with entries in R. Thus to each representation of G on the R-free R-module V of finite rank n, there is a matrix representation $\rho^* : G \longrightarrow GL(n,R)$. In particular, if $R = F$ is a field, then all finitely generated F-modules are free of finite rank. In this case the study of FG-modules, representations of G on F-modules, and matrix representations of G over F are essentially equivalent.

1.10. PROPOSITION. Let H be a subgroup of an arbitrary group G and let R be a commutative ring. If X is a generating set for H, then

$$RG\cdot I(H) = \sum_{x\in X} RG(x-1) \quad \text{and} \quad I(H)RG = \sum_{x\in X} (x-1)RG$$

Proof. We shall only establish the first equality, since the second follows by a similar argument. It is clear that $\sum_{x\in X} RG(x-1) \subseteq RG\cdot I(H)$. Thus we need only show that for any nonidentity h in H, $h-1 \in \sum_{x\in X} RG(x-1)$. Now h is a group word in the $x_i \in X$ and we proceed by induction on the length t of the word. First it is true for words of length 1 :

$$x_i-1 \in \sum_{x\in X} RG(x-1) \qquad x_i^{-1}-1 = -x_i^{-1}(x_i-1) \in \sum_{x\in X} RG(x-1)$$

Suppose it is true for words h of length t. Any word of length $t+1$ is of the form $x_i^{\pm 1}h$ and both $h-1$ and $x_i^{\pm 1}-1$ are in $\sum_{x\in X} RG(x-1)$. Hence

$$x_i^{\pm 1}h-1 = x_i^{\pm 1}(h-1) + (x_i^{\pm 1}-1) \in \sum_{x\in X} RG(x-1)$$

as desired. ∎

2. CENTRAL IDEMPOTENTS.

Throughout this section, G denotes a finite group and F a field of characteristic $p > 0$. Our aim is to provide information on support of central idempotents of the group algebra FG.

Let A be an F-algebra and define $[A,A]$, the *commutator subspace* of A, to be the F-linear span of all Lie products $[x,y] = xy-yx$ with $x,y \in A$. The following observation illustrates how the commutator subspace can be brought into argument.

2.1. **LEMMA.** Let A be an algebra over a field F of characteristic $p > 0$. If $a_1, a_2, \ldots, a_m \in A$ and if $n > 0$ is a given integer, then

$$\left(\sum_{i=1}^{m} a_i\right)^{p^n} \equiv \sum_{i=1}^{m} a_i^{p^n} \pmod{[A,A]}$$

Proof. By induction argument, we need only prove that

$$(a+b)^{p^n} \equiv a^{p^n} + b^{p^n} \pmod{[A,A]}$$

for all $a,b \in A$. Suppose first that $n = 1$. Expanding by the distributive law,

$$(a+b)^p = a^p + b^p + \sum a_1 a_2 \ldots a_p$$

where the sum is over all products $a_1 \ldots a_p$ of p terms, $a_i \in \{a,b\}$, not all equal to a or b. With each word $a_1 a_2 \ldots a_p$ associate its cyclic permutations

$$a_1 a_2 \ldots a_p \; , \; a_2 a_3 \ldots a_p a_1, \ldots, a_p a_1 \ldots a_{p-1}$$

All these products are congruent modulo $[A,A]$. This is so since for

$$x = a_{i_1} a_{i_2} \ldots a_{i_p} \; , \; y = a_{i_j} a_{i_{j+1}} \ldots a_{i_p} a_{i_1} \ldots a_{i_{j-1}}$$

we have $x - y = [\gamma,\delta] \in [A,A]$ where

$$\gamma = a_{i_1} a_{i_2} \ldots a_{i_{j-1}} \quad \text{and} \quad \delta = a_{i_j} a_{i_{j+1}} \ldots a_{i_p}$$

Therefore the sum of these cyclic permutations is $px_1 \ldots x_p$ modulo $[A,A]$, and hence it belongs to $[A,A]$. Invoking (1), it follows that

$$(2) \qquad (x+y)^p = x^p + y^p + z \qquad (z \in [A,A])$$

and hence that

$$(xy-yx)^p \equiv (xy)^p - (yx)^p \pmod{[A,A]}$$
$$\equiv [x, (yx)^{p-1} y] \pmod{[A,A]}$$

Accordingly

(3) $$s \in [A,A] \quad \text{implies} \quad s^p \in [A,A]$$

The result now follows from (2) and (3) by induction on n. ∎

The next lemma will enable us to exploit the previous result. In what follows $C\ell(G)$ denotes the set of all conjugacy classes of G.

2.2. LEMMA. $[FG,FG] = \{\Sigma x_g g \in FG \mid \sum_{g\in C} x_g = 0 \text{ for all } C \in C\ell(G)\}$

Proof. Given $C = \{g_1,\ldots,g_n\} \in C\ell(G)$, assume that $x = \sum_{i=1}^{n} \lambda_i g_i$, where $\lambda_i \in F$ and $\sum_{i=1}^{n} \lambda_i = 0$. Then $x = \sum_{i=1}^{n-1} \lambda_i(g_i - g_n)$ and so in view of the identity

$$g - tgt^{-1} = (gt^{-1})t - t(gt^{-1}) \qquad (t,g \in G)$$

we have $x \in [FG,FG]$. Since $[FG,FG]$ is spanned by all $xy-yx$ with $x,y \in G$, the result follows by virtue of the equality $xy-yx = xy-x^{-1}(xy)x$. ∎

Let H be a subgroup of G. Then the natural projection

$$\pi : FG \longrightarrow FH$$

defined by

$$\pi(\sum_{g\in G} x_g g) = \sum_{g\in H} x_g g$$

is obviously an F-linear map. In what follows, for each subset X of G, we define $X^+ \in FG$ to be the sum of all elements in X. The next result, due to Brauer, will be of fundamental importance.

2.3. THEOREM. Let P be a p-subgroup of G, let $S = C_G(P)$ and let H be a subgroup of G for which $S \subseteq H \subseteq N_G(P)$. Then the natural projection $\pi : FG \longrightarrow FS$ induces a ring homomorphism of $Z(FG)$ into $Z(FH)$ whose kernel is the F-linear span of all C^+ with $C \in C\ell(G)$ and $C \cap S = \emptyset$,

Proof. Let $C_1, C_2, \ldots, C_r$ be all conjugacy classes of G, let $X_i = C_i \cap S$ and assume that $X_k \neq \emptyset$ for $k \in \{1,\ldots,t\}$ and $X_k = \emptyset$ for $k \in \{t+1,\ldots,r\}$. By Proposition 1.7, the elements $C_1^+, C_2^+, \ldots C_r^+$ constitute an F-basis for $Z(FG)$. Note that

$$\pi\left(\sum_{i=1}^{r} \lambda_i C_i^+\right) = \sum_{i=1}^{t} \lambda_i X_i^+$$

and that $X_1,\ldots,X_t$ are mutually disjoint. Moreover, each X_i, $1 \leqslant i \leqslant t$, is a union of conjugacy classes of H. Hence π induces an F-homomorphism from $Z(FG)$ to $Z(FH)$ whose kernel is of the required form.

We are therefore left to verify that for all $i,j \in \{1,\ldots,r\}$,

$$\pi(C_i^+C_j^+) = \pi(C_i^+)\pi(C_j^+)$$

Given $s \in S$, we put $T_s = \{(x,y) \mid x \in C_i, y \in C_j \text{ and } xy = s\}$ and denote by T the union of all T_s with $s \in S$. If $T = \emptyset$, then either $X_i = \emptyset$ or $X_j = \emptyset$, in which case $\pi(C_i^+C_j^+) = 0 = \pi(C_i^+)\pi(C_j^+)$. Now suppose that $T \neq \emptyset$. Then

$$\pi(C_i^+C_j^+) = \sum_{(x,y)\in T} xy = \sum_{s\in S} \sum_{(x,y)\in T_s} xy$$

Since for all $h \in P$ and $s \in S$, $hsh^{-1} = s$, it follows that P acts as a permutation group on T_s via ${}^h(x,y) = (hxh^{-1}, hyh^{-1})$. Because char$F = p$, the sum of all elements xy where (x,y) ranges all orbits of T_s of size $\neq 1$, is equal to 0. On the other hand, an orbit of T_s has size 1 if and only if it is of the form $\{(x,y)\}$ with $x,y \in S$. The conclusion is that

$$\pi(C_i^+C_j^+) = \sum_{(x,y)\in X_i\times X_j} xy = \pi(C_i^+)\pi(C_j^+),$$

as required. ∎

In future we shall refer to the homomorphism $Z(FG) \to Z(FH)$ constructed in Theorem 2.3, as the Brauer homomorphism. We come now to the main result of this section.

2.4. THEOREM. (Osima (1955)). Let e be a nonzero central idempotent of FG. Then Supp$e$ consists of p'-elements.

Proof. (Passman(1969a)). Assume by way of contradiction that $z \in$ Supp$e$ with $z = xy = yx$ where $x \neq 1$ is a p-element and y is a p'-element. Let P be the subgroup of G generated by x. Then, by Theorem 2.3, $\pi(e)$ is a central idempotent in FS where $S = C_G(P)$. Furthermore, since $z \in S$, we also have $z \in$ Supp$\pi(e)$. Thus we may harmlessly assume that $x \in Z(G)$. Invoking

Lemma 2.2, we infer that

(*) $x \in \mathrm{Supp}v$ for all $v \in [FG,FG]$

Choose an integer n with $p^n \geq |G|$ and with $p^n \equiv 1(\mathrm{mod}q)$ and put $t = y^{-1}e$. If $t = \Sigma t_g g$, $t_g \in F$, then by Lemma 2.1,

$$t^{p^n} \equiv \Sigma t_g^{p^n} g^{p^n} \pmod{[FG,FG]}$$

Since g^{p^n} is a p'-element, it follows from (*) that $x \notin \mathrm{Supp}t^{p^n}$. On the other hand, since e is a central idempotent and since $p^n \equiv 1(\mathrm{mod}q)$, we have

$$t^{p^n} = y^{-p^n}e = y^{-1}e = t$$

But, by the definition of t, $x \in \mathrm{Supp}t$, a contradiction. So the theorem is true. ∎

Let C be a conjugacy class of G and let $g \in C$. A Sylow p-subgroup of $C_G(g)$ is called a *defect group* of C (with respect to p). Thus all defect groups of C are conjugate and so have a common order, say p^d. The integer d is called the *defect* of C. We denote by $\delta(C)$ any defect group of C. If H_1 and H_2 are subgroups of G, we write $H_1 \leq_G H_2$ (respectively, $H_1 <_G H_2$) to indicate that H_1 is conjugate in G to a subgroup of H_2 (respectively, to a proper subgroup of H_2), while $H_1 =_G H_2$ will mean that H_1 and H_2 are conjugate in G.

Let e be a block idempotent of FG and let $B = B(e)$. Since $e \in Z(FG)$, Proposition 1.7 implies that

$$\mathrm{Supp}e = C_1 \cup C_2 \cup \ldots \cup C_t$$

for some $C_i \in C\ell(G)$. The largest of the defect groups of C_i, $1 \leq i \leq r$, denoted by $\delta(e)$ (or $\delta(B)$), is called a *defect group* of e (or of B). It will be shown (Proposition 2.6) that all defect groups of e are conjugate and so have a common order, say p^d. The integer d is called the *defect* of e (or of B).

2.5. LEMMA. Let D be a p-subgroup of G and let $C_1, C_2, \ldots, C_t$ be all elements in $C\ell(G)$ with $\delta(C_i) \leqslant_G D$. Then the F-linear span $I_D(G)$ of $C_1^+, C_2^+, \ldots, C_t^+$ is an ideal of $Z(FG)$.

Proof. Fix $i \in \{1,\ldots,t\}$ and denote by C_j any conjugacy class of G. It clearly suffices to verify that $C_i^+ C_j^+ \in I_D(G)$.

We may assume that $C_i^+ C_j^+ \neq 0$ in which case we may choose $g \in \mathrm{Supp} C_i^+ C_j^+$. Because

$$C_i^+ C_j^+ = \sum_{(x,y)\in C_i \times C_j} xy$$

we have $g = uv$ for some $u \in C_i$ and $v \in C_j$. Let P be a defect group of the conjugacy class containing g and let

$$X = \{(x,y) \mid xy = g \text{ and } (x,y) \in C_i \times C_j\}$$

Then P acts on X by conjugation, so by the argument employed in the proof of Theorem 2.3, we have $C_i \cap C_G(P) \neq \emptyset$. Hence $P \leqslant_G \delta(C_i)$ and therefore $P \leqslant_G D$. Thus $C_i^+ C_j^+ \in I_D(G)$, as required. ∎

If D is a defect group of $C \in C\ell(G)$, we shall write $I[C]$ instead of $I_D(G)$. The fact that $I[C]$ is an ideal of $Z(FG)$ will be exploited in the proof of the following result.

2.6. PROPOSITION. Let e be a block idempotent of FG, and let γ be the irreducible representation of $Z(FG)$ associated with e

(i) $\mathrm{Supp}\, e = C_1 \cup C_2 \cup \ldots \cup C_t \cup C_{t+1} \cup \ldots \cup C_r$

where $C_1,\ldots,C_r$ are p-regular classes of G such that

(a) For all $i \in \{1,\ldots,t\}$, $e \in I[C_i]$ and $\delta(C_i) =_G \delta(C_j)$ for all $i,j \in \{1,\ldots,t\}$

(b) $\delta(C_k) <_G \delta(C_i)$ for all $k \in \{t+1,\ldots,r\}$ and $i \in \{1,\ldots,t\}$

(ii) If $C \in C\ell(G)$ is such that $\delta(C) <_G \delta(C_i)$ for some $i \in \{1,\ldots,t\}$, then $\gamma(C^+) = 0$

(iii) There exists $i \in \{1,\ldots,t\}$ such that $\gamma(C_i^+) \neq 0$

(iv) If $C \in C\ell(G)$ is such that $\gamma(C^+) \neq 0$, then $\delta(e) \leqslant_G \delta(C)$.

Proof. (i) Owing to Theorem 2.4, $e = \sum_{i=1}^{r} \lambda_i C_i^+$ for some nonzero λ_i in F and some p-regular classes $C_1,\dots,C_r$ of G. Since $\gamma(1) = \gamma(e) \neq 0$, there is some $k \in \{1,\dots,r\}$ with $\gamma(C_k^+) \neq 0$. But $C_k^+ \in I[C_k]$, so we must have $\gamma(I[C_k]) \neq 0$. Invoking Proposition 1.10.10(iv), we infer that $e \in I[C_k]$. By renumbering $C_1,\dots,C_r$ in such a way that $\{C_1,\dots,C_t\}$ is the subset of $\{C_1,\dots,C_r\}$ consisting of all C_i with $\delta(C_i) \underset{G}{=} \delta(C_k)$, we deduce that (a) and (b) hold.

(ii) By (i), $\delta(C) \underset{G}{<} \delta(C_k)$. Suppose that $\gamma(C^+) \neq 0$. Since $C^+ \in I[C]$ we have $\gamma(I[C]) \neq 0$ and so by Proposition 1.10.10(iv), $e \in I[C]$. Hence $\delta(C_k) \underset{G}{\leqslant} \delta(C)$, a contradiction.

(iii) This was established in (i)

(iv) By the argument of (ii), $e \in I[C]$. The desired assertion is now a consequence of the definition of $I[C]$. ■

For future use, we shall need a deep result due to Brauer (see Curtis and Reiner(1962)). The proof will be omitted since it relies on modular character theory, a topic we do not touch upon in this book.

2.7. THEOREM. Let F be an algebraically closed field of characteristic $p > 0$, let G be a group of order $p^n m$ with $(p,m) = 1$ and let B be a block of FG with defect d. Then d is the smallest integer such that p^{n-d} divides the dimensions of all the irreducible FG-modules in B.

3. THE NUMBER OF IRREDUCIBLE FG-MODULES.

Throughout this section G denotes a finite group and F a field of characteristic $p > 0$. Our goal is to determine the number of irreducible FG-modules in the case where F is a splitting field for FG.

3.1. LEMMA. Let A be a finite-dimensional F-algebra, let F be a splitting field for A and let $S = [A,A]$. Then

$$T = \{x \in A \mid x^{p^m} \in S \text{ for some integer } m \geqslant 1\}$$

is an F-subspace of A, and the number of nonisomorphic irreducible A-modules equals $\dim_F(A/T)$.

Proof. As we have seen in the proof of Lemma 2.1, $a \in [A,A]$ implies $a^p \in [A,A]$ and so $S \subseteq T$. Again, by Lemma 2.1, if $a,b \in T$ then for a sufficiently large m,

$$(a+b)^{p^m} \equiv a^{p^m} + b^{p^m} \equiv 0(\mathrm{mod} S)$$

proving that $a + b \in T$. It is clear that for all $\lambda \in F$ and $a \in T$, $\lambda a \in T$ and thus T is an F-subspace of A. Since $J(A)$ is nilpotent, it follows from the definition of T that $J(A) \subseteq T$.

Now put $\bar{A} = A/J(A)$ and $\bar{T} = T/J(A)$. Denote by r the number of irreducible A-modules. Because F is a splitting field for A, Corollary 1.10.4 tells us that

$$\bar{A} \cong \prod_{i=1}^{r} M_{n_i}(F)$$

for some positive integers $n_1,\ldots,n_r$. Next put $\bar{S} = (S+J(A))/J(A)$. Then clearly $\bar{S} = [\bar{A},\bar{A}]$ and

$$\bar{T} = \{x \in \bar{A} | x^{p^m} \in \bar{S} \text{ for some integer } m \geqslant 1\}$$

Let S_i, T_i be the analogous F-subspaces defined for the F-algebra $A_i = M_{n_i}(F)$. Then

$$\dim_F(A/T) = \dim_F(\bar{A}/\bar{T}) = \sum_{i=1}^{r} \dim_F(A_i/T_i)$$

and so we are left to verify that

$$\dim_F(A_i/T_i) = 1 \qquad (1 \leqslant i \leqslant r)$$

Let e_{st} denote the $n_i \times n_i$ matrix with (s,t)-th entry equal to 1 and all other entries 0. Then for any $s \neq t$

$$e_{st} = e_{sj}e_{jt} - e_{jt}e_{st} \in S_i$$

and

$$e_{ss} - e_{tt} = e_{st}e_{ts} - e_{ts}e_{st} \in S_i$$

Hence S_i contains the n_i^2-1 linearly independent elements $e_{st}(1 \leqslant s,t \leqslant n_i,$ $s \neq t)$ and $e_{ss} - e_{11}$ $(s = 2,\dots,n_i)$. Because $S_i \subseteq T_i$, this implies that

$$\dim_F(A_i/T_i) = \dim_F A_i - \dim_F T_i \leqslant n_i^2 - (n_i^2-1) = 1$$

On the other hand, the elements of S_i are all $n_i \times n_i$ matrices with trace 0. In particular, $e_{11} \notin S_i$. But $e_{11}^{p^m} = e_{11}$ for all $m \geqslant 1$, so $e_{11} \notin T_i$. Hence $T_i \neq A_i$ and therefore $\dim_F(A_i/T_i) = 1$. Thus the lemma is proved. ∎

We come now to the demonstration for which the present section has been developed.

3.2. THEOREM (Brauer (1935)). Let G be a finite group and let F be a field of characteristic $p > 0$ such that F is a splitting field for FG. Then the number of irreducible FG-modules is equal to the number of p-regular classes of G.

Proof. Let $S = [FG,FG]$, let $T = \{x \in FG \mid x^{p^m} \in S$ for some integer $m \geqslant 1\}$ and let $v_1,v_2,\dots,v_t$ be an F-basis of T. Choose an integer m so large that $a_i^{p^m} \in S$ for each $i \in \{1,\dots,t\}$ and p^m is at least as large as the order of a Sylow p-subgroup of G. Owing to Lemma 2.1, we have

$$\left(\sum_{i=1}^{t} \lambda_i v_i\right)^{p^m} \equiv \sum_{i=1}^{t} \lambda_i^{p^m} v_i^{p^m} \equiv 0 (\text{mod} S)$$

for all $\lambda_i \in F$. Hence $x = \Sigma x_g g \in T$ if and only if $x^{p^m} \in S$ or equivalently

(1) $\qquad x \in T$ if and only if $\Sigma x_g^{p^m} g^{p^m} \in S$

where, by our choice of m, each g^{p^m} is p-regular. Let $C_1,\dots,C_s$ be all p-regular classes of G, and put

$$C_i^* = \{g \in G \mid g^{p^m} \in C_i\} \qquad (1 \leqslant i \leqslant s)$$

Applying (1) and Lemma 2.2, we conclude that $x = \Sigma x_g g \in T$ if and only if

$$\left(\sum_{g\in C_i^*} x_g\right)^{p^m} = \sum_{g\in C_i^*} x_g^{p^m} = 0$$

for all $i \in \{1,\dots,s\}$. Thus

(2) $\qquad \Sigma x_g g \in T$ if and only if $\sum_{g\in C_i^*} x_g = 0$ for all $i \in \{1,\dots,s\}$

Finally, define $\psi : FG \longrightarrow F\times\ldots\times F$ (s copies) as the F-linear mapping given by

$$\Sigma x_g g \longmapsto (\beta_1,\ldots,\beta_s)$$

where $\beta_i = \sum_{g\in C_i^*} x_g$, $1 \leqslant i \leqslant s$. Then ψ is clearly surjective and by (2) T is the kernel of ψ. Hence, by Lemma 3.1,

$$r = \dim_F(FG/T) = s$$

and the result follows. ■

4. INDUCED MODULES.

Throughout this section, F denotes an arbitrary field, G a finite group and H a subgroup of G. If V is an FG-module, then we shall denote by V_H the FH-module obtained by the restriction of algebra; thus as an F-module, V_H equals V, but only action of FH is defined on V_H. This process will be called *restriction* and it permits us to go from any FG-module V to a uniquely determined FH-module V_H.

Let W be any FH-module. Since we may consider FH as a subalgebra of FG, we can define an FG-module structure on the tensor product $FG \otimes_{FH} W$. This is the induced module and we denote it by W^G.

We say that an FG-module V is *imprimitive* if V can be written as a direct sum $V = V_1 \oplus \ldots \oplus V_n$ of F-subspaces V_i with $n > 1$ such that for each $g \in G, V_i \mapsto gV_i$ is a permutation of the set $\{V_1,\ldots,V_n\}$. Expressed otherwise, G acts on the set $\{V_1,\ldots,V_n\}$ by the rule ${}^gV_i = gV_i$ for all $g \in G$, $i \in \{1,\ldots,n\}$. We shall refer to the set $\{V_1,\ldots,V_n\}$ as a *system of imprimitivity* for V. The following observation relates the concept of imprimitivity to that of induction.

4.1. LEMMA. (i) Let H be a proper subgroup of G, let W be an FH-module, and let $g_1,\ldots,g_n$ be a left transversal for H in G, with $g_1 \in H$. Then

$$\{g_1 \otimes W, g_2 \otimes W, \ldots\ g_n \otimes W\}$$

is a system of imprimitivity for W^G and the action of G is transitive.

(ii) Let V be an imprimitive FG-module with $\{V_1,\ldots,V_n\}$ as a system of imprimitivity and assume that G acts transitively on $\{V_1,\ldots,V_n\}$. Let H be the stabilizer of V_1 under the action of G. Then V_1 is an FH-module and $V \cong V_1^G$.

Proof. (i) By Proposition 1.4, $FG = \bigoplus_{i=1}^{n} g_i FH$ and so

$$W^G = \left(\bigoplus_{i=1}^{n} g_i FH\right) \otimes W = \bigoplus_{i=1}^{n}(g_i \otimes W)$$

Since the left multiplication by $g \in G$ induces a permutation of the set $\{g_1 \otimes W,\ldots,g_n \otimes W\}$ and $g(g_1 \otimes W) = g \otimes W$, the assertion follows.

(ii) Since G acts transitively on $\{V_1,\ldots,V_n\}$ and H is the stabilizer of V_1, it follows that $n = (G:H)$. Thus we can choose a left transversal for H in G, say $g_1,\ldots,g_n$ such that $g_i V_1 = V_i$, $1 \leqslant i \leqslant n$. Since

$$V_1^G = (g_1 \otimes V_1) \oplus (g_2 \otimes V_1) \oplus \cdots \oplus (g_n \otimes V_1)$$

and

$$V = g_1 V_1 \oplus \cdots \oplus g_n V_1$$

as F-spaces, the map $\theta : V_1^G \longrightarrow V$ given by

$$\sum_{i=1}^{n} g_i \otimes v_i \longmapsto \sum_{i=1}^{n} g_i v_i \qquad (v_i \in V_i)$$

is a vector space isomorphism. However, for each $g \in G$ and each $i \in \{1,\ldots,n\}$, there exist uniquely determined $h \in H$ and $j \in \{1,\ldots,n\}$ such that $gg_i = g_j h$. Thus, for all $v \in V_1$, we have

$$\begin{aligned}\theta(g(g_i \otimes v)) &= \theta(gg_i \otimes v) = \theta(g_j \otimes hv)\\ &= g_j hv = g(g_i v)\\ &= g\ \theta(g_i \otimes v)\end{aligned}$$

as required. ∎

Let H be a subgroup of G, let U be a submodule of an FH-module V and let $j : U \longrightarrow V$ be the inclusion map. Since FG is a free FH-module

(Proposition 1.4), it follows from Proposition 1.8.5 that FG is a flat FH-module. Hence the canonical homomorphism

$$1 \otimes j : U^G \longrightarrow V^G$$

is injective. For this reason, from now on we shall identify U^G with its image in V^G. The following lemma collects together some elementary properties of induced modules.

4.2. LEMMA. Let V_1 and V_2 be submodules of an FH-module V.

(i) $V_1^G \subseteq V_2^G$ if and only if $V_1 \subseteq V_2$

(ii) $V_1^G = V_2^G$ if and only if $V_1 = V_2$

(iii) $(V_1+V_2)^G = V_1^G + V_2^G$

(iv) $(V_1 \cap V_2)^G = V_1^G \cap V_2^G$

(v) If $V = V_1 \oplus V_2$, then $V^G = V_1^G \oplus V_2^G$

Proof. We first observe that (ii) is a consequence of (i), that (v) follows from (iii) and (iv), and that (iv) is a consequence of the fact that FG is a free FH-module. To prove (i), observe that $V_1 \subseteq V_2$ obviously implies that $V_1^G \subseteq V_2^G$. Let T be a transversal for H in G containing 1. Because

$$V^G = \bigoplus_{t \in T} t \otimes V$$

each element in V^G can be uniquely written in the form $\sum_{t \in T} t \otimes v_t$ with $v_t \in V$. Therefore, if $v_1 \in V_1$ is such that $1 \otimes v_1 \in V_2^G$, then $v_1 \in V_2$. Thus $V_1^G \subseteq V_2^G$ implies $V_1 \subseteq V_2$, proving (i).

To prove (iii), observe that $V_1^G + V_2^G \subseteq (V_1+V_2)^G$ by virtue of (i). The opposite containment being a consequence of the equality

$$t \otimes (v_1+v_2) = t \otimes v_1 + t \otimes v_2 \qquad (v_1 \in V_1, v_2 \in V_2, t \in T)$$

the result follows. ■

We next provide some important formal properties of induced modules.

4.3. PROPOSITION. Let $A \subseteq B$ be left ideals of FH. Then

(i) $(B/A)^G \cong FG \cdot A/FG \cdot B$

(ii) Let I be a right ideal of FG such that $I \subseteq FG \cdot A$. If V is an FH-module, then there is a canonical injective map

$$\psi : I(V^G) \longrightarrow (AV)^G$$

Proof. All tensor products are to be taken over FH.

(i) Consider the diagram

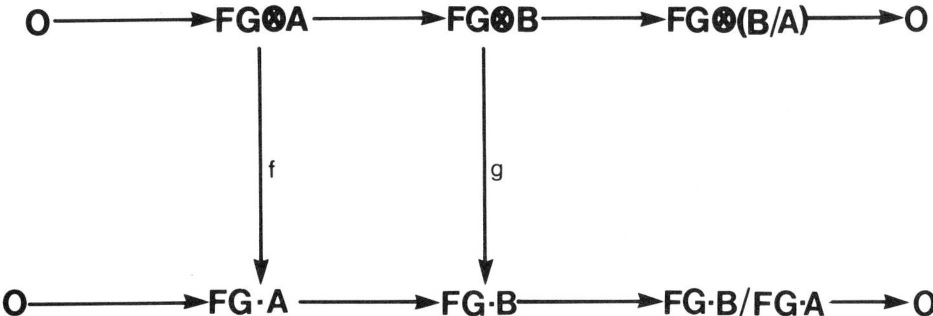

where all maps are canonical. Then the diagram is commutative, and since FG is a flat right FH-module, both rows are exact, and f,g are isomorphisms. This proves the required assertion.

(ii) Let μ denote the canonical map $V \longrightarrow V/AV$. Then the sequence

$$0 \longrightarrow FG \otimes AV \longrightarrow FG \otimes V \xrightarrow{1\otimes\mu} FG \otimes (V/AV) \longrightarrow 0$$

is exact. Let $x \in I, y \in FG$, and $v \in V$. Then $xy \in I \subseteq FG \cdot A$, so xy can be written in the form $xy = y_1a_1 + \dots + y_na_n$ with $y_i \in FG, a_i \in A$. Thus

$$\begin{aligned}(1 \otimes \mu)(x(y \otimes v)) &= (1 \otimes \mu)(xy \otimes v) = xy \otimes (v+AV) \\ &= y_1 \otimes a_1(v+AV) + \dots + y_n \otimes a_n(v+AV) \\ &= 0\end{aligned}$$

and therefore $I(FG \otimes V) \subseteq \text{Ker}(1 \otimes \mu)$, which finishes the proof. ■

4.4. PROPOSITION. Let H be a subgroup of G and let $U \xrightarrow{\lambda} V \xrightarrow{\mu} W$ be a sequence of homomorphisms of FH-modules. Then the following properties hold:

(i) The sequence

$$0 \longrightarrow U \xrightarrow{\lambda} V \xrightarrow{\mu} W \longrightarrow 0 \qquad (*)$$

is exact if and only if the corresponding sequence of FG-modules

$$0 \longrightarrow U^G \xrightarrow{1\otimes\lambda} V^G \xrightarrow{1\otimes\mu} W^G \longrightarrow 0 \qquad (**)$$

is exact

(ii) Assume that the sequence (*) is exact. Then (*) splits if and only if (**) splits.

(iii) If U is a submodule of V, then $V^G/U^G \cong (V/U)^G$

Proof. (i) That exactness of (*) implies that of (**) is a consequence of the fact that FG is a flat FH-module. Conversely, suppose that (**) is exact. Then

$$\mathrm{Ker}(1 \otimes \mu) = (\ker\mu)^G \quad \text{and} \quad Im(1 \otimes \lambda) = (Im\lambda)^G$$

Since $\mathrm{Ker}(1 \otimes \mu) = \mathrm{Im}(1 \otimes \lambda)$, it follows from Lemma 4.2(ii) that $\mathrm{Ker}\mu = Im\lambda$. A similar argument shows that $\mathrm{Ker}(1 \otimes \lambda) = 0$ implies $\mathrm{Ker}\lambda = 0$ and that $Im(1 \otimes \mu) = W^G$ implies $Im\mu = W$. Thus the sequence (*) is also exact.

(ii) Suppose that $0 \longrightarrow U \xrightarrow{\lambda} V \xrightarrow{\mu} W \rightarrow 0$ is a split exact sequence and let $\gamma : W \longrightarrow V$ be a splitting homomorphism. Then $1 \otimes \gamma : W^G \longrightarrow V^G$ is obviously a splitting homomorphism for the induced sequence

Conversely, assume that the exact sequence

$$0 \longrightarrow U^G \xrightarrow{1\otimes\lambda} V^G \xrightarrow{1\otimes\mu} W^G \longrightarrow 0$$

of FG-modules splits. Let T denote a transversal for H in G containing 1. Consider the mappings

$$\begin{cases} W \xrightarrow{\sigma} W^G \\ w \longmapsto 1 \otimes w \end{cases} \qquad\qquad \begin{cases} V^G \xrightarrow{\tau} V \\ \sum\limits_{t\in T} t \otimes v_t \longmapsto v_1 \end{cases}$$

Then σ and τ are well defined FH-linear maps. Let $\psi : W^G \longrightarrow V^G$ be a splitting homomorphism and let $w \in W$. Then $\psi(1 \otimes w) = \sum\limits_{t\in T} t \otimes v_t, v_t \in V$, and

$$(1 \otimes \mu)\psi(1 \otimes w) = \sum_{t \in T} t \otimes \mu(v_t) = 1 \otimes w$$

Thus $w = \mu(v_1)$ and therefore

$$\begin{aligned}(\mu\tau\psi\sigma)(w) &= (\mu\tau\psi)(1 \otimes w) \\ &= (\mu\tau)(\sum_{t \in T} t \otimes v_t) \\ &= \mu(v_1) = w\end{aligned}$$

It follows that $\tau\psi\sigma : W \longrightarrow V$ is a splitting homomorphism, as required.

(iii) Direct consequence of (i) applied to the natural exact sequence $0 \longrightarrow U \longrightarrow V \longrightarrow V/U \longrightarrow 0$. ■

As an immediate consequence, we derive

4.5. COROLLARY. Let V be an FH-module. If V^G is completely reducible, then so is V.

Another useful property of induced modules is given by

4.6. PROPOSITION (Transitivity of the induction). If L, H are subgroups of G with $H \subseteq L$, then for any FH-module V,

$$(V^L)^G \cong V^G$$

Proof. The required assertion follows from

$$\begin{aligned}(V^L)^G &= FG \otimes_{FL} (FL \otimes_{FH} V) \cong (FG \otimes_{FL} FL) \otimes_{FH} V \\ &\cong FG \otimes_{FH} V = V^G \quad ■\end{aligned}$$

Let S be a subring of a ring R and let M be a left S-module. Given $r \in R, \phi \in \mathrm{Hom}_S(R,M)$, define $r\phi : R \longrightarrow M$ by $(r\phi)(x) = \phi(xr)$. It is evident that the additive group $\mathrm{Hom}_S(R,M)$ becomes a (left) R-module. The following result illustrates how a module of this type can be brought into argument.

4.7. PROPOSITION. Let V be an FH-module. Then

$$\mathrm{Hom}_{FH}(FG,V) \cong V^G \qquad \text{as } FG\text{-modules}$$

Proof. Let $\{g_1,\ldots,g_n\}$ be a left transversal for H in G, and for any $\phi \in \operatorname{Hom}_{FH}(FG,V)$, define $\theta(\phi) \in V^G$ by $\theta(\phi) = \sum_{i=1}^{n} g_i \otimes \phi(g_i^{-1})$. Then θ is obviously an F-homomorphism. Assume that $\sum_{i=1}^{n} g_i \otimes \phi(g_i^{-1}) = 0$. Then $\phi(g_i^{-1}) = 0$ for all $i \in \{1,2,\ldots,n\}$. Since $\{g_1^{-1},\ldots,g_n^{-1}\}$ is a right transversal for H in G, given $g \in G$, we have $g = hg_i^{-1}$ for some $h \in H$. Hence $\phi(g)=h\phi(g_i^{-1})=0$ and thus θ is injective. Because FG is a free left FH-module with the elements $g_1^{-1},\ldots,g_n^{-1}$ as a basis, for every set $\{v_i \in V \mid 1 \leqslant i \leqslant n\}$ there is a $\phi \in \operatorname{Hom}_{FH}(FG,V)$ such that $\phi(g_i^{-1}) = v_i, 1 \leqslant i \leqslant n$. Thus θ is an F-isomorphism.

Given $g \in G$, there is an $h_i \in H$ and a permutation $i \mapsto \pi(i)$ of $\{1,\ldots,n\}$ such that $g_i^{-1}g = h_i g_{\pi(i)}^{-1}$ and hence that $gg_{\pi(i)} = g_i h_i$. Thus

$$\begin{aligned}
\theta(g\phi) &= \sum_{i=1}^{n} g_i \otimes (g\phi)g_i^{-1} = \sum_{i=1}^{n} g_i \otimes \phi(g_i^{-1}g) \\
&= \sum_{i=1}^{n} g_i \otimes h_i\phi(g_{\pi(i)}^{-1}) = \sum_{i=1}^{n} g_i h_i \otimes \phi(g_{\pi(i)}^{-1}) \\
&= \sum_{i=1}^{n} gg_{\pi(i)} \otimes \phi(g_{\pi(i)}^{-1}) = g\sum_{i=1}^{n} g_{\pi(i)} \otimes \phi(g_{\pi(i)}^{-1}) \\
&= g\theta(\phi)
\end{aligned}$$

as asserted. ∎

Let V be an FH-module. Then the mapping

$$\begin{cases} V \xrightarrow{\ f\ } V^G \\ v \mapsto 1 \otimes v \end{cases}$$

is an injective FH-homomorphism. In what follows we shall refer to f as the *canonical injection.*

4.8. PROPOSITION (Universal characterization of induced modules). Let H be a subgroup of G, let V be an FH-module, and let $f : V \longrightarrow V^G$ be the canonical injection. Then, for any FG-module W and any $\phi \in \operatorname{Hom}_{FH}(V,W_H)$ there exists one and only one $\psi \in \operatorname{Hom}_{FG}(V^G,W)$ which renders commutative the following diagram:

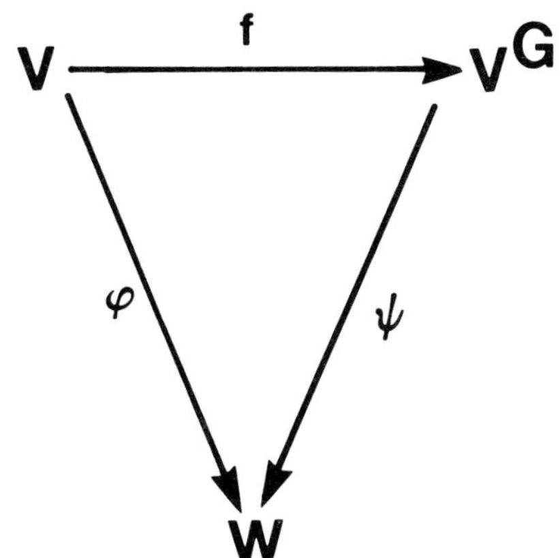

Proof. Since $g \otimes v = g(1 \otimes v), g \in G, v \in V$, any FG-homomorphism $V^G \longrightarrow W$ is uniquely determined by its restriction to $1 \otimes V$. Hence there exists at most one such ψ. Let T be a transversal for H in G containing 1 and let $v_1, \ldots, v_n$ be an F-basis of V. Then the elements $\{t \otimes v_i \mid t \in T, 1 \leqslant i \leqslant n\}$ form an F-basis of V^G. Let ψ be the F-homomorphism $V^G \longrightarrow W$ determined by $\psi(t \otimes v_i) = t\phi(v_i)$. Then $\psi(t \otimes v) = t\phi(v)$ for any $v \in V$. Let $g \in G$ and write $gt = t_1 h$ for some $t_1 \in T, h \in H$. Then we must have

$$\begin{aligned}\psi(g(t \otimes v)) &= \psi(gt \otimes v) = \psi(t_1 \otimes hv)\\ &= t_1\phi(hv) = t_1 h\phi(v) = gt\phi(v)\\ &= g\psi(t \otimes v)\end{aligned}$$

whence ψ is an FG-homomorphism. Since $\psi \circ f = \phi$, the result follows. ∎

We are now ready to prove the following important result.

4.9. THEOREM. Let H be a subgroup of a group G, let V be an FH-module, and let W be an FG-module. Then

(i) $\operatorname{Hom}_{FG}(V^G, W) \cong \operatorname{Hom}_{FH}(V, W_H)$ as F-spaces

(ii) $\operatorname{Hom}_{FG}(W, V^G) \cong \operatorname{Hom}_{FH}(W_H, V)$ as F-spaces

Proof. (i) Let $f : V \longrightarrow V^G$ be the canonical injection. Given $\psi \in \operatorname{Hom}_{FG}(V^G, W)$, we have $\psi f \in \operatorname{Hom}_{FH}(V, W_H)$ and the map

$$\begin{cases} \operatorname{Hom}_{FG}(V^G,W) \xrightarrow{\theta} \operatorname{Hom}_{FH}(V,W_H) \\ \psi \longmapsto \psi\circ f \end{cases}$$

is an injective F-homomorphism. If $\phi \in \operatorname{Hom}_{FH}(V,W_H)$, then by Proposition 4.8 there exists a unique $\psi \in \operatorname{Hom}_{FG}(V^G,W)$ such that $\phi = \psi f$. Thus θ is an isomorphism.

(ii) Owing to Proposition 4.7, we need only verify that

$$\operatorname{Hom}_{FG}(W, \operatorname{Hom}_{FH}(FG,V)) \cong \operatorname{Hom}_{FH}(W_H,V) \qquad \text{as } F\text{-spaces}$$

Given $\phi \in \operatorname{Hom}_{FG}(W,\operatorname{Hom}_{FH}(FG,V))$, define $\phi^* \in \operatorname{Hom}_{FH}(W_H,V)$ by

$$\phi^*(w) = [\phi(w)](1) \qquad (w \in W)$$

Then the map $\phi \mapsto \phi^*$ is an injective F-homomorphism. If $\psi \in \operatorname{Hom}_{FH}(W_H,V)$, then the map $\phi : W \longrightarrow \operatorname{Hom}_{FH}(FG,V)$ defined by

$$[\phi(w)](x) = \psi(xw)$$

is an FG-homomorphism. Since $\phi^* = \psi$, the assertion follows. ■

Let U and V be any FG-modules. Then the *intertwining number* for U and V is defined to be

$$i(U,V) = \dim_F \operatorname{Hom}_{FG}(U,V)$$

Let U be a completely reducible FG-module, and let V be an irreducible FG-module. We say that a nonnegative integer m is the *multiplicity* of V as an irreducible constituent of U if a decomposition of U into irreducible submodules contains exactly m submodules isomorphic to V.

Let V be an FG-module. In what follows we shall refer to the factor module $V/J(V)$ as the *head* of V.

4.10. THEOREM. (Nakayama Reciprocity). Let F be an arbitrary field and let H be a subgroup of G. Suppose further that V is an irreducible FH-module and W is an irreducible FG-module. Let m be the multiplicity of W as an irreducible constituent in the head of V^G and let n be the multiplicity of V as an irreducible constituent in the socle of W_H. Then

$$m\, i(W,W) = n\, i(V,V)$$

In particular,

(i) If F is a splitting field for FG and FH, then $m = n$

(ii) V is isomorphic to a submodule of W_H if and only if W is isomorphic to a factor module of V^G.

Proof. We first note that $\operatorname{Hom}_{FG}(U,W) = 0$ for all irreducible FG-modules U with $U \not\cong W$. Since for any FG-homomorphism $f : V^G \longrightarrow W$, $J(V^G) \subseteq \operatorname{Ker} f$ we have

$$\begin{aligned}\operatorname{Hom}_{FG}(V^G,W) &\cong \operatorname{Hom}_{FG}(V^G/J(V^G),W)\\ &\cong \operatorname{Hom}_{FG}(W,W) \oplus \dots \oplus \operatorname{Hom}_{FG}(W,W) \qquad (m \text{ times})\end{aligned}$$

Thus $i(V^G,W) = m i(W,W)$. Since the image of any FH-homomorphism $V \longrightarrow W_H$ is contained in the socle S of W_H, we have

$$\begin{aligned}\operatorname{Hom}_{FH}(V,W_H) &\cong \operatorname{Hom}_{FH}(V,S)\\ &\cong \operatorname{Hom}_{FH}(V,V) \oplus \dots \oplus \operatorname{Hom}_{FH}(V,V) \qquad (n \text{ times})\end{aligned}$$

Thus $i(V,W_H) = n\, i(V,V)$ and the result follows by virtue of Theorem 4.9(i). ∎

5. RELATIVE PROJECTIVE AND INJECTIVE MODULES.

Throughout this section, F denotes an arbitrary field, G a finite group and H a subgroup of G.

An FG-module V is said to be *H-projective* if every exact sequence of FG-modules

$$0 \longrightarrow U \longrightarrow W \longrightarrow V \longrightarrow 0$$

for which the associated sequence of FH-modules

$$0 \longrightarrow U_H \longrightarrow W_H \longrightarrow V_H \longrightarrow 0$$

splits is also a split exact sequence of FG-modules. The FG-modules V is called H-*injective* if every exact sequence of FG-modules

$$0 \longrightarrow V \longrightarrow W \longrightarrow U \longrightarrow 0$$

for which

$$0 \longrightarrow V_H \longrightarrow W_H \longrightarrow U_H \longrightarrow 0$$

is a split exact sequence of FH-modules, is also a split exact sequence of FG-modules. Thus an FG-module is projective (respectively, injective) if and only if it is 1-projective (respectively, 1-injective).

5.1. LEMMA. Let T be a transversal for H in G, let V and W be FG-modules, and let $f \in \mathrm{Hom}_{FH}(V_H, W_H)$. Then the map $f^* : V \to W$ defined by

$$f^*(v) = \sum_{t \in T} (tft^{-1})v$$

is an FG-homomorphism which is independent of the choice of T.

Proof. Assume that $s_t = tu_t$ for each $t \in T$ with $u_t \in H$. Then

$$\begin{aligned}\sum_{t \in T} (s_t f s_t^{-1})v &= \sum_{t \in T} tu_t f u_t^{-1} t^{-1} v \\ &= \sum_{t \in T} tu_t u_t^{-1} f t^{-1} v \\ &= f^*(v)\end{aligned}$$

proving that f^* is independent of the choice of T. It therefore follows that for all $g \in G$.

$$\begin{aligned}f^*(gv) &= \sum_{t \in T} (tft^{-1})(gv) = \sum_{t \in T} g(g^{-1}t)f(t^{-1}g)v \\ &= \sum_{t \in T} g(g^{-1}t)f(g^{-1}t)^{-1}v \\ &= gf^*(v)\end{aligned}$$

as required. ∎

5.2. LEMMA. Let T be a transversal for H in G, let V be an FG-module, and let $f : (V_H)^G \to V$ be given by

$$f\left(\sum_{t \in T} t \otimes v_t\right) = \sum_{t \in T} tv_t \qquad (v_t \in V)$$

Then f is a surjective FG-homomorphism and $\text{Ker}f$ is a direct summand of $(V_H^G)_H$.

Proof. It is clear that f is a surjective F-homomorphism. To prove that f is in fact an FG-homomorphism, fix $g \in G$. Then for each $t \in T$ there exists $t' \in T$ and $u_t \in H$ such that $gt = t'u_t$. It follows that

$$\begin{aligned} g(\sum_{t\in T} t \otimes v_t) &= \sum_{t\in T} gt \otimes v_t = \sum_{t\in T} t'u_t \otimes v_t \\ &= \sum_{t\in T} t' \otimes u_t v_t \end{aligned}$$

and so

$$\begin{aligned} f(g(\sum_{t\in T} t \otimes v_t)) &= \sum_{t\in T} t'u_t v_t = \sum_{t\in T} gtv_t \\ &= gf(\sum_{t\in T} t \otimes v_t), \end{aligned}$$

proving that f is a surjective FG-homomorphism. Let $\varepsilon : V_H \longrightarrow (V_H)^G$ be the canonical injection. Obviously $\varepsilon(V_H) \cap \text{Ker}f = 0$. Because

$$\sum_{t\in T} t \otimes v_t - 1 \otimes (\sum_{t\in T} tv_t) \in \text{Ker}f$$

we conclude that

$$((V_H)^G)_H = \text{Ker}f \oplus \varepsilon(V_H)$$

as asserted. ∎

5.3. LEMMA. Let T be a transversal for H in G, let V be an FG-module, and let $f : V \longrightarrow (V_H)^G$ be defined by

$$f(v) = \sum_{t\in T} t \otimes t^{-1}v$$

Then f is an injective FG-homomorphism which is independent of the choice of the transversal T and $(f(V))_H$ is a direct summand of $((V_H)^G)_H$.

Proof. Assume that T' is another transversal for H in G. Then, for each $t \in T$, there exists $u_t \in H$ and $t' \in T'$ such that $t = t'u_t$. Hence

$$\sum_{t\in T} t \otimes t^{-1}v = \sum_{t\in T} t'u_t \otimes u_t^{-1}(t')^{-1}v$$

$$= \sum_{t'\in T'} t' \otimes (t')^{-1}v,$$

proving that f is independent of the choice of the transversal.

It follows that for all $g \in G$,

$$f(gv) = \sum_{t\in T} t \otimes t^{-1}(gv) = g \sum_{t\in T} (g^{-1}t) \otimes (t^{-1}g)v$$
$$= gf(v)$$

so that f is an FG-homomorphism. Since

$$(V_H)^G = \bigoplus_{t\in T} t \otimes V_H$$

f is an injection. Now $W' = \bigoplus_{t\in T, t\notin H} t \otimes V_H$ is obviously an FH-submodule of $((V_H)^G)_H$, and we show that

$$((V_H)^G)_H = W' \oplus (f(V))_H$$

If $f(v) = \sum_{t\in T} t \otimes t^{-1}v \in f(V) \cap W'$, then $v = 0$. On the other hand, if $t' \in T \cap H$, we have

$$\sum_{t\in T} t \otimes v_t - \sum_{t\in T} t \otimes t^{-1}t'v_{t'} \in W'$$

So the lemma is verified. ∎

The next result establishes several important characterizations of H-projective and H-injective modules.

5.4. THEOREM. (Higman(1954)). Let V be an FG-module and let T be a transversal for H in G containing 1. Then the following statements are equivalent:

(i) V is H-projective

(ii) V is isomorphic to a direct summand of $(V_H)^G$

(iii) V is isomorphic to a direct summand of W^G, where W is an FH-module

(iv) There exists $\psi \in \mathrm{End}_{FH}(V_H)$ such that

$$\Big(\sum_{t\in T} t\psi t^{-1}\Big)v = v \qquad \text{for all } v \in V$$

(v) V is H-injective

Proof. (i) $\Rightarrow$ (ii): By Lemma 5.2, there exists an exact sequence of FG-modules $0 \longrightarrow U \longrightarrow (V_H)^G \xrightarrow{f} V \longrightarrow 0$ such that the associated sequence of FH-submodules splits. By hypothesis, V is H-projective, so $\mathrm{Ker} f$ is a direct summand of $(V_H)^G$. Thus

$$(V_H)^G = \mathrm{Ker} f \oplus V'$$

where $V' \cong V$.

(ii) $\Rightarrow$ (iii): Obvious

(iii) $\Rightarrow$ (iv): We first prove that any FG-module V of the form $V = W^G$, where W is an FH-module, satisfies (iv). Define $\psi : V \longrightarrow V$ by

$$\psi\Big(\sum_{t\in T} t \otimes w_t\Big) = 1 \otimes w_1 \qquad (w_t \in W)$$

It is easily checked that ψ is an FH-homomorphism. Moreover, for all $w \in W$, we have

$$\Big(\sum_{t\in T} t\psi t^{-1}\Big)(t' \otimes w) = \sum_{t\in T} t\psi(t^{-1}t' \otimes w) = t' \otimes w$$

Thus ψ satisfies (iv). Turning to the general case, we may harmlessly assume that

$$W^G = V \oplus V' \qquad \text{(direct sum of } FG\text{-modules)}$$

Let $\pi : W^G \longrightarrow V$ be the projection map and let ψ satisfy (iv) with respect to W^G. Then $\pi\circ\psi$ induces an FH-homomorphism $V \longrightarrow V$, and for all $v \in V$, we have

$$v = \pi\Big(\sum_{t\in T} t\psi t^{-1} v\Big) = \Big(\sum_{t\in T} t(\pi\cdot\psi)t^{-1}\Big)v$$

since π is an FG-homomorphism. The desired implication follows.

(iv) $\Rightarrow$ (v): Suppose that V is an FG-submodule of the FG-module U and that $U_H = V_H \oplus W$ for some FH-submodule W. Denote by π the projection of U_H onto V_H and let $\phi \in \mathrm{Hom}_F(U,V)$ be defined by

$$\phi(u) = \sum_{t\in T} (t\psi\circ\pi t^{-1})u$$

where $\psi \in \operatorname{End}_{FH}(V_H)$ satisfies (iv). If $v \in V$, then $\phi(v) = v$ and thus $U = V \oplus \operatorname{Ker}\phi$. But, by Lemma 5.1, ϕ is an FG-homomorphism, so $\operatorname{Ker}\phi$ is an FG-submodule. Hence V is H-injective.

(v) ⇒ (i): Assume that V is H-injective. Then, by Lemma 5.3, V is isomorphic to a direct summand of $(V_H)^G$. Invoking implication (iii) ⇒ (iv), we infer that there exists $\psi \in \operatorname{End}_{FH}(V_H)$ such that

$$(\sum_{t\in T} t\psi t^{-1})v = v$$

Assume that $f : V \longrightarrow V$ is a surjective FG-homomorphism for which

$$U = \operatorname{Ker}f \oplus W \qquad \text{(direct sum of } FH\text{-modules)}$$

The restriction $f_1 = f|W$ is an FH-isomorphism of W onto V_H. Set $\phi = f_1^{-1}$ and define $\theta : U \longrightarrow U$ by

$$\theta(u) = \sum_{t\in T} t\phi\psi f t^{-1}u$$

Then, by Lemma 5.1, θ is an FG-homomorphism and

$$\begin{aligned}(f\circ\theta)(u) &= \sum_{t\in T} f t\phi\psi f t^{-1}u \\ &= \sum_{t\in T} t f\phi\psi t^{-1}u \\ &= \sum_{t\in T} t\psi t^{-1} f u \\ &= f(u)\end{aligned}$$

proving that $U = \theta(U) + \operatorname{Ker}f$. Finally, if $\theta(u) \in \theta(U) \cap \operatorname{Ker}f$, then

$$\theta(u) = \sum_{t\in T} t\phi\psi f t^{-1}u = \sum_{t\in T} t\phi\psi t^{-1} f u = 0$$

Thus $U = \theta(U) \oplus \operatorname{Ker}f$ and the result follows. ■

5.5. COROLLARY. (Higman, 1954). Assume that the characteristic of F does not divide the index of H in G. Then any FG-module is H-projective and H-

injective.

Proof. Let V be an FG-module and let $\psi : V \longrightarrow V$ be defined by

$$\psi(v) = (G:H)^{-1}v$$

Then ψ is an FH-homomorphism and for all $v \in V$

$$(\sum_{t\in T} t\psi t^{-1})v = v$$

where T is a transversal for H in G. Now apply Theorem 5.4. ∎

5.6. COROLLARY. Let N be a normal subgroup of G and let V be an irreducible FN-module. If the characteristic of F does not divide the index of N in G, then V^G is completely reducible.

Proof. Let W be a submodule of V^G and let T be a transversal for N in G. Since

$$(V^G)_N = \bigoplus_{t\in T} t \otimes V$$

it follows that $(V^G)_N$ is completely reducible. Hence W_N is completely reducible, and thus W_N is a direct summand of $(V^G)_N$. By Corollary 5.5, W is N-injective. Consequenctly, W is a direct summand of V^G and the result follows. ∎

Let U and V be FG-modules with F-bases $u_1,\ldots,u_n$ and $v_1,\ldots,v_m$, respectively. Then $U \otimes_F V$ is a vector space over F with an F-basis $\{u_i \otimes v_j | 1 \leqslant i \leqslant n,\ 1 \leqslant j \leqslant m\}$. We can define an action of G on $U \otimes_F V$ by

$$g(u_i \otimes v_j) = gu_i \otimes gv_j$$

for all i,j and $g \in G$, and then extend to $U \otimes_F V$ and FG by linearity. The FG-module $U \otimes_F V$ is called the (inner) *tensor product* of U and V .

5.7. PROPOSITION. Let $S \subseteq H$ be subgroups of G and let V,W be FG-modules.

(i) If V is H-projective and W_H is S-projective, then $V \otimes_F W$ is S-projective

(ii) If W is projective, then $V \otimes_F W$ is also projective.

Proof. (i) Let X be a transversel for H in G and let Y be a transversal for S in H. Then XY is a transversal for S in G. By Theorem 5.4, there exist

$$\gamma \in \operatorname{End}_{FH}(V_H) \quad \text{and} \quad \delta \in \operatorname{End}_{FS}(W_S)$$

such that

$$\left(\sum_{x \in X} x\gamma x^{-1}\right)v = v \quad \text{and} \quad \sum_{y \in Y} (y\delta y^{-1})w = w$$

for all $v \in V, w \in W$. Then $\gamma \otimes \delta \in \operatorname{End}_{FS}((V \otimes_F W)_S)$ and a straight-forward calculation shows that

$$\left(\sum_{x \in X, y \in Y} xy(\gamma \otimes \delta)y^{-1}x^{-1}\right)(v \otimes w) = v \otimes w$$

for all $v \in V, w \in W$. The desired conclusion is now a consequence of Theorem 5.4.

(ii) Apply (i) for the case $H = G$ and $S = 1$. ∎

The following lemma collects together some simple properties of H-projective modules.

5.8. LEMMA. Let $S \subseteq H$ be subgroups of G and let V be an FG-module.

(i) Let $V = \bigoplus_{i=1}^{n} V_i$, where V_i is an FG-module, $1 \leqslant i \leqslant n$. Then V is H-projective if and only if each V_i is H-projective

(ii) If V is S-projective, then V is also H-projective

(iii) If V is H-projective and V_H is S-projective, then V is S-projective

(iv) For any FH-module W, the FG-module W^G is H-projective

(v) If W is an S-projective FH-module, then any FG-module which is a direct summand of W^G is S-projective.

Proof. (i) Assume that V is H-projective. Then there exists $\psi \in \operatorname{End}_{FH}(V_H)$ which satisfies Theorem 5.4(iv). Let π_i, $1 \leqslant i \leqslant n$, be the projection of V onto V_i defined by the decomposition $V = \oplus V_i$. Then, as in the proof of implication (ii) ⇒ (iii) of Theorem 5.4, $\pi_i \circ \psi$ induces an FH-homomorphism of V_i onto V_i, which also satisfies Theorem 8.4(iii). Thus each V_i is H-projective. Conversely, if each $\psi_i \in \operatorname{End}_{FH}(V_i)_H$ satisfies Theorem 5.4.(iv), then $\psi \in \operatorname{End}_{FH}(V_H)$

defined by $\psi|V_i = \psi_i$, obviously satisfies the same condition. Hence V is H-projective.

(ii) Let $0 \longrightarrow U \longrightarrow W \longrightarrow V \longrightarrow 0$ be an exact sequence of FG-modules such that the associated sequence $0 \to U_H \to W_H \to V_H \to 0$ splits. Then the sequence $0 \to U_S \to W_S \to V_S \to 0$ splits. Since V is S-projective, the sequence $0 \to U \to W \to V \to 0$ must also split. Thus V is H-projective.

(iii) By Theorem 5.4, $(V_H)^G = V' \oplus X$ and $(V_S)^H = V'' \oplus Y$, where $V' \cong V$ and $V'' \cong V_H$. Thus $V_S^G \cong (V_H)^G \oplus Y^G = V' \oplus X \oplus Y^G$, so V is isomorphic to a direct summand of V_S^G. Now apply Theorem 5.4.

(iv) This was established in the course of the proof of Theorem 8.4.

(v) By Theorem 5.4, $W_S^H = W' \oplus X$ where $W' \cong W$, so that $W_S^G \cong W^G \oplus X^G$. By (iv), W_S^G is S-projective and therefore, by (i), W^G is S-projective. Hence any direct summand of W^G is S-projective, again by appealing to (i). ∎

5.9. THEOREM. Let H be a subgroup of G.

(i) If V is a projective FG-module, then V_H is a projective FH-module. The converse is true, provided $\text{char}F \nmid (G:H)$.

(ii) If V is an FH-module, the induced module V^G is projective if and only if V is projective.

(iii) Let $\text{char}F = p > 0$, let P be a Sylow p-subgroup of G, and let V be an FG-module. Then V is projective if and only if V_P is projective.

Proof. (i) Let V be a projective FG-module. Then V_H is a projective FH-module, by virtue of Lemma 5.8(ii). Conversely, assume that V is an FG-module such that V_H is projective and let $\text{char}F \nmid (G:H)$. Let

$$0 \longrightarrow U \longrightarrow W \longrightarrow V \longrightarrow 0$$

be an exact sequence of FG-modules. Then the associated sequence

$$0 \longrightarrow U_H \longrightarrow W_H \longrightarrow V_H \longrightarrow 0$$

splits, since V_H is projective. Owing to Corollary 5.5, V is H-projective; hence the sequence

$$0 \longrightarrow U \longrightarrow W \longrightarrow V \longrightarrow 0$$

also splits. Consequently, V is a projective FG-module.

(ii) Let V be an FH-module and let T be a transversal for H in G. Then

$$(V^G)_H = 1 \otimes V \oplus \sum_{\substack{t \in T \\ t \notin H}} t \otimes V \qquad \text{(direct sum of } FH\text{-modules)} \qquad (14)$$

where, of course, $1 \otimes V$ is identifiable with V. Assume now that V^G is projective. Then, by (i), $(V^G)_H$ is also projective. Hence, by (14) and Lemma 5.8(i), V is projective. Conversely, suppose that V is projective. If V is a free FH-module, then the FG-isomorphism $FG \otimes_{FH} FH \cong FG$ implies that V^G is a free FG-module. Hence V^G is projective. In the general case, we have $W \cong V \oplus U$ for some free FH-module W and some FH-module U. Then $W^G \cong V^G \oplus U^G$ and hence V^G is projective by Lemma 5.8(i).

(iii) Direct consequence of (i). ∎

6. VERTICES OF FG-MODULES.

Throughout this section, G denotes a finite group and F a field of characteristic $p > 0$ dividing the order of G.

Let V and W be FG-modules. We say that V is a *component* of W if V is isomorphic to a direct summand of W.

Assume that V is an indecomposable FG-module. Then a subgroup Q of G is said to be a *vertex* of V if the following two properties hold:

(i) V is a component of $(V_Q)^G$

(ii) V is not a component of $(V_H)^G$ for any proper subgroup H of Q.

Hence 1 is the vertex of V if and only if V is a principal indecomposable module. Observe that if $p \nmid |G|$, then by Maschke's theorem all indecomposable FG-modules have vertex 1. In order to avoid trivial statements, we made the restriction that p divides $|G|$.

Owing to Theorem 5.4, (i) and (ii) are equivalent to the condition that V is Q-projective, but not H-projective for any proper subgroup H of Q. Observe

also that by Theorem 5.4, (i) is equivalent to

(i') V is a component of U^G for a suitable FQ-module U

Assume that Q is a vertex of V and write

$$V_Q = U_1 \oplus \ldots \oplus U_r$$

as a sum of indecomposable modules. Then

$$(V_Q)^G = U_1^G \oplus \ldots \oplus U_r^G$$

and hence, by the Krull-Schmidt theorem, V is a component of U^G for some indecomposable FQ-module U. Any such module U is called a *source* of V. Our aim in this section is to prove that the set of vertices of V is a class of conjugate p-subgroups of G.

Let W be any FH-module, where H is a subgroup of G. For any $g \in G$, define the $F(gHg^{-1})$-module ${}^gW$ as follows: ${}^gW$ has the same underlying space as W and the elements $x, x \in gHg^{-1}$ act according to the rule

$$x * w = (g^{-1}xg)w$$

6.1. THEOREM (Mackey decomposition). Let H and L be subgroups of G, let T be a set of double coset representatives for (L,H) in G and let V be an FH-module. For each $t \in T$, let V_t denote the restriction of ${}^tV$ to $F(tHt^{-1} \cap L)$. Then

$$(V^G)_L \cong \bigoplus_{t \in T} (V_t)^L \qquad \text{(as } FL\text{-modules)}$$

Proof. Choose a left transversal $g_1 = 1, g_2, \ldots, g_n$ for H in G. Then

$$V^G = \bigoplus_{i=1}^{n} (g_i \otimes V) \qquad \text{(as } F\text{-spaces)}$$

Since $g_i(1 \otimes V) = g_i \otimes V$ and since $V \cong 1 \otimes V$, we have

$$g_i \otimes V \cong {}^{g_i}V \qquad \text{as } F(g_iHg_i^{-1})\text{-modules} \qquad (*)$$

Put $S = \{g_1 \otimes V, g_2 \otimes V, \ldots, g_n \otimes V\}$. Then G, and in particular L, acts on S. Furthermore, $g_i \otimes V$ and $g_j \otimes V$ lie in the same L-orbit if and only if

g_i and g_j belong to the same double (L,H)-coset.

Given $t \in T$, denote by W_t the sum of those $g_j \otimes V$ for which $g_j \in LtH$. Then each W_t is an FL-module and

$$(V^G)_L = \bigoplus_{t\in T} W_t$$

It therefore suffices to verify that $W_t \cong (V_t)^L$. Let $X \subseteq \{g_1,\ldots,g_n\}$ be such that $LtH = \bigcup_{x\in X} xH$ so that $W_t = \bigoplus_{x\in X} (x \otimes V)$. We may harmlessly assume that $t \in X$. Now L acts transitively on the set $\{x \otimes V | x \in X\}$ and under this action the stabilizer of $t \otimes V$ is

$$\{\ell \in L | \ell t \otimes V = t \otimes V\} = \{\ell \in L | t^{-1}\ell t \in H\} = tHt^{-1} \cap L$$

Thus, by (*) and Lemma 4.1, $W_t \cong (V_t)^L$ and the result follows. ∎

6.2. LEMMA. Let H and L be subgroups of G and let W be an FG-module such that W is a component of V^G for some FH-module V. Write

$$W_L = W_1 \oplus \ldots \oplus W_n$$

as a direct sum of indecomposable FL-modules. Then there exist elements $g_1,\ldots,g_n$ of G such that for each $i \in \{1,\ldots,n\}$, W_i is a component of V_i^L, where V_i is the restriction of ${}^{g_i}V$ to $F(g_iHg_i^{-1} \cap L)$. In particular,

$$W_i \text{ is } (g_iHg_i^{-1} \cap L)\text{-projective}$$

Proof. Denote by T a set of double coset representatives for (L,H) in G and for each $t \in T$, let V_t be the restriction of ${}^tV$ to $F(tHt^{-1} \cap L)$. Then, by Theorem 6.1,

$$(V^G)_L \cong \bigoplus_{t\in T} (V_t)^L$$

Because W is a component of V^G, it follows that each W_i is a component of $(V^G)_L$. Hence the required assertion is a consequence of the indecomposability of W_i and the Krull-Schmidt theorem. ∎

6.3. LEMMA. Let W be an indecomposable FG-module and let Q be a vertex of W.

(i) For any $g \in G$, gQg^{-1} is also a vertex of W

(ii) If W is a component of $(W_L)^G$ for some subgroup L of G, then $L \supseteq gQg^{-1}$ for some $g \in G$.

Proof. (i) Let H be a subgroup of G. Then it is easy to verify that

$$(W_H)^G \cong (W_{gHg^{-1}})^G \qquad \text{for all } g \in G$$

Hence, if W is H-projective, then for all $g \in G$, W is gHg^{-1}-projective. In particular, for all $g \in G$, W is gQg^{-1}-projective. Now any subgroup of gQg^{-1} is of the form gQ_1g^{-1} for some subgroup Q_1 of Q. If W is gQ_1g^{-1}-projective, then by the above, W is Q_1-projective. Since Q is a vertex of W, we have $Q_1 = Q$ and so $gQ_1g^{-1} = gQg^{-1}$. This proves that gQg^{-1} is a vertex of W.

(ii) Owing to Lemma 6.2, $W_L = W_1 \oplus \ldots \oplus W_n$, where each W_i is an indecomposable FL-module which is $(g_iQg_i^{-1} \cap L)$-projective for some $g_i \in G$. In particular, W is a component of $W_1^G \oplus \ldots \oplus W_n^G$ and so, by the Krull-Schmidt theorem, W is a component of W_i^G for some i. Invoking Lemma 5.8(v), we infer that W is $(g_iQg_i^{-1} \cap L)$-projective. But, by (i), $g_iQg_i^{-1}$ is a vertex of W, hence $g_iQg_i^{-1} = g_iQg_i^{-1} \cap L$ and the result follows. ■

We are now ready to prove

6.4. THEOREM. Let W be an indecomposable FG-module with vertex Q. Then Q is a p-subgroup of G and a subgroup H of G is a vertex of W if and only if H is conjugate to Q in G.

Proof. Let P be a Sylow p-subgroup of G. By Corollary 5.5, W is P-projective, so W is a component of $(W_P)^G$. Applying Lemma 6.3(ii), we deduce that $P \supseteq gQg^{-1}$ for some $g \in G$. Hence Q is a p-subgroup of G. By Lemma 6.3(i), it suffices to show that if H is a vertex of W, then H is conjugate to Q. So suppose that H is a vertex of W. Then, by Lemma 6.3(ii), $H \supseteq gQg^{-1}$ for some $g \in G$. Because gQg^{-1} is a vertex of W, W is gQg^{-1}-projective and thus $H = gQg^{-1}$. ■

3
The Jacobson radical of group algebras: Foundations of the theory

The aim of this chapter is to present a number of general results concerning the Jacobson radical of group algebras. The material discussed constitutes the foundations of the theory and will be frequently applied in our subsequent investigations. After presenting some properties of elementary nature, we examine the Jacobson radical of direct products. We then provide necessary and sufficient conditions for an element $x = \Sigma x_g g \in FG$ to be in $J(FG)$. These conditions are subsequently simplified in the special case where G is p-solvable. As an application, we characterize the elements of $J(Z(FG))$. Further topics include Frobenius groups and the dimensions of $J(FG)$ and $J(Z(FG))$. Next we establish an important decomposition theorem due to Morita and provide a number of applications. Among other topics, we examine criteria for the commutativity of $J(FG)$ and $J(B_o)$, where B_o is the principal block of FG. The chapter ends with a beautiful result due to Morita, which provides circumstances under which $J(FG)$ is expressible as a principal left and right ideal.

1. ELEMENTARY PROPERTIES.

The aim of this section is to present a number of results of elementary nature which will be frequently applied in subsequent investigations. *Throughout* G *denotes a finite group and* F *a field of characteristic* $p > 0$. All vector spaces over F are assumed to be finite-dimensional.

1.1. PROPOSITION. Assume that P is a normal p-subgroup of G. Then

(i) $FG \cdot I(P) \subseteq J(FG)$

(ii) $J(F(G/P)) \cong J(FG)/FG \cdot I(P)$

(iii) If P is a Sylow p-subgroup of G, then $J(FG) = FG \cdot I(P)$.

Proof. (i) Given $x \in P$, we have $x^{p^n} = 1$ for some $n \geqslant 1$. Then

$$(x-1)^{p^n} = x^{p^n} - 1 = 0$$

which proves, by Proposition 1.10.5, that $I(P)$ is nilpotent, since the elements $x-1, x \neq 1, x \in P$, constitute an F-basis for $I(P)$. Because $P \lhd G$, we have

$$FG \cdot I(P) = I(P) \cdot FG$$

and hence

$$(FG \cdot I(P))^n = FG \cdot I(P)^n \qquad \text{for all } n \geqslant 1$$

Thus $FG \cdot I(P)$ is a nilpotent ideal of FG and so $FG \cdot I(P) \subseteq J(FG)$.

(ii) Direct consequence of (i), Proposition 2.1.2 and Corollary 1.6.17(ii)

(iii) If P is a Sylow p-subgroup of G, then by Maschke's theorem $J(F(G/P)) = 0$. Now apply (ii). ■

1.2. COROLLARY. If G is a p-group, then $J(FG) = I(G)$ and FG is a local ring. In particular, 0 and 1 are the only idempotents of FG.

Proof. By Proposition 1.1(iii), $J(FG) = I(G)$. Since $FG/I(G) \cong F$, we conclude that FG is a local ring. ■

1.3. LEMMA. Let H be a subgroup of G and let T be a right tranversal for H in G.

(i) If I is a left ideal of FH with F-basis X, then $\{tx \mid t \in T, x \in X\}$ is an F-basis of $FG \cdot I$. In particular,

$$\dim_F FG \cdot I = |G:H| \dim_F I$$

(ii) The elements $\{t(h-1) \mid t \in T, h \in H, h \neq 1\}$ constitute an F-basis of $FG \cdot I(H)$

Proof. (i) For any $t \in T, \{tx \mid x \in X\}$ is an F-basis of $t \cdot I$. On the other hand, by Proposition 2.1.4, $FG \cdot I = \bigoplus_{t \in T} t \cdot I$, proving the assertion.

(ii) Direct consequence of (i) and the fact that $\{h-1 \mid h \in H, h \neq 1\}$ is an F-basis of $I(H)$. ■

1.4. COROLLARY. Assume that a Sylow p-subgroup P of G is normal. Then, for any transversal T of P in G, the set

$$\{t(x-1) \mid t \in T \; x \in P \; x \neq 1\}$$

constitute an F-basis for $J(FG)$. In particular,

$$\dim_F J(FG) = |G| - |G:P|$$

Proof. Owing to Proposition 1.1(iii), $J(FG) = FG \cdot I(P)$. Now apply Lemma 1.3(ii). ■

1.5. COROLLARY. Assume that P is a normal p-subgroup of G. Then

$$\dim_F J(FG) = \dim_F J(F(G/P)) + |G| - |G:P|$$

Proof. By Proposition 1.1(ii),

$$\dim_F J(FG) = \dim_F J(F(G/P)) + \dim_F FG \cdot I(P)$$

Invoking Lemma 1.3(ii), we also have $\dim_F FG \cdot I(P) = |G| - |G:P|$, proving the assertion. ■

A subgroup H of G is said to be a *p-complement* if $G = H \cdot P$ with $H \cap P = 1$, where $P \in \mathrm{Syl}_p(G)$. Since $|G| = |H||P|$, the p-complement H is always a p'-group.

1.6. PROPOSITION. Let H be a subgroup of G.

(i) $J(FG) \cap FH \subseteq J(FH)$

(ii) $\dim_F J(FG) \leqslant \dim_F J(FH) + |G| - |H|$

(iii) If G has a p-complement, then $\dim_F J(FG) \leqslant |G| - |G:P|$

Proof. (i) Direct consequence of the fact that $J(FG) \cap FH$ is a nilpotent ideal of FH.

(ii) By (i), we have

$$\begin{aligned}\dim_F J(FG) &= \dim_F (J(FG)/(J(FG) \cap FH)) + \dim_F (J(FG) \cap FH) \\ &\leqslant \dim_F (FG/FH) + \dim_F J(FH) \\ &= |G| - |H| + \dim_F J(FH)\end{aligned}$$

as required.

(iii) Let H be a p-complement. Since H is a p'-group, $J(FH) = 0$ by Maschke's theorem. Now apply (ii). ∎

In what follows, we denote by $O_p(G)$ the maximal normal p-subgroup of G.

1.7. PROPOSITION. $G \cap (1 + J(FG)) = O_p(G)$

Proof. That $O_p(G) \subseteq G \cap (1 + J(FG))$ is a consequence of Proposition 1.1(i). Conversely, assume that $g \in G \cap (1 + J(FG))$. Then $g-1 \in J(FG)$ is nilpotent. Hence, there exists $n \geq 1$ such that $(g-1)^{p^n} = 0$. But $(g-1)^{p^n} = g^{p^n} - 1$, so $g^{p^n} = 1$ which proves that $G \cap (1 + J(FG))$ is a p-group. Thus $G \cap (1 + J(FG)) \subseteq O_p(G)$ and the result follows. ∎

1.8. PROPOSITION. Let N be a normal subgroup of G of index n.

(i) (Passman (1970a)). $J(FG)^n \subseteq FG \cdot J(FN) \subseteq J(FG)$

(ii) (Villamayor (1959)) If $p \nmid n$, then

$$J(FG) = FG \cdot J(FN) = J(FN)FG$$

and, in particular,

$$\dim_F J(FG) = |G:N| \dim_F J(FN)$$

(iii) Willems (1976). If $J(FG) = FG \cdot J(FN)$, then $p \nmid n$.

Proof. (i) Since conjugation by $g \in G$ is an automorphism of FN, we have $g^{-1}J(FN)g = J(FN)$. Thus $FG \cdot J(FN) = J(FN) \cdot FG$ and therefore

$$(FG \cdot J(FN))^n = FG \cdot J(FN)^n \qquad \text{for all } n \geq 1$$

It follows that $FG \cdot J(FN)$ is a nilpotent ideal and so $FG \cdot J(FN) \subseteq J(FG)$.

Let V be an irreducible FN-module and let $\{g_1, \ldots, g_n\}$ be a transversal for N in G. Then

$$(V^G)_N = \bigoplus_{i=1}^{n} g_i \otimes V \qquad (*)$$

where each $g_i \otimes V$ is an irreducible FN-module. Hence $(V^G)_N$ has a composition series of length n. Since any chain of FG-submodules of V^G restricts to a chain of FN-submodules of $(V^G)_N$, we infer that V^G has a composition series of length $m \leqslant n$. Hence

$$J(FG)^m V^G = 0$$

by Proposition 1.6.33(i). Now FG is a free FN-module with the elements $g_1,\ldots,g_n$ as a basis. Therefore, given $x \in J(FG)^m$, there exist unique $y_1,\ldots,y_n$ in FN such that

$$x = g_1 y_1 + \ldots + g_n y_n$$

Since for any $v \in V, x\,(1 \otimes v) = 0$ we have

$$g_1 \otimes y_1 v + \ldots + g_n \otimes y_n v = 0$$

and so, by (*), $y_i v = 0, 1 \leqslant i \leqslant n$. It follows that $y_i \in J(FN)$ and hence that $x \in FG{\cdot}J(FN)$. This proves that $J(FG)^m \subseteq FG{\cdot}J(FN)$ and since $J(FG)^n \subseteq J(FG)^m$, we finally deduce that

$$J(FG)^n \subseteq FG{\cdot}J(FN)$$

as required.

(ii) Assume that $p \nmid n$. Let V be an irreducible FN-module and let $x \in J(FG)$. Then V^G is completely reducible (Corollary 2.5.6), hence $xV^G = 0$ and, by the argument of (i), $x \in FG{\cdot}J(FN)$. Thus $J(FG) \subseteq FG{\cdot}J(FN)$ and therefore, by (i), $J(FG) = FG{\cdot}J(FN)$. The final assertion follows by virtue of Lemma 1.3(i).

(iii) We may view $F(G/N)$ as an FG-module by means of $x(gN) = xgN$, for all $x,g \in G$. Fix $x \in J(FN)$. Since $J(FN) \subseteq I(N)$, $x = \sum_{a\in N} x_a(a-1)$ for some $x_a \in F$. Hence, for all $g \in G$,

$$x(gN) = \sum_{a\in A} x_a(a-1)gN = \sum_{a\in A} x_a(agN-gN) = 0$$

Thus by our assumption

$$J(FG)F(G/N) = FG{\cdot}J(FN)F(G/N) = 0$$

Hence $F(G/N)$ is a completely reducible FG-module. Therefore $F(G/N)$ is also a completely reducible $F(G/N)$-module. The desired conclusion is now a consequence of Proposition 2.1.8. ■

Let p be a prime. A group G is said to be *p-solvable* if the composition factors of G are either p-groups or p'-groups. It is an immediate consequence of the definition that:

(i) Any solvable group is p-solvable

(ii) Any extension of a p-solvable group by a p-solvable group is p-solvable.

(iii) Subgroups and homomorphic images of p-solvable groups are p-solvable.

1.9. PROPOSITION (Passman(1970a)). Let N be a normal subgroup of G such that G/N is p-solvable. If $|G/N| = np^a$ where $(n,p) = 1$, then

$$J(FG)^{p^a} \subseteq FG\cdot J(FN) \subseteq J(FG)$$

In particular, if G is p-solvable of order np^a where $(n,p) = 1$, then

$$J(FG)^{p^a} = 0$$

Proof. Owing to Proposition 1.8(i), we need only verify that

$$J(FG)^{p^a} \subseteq FG\cdot J(FN)$$

We proceed by induction on $|G/N|$. If G/N is a p-group, the result follows from Proposition 1.8(i) and if G/N is a p'-group, the result follows from Proposition 1.8(ii). If neither of these cases occurs, then G/N must have a proper normal subgroup M/N. Write $|G/M| = rp^b, |M/N| = sp^c$ with $(r,p) = (s,p) = 1$. By induction

$$J(FG)^{p^b} \subseteq FG\cdot J(FM)$$

and

$$J(FM)^{p^c} \subseteq FM\cdot J(FN)$$

Since $FG\cdot J(M) = J(M)\cdot FG$, we infer that

$$J(FG)^{p^b p^c} \subseteq (FG)(FM)(J(FN)) = FG\cdot J(FN)$$

Since $p^b p^c = p^a$ the result follows. ■

1.10. LEMMA. Let P be a normal p-subgroup of G. Then the kernel of the Brauer homomorphism

$$\pi : Z(FG) \longrightarrow Z(FG)$$

is a nilpotent ideal.

Proof. Let $C \in C\ell(G)$ be such that $C \cap C_G(P) = \emptyset$. Invoking Theorem 2.2.3, it suffices to verify that C^+ is nilpotent. Since $FG \cdot I(P)$ is a nilpotent ideal of FG (Proposition 1.1(i)), the latter will follow provided we show that $C^+ \in FG \cdot I(P)$.

The group P acts by conjugation as a permutation group of the set C. Because $C \cap C_G(P) = \emptyset$, the size of each orbit under this action is divisible by p. But for all $x \in P$ and $g \in C$,

$$x^{-1}gx - g = x^{-1}g(x-1) + (x^{-1}-1)g \in FG \cdot I(P)$$

Hence $C^+ \in FG \cdot I(P)$ and the result follows. ■

1.11. LEMMA. Let P be a normal p-subgroup of G and let e be a central idempotent of FG. Then $e \in FC_G(P)$.

Proof. We may harmlessly assume that e is a block idempotent of FG. Let $e_1, \ldots, e_n$ be all block idempotents of FG and let

$$\pi : Z(FG) \longrightarrow Z(FG)$$

be the Brauer homomorphism. Then

$$1 = \pi(e_1) + \ldots + \pi(e_n) \quad \text{and} \quad \pi(e_i)\pi(e_j) = 0 \qquad \text{for } i \neq j$$

Furthermore, by Lemma 1.10, $\mathrm{Ker}\pi$ is a nilpotent ideal, so each $\pi(e_i)$ is a nonzero idempotent. Hence $\pi(e_1), \ldots, \pi(e_n)$ are all block idempotents of FG and, in particular, $e \in \mathrm{Im}\pi$. Since for any $x \in \mathrm{Im}\pi$, $\mathrm{Supp} \subseteq C_G(P)$, the result follows. ■

We next present a criterion for FG to be indecomposable. First, however, we must develop our vocabulary.

In what follows we write $O_{p'}(G)$ for the maximal normal p'-subgroup of G. The inverse image of $O_p(G/O_{p'}(G))$ in G is denoted by $O_{p',p}(G)$. A group G is called *p-constrained* if the inverse image of $C_{G/O_{p'}(G)}(O_{p',p}(G)/O_{p'}(G))$ in G is contained in $O_{p',p}(G)$. Thus, if G is p-constrained and $O_{p'}(G) = 1$, then

$$C_G(O_p(G)) \subseteq O_p(G)$$

It is well-known (see Huppert and Blackburn (1982a)) that any p-solvable group is p-constrained.

An obvious necessary condition for FG to be indecomposable is that $O_{p'}(G) = 1$. Indeed, if $m = |O_{p'}(G)| > 1$ then

$$e = m^{-1} \sum_{x \in O_{p'}(G)} x$$

is a nontrivial central idempotent. The following result provides a class of groups for which the converse is also true.

1.12. PROPOSITION. Let G be p-constrained. Then FG is indecomposable if and only if $O_{p'}(G) = 1$.

Proof. By the foregoing remark, we need only prove that if $O_{p'}(G) = 1$, then FG is indecomposable. So assume that $O_{p'}(G) = 1$ and set $P = O_p(G)$. Since G is p-constrained, we must have $C_G(P) \subseteq P$. Hence, by Lemma 1.11, given a central idempotent e of FG, we have $e \in FC_G(P) \subseteq FP$. Invoking Corollary 1.2, we deduce that $e = 0$ or $e = 1$, as desired. ■

Having discovered a class of groups G for which FG is indecomposable, we now prove

1.13. PROPOSITION. Let $G \neq 1$ be a group such that FG is indecomposable (e.g. G is p-constrained, $O_{p'}(G) = 1$ as $G \neq 1$). If $C_1, C_2, \ldots, C_r$ are all conjugacy classes of G distinct from 1, then the elements

$$C_i^+ - |C_i| \cdot 1 \qquad (1 \leqslant i \leqslant r)$$

constitute an F-basis for $J(Z(FG))$.

Proof. Owing to Proposition 2.10.8(iv), $Z(FG)$ is a local ring. Let

$$\pi : Z(FG) \longrightarrow F$$

be the restriction to $Z(FG)$ of the augmentation map $FG \longrightarrow F$. Then π is a homomorphism and hence $J(Z(FG)) = \text{Ker}\pi$. The elements $C_i^+ - |C_i|\cdot 1$, $1 \leqslant i \leqslant r$, are linearly independent and have augmentation zero. Hence their linear span has dimension r and is contained in $J(Z(FG))$. The desired conclusion is therefore a consequence of the fact $\dim_F Z(FG) = r+1$. ■

Example. Let $G = S_4$ and let $p = 2$. Then G is p-constrained and $O_{p'}(G)=1$. Put $x = (1234)$, $y = (123)$, $z = (12)$ and $u = (12)(34)$ and denote by C_x, C_y, C_z and C_u the corresponding conjugacy classes. Then C_x, C_y, C_z, C_u are all conjugacy classes of G distinct from 1 and

$$|C_x| = 6\ ,\ |C_y| = 8\ ,\ |C_z| = 6 \ \text{ and } \ |C_u| = 3$$

Hence the elements

$$C_x^+\ ,\ C_y^+\ ,\ C_z^+\ ,\ C_u^+ + 1$$

constitute an F-basis for $J(Z(FG))$. ■

We next examine the behaviour of $J(FG)$ under extension of ground field.

Let E be a field and K a subfield; we write E/K for the field E considered as an extension of K. A polynomial f over a field K is called *separable* over K if it is a product of irreducible polynomials in $K[X]$, all of which have only simple roots (in a splitting field of f). An element of E/K is *separable* over K if it is algebraic over K and if its minimal polynomial over K is separable. An extension E/K is a *separable extension* if every element of E is separable over K.

A field K is said to be *perfect* if either char$K = 0$ or char$K = p > 0$ and $K^p = K$.

Let A denote a finite-dimensional algebra over a field K. An A-module V is said to be separable if V_E is a completely reducible A_E-module for every field extension E of K. Thus a separable module is completely reducible; the converse need not be true as we shall see below.

For convenience of reference, we now quote the following two standard facts:

1.14. PROPOSITION. A field K is perfect if and only if every algebraic extension of K is separable.

Proof. See Jacobson(1964). ∎

1.15. PROPOSITION. Let V be a completely reducible A-module. Then V is separable if and only if for every irreducible submodule W of V the centre of the division algebra $\mathrm{End}_A(W)$ is a finite separable field extension of F.

Proof. See Bourbaki(1958). ∎

The algebra A is said to be *separable* if the left regular module ${}_AA$ is separable. Expressed otherwise, A is separable if for every field extension E of K, A_E is a semisimple E-algebra.

1.16. PROPOSITION. The following conditions are equivalent:

(i) $A/J(A)$ is separable

(ii) For every field extension E/K, $J(A_E) = J(A)_E$

(iii) For every field extension E/K, $(A/J(A))_E \cong A_E/J(A_E)$

In particular, if $A/J(A)$ is separable, then so is $A_E/J(A_E)$ for any field extension E/K.

Proof. Direct consequence of Lemma 1.10.1 (iii),(iv). ∎

Let A be an algebra over a field K and let L be a subfield of K. We say that A is *definable* over L if there exists a K-basis X of A such that for all $x_1, x_2 \in X$, x_1x_2 is an L-linear combination of the elements of X. Expressed otherwise, A is definable over L if $A \cong K \otimes_L A_1$ for some L-algebra A_1.

1.17. PROPOSITION. Let A be an algebra over a field K and assume that A is definable over a perfect subfield of K. Then $A/J(A)$ is separable

Proof. By hypothesis, $A \cong K \otimes_L A_1$ for some L-algebra A_1 and some perfect subfield L of K. By Proposition 1.16, we need only prove that $A_1/J(A_1)$ is a separable L-algebra. The latter being a consequence of Propositions 1.14 and 1.15, the result follows. ■

1.18. COROLLARY. Let K be an arbitrary field. Then the K-algebras $KG/J(KG)$ and $Z(KG)/J(Z(KG))$ are separable. In particular, for every field extension E of K,

$$J(EG) = E \otimes_K J(KG) \quad \text{and} \quad J(Z(EG)) = E \otimes_K J(Z(KG))$$

Proof. Let L be the prime subfield of K. Then L is perfect and

$$KG \cong K \otimes_L LG \quad \text{and} \quad Z(KG) \cong K \otimes_L Z(LG)$$

Now apply Propositions 1.17 and 1.16. ■

1.19. PROPOSITION. Let A be an algebra over a field K and assume that $A/J(A)$ is separable. Then for any field extension L of K and any irreducible A-module V,

$$P(V_L) = P(V)_L \quad \text{and, in particular,} \quad \dim_K P(V) = \dim_L P(V_L)$$

Proof. Since $P(V)$ is a projective A-module, $P(V)_L$ is a projective A_L-module. By Lemma 1.10.1(v),

$$V_L \cong (P(V)/J(P(V)))_L \cong P(V)_L/J(P(V))_L$$

Since $A/J(A)$ is separable, V_L is a completely reducible A_L-module and so $J(V_L) = 0$. Applying Lemma 1.10.1(vi) and Corollary 1.6.5, we conclude that $J(P(V))_L = J(P(V)_L)$. Hence, by Theorem 1.9.3, $P(V)_L$ is a projective cover of V_L. ■

1.20. PROPOSITION. Let G be p-constrained, let $N = O_{p'}(G)$ and let

$$e = |N|^{-1} \sum_{x \in N} x$$

Then $FGe \cong F(G/N)$ is the principal block of FG.

Proof. By Proposition 2.1.9, e is a central idempotent of FG such that $FGe \cong F(G/N)$. Since G is p-constrained, so is $\bar{G} = G/N$. But $O_{p'}(\bar{G}) = 1$ and hence, by Proposition 1.12, $F\bar{G}$ is indecomposable. This proves that e is a block idempotent of FG. Finally, since e acts as identity on the trivial FG-module, the result follows. ■

2. DIRECT PRODUCTS.

Throughout this section, G denotes a finite group and F a field of characteristic $p > 0$. Our aim is to investigate $J(FG)$ for the case $G = G_1 \times G_2$ for some $G_i \lhd G$, $i = 1,2$.

Let A and B be (finite-dimensional) algebras over a field K. As is customary, we shall identify the algebras A and B with their images $A \otimes 1$ and $1 \otimes B$ in $A \otimes_K B$. With these identifications, A and B become commuting subalgebras which generate $A \otimes_K B$.

Let V and W be two vector spaces over K. Given $f \in \text{End}_K(V)$, $g \in \text{End}_K(W)$, one can define an element $f \otimes g$ of $\text{End}_K(V \otimes W)$ by setting

$$(f \otimes g)(v \otimes w) = f(v) \otimes g(w) \qquad (v \in V, w \in W)$$

If $f : A \to \text{End}_K(V)$ and $g : B \to \text{End}_K(W)$ are algebra homomorphisms, then

$$\begin{cases} A \otimes_K B \longrightarrow \text{End}_K(V \otimes W) \\ a \otimes b \longmapsto f(a) \otimes g(b) \end{cases}$$

is also an algebra homomorphism. This means that if V and W are A and B-modules, respectively, then $V \otimes_K W$ is an $A \otimes_K B$-module by means of

$$(a \otimes b)(v \otimes w) = av \otimes bw \qquad (a \in A, b \in B, v \in V, w \in W)$$

Let G and H be groups and let V and W be KG and KH-modules, respectively. Then the *outer tensor product* $V \# W$ of V and W is a $K(G \times H)$-module whose underlying vector space is $V \otimes_K W$ with the module operation given by

$$(g,h)(v \otimes w) = gv \otimes hw \qquad (g \in G, h \in H, v \in V, w \in W)$$

On the other hand, as we have seen above, the vector space $V \otimes_K W$ is an $KG \otimes_K KH$-

module by means of

$$(g \otimes h)(v \otimes w) = gv \otimes hw \qquad (g \in G, h \in H, v \in V, w \in W)$$

Hence if we identify $KG \otimes_K KH$ with $K(G \times H)$ (see Proposition 2.1.6), $V \otimes_K W$ is nothing else but the outer tensor product of V and W.

2.1. THEOREM. Let A and B be algebras over an algebraically closed field K, and let $C = A \otimes_K B$. If $V_1, \ldots, V_r$ and $W_1, W_2, \ldots, W_t$ are all nonisomorphic irreducible A and B-modules, respectively, then $V_i \otimes_K W_j, 1 \leqslant i \leqslant r, 1 \leqslant j \leqslant t$, are all nonisomorphic irreducible C-modules.

Proof. Let $f_i : A \longrightarrow \mathrm{End}_K(V_i)$ and $g_j : B \longrightarrow \mathrm{End}_K(W_j)$ be the homomorphisms associated with the modules V_i and W_j, respectively. It is a consequence of Corollary 1.10.4 that each f_i and g_j is surjective; hence so is $f_i \otimes g_j$ and therefore $V_i \otimes_K W_j$ is an irreducible C-module.

Assume now that U is an irreducible C-module and that $\psi : C \longrightarrow \mathrm{End}_K(U)$ is the homomorphism associated with U. Because the C-module U is finite-dimensional, we may choose an irreducible A-submodule V. Let L be the space of all A-homomorphisms from V to U. Then $L \neq 0$ and it is easy to verify that L is a B-module under the action

$$b\ell = \psi(b) \circ \ell \qquad (b \in B, \ell \in L)$$

Let W be an irreducible B-submodule of L and let

$$\phi : V \otimes_K W \longrightarrow U$$

be the F-linear map defined by $\phi(v \otimes w) = w(v)$; the map ϕ is in fact a C-homomorphism since for all $a \in A, b \in B, v \in V, w \in W$,

$$\begin{aligned}\phi((a \otimes b)(v \otimes w)) &= \phi((av)\otimes(bw)) = (bw)(av) = (\psi(b)\circ w)(av)\\ &= \psi(b)\psi(a)w(v) = \psi(ab)w(v) = (a \otimes b)w(v)\\ &= (a \otimes b)\phi(v \otimes w)\end{aligned}$$

Now U is irreducible, by hypothesis, and $V \otimes_K W$ is irreducible by the foregoing.

Since $\phi \neq 0$, we infer that $U \cong V \otimes_K W$.

We are therefore left to verify that the isomorphism class of $V_i \otimes_K W_j$ determines the isomorphism classes of both V_i and W_j. Let M_i be an irreducible A-submodule of $V_i \otimes_K W_j$. For any $b \in B, bM_i$ is also an irreducible A-submodule and obviously, $M_i \cong bM_i$. Because

$$V_i \otimes_K W_j = \sum_{b \in B} bM_i ,$$

it follows that as an A-module, $V_i \otimes_K W_j$ is a direct sum of copies of M_i. Since V_i is identifiable with an irreducible A-submodule of $V_i \otimes_K W_j$, we deduce from the Krull-Schmidt theorem that $V_i \cong M_i$. Thus $V_i \otimes_K W_j$ determines the isomorphism class of V_i. A similar argument shows that $V_i \otimes_K W_j$ determines the isomorphism class of W_j, so the theorem is true. ■

2.2. COROLLARY. Let K be an algebraically closed field, let G be the (internal) direct product of G_1 and G_2 and let $V_1, V_2, \ldots, V_r$ and $W_1, W_2, \ldots, W_t$ be all nonisomorphic irreducible KG_1 and KG_2-modules, respectively Then, for each $i \in \{1, \ldots, r\}, j \in \{1, \ldots, t\}$, $V_i \otimes_K W_j$ is a KG-module via

$$(g_1 g_2)(v_i \otimes w_j) = g_1 v_i \otimes g_2 w_j \qquad (v_i \in V_i, w_j \in W_j, g_1 \in G_1, g_2 \in G_2)$$

Furthermore, the $V_i \otimes_K W_j$ $(1 \leqslant i \leqslant r,\ 1 \leqslant j \leqslant t)$ are all nonisomorphic irreducible KG-modules.

Proof. Direct consequence of Theorem 2.1 and the fact (see Proposition 2.1.6) that the map $g_1 g_2 \mapsto g_1 \otimes g_2, g_1 \in G_1, g_2 \in G_2$, induces an isomorphism $KG \to KG_1 \otimes_K KG_2$ of K-algebras. ■

Let I be a nilpotent ideal of a ring R. By the *nilpotency index* of I we understand the least positive integer n such that $I^n = 0$. In what follows, we write $t(G)$ for the nilpotency index of $J(FG)$. For the sake of convenience, we put

$$J(FG)^0 = FG$$

We are now ready to provide information on $J(F(G_1 \times G_2))$. Since the corresponding result entails a fair amount of notation, it will be especially useful to pause to assemble most of it in one place.

G is the (internal) direct product of G_1 and G_2.

$r_i = \dim_F J(FG_1)^i \quad (i \geq 0)$

$r'_i = \dim_F J(FG_2)^i \quad (i \geq 0)$

s is a fixed integer such that $1 \leq s < t(G)$

$T_i = \{a_{ij} | 1 \leq j \leq r_i - r_{i+1}\}$ is a subset of $J(FG_1)^i$ which forms an F-basis of $J(FG_1)^i$ modulo $J(FG_1)^{i+1}$, $0 \leq i < s$ (If $J(FG_1)^i = 0$, then we set $T_i = \emptyset$ and $r_i = 0$).

$T_s = \{a_{sk} | 1 \leq k \leq r_s\}$ is an F-basis of $J(FG_1)^s$ (If $J(FG_1)^s = 0$, then we set $T_s = \emptyset$ and $r_s = 0$)

$T'_k = \{a'_{k\ell} | 1 \leq \ell \leq r'_k - r'_{k+1}$ and $T'_s = \{a'_{s\ell} | 1 \leq \ell \leq r'_s\}$ are defined similarly for FG_2.

2.3. THEOREM (Loncour (1971), Motose(1974b)). With the notation above, the following properties hold:

(i) $\dim_F J(FG)^s = \sum_{i=0}^{s} r_i r'_{s-i} - \sum_{i=1}^{s} r_i r'_{s-i+1}$

(ii) $J(FG)^s = \sum_{i=0}^{s} J(FG_1)^i J(FG_2)^{s-i}$ and, in particular,

$$J(FG) = J(FG_1)FG_2 + J(FG_2)FG_1$$

(iii) $t(G) = t(G_1) + t(G_2) - 1$

(iv) $X = \bigcup_{i+k \geq s} T_i T'_k$ form an F-basis for $J(FG)^s$, where

$$T_i T'_k = \{tt' | t \in T_i, t' \in T'_k\}$$

Proof. (i) and (ii). We first consider the case where $s = 1$. Let E be an algebraic closure of F and let $\{U_i | 1 \leq i \leq a\}$ and $\{V_j | 1 \leq j \leq b\}$ be the set of all irreducible EG_1 and EG_2-modules, respectively. By Corollary 2.2, $\{U_i \otimes V_j | 1 \leq i \leq a, 1 \leq j \leq b\}$ is the set of all irreducible LG-modules. Thus

$$r_1 r'_0 + r_0 r'_1 - r_1 r'_1 = r'_0(r_0 - \sum_{i=1}^{a} (\dim_L U_i)^2) + r_0(r'_0 - \sum_{i=1}^{b} (\dim_L V_i)^2)$$
$$- (r_0 - \sum_{i=1}^{a} (\dim_L U_i)^2)(r'_0 - \sum_{i=1}^{b} (\dim_L V_i)^2)$$

$$= r_o r'_o - \sum_{i=1}^{a} \sum_{j=1}^{b} (\dim_L U_i)^2 (\dim_L V_j)^2$$

$$= \dim_L J(LG)$$

$$= \dim_F J(FG) \qquad \text{(by Corollary 1.17)}$$

which proves (i) for the case $s = 1$.

To prove (ii), observe that

$$J(FG_1)FG_2 \cap J(FG_2)FG_1 = J(FG_1)J(FG_2)$$
$$\cong J(FG_1) \otimes_F J(FG_2)$$

It therefore follows that

$$\dim_F(J(FG_1)FG_2 + J(FG_2)FG_1) = \dim_F J(FG_1)FG_2 + \dim_F J(FG_2)FG_1$$
$$- \dim_F(J(FG_1)FG_2 \cap J(FG_2)FG_1)$$
$$= r_1 r'_o + r_o r'_1 - r_1 r'_1$$
$$= \dim_F J(FG)$$

Since $J(FG_1)FG_2 + J(FG_2)FG_1 \subseteq J(FG)$, we conclude that

$$J(FG) = J(FG_1)FG_2 + J(FG_2)FG_1$$

which proves (ii) for the case $s = 1$.

Turning to the general case, we have

$$J(FG)^s = (J(FG_1)FG_2 + J(FG_2)FG_1)^s$$
$$= \sum_{i=o}^{s} J(FG_1)^i J(FG_2)^{s-i}$$

proving (ii). The property (i) now follows from (ii) upon noting that

$$\sum_{k=i}^{s} J(FG_1)^k J(FG_2)^{s-k} \cap J(FG_1)^{i-1} J(FG_2)^{s-i+1} = J(FG_1)^i J(FG_2)^{s-i+1}$$

(iii) Direct consequence of (ii).

(iv) Since $|X| = \dim_F J(FG)^s$, we need only verify that X is linearly independent over F. To this end, assume that

$$\sum_{i,j,k,\ell} \lambda_{ijk\ell} a_{ij} a'_{k\ell} = 0$$

where the $\lambda_{ijk\ell}$ are elements of F. Setting

$$\mu_m = \sum_{i,j,\ell} \lambda_{ijm\ell} a_{ij} a'_{m\ell} \quad \text{and} \quad \beta_n = \sum_{q=n}^{s} \mu_q ,$$

it follows that

$$\beta_n + \sum_{q<n} \mu_q = 0$$

and that

$$\mu_m \in J(FG_1)^{s-m} J(FG_2)^m , \ \beta_n \in FG_1 \cdot J(FG_2)^n$$

Now

$$\mu_o = -\beta_1 \in J(FG_1)^s FG_2 \cap J(FG_2) FG_1 = J(FG_1)^s J(FG_2)$$

and the map

$$\begin{cases} (J(FG_1)^s \otimes FG_2)/(J(FG_1)^s \otimes J(FG_2)) \longrightarrow J(FG_1)^s \otimes [FG_2/J(FG_2)] \\ \qquad x \otimes y + J(FG_1)^s \otimes J(FG_2) \longmapsto x \otimes (y + J(FG_2)) \end{cases}$$

is an isomorphism. Hence all the coefficients of μ_o are zero and thus $\mu_1 = -\beta_2$ is an element of

$$(J(FG_1)^{s-1} J(FG_2)) \cap (FG_1 \cdot J(FG_2)^2) = J(FG_1)^{s-1} J(FG_2)^2$$

Invoking the natural isomorphism

$$(J(FG_1)^{s-1} \otimes J(FG_2))/(J(FG_1)^{s-1} \otimes J(FG_2)^2) \cong J(FG_1)^{s-1} \otimes [J(FG_2)/J(FG_2)^2]$$

we conclude that all coefficients of μ_1 are zero. Repeating the above procedure, we obtain eventually $\lambda_{ijk\ell} = 0$ for all i,j,k,ℓ. This completes the proof of the theorem. ■

2.4. COROLLARY. For any groups $G_1, G_2, \ldots, G_k$,

$$t(G_1 \times G_2 \times \ldots \times G_k) = \sum_{i=1}^{k} t(G_i) - (k-1)$$

Proof. Apply induction on k and Theorem 2.3(iii). ■

2.5. COROLLARY. Let $G = \mathbb{Z}_{p^{n_1}} \times \mathbb{Z}_{p^{n_2}} \times \ldots \times \mathbb{Z}_{p^{n_k}}$ for some positive integers n_i, $1 \leqslant i \leqslant k$. Then

$$t(G) = \sum_{i=1}^{k} p^{n_i} - (k-1)$$

Proof. Let $G = \langle g \rangle$ be a cyclic group of order p^n. Owing to Corollary 2.4, we need only verify that

$$t(G) = p^n$$

By Proposition 1.1, $J(FG) = I(G)$ and so $J(FG)$ is the F-linear span of all elements g^i-1, $1 \leqslant i < p^n$. Since

$$g^i-1 = (g-1)(1+g+\ldots+g^{i-1})$$

$J(FG) = FG(g-1)$. The desired conclusion is therefore a consequence of the equality

$$g^{p^k}-1 = (g-1)^{p^k}$$

applied for $k \leqslant n$. ■

3. A CHARACTERIZATION OF ELEMENTS OF $J(FG)$: THE GENERAL CASE

Let G be a finite group and let F be an arbitrary field of characteristic $p > 0$. The problem that motivates this section is to discover necessary and sufficient conditions for an element $x = \Sigma x_g g \in FG$ to be in $J(FG)$. Under certain additional hypotheses (e.g. when G is p-solvable), these conditions will be simplified in Sec. 5.

To accomplish our aim, we shall have to set up some machinery pertaining to the theory of Frobenius and symmetric algebras. Many of the results presented below will only be needed in subsequent investigations. We present them here in order not to interrupt future discussions at an awkward stage. Our method borrows heavily from an important paper due to Kulshammer (1981a).

Throughout this section, unless explicitely stated otherwise, A denotes a finite-dimensional algebra over an arbitrary field F and $A^* = \text{Hom}_F(A,F)$. The left (right) A-module A means the regular left (right) A-module.

Given a subset X of A, define AX and XA by

$$AX = \{\Sigma a_i x_i | a_i \in A, x_i \in X\}$$

$$XA = \{\Sigma x_i a_i | a_i \in A, x_i \in X\}$$

If $X = \{a\}$ then we shall write Aa and aA instead of AX and XA, respectively. A pair (A,ψ), where $\psi \in A^*$, is said to be a *Frobenius algebra* if for all $a \in A$ the following conditions are equivalent:

(i) $\psi(aA) = 0$ (ii) $\psi(Aa) = 0$ (iii) $a = 0$

We shall refer to A itself as a *Frobenius algebra* if there exists $\psi \in A^*$ such that the pair (A,ψ) is a Frobenius algebra.

A pair (A,ψ), where $\psi \in A^*$, is said to be a *symmetric algebra* if (A,ψ) is a Frobenius algebra such that

$$\psi(xy) = \psi(yx) \qquad \text{for all } x,y \in A$$

The algebra A is called a *symmetric algebra* if there exists $\psi \in A^*$ such that (A,ψ) is a symmetric algebra.

Let $f : A \times A \longrightarrow F$ be a bilinear form. Then f is called *nonsingular* if the following conditions are equivalent.

(i) $f(a,A) = 0$ (ii) $f(A,a) = 0$ (iii) $a = 0$

We shall refer to f as being *associative* if

$$f(xy,z) = f(x,yz) \qquad \text{for all } x,y,z \in A$$

and *symmetric* if

$$f(x,y) = f(y,x) \qquad \text{for all } x,y \in A$$

3.1. LEMMA. (i) Assume that (A,ψ) is a Frobenius algebra. Then the map

$f : A \times A \longrightarrow F$ defined by $f(x,y) = \psi(xy)$ is a nonsingular associative F-bilinear form. Furthermore, if (A,ψ) is a symmetric algebra, then f is also symmetric.

(ii) Assume that $f : A \times A \longrightarrow F$ is a nonsingular associative F-bilinear form and let $\psi : A \longrightarrow F$ be defined by $\psi(x) = f(x,1)$. Then (A,ψ) is a Frobenius algebra. Furthermore, if f is symmetric then (A,ψ) is a symmetric algebra.

Proof. (i) For any $x,y,z \in A$, we have

$$f(x+y,z) = \psi((x+y)z) = \psi(xz)+\psi(yz) = f(x,z)+f(y,z)$$

and similarly $f(x,y+z) = f(x,y)+f(x,z)$. Given $\lambda \in F$, we also have

$$f(\lambda x,y) = \psi(\lambda xy) = \lambda\psi(xy) = \lambda f(x,y)$$

and similarly $f(x,\lambda y) = \lambda f(x,y)$. Hence f is an F-bilinear form on A.

Suppose that $f(x,y) = 0$ for all $y \in A$. Then $\psi(xA) = 0$ and therefore $x = 0$. A similar argument shows that if $f(x,y) = 0$ for all $x \in A$, then $y=0$, proving that f is nonsingular. For all $x,y,z \in A$, we have

$$f(xy,z) = \psi((xy)z) = \psi(x(yz)) = f(x,yz),$$

proving that f is associative. Because $\psi(xy) = \psi(yx)$ implies $f(x,y)=f(y,x)$, the assertion follows.

(ii) It is clear that $\psi \in A^*$. Let $a \in A$ be such that $\psi(aA) = 0$. Then $f(ab,1) = 0$ for all $b \in A$. However, $f(ab,1) = f(a,b)$, so the nonsingularity of f implies that $a = 0$. A similar argument shows that if $\psi(Aa) = 0$ then $a = 0$, proving that (A,ψ) is a Frobenius algebra. If f is symmetric, then for all $x,y \in A$

$$\psi(xy) = f(xy,1) = f(x,y) = f(y,x) = f(yx,1) = \psi(yx)$$

as asserted. ∎

Note that A^* is an (A,A)-bimodule via

$$(af)(x) = f(xa)$$
$$(fa)(x) = f(ax)$$

for all $x,a \in A$ and $f \in A^*$.

3.2. LEMMA. (i) Assume that (A,ψ) is a Frobenius algebra. Then the map

$$f : A \longrightarrow A^*$$

defined by $f(a)(x) = \psi(xa)$ for all $x,a \in A$, is an isomorphism of left A-modules. Furthermore, if (A,ψ) is a symmetric algebra, then f is an isomorphism of (A,A)-bimodules.

(ii) Assume that $f : A \longrightarrow A^*$ is an isomorphism of left A-modules and set $\psi(a) = f(1)(a)$. Then (A,ψ) is a Frobenius algebra. Furthermore, if f is an isomorphism of (A,A)-bimodules, then (A,ψ) is a symmetric algebra.

Proof. (i) It is obvious that for all $a \in A, f(a) \in A^*$ and that

$$f(x+y) = f(x)+f(y)$$

for all $x,y \in A$. If $r,x \in A$, then

$$f(ra)(x) = \psi(xra) = f(a)(xr) = (rf(a))(x),$$

proving that f is a homomorphism of left A-modules. By the definition of ψ, f is injective and since $\dim_F A = \dim_F A^*$, f is a bijection.

Assume that (A,ψ) is a symmetric algebra. Then, for all $r,x \in A$, we have

$$f(ar)(x) = \psi(xar) = \psi(rxa) = f(a)(rx) = (f(a)r)(x),$$

as desired.

(ii) It is clear that $\psi \in A^*$. Suppose that $\psi(ax) = 0$ for all $x \in A$. Then

$$0 = f(1)(ax) = f(x)(a)$$

for all $x \in X$ and hence $a = 0$. A similar argument proves that if $\psi(xa) = 0$ for all $x \in A$, then $a = 0$. Hence (A,ψ) is a Frobenius algebra.

Assume that f is an isomorphism of (A,A)-bimodules. Then, for all $x,y \in A$, we have

$$\psi(xy) = f(1)(xy) = f(x)(y) = f(1)(yx) = \psi(yx),$$

as required. ■

3.3. COROLLARY. The following conditions are equivalent:

(i) A is a Frobenius (respectively, symmetric) algebra

(ii) There exists a nonsingular associative F-bilinear form $f : A \times A \longrightarrow F$ (respectively, there exists a nonsingular, associative, symmetric F-bilinear form $f : A \times A \longrightarrow F$)

(iii) $A \cong A^*$ as left modules (respectively, $A \cong A^*$ as (A,A)-bimodules)

Proof. Direct consequence of Lemmas 3.1 and 3.2. ■

3.4. LEMMA. Let $A_1, A_2, \ldots, A_n$ be F-algebras. Then $\prod_{i=1}^{n} A_i$ is a Frobenius (respectively, symmetric) algebra if and only if each A_i is a Frobenius (respectively, symmetric) algebra.

Proof. Suppose that $\psi_i \in A_i^*$ is such that (A_i, ψ_i) is a Frobenius algebra $1 \leqslant i \leqslant n$. Put $A = \prod_{i=1}^{n} A_i$ and define $\psi : A \longrightarrow F$ by

$$\psi(a_1, \ldots, a_n) = (\psi_1(a_1), \ldots, \psi_n(a_n))$$

Then (A,ψ) is obviously a Frobenius algebra. Conversely, if (A,ψ) is a Frobenius algebra, then so is (A_i, ψ_i), $1 \leqslant i \leqslant n$, with ψ_i being the projection of ψ onto A_i. Finally, for $a = (a_1, \ldots, a_n)$ and $b = (b_1, \ldots, b_n)$, $\psi(ab) = \psi(ba)$ if and only if $\psi_i(a_i b_i) = \psi_i(b_i a_i)$ for all $i \in \{1, \ldots, n\}$. Thus the lemma is proved. ■

We have now accumulated all the information necessary to prove the following important result.

3.5. THEOREM. (Eilenberg and Nakayma (1955)).

(i) If A_1 and A_2 are Frobenius (or symmetric) F-algebras, then so is $A = A_1 \otimes_F A_2$

(ii) If A is a semisimple F-algebra, then A is symmetric

Proof. (i) Suppose that A_1 and A_2 are Frobenius F-algebras. Then, by corollary 3.3, there exists an isomorphism $f_i : A_i \longrightarrow A_i^*$ of left A_i-modules, $i = 1,2$. Consider the map $f : A \longrightarrow A^*$ defined by

$$[f(a_1 \otimes a_2)](b_1 \otimes b_2) = [f_1(a_1)(b_1)][f_2(a_2)(b_2)] \qquad (a_1, b_1 \in A_1, a_2, b_2 \in A_2)$$

A routine calculation shows that f is an isomorphism of left A-modules. Moreover, if each f_i is an (A_i,A_i)-isomorphism, then f is an (A,A)-isomorphism. The desired assertion is now a consequence of Corollary 3.3.

(ii) We first assume that $A = M_n(F)$ for some $n \geqslant 1$. Then, by looking at the trace map, we conclude that A is symmetric. If D is a division algebra over F, then

$$D \underset{F}{\otimes} M_n(F) \cong M_n(D)$$

Invoking (i) and Lemma 1.4, we may therefore assume that A is a division algebra. Then, for any nonzero ψ in $A^*, (A,\psi)$ is a Frobenius algebra. To prove that (A,ψ) is a symmetric algebra, it suffices to verify that $\psi([A,A]) = 0$ for some nonzero ψ in A^*. The latter will follow provided we show that $[A,A] \neq A$. The last statement being independent of the ground field, we may assume that F is the centre of A.

Let E be a field extension of F which is a splitting field for A_E. Then $A_E \cong M_n(E)$ for some $n \geqslant 1$. Since each matrix in $[M_n(E),M_n(E)]$ has trace 0, we have $[A_E,A_E] \neq A_E$. But $[A_E,A_E] = [A,A]_E$, so $[A,A] \neq A$, as required. ■

For a given subset X of A, we denote by $\ell(X)$ and $r(X)$ the *left* and *right annihilators* of X defined by

$$\ell(X) = \{a \in A \mid aX = 0\}$$
$$r(X) = \{a \in A \mid Xa = 0\}$$

It is obvious that $\ell(X)$ and $r(X)$ are left and right ideals of A, respectively. Moreover, if X is a left (respectively, right) ideal of A, then $\ell(X)$ (respectively, $r(X)$) is a two-sided ideal of A.

Suppose that (A,ψ) is a Frobenius algebra and that X is a subset of A. Define the subsets ${}^{\perp}X$ and $X^{\perp}$ of A by

$${}^{\perp}X = \{a \in A \mid \psi(aX) = 0\}$$
$$X^{\perp} = \{a \in A \mid \psi(Xa) = 0\}$$

It is plain that if X is a subspace of A, then so are ${}^{\perp}X$ and $X^{\perp}$.

3.6. THEOREM. Let (A,ψ) be a Frobenius algebra and let X be an F-subspace of A.

(i) $({}^{\perp}X)^{\perp} = {}^{\perp}(X^{\perp}) = X$ and $\dim_F X^{\perp} = \dim_F {}^{\perp}X = \dim_F A - \dim_F X$

(ii) If X is a left (respectively, right) ideal of A, then $r(X) = X^{\perp}$ (respectively, $\ell(X) = {}^{\perp}X$). In particular, for every idempotent e of A.

$$(Ae)^{\perp} = (1-e)A\ , \quad A/(Ae)^{\perp} \cong eA \quad \text{as right } A\text{-modules and } \dim_F eA = \dim_F Ae.$$

(iii) If X is a left ideal of A, then

$$\dim_F X + \dim_F r(X) = \dim_F A \quad \text{and} \quad \ell(r(X)) = X$$

If X is a right ideal of A, then

$$\dim_F X + \dim_F \ell(X) = \dim_F A \quad \text{and} \quad r(\ell(X)) = X$$

(iv) The mapping $X \mapsto r(X)$ is a duality of the lattice of left ideals of A onto the lattice of right ideals, that is

$$r(X_1+X_2) = r(X_1) \cap r(X_2) \quad \text{and} \quad r(X_1 \cap X_2) = r(X_1) + r(X_2)$$

for all left ideals X_1, X_2 of A.

Similarly, $X \mapsto \ell(X)$ is a duality of the lattice of right ideals of A onto the lattice of left ideals of A.

(v) If (A,ψ) is symmetric, then

$$r(X) = \ell(X)$$

for any two-sided ideal X of A.

(vi) If (A,ψ) is symmetric and F is a splitting field for A, then the Cartan matrix of A is symmetric.

Proof. (i) Fix an F-basis $\{a_1,\ldots,a_n\}$ of A such that $\{a_1,\ldots,a_m\}, m \leqslant n$ is an F-basis of X. The map

$$\begin{cases} A \longrightarrow F\times F\times\ldots\times F & (n \text{ times}) \\ a \longmapsto (\psi(a_1a),\psi(a_2a),\ldots,\psi(a_na)) & \end{cases}$$

is an F-isomorphism such that the image of $X^{\perp}$ consists of all $(\lambda_1,\ldots,\lambda_n)$ with $\lambda_1 = \lambda_2 = \ldots = \lambda_m = 0$. Hence $\dim_F X^{\perp} = \dim_F A - \dim_F X$ and, by a similar argument, $\dim_F {}^{\perp}X = \dim_F A - \dim_F X$. It follows that

$$\dim_F ({}^{\perp}X)^{\perp} = \dim_F X$$

Because $({}^{\perp}X)^{\perp} \supseteq X$, we infer that $({}^{\perp}X)^{\perp} = X$, as required. The second equality is proved similarly.

(ii) Suppose that X is a right ideal of A. Then $a \in {}^{\perp}X$ if and only if $\psi(aX) = 0$ or if and only if $\psi(aXA) = 0$. Since the latter equality is equivalent to $aX = 0$, we have $\ell(X) = {}^{\perp}X$. The equality $r(X) = X^{\perp}$ for a left ideal X of A follows by the same argument. Now for all $x \in A$,

$$x \in (Ae)^{\perp} \Leftrightarrow Aex = 0 \Leftrightarrow ex = 0 \Leftrightarrow x \in (1-e)A,$$

so

$$(Ae)^{\perp} = (1-e)A \quad \text{and} \quad A/(Ae)^{\perp} \cong eA$$

Finally $\dim_F eA = \dim_F Ae$, by applying (i).

(iii) Direct consequence of (i) and (ii).

(iv) It is clear that

$$r(X_1+X_2) = r(X_1) \cap r(X_2) \quad \text{and} \quad \ell(X_1+X_2) = \ell(X_1) \cap \ell(X_2)$$

Invoking (iii), we conclude that

$$\begin{aligned}\ell(X_1 \cap X_2) &= \ell(r(\ell(X_1)) \cap r(\ell(X_2))) = \ell(r(\ell(X_1)) + \ell(X_2))) \\ &= \ell(X_1) + \ell(X_2)\end{aligned}$$

The corresponding assertion for left ideals follows similarly.

(v) Assume that (A,ψ) is a symmetric algebra. Then ${}^{\perp}X = X^{\perp}$ for any subset X of A. Now apply (ii).

(vi) Assume that (A,ψ) is symmetric and that F is a splitting field for A. Let $e_1,\ldots,e_n$ be a complete set of orthogonal primitive idempotents of A and let $C = (c_{ij})$ be the Cartan matrix of A. Then

$$A = \bigoplus_{i=1}^{n} Ae_i = \bigoplus_{i=1}^{n} e_iA$$

and therefore

$$A = \bigoplus_{i,j=1}^{n} e_jAe_i \qquad \text{(direct sum of } F\text{-spaces)}$$

Fix $j,k \in \{1,\dots,n\}$ with $j \neq k$. Then

$$\bigoplus_{(k,s)\neq(j,i)} e_kAe_s \subseteq (e_jAe_i)^{\perp}$$

and so, applying (i) and Proposition 1.10.16(i), we have

$$\begin{aligned} c_{ji} &= \dim_F e_iAe_j = \dim_F A - \dim_F (e_iAe_j)^{\perp} \\ &\leqslant (\dim_F A)\,(\sum_{(k,s)\neq(j,i)} \dim_F e_kAe_s) \\ &= \dim_F e_jAe_i \\ &= c_{ij} \end{aligned}$$

A similar argument shows that $c_{ij} \leqslant c_{ji}$ and the result follows. ∎

We next concentrate on factor algebras and prove the following result.

3.7. THEOREM (Nakayama(1939)). Let X be an ideal of A.

(i) Assume that (A,ψ) is a Frobenius algebra. Then A/X is a Frobenius algebra if and only if there exists $a \in A$ such that $r(X) = aA = Aa$.

(ii) Assume that (A,ψ) is a symmetric algebra. Then A/X is a symmetric algebra if and only if there exists $a \in Z(A)$ such that $r(X) = aA$.

Proof. (i) Suppose that $(A/X,\mu)$ is a Frobenius algebra, and let $\pi : A \longrightarrow A/X$ be the natural map. Then, by Lemma 3.2(i) and Corollary 3.3, there exists $a \in A$ such that $\mu\pi = \psi_a$ where $\psi_a(b) = \psi(ba)$ for all $b \in A$. We shall show that $r(X) = aA = Aa$. Since $\psi_a(X) = \psi(Xa) = 0$, we have $a \in X^{\perp}$. Therefore $aA \subseteq X^{\perp}$ and, by Theorem 3.6(ii), $aA \subseteq r(X)$. Suppose that $b \in \ell(aA)$ or $b \in \ell(Aa)$. Then $ba = 0$ and hence

$$\mu((A/X)(b+X)) = (\mu\pi)(Ab) = \psi_a(Ab) = \psi(Aba) = 0$$

Thus $b \in X$ and so both $\ell(aA)$ and $\ell(Aa)$ are in X. Applying Theorem 3.6(iii), we infer that $aA = r(\ell(aA)) \supseteq r(X)$, proving that $r(X) = aA$. Consequently,

$Aa \subseteq r(X)$ and therefore $\ell(Aa) \supseteq X$. It follows that $X = \ell(Aa)$ and thus $r(X) = r\ell(Aa) = Aa$.

Conversely, assume that $r(X) = aA = Aa$ for some $a \in A$. Define

$$\mu : A/X \longrightarrow F$$

by

$$\mu(b+X) = \psi(ba) \qquad \text{for all } b \in A$$

Because $a \in r(X)$, μ is well defined and obviously an F-linear map. Suppose that $b \in A$ is such that $\mu((A/X)(b+X)) = 0$. Then $\psi(Aba) = 0$, so $ba = 0$ and thus $b \in \ell(aA)$. But $r(X) = aA$ implies that $X = \ell(r(X)) = \ell(aA)$; hence $b \in X$. A similar argument shows that $\mu((b+X)(A/X)) = 0$ implies $b \in X$, proving that A/X is a Frobenius algebra.

(ii) Suppose that A/X is symmetric. By (i), $r(X) = aA$ for some $a \in A$. On the other hand,

$$\psi(xya) = \psi_a(xy) = \psi_a(yx) = \psi(yxa) = \psi(xay)$$

for all $x,y \in A$. Hence $ay = ya$ for all $y \in A$, and so $a \in Z(A)$.

Conversely, suppose that $a \in Z(A)$. Then, by (i), A/X is obviously symmetric. So the theorem is proved. ■

We next exhibit some important properties of socles and commutator subspaces of Frobenius and symmetric algebras. First, however, it will be convenient to recall the following piece of information.

Let V be a left A-module. Then the socle of V, written $\mathrm{Soc}V$, is defined as the sum of all irreducible submodules of V. The socle of a right A-module is defined similarly. The left and right socles of A are defined as the socles of ${}_AA$ and A_A, respectively. In case the left and right socles of A coincides, we shall write $\mathrm{Soc}A$ to denote both of them.

3.8. THEOREM. Assume that A is a Frobenius algebra.

(i) The left and right socles of A coincide. Furthermore,

$$\dim_F \mathrm{Soc}A = \dim_F (A/J(A))$$

(ii) For each primitive idempotent $e \in A$, $(\mathrm{Soc}A)e$ is the unique irreducible submodule of Ae. In particular, $\mathrm{Soc}(Ae) = (\mathrm{Soc}A)e$

(iii) If A is symmetric, then

(a) There exists $z \in Z(A)$ such that $\mathrm{Soc}A = Az$ and $J(A) = \ell(z)$

(b) For any primitive idempotent $e \in A$, $\mathrm{Soc}(Ae) \cong Ae/J(A)e$

(c) For any irreducible A-module V, V is isomorphic to a submodule of $P(V)$.

Proof. (i) and (ii): Setting $J = J(A)$, it follows from Proposition 1.6.29 that $r(J)$ is the left socle of A. By a similar argument, $\ell(J)$ is the right socle of A. Let e be a primitive idempotent of A. Then Je is the unique maximal submodule of Ae, and the quotient Ae/Je is an irreducible module (Propositions 1.9.4 and 1.6.36). By Theorem 3.6, the map $X \mapsto X^{\perp}$ is an inclusion-reversing bijection between the set of left ideals X of A contained in Ae, and the set of right ideals of A containing $(Ae)^{\perp}$. We thus obtain an inclusion-reversing bijection $X \mapsto X^{\perp}/(Ae)^{\perp}$ between the set of submodules of Ae and the set of submodules of $A/(Ae)^{\perp}$. Again, applying Theorem 3.6, we obtain an inclusion-reversing bijection $X \mapsto eX^{\perp}$ between the set of submodules X of Ae and the set of submodules of eA. In particular, $e(Je)^{\perp}$ is the unique irreducible submodule of eA. Further, for $y \in A$, we have

$$y \in (Je)^{\perp} \Leftrightarrow Jey = 0 \Leftrightarrow ey \in r(J)$$

which implies that $e(Je)^{\perp} = er(J)$. Thus $er(J)$ is the unique irreducible submodule of eA.

Because J annihilates all irreducible A-modules, we have $er(J)J = 0$. Thus $r(J)J$ is annihilated by all primitive idempotents $e \in A$. The latter implies that $r(J)J = 0$, and therefore $r(J) \subseteq \ell(J)$. The reverse inclusion holds by symmetry. The furthermore assertion being a consequence of Theorem 3.6(i), property (i) is established. The same argument (reversing left and right) also proves (ii).

(iii) Note that (a) is a consequence of Theorems 3.5(ii), 3.6(iii) and 3.7(ii).

To prove (b), let $\psi \in A^*$ be such that (A,ψ) is a symmetric algebra. By the foregoing, $r(J)e$ is the unique irreducible submodule of Ae and $Jr(J)e = 0$. Hence every $f \in \mathrm{Hom}_A(Ae,r(J)e)$ induces a map $\lambda \in \mathrm{Hom}_A(Ae/Je,r(J)e)$. Since $\mathrm{Hom}_A(Ae/Je,r(J)e) \cong er(J)e$ as additive groups (Lemma 1.5.5), we are left to verify that $er(J)e \neq 0$.

Assume by way of contradiction that $er(J)e = 0$. Then

$$0 = \psi(er(J)e) = \psi(r(J)e) = \psi(Ar(J)e)$$

and hence $r(J)e = 0$, which is impossible. This proves (b).

To prove (c), observe that $V \cong Ae/J(A)e$ and $P(V) \cong Ae$ for some primitive idempotent e of A (Proposition 1.9.4). By (b), $V \cong \mathrm{Soc}(Ae) \subseteq Ae$ and the result follows. ■

As a preliminary to the next result, we prove

3.9. LEMMA. (i) Assume that (A,ψ) is a Frobenius algebra. Then, for any subset X of A, $\ell(X)$ is the largest left ideal of A contained in ${}^{\perp}X$

(ii) Assume that (A,ψ) is a symmetric algebra. Then $Z(A) = [A,A]^{\perp}$

Proof. (i) By the definition of ${}^{\perp}X$, we have $\ell(X) \subseteq {}^{\perp}X$. Let I be any left ideal of A contained in ${}^{\perp}X$. Then

$$\psi(AIX) = \psi(IX) = 0$$

so $IX = 0$ and thus $I \subseteq \ell(X)$, as required.

(ii) For any $z \in A$, $z \in [A,A]^{\perp}$ if and only if for all $x,y \in A$

$$\psi((xy-yx)z) = \psi(xyz) - \psi(yxz) = \psi(xyz) - \psi(xzy) = 0$$

or if and only if

$$\psi(A(yz-zy)) = 0 \qquad \text{for all } y \in A$$

Since the latter is equivalent to $yz-zy = 0$ for all $y \in A$, the result is established. ■

From now on, we assume that $\mathrm{char}F = p > 0$ *and set*

$$T(A) = \{a \in A \mid a^{p^n} \in [A,A] \text{ for some } n \geqslant 1\}$$

Then $T(A)$ is an F-subspace of A containing $[A,A]$. We are now ready to prove the following important result.

3.10. THEOREM (Külshammer(1981a)). With the notation above, the following properties hold:

(i) $T(A) = [A,A] + J(A)$ and $J(A)$ is the largest left ideal of A contained in $T(A)$.

(ii) If A is a Frobenius algebra, then $\text{Soc}A = T(A)^{\perp}A$

(iii) If A is a symmetric algebra, then

(a) $T(A)^{\perp} = Z(A) \cap \text{Soc}A$ and, in particular, $T(A)^{\perp}$ is an ideal of $Z(A)$

(b) $\text{Soc}A = (Z(A) \cap \text{Soc}A)A$

(c) $(T(A)^{\perp})^2 = Z(B_1) \oplus \ldots \oplus Z(B_r)$, where $B_1,\ldots,B_r$ are all blocks of A which are simple F-algebras.

(d) If F is a splitting field for A, then the number of blocks of A that are simple algebras is equal to $\dim_F (T(A)^{\perp})^2$

Proof. (i) We first treat the case where A is a simple algebra. Since for any $z \in Z(A)$ and $x,y \in A$,

$$(xy-yx)z = x(yz) - (yz)x \in [A,A]$$

both $[A,A]$ and $T(A)$ can be regarded as algebras over the field $Z(A)$. We may therefore assume that $Z(A) = F$, i.e. that A is a central simple F-algebra. Let E be the algebraic closure of F so that

$$A_E = E \otimes_F A \cong M_n(E)$$

for some $n \geqslant 1$. Then clearly $T(A_E) = [A_E,A_E]$ and

$$\dim_E T(A_E) = \dim_E A_E - 1$$

Because

$$\dim_F [A,A] = \dim_E [A_E,A_E] \text{ and } \dim_F A = \dim_E A_E$$

we infer that $\dim_F[A,A] = \dim_F A - 1$. If it were true that $A = T(A)$, then since $T(A_E)$ is an E-subspace of A_E we would have $A_E = T(A_E)$, a contradiction. Hence $A \neq T(A)$ and therefore, by the above, $T(A) = [A,A]$. Since A is simple, (A,ψ) is a symmetric algebra for some $\psi \in A^*$ (Theorem 3.5(ii)). But then $[A,A] \subseteq \operatorname{Ker}\psi$ and hence $T(A)$ contains no nonzero left ideals of A. This proves the case where A is a simple algebra.

Turning to the general case, set $\bar{A} = A/J(A)$ and write

$$\bar{A} = A_1 \oplus \ldots \oplus A_m$$

where the A_i are pairwise orthogonal simple F-algebras. Then, $T(A_i) = [A_i,A_i]$, $1 \leqslant i \leqslant m$, by the special case proved above, and thus

$$\begin{aligned}([A,A] + J(A))/J(A) &= [\bar{A},\bar{A}] = [A_1,A_1] \oplus \ldots \oplus [A_m,A_m] \\ &= T(A_1) \oplus \ldots \oplus T(A_m) \\ &= T(\bar{A})\end{aligned}$$

Obviously $[A,A] + J(A) \subseteq T(A)$ and $T(A)/J(A) \subseteq T(\bar{A})$. Consequently, $T(\bar{A}) = T(A)/J(A)$ and so $T(A) = [A,A] + J(A)$.

Assume that I is a left ideal of A contained in $T(A)$. Then $(I+J(A))/J(A)$ is a left ideal of $\bar{A}$ contained in $T(\bar{A})$ and hence in $[\bar{A},\bar{A}]$. Since $\bar{A}$ is semisimple, $\bar{A}$ is a symmetric algebra (Theorem 3.5(iii)). Thus, by the previous argument, $[\bar{A},\bar{A}]$ contains no nonzero left ideals of $\bar{A}$. Hence $I \subseteq J(A)$ and the required assertion follows.

(ii) Assume that A is a Frobenius algebra. Thanks to Theorem 3.6, we have

$$^{\perp}(T(A)^{\perp}) = T(A) \tag{1}$$

$$r[\ell(T(A)^{\perp}A)] = T(A)^{\perp}A \tag{2}$$

By (i) and (1), $J(A)$ is the largest left ideal of A contained in $^{\perp}(T(A)^{\perp})$. Applying Lemma 3.9(i), we conclude that

$$J(A) = \ell(T(A)^{\perp}) = \ell(T(A)^{\perp}A) \tag{3}$$

Invoking (2) and (3), it follows that

$$\mathrm{Soc}A = r(J(A)) = T(A)^{\perp}A,$$

as required.

(iii) Assume that A is a symmetric algebra. By (ii) and Lemma 3.9(ii), to prove (a) we need only verify that

$$T(A)^{\perp} = [A,A]^{\perp} \cap T(A)^{\perp}A$$

Because $\ell(T(A)^{\perp}A)$ is a left ideal of A, we have

$$[\ell(T(A)^{\perp}A)]^{\perp} = r[\ell(T(A)^{\perp}A] \tag{4}$$

Consequently,

$$\begin{aligned} T(A)^{\perp} &= [A,A]^{\perp} \cap J(A)^{\perp} && \text{(by (ii))} \\ &= [A,A]^{\perp} \cap [\ell(T(A)^{\perp}A)] && \text{(by (3))} \\ &= [A,A]^{\perp} \cap r[\ell(T(A)^{\perp}A)] && \text{(by (4))} \\ &= [A,A]^{\perp} \cap T(A)^{\perp}A, && \text{(by (2))} \end{aligned}$$

proving (a). Property (b) is a consequence of (a) and (ii), while (d) follows from (c). We are therefore left to verify (c).

To prove (c), we may harmlessly assume that A is indecomposable, in which case it suffices to verify that

$$[T(A)^{\perp}]^2 = \begin{cases} 0 & \text{if } J(A) \neq 0 \\ Z(A) & \text{if } J(A) = 0 \end{cases}$$

Suppose first that $J(A) = 0$. Then $1 \in \mathrm{Soc}A \cap Z(A) = T(A)^{\perp}$. But $T(A)^{\perp}$ is an ideal of $Z(A)$, so $T(A)^{\perp} = Z(A)$ and thus $[T(A)^{\perp}]^2 = Z(A)$. Next assume that $J(A) \neq 0$. Then $1 \notin \mathrm{Soc}A$ and so $1 \in T(A)^{\perp}$. Taking into account that $Z(A)$ is a local algebra, we have

$$T(A)^{\perp} \subseteq J(Z(A)) \subseteq J(A)$$

However, by (a) $T(A)^{\perp} \subseteq \mathrm{Soc}A$ and thus

$$[T(A)^{\perp}]^2 \subseteq J(A)\mathrm{Soc}A = 0$$

as required. ■

We have completed the development of the general machinery, and we are confronted with the problem of characterizing the elements of $J(FG)$, where F is a field of characteristic $p > 0$.

Define a map

$$tr : FG \longrightarrow F$$

called the *trace*, by

$$tr(\sum_{g\in G} x_g g) = x_1 \qquad (x_g \in F)$$

The following observation will enable us to take full advantage of the results so far obtained.

3.11. LEMMA. (i) (FG,tr) is a symmetric algebra

(ii) Let e be an idempotent of FG and let tr' be the restriction of tr to $eFGe$. Then $(eFGe,tr')$ is a symmetric algebra.

Proof. (i) It is clear that $tr \in \mathrm{Hom}_F(FG,F)$. Fix $x = \sum_{g\in G} x_g g$ and $y = \sum_{g\in G} y_g g$ in FG. Then we have

$$tr(xy) = \sum_{g\in G} x_g y_{g^{-1}} = \sum_{g\in G} y_g x_{g^{-1}} = tr(yx)$$

and

$$tr(g^{-1}x) = x_g \qquad \text{for all } g \in G$$

In particular, if $tr(FGx) = 0$ then $x = 0$ as required.

(ii) Assume that $y \in eFGe$ is such that $tr'(eFGey) = 0$. Then $y = eye = ye$ and

$$0 = tr(eFGey) = tr(eFGeye) = tr(FGeye) = tr(FGy),$$

proving that $y = 0$. So the lemma is true. ■

For any subset X of G, we write

$$X^{-1} = \{x^{-1} | x \in X\}, \quad X^{+} = \sum_{x\in X} x$$

and denote by FX the F-linear span of X.

Let $C_1 = \{1\}, C_2, \dots, C_r$ be all p-regular classes of G. By a *p-regular section of G associated with C_i*, we understand the set

$$S_i = \{g \in G \mid g_{p'} \in C_i\} \qquad (1 \leqslant i \leqslant r)$$

It is an easy consequence of the definition that

(i) $G = \bigcup_{i=1}^{r} S_i$ with $S_i \cap S_j = \emptyset$ for $i \neq j$, each S_i is a union of conjugacy classes of G and S_1 is a union of conjugacy classes of p-elements of G.

(ii) An element $C \in C\ell(G)$ is a p-regular section if and only if C is of p-defect zero.

3.12. LEMMA. Let X be a subset of G and let $u = \sum_{g \in G} u_g g \in FG$.

(i) $tr(uX^+) = \sum_{g \in X^{-1}} u_g$ and, in particular, $tr(uS_i^+) = \sum_{g \in S_i^{-1}} u_g$

(ii) $\ell(X^+) = \{\sum_{g \in G} u_g g \in FG \mid \sum_{g \in X^{-1}} u_{tg} = 0$ for all $t \in G\}$.

Proof. By the definition of the product of elements in FG, we have

$$uX^+ = \sum_{t \in G} \left(\sum_{g \in X} u_{tg^{-1}}\right) t$$

Hence, by the definition of the trace map, $tr(uX^+) = \sum_{g \in X} u_{g^{-1}} = \sum_{g \in X^{-1}} u_g$. Also $uX^+ = 0$ if and only if for all $t \in G$

$$\sum_{g \in X} u_{tg^{-1}} = \sum_{g \in X^{-1}} u_{tg} = 0$$

as required. ■

The following result will enable us to take full advantage of Theorem 3.10(iii).

3.13. LEMMA. Let $S_1, S_2, \dots, S_r$ be all p-regular sections of G. Then

(i) $T(FG) = \{\sum_{g \in G} x_g g \in FG \mid \sum_{g \in S_i} x_g = 0$ for all $i \in \{1, \dots, r\}\}$

(ii) $T(FG)^{\perp} = \sum_{i=1}^{r} FS_i^+$

Proof. (i) We first observe that there exists $n \geqslant 1$ such that

(*) $\qquad g^{p^m} = g_{p'}$ for all $m \equiv 0 \pmod{n}$ and all $g \in G$

Indeed, write $|G| = p^a k$ with $(p,k) = 1$. Because $(p,k) = 1$, there exists $n \in \mathbb{N}$ such that $p^n \equiv 1 \pmod{k}$. Replacing n by its multiple, if necessary, we may assume that $n \geqslant a$. Suppose that m is divisible by n and let $g \in G$. Then $g_p^{p^m} = 1$ and, since $p^m \equiv 1 (\text{mod } k)$, $g_{p'}^{p^m} = g_{p'}$. Thus

$$g^{p^m} = g_p^{p^m} g_{p'}^{p^m} = g_{p'} \ ,$$

proving (*).

Let $x = \sum_{g \in G} x_g g \in FG$. Then, by (*), we may choose $m \geqslant 1$ such that $g^{p^m} = g_{p'}$ for all $g \in G$ and such that $x \in T(FG)$ if and only if $x^{p^m} \in [FG,FG]$.

Owing to Lemma 2.2.1,

$$x^{p^m} \equiv \sum_{g \in G} x_g^{p^m} g^{p^m} \ (\text{mod}[FG,FG]$$

and thus $x \in T(FG)$ if and only if $\sum_{g \in G} x_g^{p^m} g_{p'} \in [FG,FG]$. It follows, from Lemma 2.2.2, that $x \in T(FG)$ if and only if $\sum_{g \in G_i} x_g^{p^m} = 0$ for all $i \in \{1,\dots,r\}$. Since

$$\sum_{g \in S_i} x_g^{p^m} = (\sum_{g \in S_i} x_g)^{p^m} \ ,$$

the required assertion follows.

(ii) By Lemma 3.12(i) and (i), $\sum_{i=1}^{r} FS_i^+ \subseteq T(FG)^\perp$. Conversely, suppose that $x = \sum_{g \in G} x_g g \in T(FG)^\perp$ and let $a,b \in S_i$. Then a^{-1} and b^{-1} belong to the p-regular section S_i^{-1} and hence, by (i), $a^{-1} - b^{-1} \in T(FG)$. But then $tr(xa^{-1} - xb^{-1}) = 0$ which ensures that

$$x_a = tr(xa^{-1}) = tr(xb^{-1}) = x_b$$

proving (ii). ■

We are now ready to prove the following important result.

3.14. THEOREM (Brauer (1955), Tsushima (1978a)). Let F be a field of characteristic $p > 0$ and let $S_1, S_2, \dots, S_r$ be all p-regular sections of G. Then

$$\text{Soc}FG = \sum_{i=1}^{r} FG \cdot S_i^+ \quad \text{and} \quad J(FG) = \bigcap_{i=1}^{r} \ell(S_i^+)$$

Proof. By Theorem 3.10(iii)(a),(b),

$$\text{Soc}FG = FG \cdot T(FG)^\perp$$

Invoking Lemma 3.13(ii), we conclude that $\mathrm{Soc}FG = \sum_{i=1}^{r} FG\cdot S_i^+$. It follows from Proposition 1.6.30 and Theorem 3.8(iii) that

$$\begin{aligned} J(FG) &= \ell(\mathrm{Soc}FG) = \ell\left(\sum_{i=1}^{r} FG\cdot S_i^+\right) \\ &= \bigcap_{i=1}^{r} \ell(FG\cdot S_i^+) \end{aligned}$$

The desired conclusion is now a consequence of the equality $\ell(FG\cdot S_i^+) = \ell(S_i^+)$. ■

We have now come to the demonstration for which the present section has been developed.

3.15. COROLLARY. Let G be a finite group, let F be a field of characteristic $p > 0$ and let $S_1, S_2, \ldots, S_r$ be all p-regular sections of G. Then

$$J(FG) = \{\sum_{g\in G} x_g g \in FG \mid \sum_{s\in S_i} x_{gs} = 0 \text{ for all } g \in G \text{ and all } i \in \{1,\ldots,r\}\}$$

Proof. Apply Theorem 3.14 and Lemma 3.12(ii) together with the fact that $S_i \mapsto S_i^{-1}$ is a permutation of the set $\{S_1,\ldots,S_r\}$. ■

4. CONLON'S THEOREM, FONG'S DIMENSION FORMULA AND RELATED RESULTS

Our main goal in this section is to investigate the decomposition of induced modules from normal subgroups and the dimensions of projective covers of irreducible FG-modules . We do this for use in the sequel where we will pursue our study of $J(FG)$. As a preparation for the proof of the main theorems, we prove a number of important results which will be frequently applied at a later stage.

Throughout, G denotes a finite group, N a normal subgroup of G and F a field. All modules are assumed to be finitely generated.

Let V be an FN-module and let $g \in G$. We define ${}^gV$ to be the FN-module whose underlying space is V and on which the elements $n \in N$ act according to the rule

$$n * v = (g^{-1}ng)v$$

We say that two FN-modules V_1 and V_2 are *G-conjugate* if $V_1 \cong {}^gV_2$ for some $g \in G$. It is clear that G-conjugacy is an equivalence relation, so the set of FN-modules is a disjoint union of G-conjugacy classes. A straightforward verification shows that V is irreducible (completely reducible, indecomposable) if and only if ${}^gV$ is irreducible (completely reducible, indecomposable).

Given an FN-module V, consider the set $H = \{g \in G \mid V \cong {}^gV\}$. Then H is a subgroup of G containing N. We shall refer to H as the *inertia group* of V; in case $H = G$ we shall say that V is *G-invariant*.

4.1. THEOREM (Clifford(1937)). Let V be an irreducible FG-module and let U be an irreducible FN-submodule of V_N. Let H and W denote, respectively, the inertia group of U and the sum of all submodules of V_N isomorphic to U. Then there exists a positive integer e such that

(i) $V_N \cong e({}^{g_1}U \oplus \dots \oplus {}^{g_n}U)$, where $\{1 = g_1, g_2, \dots, g_n\}$ is a left transversal for H in G and ${}^{g_1}U, \dots, {}^{g_n}U$ are pairwise nonisomorphic irreducible FN-modules. In particular, V_N is completely reducible.

(ii) W is an irreducible FH-module such that

$$W_N \cong eU \quad \text{and} \quad V \cong W^G$$

(Here, of course, eU means a direct sum of e copies of U).

Proof. First of all, it is clear that $V = \sum_{g\in G} gU$ and that gU is an FN-module isomorphic to ${}^gU$. Therefore each gU is an irreducible FN-module. Thus V_N is a sum of irreducible FN-modules gU. It follows, from the definition of ${}^gV$, that ${}^x({}^yV) \cong {}^{xy}V$ for all $x, y \in G$ and so ${}^xU \cong {}^yU$ if and only if $xH = yH$. Setting $W_i = \sum_{g\in g_iH} gU$, $1 \leqslant i \leqslant n$, we deduce from Proposition 1.3.6, that

$$V_N = W_1 \oplus \dots \oplus W_n \qquad (W_1 = W)$$

and that

$$W_i \cong e_i\,{}^{g_i}U \qquad 1 \leqslant i \leqslant n$$

for some positive integer e_i.

Fix $g \in G, i \in \{1,2,\dots,n\}$, and write $gg_i = g_js$ for some $j \in \{1,\dots,n\}$ and $s \in H$. Then

$$gW_i = g\sum_{h\in H} g_ihU = \sum_{h\in H} g_jshU = W_j$$

and obviously $x(yW_i) = (xy)W_i$, for all $x,y \in G$. Thus the group G acts on the set $\{W_1,\ldots,W_n\}$ transitively and H is the stabilizer of W under this action. It follows that $\dim_F W_i = \dim_F W$ and hence that all the e_i are equal. Invoking Lemma 2.4.1(ii), we also deduce that W is an FH-module and $V \cong W^G$. Finally, because V is irreducible, so is W and the result follows. ∎

In order to make further progress, we need to introduce the notions of projective representations and twisted group algebras.

Let F^* denote the multiplicative group of F and let $Z^2(G,F^*)$ be the set of all functions

$$\alpha : G \times G \longrightarrow F^*$$

which satisfy the following identities

$$\alpha(x,1) = \alpha(1,x) = 1 \qquad (x \in G) \tag{1}$$

$$\alpha(x,y)\alpha(xy,z) = \alpha(y,z)\alpha(x,yz) \qquad (x,y,z \in G) \tag{2}$$

We shall refer to the elements of $Z^2(G,F^*)$ as *cocycles*. Given $\alpha,\beta \in Z^2(G,F^*)$, define $\alpha\beta$ by the rule

$$(\alpha\beta)(x,y) = \alpha(x,y)\beta(x,y) \qquad (x,y \in G)$$

It is then obvious that $\alpha\beta$ is also a cocycle and that $Z^2(G,F^*)$ becomes an abelian group.

Let $t : G \to F^*$ be such that $t(1) = 1$ and let $\delta t : G \times G \to F^*$ be defined by

$$(\delta t)(x,y) = t(x)t(y)t(xy)^{-1} \qquad (x,y \in G)$$

We shall refer to δt as a *coboundary*. It is routine to verify that the set $B^2(G,F^*)$ of all coboundaries constitutes a subgroup of $Z^2(G,F^*)$. The factor group

$$H^2(G,F^*) = Z^2(G,F^*)/B^2(G,F^*)$$

is said to be the *second cohomology group of* G *over* F^*. The elements of

$H^2(G,F^*)$ are called *cohomology classes;* any two cocycles contained in the same cohomology class are said to be *cohomologous*. Given $\alpha \in Z^2(G,F^*)$, we shall write $\bar{\alpha}$ for the cohomology class containing α.

One can easily check that if the requirement (1) is dropped, then the resulting cohomology group will be isomorphic to the one introduced.

Let V be a vector space over F. A mapping $\rho : G \longrightarrow GL(V)$ is called a *projective representation* of G over F if there exists a mapping $\alpha : G \times G \longrightarrow F^*$ such that

$$\rho(x)\rho(y) = \alpha(x,y)\rho(xy) \qquad \text{for all } x,y \in G \tag{3}$$

$$\rho(1) = 1_V \tag{4}$$

Thus an ordinary representation is a projective representation with $\alpha(x,y) = 1$ for all $x,y \in G$. If we identify $GL(V)$ with $GL(n,F), n = \dim_F V$, the resulting map is called a *projective matrix representation* of G over F. As in the case of ordinary representations, we shall treat the terms "projective representation" and "projective matrix representation" as interchangeable.

In view of the associativity of the multiplication in G, conditions (3) and (4) imply that $\alpha \in Z^2(G,F^*)$. To stress the dependence of ρ on V and α, we shall often refer to ρ as an *α-representation on the space* V.

Two projective representations $\rho_i : G \rightarrow GL(V_i), i = 1,2$ are said to be *linearly equivalent* if there exists a vector space isomorphism $f : V_1 \longrightarrow V_2$ such that

$$\rho_2(g) = f\rho_1(g)f^{-1} \tag{5}$$

for all $g \in G$. The projective representation ρ on the space V is called *irreducible* if 0 and V are the only subspaces of V which are sent into themselves by all the transformations $\rho(g), g \in G$. Let $\rho_i : G \rightarrow GL(V_i), i = 1,2$, be two α-representations. Then the map

$$\rho_1 + \rho_2 : G \longrightarrow GL(V_1 \oplus V_2)$$

defined by

$$(\rho_1+\rho_2)(g)(v_1+v_2) = \rho_1(g)v_1 + \rho_2(g)v_2 \qquad (v_i \in V_i)$$

is easily seen to be again an α-representation. We warn the reader that unlike the situation for ordinary representations, there is no natrual way to define a sum of an α-representation and β-representation when $\alpha \neq \beta$. An α-representation ρ on the space V is *completely reducible* if for any subspace W invariant under all transformations $\rho(g), g \in G$, there exists another such subspace W' with $V = W \oplus W'$.

Ordinary representations of G over F can be interpreted as FG-modules. The same situation prevails for projective representations in which the role of the group algebra FG plays the twisted group algebra defined below.

Given $\alpha \in Z^2(G,F^*)$, denote by $F^\alpha G$ the vector space over F with basis $\{\bar{g} | g \in G\}$ which is in one-to-one correspondence with G. Define multiplication in $F^\alpha G$ distributively using for all $x,y \in G$

$$\bar{x}\bar{y} = \alpha(x,y)\overline{xy}$$

Then $F^\alpha G$ becomes an F-algebra to which we refer as the *twisted group algebra* G over F. Note that if $\alpha(x,y) = 1$ for all $x,y \in G$, then $F^\alpha G \cong FG$. More generally, we have:

4.2. LEMMA. $F^\alpha G \cong FG$ if and only if α is a coboundary.

Proof. Assume that $F^\alpha G \cong FG$. Since FG admits an F-algebra homomorphism into F, so does $F^\alpha G$, say $f : F^\alpha G \to F$. Then, for all $x,y \in G$,

$$f(\bar{x})f(\bar{y}) = f(\bar{x}\bar{y}) = f(\alpha(x,y)\overline{xy}) = \alpha(x,y)f(\overline{xy})$$

and so $\alpha = \delta t$, where $t : G \to F^*$ is defined by $t(g) = f(\bar{g})$. Conversely, assume that $\alpha = \delta t$ for some $t : G \to F^*$ with $t(1) = 1$. Then the map $\psi : F^\alpha G \to FG$, which is the extension of $\bar{g} \to t(g)g$ by F-linearity gives the desired isomorphism. ■

4.3. LEMMA. Let A be an F-algebra and let f be a mapping of G into the unit group of A which satisfies

$$f(x)f(y) = \alpha(x,y)f(xy) \qquad (x,y \in G)$$

Then the mapping $f^* : F^{\alpha}G \longrightarrow A$ defined by $f^*(\Sigma\lambda_g\bar{g}) = \Sigma\lambda_g f(g)$ is a homomorphism of F-algebras.

Proof. Since f^* is the extension of $\bar{g} \longmapsto f(g)$ by F-linearity, f^* is a vector space homomorphism. To prove that f^* preserves multiplication, it suffices to check it on the basis elements $\bar{g}, g \in G$. Since

$$f^*(\bar{x}\,\bar{y}) = f^*(\alpha(x,y)\overline{xy}) = \alpha(x,y)f(xy) = f(x)f(y) = f^*(\bar{x})f^*(\bar{y}),$$

the result follows. ■

The next result shows that the study of α-representations is equivalent to the study of $F^{\alpha}G$-modules.

4.4. PROPOSITION. There is a bijective correspondence between α-representations of G and $F^{\alpha}G$-modules. This correspondence preserves sums and maps bijectively linearly equivalent (irreducible, completely reducible) representations into isomorphic (irreducible, completely reducible) modules.

Proof. Let ρ be an α-representation of G on the space V. Due to Lemma 4.3, we can define a homomorphism $f : F^{\alpha}G \longrightarrow \mathrm{End}_F(V)$ by setting $f(\bar{g}) = \rho(g)$ and extending by linearity. Thus V becomes an $F^{\alpha}G$-module by setting

$$(\Sigma\lambda_g\bar{g})v = \Sigma\lambda_g\rho(g)v \qquad (\lambda_g \in F, g \in G, v \in V)$$

Conversely, given an $F^{\alpha}G$-module V, and hence a homomorphism $f : F^{\alpha}G \longrightarrow \mathrm{End}_F(V)$, define $\rho(g) = f(\bar{g})$. Then $\rho(g) \in GL(V)$ since $\bar{g}$ is a unit of $F^{\alpha}G$. Moreover,

$$\rho(x)\rho(y) = f(\bar{x})f(\bar{y}) = f(\bar{x}\,\bar{y}) = f(\alpha(x,y)\overline{xy}) = \alpha(x,y)\rho(xy)$$

so that ρ is an α-representation on V. This sets up a bijective correspondence between α-representations and $F^{\alpha}G$-modules.

A subspace W of V is invariant under all $\rho(g)$ if and only if W is an $F^{\alpha}G$-submodule of V. Therefore the correspondence preserves sums and maps bijectively irreducible (completely reducible) representations into irreducible (completely reducible) modules.

We next observe that an F-isomorphism $f : V_1 \longrightarrow V_2$ of $F^{\alpha}G$-modules is an $F^{\alpha}G$-isomorphism if and only if $\bar{g}f(v) = f(\bar{g}v)$ for all $g \in G, v \in V_1$. Assume now that $\rho_i : G \longrightarrow GL(V_i), i = 1,2,$ are two α-representations. Then ρ_1 is linearly equivalent to ρ_2 if and only if there is an F-isomorphism $f: V_1 \longrightarrow V_2$ such that $\rho_2(g)f = f\rho_1(g)$, for all $g \in G$. The latter is equivalent to $\rho_2(g)f(v) = f\rho_1(g)v$ or to $\bar{g}f(v) = f(\bar{g}v)$, for all $g \in G, v \in V_1$. Hence two α-representations are linearly equivalent if and only if the corresponding modules are isomorphic. ∎

Given $\alpha \in Z^2(G/N,F^*)$, we denote by $\mathrm{inf}\alpha$ the element of $Z^2(G,F^*)$ defined by

$$(\mathrm{inf}\alpha)(x,y) = \alpha(xN,yN) \qquad \text{for all } x,y \in G$$

Then the map $\alpha \mapsto \mathrm{inf}\alpha$ induces a homomorphism $H^2(G/N,F^*) \longrightarrow H^2(G,F^*)$, called the *inflation map*. Note that since $(\mathrm{inf}\alpha)(x,y) = 1$ for all $x,y \in N$, we may embed FN in $F^{\beta}G$ by $n \mapsto \bar{n}$, for all $n \in N$, where $\beta = \mathrm{inf}\alpha$ or $\beta = (\mathrm{inf}\alpha)^{-1}$.

4.5. THEOREM (Clifford(1937)). Let V be an absolutely irreducible G-invariant FN-module. Then there exists $\omega = \omega_G(V) \in Z^2(G/N,F^*)$ such that for $\gamma = \mathrm{inf}\omega$, V can be extended to an $F^{\gamma}G$-module, i.e. there exists an $F^{\gamma}G$-module W such that $W_N \cong V$. Furthermore, if ω is a coboundary, then V can be extended to an FG-module.

Proof. Let Γ be the representation of N afforded by V. To prove the first statement, it suffices (by Proposition 4.4) to exhibit $\gamma \in Z^2(G,F^*)$, whose values are constant on the cosets of N, and a γ-representation ρ of G such that $\rho(n) = \Gamma(n)$ for all $n \in N$. By hypothesis, for each $g \in G$ there exists a nonsingular matrix $L(g)$ such that

$$\Gamma(g^{-1}ng) = L(g)^{-1}\Gamma(n)L(g) \qquad \text{for all } n \in N$$

We may, of course, assume that $L(1) = 1$. Let T be a transversal for N in G containing 1 and set

$$\rho(tn) = L(t)\Gamma(n) \qquad (t \in T, n \in N)$$

Then, for all $g \in G, n \in N$,

$$\Gamma(g^{-1}ng) = \rho(g)^{-1}\Gamma(n)\rho(g) \quad \text{and} \quad \rho(n) = \Gamma(n) \tag{5}$$

$$\rho(gn) = \rho(g)\Gamma(n) \tag{6}$$

From Schur's lemma, we readily verify that $\rho(x)\rho(y)$ differs from $\rho(xy)$ only by a scalar :

$$\rho(x)\rho(y) = \gamma(x,y)\rho(xy) \qquad (x,y \in G)$$

with $\gamma(x,y) \in F^*$. Thus ρ is a γ-representation of G and hence $\gamma \in Z^2(G,F^*)$. Furthermore, since $L(1) = 1$, we have $\rho(n) = \Gamma(n)$ for all $n \in N$.

Given $x,y \in G$ and $n_1,n_2 \in N$, we have

$$\begin{aligned} \rho(xn_1)\rho(yn_2) &= \rho(x)\Gamma(n_1)\rho(y)\Gamma(n_2) && \text{(by (6))} \\ &= \rho(x)\rho(y)\Gamma(y^{-1}n_1y)\Gamma(n_2) && \text{(by (5))} \\ &= \gamma(x,y)\rho(xy)\Gamma(y^{-1}n_1yn_2) && \\ &= \gamma(x,y)\rho(xn_1yn_2) && \text{(by (6))} \end{aligned}$$

whence $\gamma(xn_1,yn_2) = \gamma(x,y)$.

By the foregoing, we may write $\gamma = \inf(\omega)$ for some $\omega \in Z^2(G/N,F^*)$. Assume that ω is a coboundary, say $\omega(xN,yN) = t(xN)t(yN)t(xyN)^{-1}$ for some $t : G/N \longrightarrow F^*$ with $t(N) = 1$. Then, setting $\rho^*(g) = t(gN)^{-1}\rho(g)$, it is immediate that ρ^* is an ordinary representation of G such that $\rho^*(n)=\rho(n)=\Gamma(n)$ for all $n \in N$. This completes the proof of the theorem. ■

In view of Theorem 4.5, the extendibility of V is governed by $\omega_G(V) \in Z^2(G/N,F^*)$. For this reason we shall refer to $\omega_G(V)$ as the *obstruction to the extension* of V to FG.

4.6. LEMMA. Let A be a finite-dimensional F-algebra, let V be an absolutely irreducible A-module, and let $W = U \otimes_F V$, where U is some finite-dimensional vector space over F. For each $a \in A$, let $\phi_a \in \mathrm{End}_F V$ be defined by $\phi_a(v)=av$, $v \in V$.

(i) If $\theta \in \mathrm{End}_F W$ is such that for all $a \in A$, $\theta(1 \otimes \phi_a) = (1 \otimes \phi_a)\theta$, then

$\theta = \psi \otimes 1$ for some $\psi \in \mathrm{End}_F U$

(ii) If $\theta \in GL(W)$ and $\tau \in GL(V)$ are such that for all $a \in A$, $\theta^{-1}(1 \otimes \phi_a)\theta = 1 \otimes \tau^{-1}\phi_a\tau$, then $\theta = \psi \otimes \tau$ for some $\psi \in GL(U)$.

Proof. (i) Let $\{u_1,\ldots,u_m\}$ be an F-basis for U. Then, for all $v \in V$, we have

$$\theta(u_i \otimes v) = \sum_{j=1}^{m} u_j \otimes \theta_{ji}(v) \tag{7}$$

for some $\theta_{ji} \in \mathrm{End}_F V$. For all $a \in A$, we have

$$\begin{aligned}\theta(1 \otimes \phi_a)(u_i \otimes v) &= \theta(u_i \otimes \phi_a(v)) = \sum_{j=1}^{m} u_j \otimes \theta_{ji}\phi_a(v)\\ &= (1 \otimes \phi_a)\theta(u_i \otimes v) = \sum_{j=1}^{m} u_j \otimes \phi_a\theta_{ji}(v)\end{aligned}$$

Because the $\{u_j\}$ are linearly independent, we deduce that

$$\phi_a\theta_{ji}(v) = \theta_{ji}\phi_a(v) \qquad (a \in A, v \in V, 1 \leqslant i,j \leqslant m)$$

or, equivalently, that $\theta_{ji} \in \mathrm{End}_A V$. Since V is absolutely irreducible, we have $\theta_{ji} = \lambda_{ji} \cdot 1_V$ for some $\lambda_{ji} \in F$. Define $\psi \in \mathrm{End}_F U$ by

$$\psi(u_i) = \sum_{j=1}^{m} \lambda_{ji} u_j \qquad (1 \leqslant i \leqslant m)$$

Then $\psi \in GL(U)$ and, by (7), $\theta = \psi \otimes 1$, proving (i).

(ii) Put $\gamma = 1 \otimes \tau$. Then

$$\theta\gamma^{-1}(1 \otimes \phi_a)\gamma\theta^{-1} = \theta(1 \otimes \tau^{-1}\phi_a\tau)\theta^{-1} = 1 \otimes \phi_a$$

and thus, by (i), $\theta\gamma^{-1} = \psi \otimes 1$ for some $\psi \in GL(U)$. Hence $\theta = (\psi \otimes 1)\gamma = \phi \otimes \tau$ as we wished to show. ■

Suppose that $\rho_i : G \longrightarrow GL(V_i), i = 1,2,$ is an α_i-representation. Consider the map

$$\rho_1 \otimes \rho_2 : G \longrightarrow GL(V_1 \otimes V_2)$$

defined by

$$(\rho_1 \otimes \rho_2)(g) = \rho_1(g) \otimes \rho_2(g)$$

Then $\rho_1 \otimes \rho_2$ is obviously an $\alpha_1\alpha_2$-representation. We shall refer to $\rho_1 \otimes \rho_2$ as the *(inner) tensor product* of the projective representations ρ_1 and ρ_2. In module-theoretic language, the (inner) tensor product can be defined as follows. Let V and W be $F^{\alpha}G$ and $F^{\beta}G$-modules, respectively. Then the vector space $V \otimes_F W$ is an $F^{\alpha\beta}G$-module where the action of the elements $\bar{g}, g \in G$, is defined by

$$\bar{g}(v \otimes w) = \bar{g}v \otimes \bar{g}w \qquad (v \in V, w \in W)$$

and then extended to $V \otimes_F W$ and $F^{\alpha\beta}G$ by F-linearity.

Let $\alpha \in Z^2(G/N, F^*)$ and let V be an $F^{\alpha}(G/N)$-module. Then one can form an $F^{\mathrm{inf}(\alpha)}G$-module $\mathrm{inf}(V)$ whose underlying space is V on which the elements $\bar{g}, g \in G$, act according to the rule:

$$\bar{g}v = \overline{gN}v \qquad (v \in V)$$

In the future we shall refer to $\mathrm{inf}(V)$ as being *inflated* from V.

4.7. THEOREM. (Clifford(1937)). Let F be an algebraically closed field, let M be an irreducible FG-module and let V be an irreducible constituent of M_N. Let H be the inertia group of V, let $\omega = \omega_H(V) \in Z^2(H/N, F^*)$ be an obstruction to the extension of V to FH and let $\mathrm{ext}(V)$ denote any extension of V to $F^{\mathrm{inf}\omega}H$ (by Theorem 4.5, such extension always exists). Then there exists an irreducible $F^{\omega^{-1}}(H/N)$-module U such that

$$M \cong (\mathrm{inf}(U) \otimes \mathrm{ext}(V))^G$$

Proof. Owing to Theorem 4.1(ii), we may harmlessly assume that $H = G$ in which case V is G-invariant. By Theorem 4.1(i), M_N is a direct sum of copies of V. We may therefore assume that $M = W \otimes V$, where W is an F-space such that for all $n \in N, w \in W$, and $v \in V$,

$$n(w \otimes v) = w \otimes nv$$

Put $\beta = \mathrm{inf}\omega, \gamma = \mathrm{inf}(\omega^{-1})$ and $S = \mathrm{ext}(V)$. If $\eta : N \longrightarrow GL(V), \rho : G \longrightarrow GL(M)$ and $\tau : G \rightarrow GL(S)$ are the respresentations afforded by V, M, and S, respectively, then

$$\rho(n) = 1 \otimes \eta(n) \quad \text{and} \quad \tau(n) = \eta(n)$$

for all $n \in N$. Also if $ng = gn_1$ for some $n, n_1 \in N$ and $g \in G$, then $\tau(n)\tau(g) = \tau(g)\tau(n_1)$ since $\beta(n,g) = \beta(g,n) = 1$ for all $n \in N, g \in G$. Hence

$$\tau(g^{-1}ng) = \tau(g)^{-1}\tau(n)\tau(g)$$

and therefore

$$\begin{aligned}\rho(g)^{-1}(1 \otimes \eta(n))\rho(g) &= \rho(g^{-1}ng) = 1 \otimes \eta(g^{-1}ng)\\ &= 1 \otimes \tau(g^{-1}ng)\\ &= 1 \otimes \tau(g)^{-1}\tau(n)\tau(g)\end{aligned}$$

Thus, by Lemma 4.6(ii), for each $g \in G$ there exists $\psi(g) \in GL(W)$ such that

$$\rho(g) = \psi(g) \otimes \tau(g)$$

Setting $\bar{g}w = \psi(g)w, w \in W$, it follows that W is an $F^{\alpha}G$-module such that

$$M \cong W \otimes \text{ext}V$$

Furthermore, because M is irreducible, W must also be irreducible. Finally, taking into account that $\psi(n) = 1$ for all $n \in N$, we deduce that $W \cong \inf(U)$ for some irreducible $F^{\omega^{-1}}(H/N)$-module U. This completes the proof of the theorem. ■

In order to apply Theorem 4.7, we need the following simple observation

4.8. LEMMA. Let F be an algebraically closed field and let G be a group of order n.

(i) $H^2(G,F^*)$ is a finite group all elements of which have order dividing n.

(ii) If $\alpha \in Z^2(G,F^*)$ and m is the order of the cohomology class of α, then α is cohomologous to a cocycle of order m.

(iii) If $\text{char}F = p > 0$, then p does not divide the order of $H^2(G,F^*)$.

Proof. We first note that (iii) is a consequence of (ii). Assume that $\alpha \in Z^2(G,F^*)$. Because $\alpha(x,y)\alpha(xy,z) = \alpha(y,z)\alpha(x,yz)$, we have

$$\prod_{x\in G}\alpha(y,z)\alpha(x,yz) = \prod_{x\in G}\alpha(x,y)\alpha(xy,z)$$

Setting $t(y) = \prod_{x\in G}\alpha(x,y)$ it now follows that

$$\alpha(y,z)^n t(yz) = t(y)t(z) \qquad (x,y,z \in G)$$

Thus $\alpha^n = \delta t$, proving that each element of $H^2(G,F^*)$ has order dividing n. To prove that $H^2(G,F^*)$ is finite, it suffices to verify (ii). Indeed, if (ii) is true, then $H^2(G,F^*)$ can be regarded as a subset of all mappings from $G \times G$ to the group of n-th roots of 1 in F. But the latter is finite, hence $H^2(G,F^*)$ is finite.

To prove (ii), write $\alpha(x,y)^m = t(x)t(y)t(xy)^{-1}, x,y \in G$, for some $t : G \longrightarrow F^*$. Because F is algebraically closed, for any $x \in G$ there exists $\mu(x) \in F^*$ such that $\mu(x)^m = t(x)^{-1}$. Setting $\beta = \alpha(\delta\mu)$, we deduce that $\beta^m = 1$, as required.

Assume that $\mathrm{char}F = p > 0$. Then, for all $a,b \in F$, $(a-b)^p = a^p - b^p$, so $a^p = b^p$ implies $a=b$. Suppose now that $\alpha \in Z^2(G,F^*)$ is such that

$$\alpha(x,y)^p = t(x)t(y)t(xy)^{-1}, \ x,y \in G,$$

for some $t : G \longrightarrow F^*$. Then $\alpha(x,y)^p = \mu(x)^p \mu(y)^p \mu(xy)^{-p}$ for some $\mu : G \longrightarrow F^*$ and hence $\alpha = \delta\mu$, as required. ■

4.9. LEMMA. Let F be an algebraically closed field of characteristic $p > 0$ and let $\alpha \in Z^2(G,F^*)$. Then $F^\alpha G \cong FG$ under either of the following hypotheses:

(i) G is a p-group

(ii) For any prime $q \neq p$, the Sylow q-subgroups of G are cyclic or generalized quaternion

(iii) G is a cyclic extension of a p-group

Proof. (i) Direct consequence of Lemmas 4.2 and 4.8.

(ii) Let q be a prime and let S be a Sylow q-subgroup of G. It is a standard fact of cohomology theory (e.g. see M Hall(1959)) that the q-component of $H^2(G,F^*)$ is isomorphic to a subgroup of $H^2(S,F^*)$. If S is generalized quaternion, then $H^2(S,F^*) = 1$ (see Karpilovsky (1985)). Hence, by Lemma 4.8, we may assume that G is a cyclic q-group and $q \neq p$. Let g be a generator of

G, say of order m. Then $\bar{g}^m = \lambda\cdot\bar{1}$ for some $\lambda \in F^*$. Choose any $\mu \in F$ with $\mu^m = \lambda^{-1}$ so that $(\mu\bar{g})^m = \bar{1}$. Then the elements $\bar{1}, \mu\bar{g}, \ldots, (\mu\bar{g})^{m-1}$ form an F-basis of $F^{\alpha}G$. Thus $F^{\alpha}G \cong FG$.

(iii) Direct consequence of (ii). ■

4.10. LEMMA. Let F be an algebraically closed field such that $\mathrm{char}F \nmid |G|$ and let V be an irreducible $F^{\alpha}G$-module. Then $\dim_F V$ divides the order of G.

Proof. By making a diagonal change of basis $\{\bar{g} | g \in G\}$ of $F^{\alpha}G$, we may replace α by any cohomologous cocycle. Thus, by Lemma 4.8 (ii), we may assume that α is of finite order, say m. Let ε be a primitive m-th root of 1 in F and let $G^* = \langle \varepsilon^i\bar{g} | g \in G, 1 \leq i \leq m\rangle$. Then the map $f : G^* \longrightarrow G$ defined by $f(\varepsilon^i\bar{g}) = g$ is a surjective homomorphism whose kernel is a central subgroup $\langle\varepsilon\rangle$. Furthermore, if $\rho : G \to GL(V)$ is an α-representation of G afforded by V, then $\rho^* : G \longrightarrow GL(V)$ defined by $\rho^*(\varepsilon^i\bar{g}) = \varepsilon^i\rho(g)$ is easily seen to be an irreducible representation of G^*. Since $\mathrm{char}F \nmid |G^*|$ and $G^*/\langle\varepsilon\rangle \cong G$, the result follows by appealing to the following standard fact (see Curtis and Reiner (1981)):

If F is algebraically closed, $\mathrm{char}F \nmid |G|$ and Z is a central subgroup of G, then the dimensions of irreducible FG-modules divide $(G:Z)$. ■

4.11. LEMMA. Let F be an algebraically closed field. Then, for any $G \neq 1$, $F^{\alpha}G$ is a local ring if and only if $\mathrm{char}F = p > 0$ and G is a p-group.

Proof. Assume that G is a p-group and that $\mathrm{char}F = p > 0$. Then, by Lemmas 4.2 and 4.8, $F^{\alpha}G \cong FG$ and hence, by Corollary 1.2, $F^{\alpha}G$ is a local ring. Conversely, assume that $F^{\alpha}G$ is a local ring and assume that G has an element g of prime order $p \neq \mathrm{char}F$. It follows from Lemma 4.9 that for $H = \langle g\rangle$, $F^{\alpha}H \cong FH$ is semisimple. Hence

$$F^{\alpha}H \cong F \times F \times \ldots \times F \qquad (p \text{ times})$$

proving that $F^{\alpha}H$ has nontrivial idempotents. But then the same is true of $F^{\alpha}G$, a contradiction. So the Lemma is proved. ■

We are now ready to prove the following important result.

4.12. THEOREM. Let N be a normal subgroup of G, let F be an algebraically closed field and let V be an irreducible FG-module. If $\mathrm{char}F = p > 0$ divides $(G:N)$, assume that G/N is p-solvable. Then $\dim_F V$ divides $(G:N)d$, where d is the dimension of an irreducible constituent of V_N.

Proof. We first show that the result is true under either of the following hypotheses:

(i) $\mathrm{char}F \nmid (G:N)$

(ii) $\mathrm{char}F = p > 0$ and G/N is a p-group

Let W be an irreducible constituent of V_N, let H be the inertia group of W, and let $\omega \in Z^2(H/N,F^*)$ be an obstruction to the extension of W to FH. By Theorem 4.7, there exists an irreducible $F^{\omega^{-1}}(H/N)$-module S such that

$$V \cong (\inf(S) \otimes \mathrm{ext}(W))^G$$

Since $\dim_F S = \dim_F \inf(S)$ and $\dim_F W = \dim_F \mathrm{ext}(W)$, we have

$$\dim_F V = (\dim_F S)(\dim_F W)(G:H)$$

In case (i) $\dim_F S$ divides $(H:N)$, by Lemma 4.9, so $\dim_F V$ divides $(G:N)\dim_F W$. In case (ii), $F^{\omega^{-1}}(H/N)$ is a local ring, by Lemma 4.11. Hence $\dim_F S = 1$ and therefore $\dim_F V$ again divides $(G:N)\dim_F W$.

Turning to the general case, we use induction on $|G|$. So assume that the result is true for groups of lower order than G. By the above, we may assume that $\mathrm{char}F = p > 0$ divides $(G:N)$, in which case G/N is p-solvable, by hypothesis. Since G/N is p-solvable there exists a proper normal subgroup M of G containing N and such that G/M is either a p or p'-group. Let W be an irreducible constituent of V_M and let S be an irreducible constituent of W_N. Then S is obviously an irreducible constituent of V_N. By the foregoing,

$$\dim_F V \text{ divides } (G:M)\dim_F W$$

and by induction hypothesis

$$\dim_F W \text{ divides } (M:N)\dim_F S$$

Hence

$$\dim_F V \text{ divides } (G:M)(M:N)\dim_F S = (G:N)\dim_F S$$

as asserted. ■

4.13. COROLLARY. (Dade(1968), Swan(1963)). Let F be an algebraically closed field of characteristic $p > 0$ and let A be an abelian normal subgroup of a p-solvable group G. The the dimensions of irreducible FG-modules divide $(G:A)$.

Proof. Direct consequence of Theorem 4.12 and the fact that dimensions of all irreducible FA-modules are equal to 1. ■

We next concentrate on decomposition of induced modules from normal subgroups.

4.14. THEOREM(Ward(1968), Willems(1976)). Let F be an arbitrary field, let V be an indecomposable FN-module, and let H be the inertia group of V. Suppose that

$$V^H = V_1 \oplus \dots \oplus V_s \tag{8}$$

where the V_i are indecomposable FH-modules. Then

(i) $V^G = V_1^G \oplus V_2^G \oplus \dots \oplus V_s^G$, where the V_i^G are indecomposable FG-modules such that $V_i^G \cong V_j^G$ implies $V_i \cong V_j$.

(ii) If V_i is irreducible for some i, then V_i^G and V are irreducible.

Proof. (i) The equality $V^G = V_1^G \oplus \dots \oplus V_s^G$ follows from the fact that induction is transitive and preserves direct sums. To prove that each V_i^G is indecomposable, assume that $V_i^G = X \oplus Y$ is a direct decomposition. Because V_i is identifiable with a direct summand of $(V_i^G)_H$, we may assume, by the Krull-Schmidt theorem, that V_i is a direct summand of X_H, say

$$X_H = V_i \oplus X' \tag{9}$$

Next we observe that $(V^H)_N$ is the direct sum of isomorphic FN-modules of the form $h \otimes V$ with $h \in H$. Applying the Krull-Schmidt theorem, we obtain from (8)

$$(V_i)_N \cong m_i V \qquad (1 \leqslant i \leqslant s) \tag{10}$$

for some positive integer m_i. Thus, for all $g \in G$, there is an FN-isomorphism

$$g \otimes V_i \cong m_i(g \otimes V) \qquad (1 \leqslant i \leqslant s)$$

which in turn yields

$$(V_i^G)_N \cong (\bigoplus_{t \in T} t \otimes V_i)_N \cong \bigoplus_{t \in T} m_i(t \otimes V) \qquad (1 \leqslant i \leqslant s) \qquad (11)$$

where T is a transversal for H in G containing 1.

Restricting (9) to N, we see by (10) that

$$X_N \cong m_i V \oplus X'_N$$

However X is an FG-module, so

$$X_N = tX_N \cong m_i(t \otimes V) \oplus (t \otimes X'_N) \qquad (t \in T)$$

proving that for all $t \in T$, $t \otimes V$ appears as a direct summand of X_N with multiplicity at least m_i. Taking into account that for distinct $t_1, t_2 \in T$, $t_1 \otimes V$ and $t_2 \otimes V$ are nonisomorphic FN-modules, we infer from (ii) that $(V_i^G)_N$ is isomorphic to a direct summand of X_N. Thus $\dim_F V_i^G \leqslant \dim_F X$ and therefore V_i^G is indeed indecomposable.

We are left to prove that $V_i^G \cong V_j^G$ implies $V_i \cong V_j$. Assume by way of contradiction that $V_i^G \cong V_j^G$ but $V_i \not\cong V_j$. Since V_i is identifiable with a direct summand of $(V_i^G)_H$, we may write

$$(V_i^G)_H = V_i \oplus V'_i \quad \text{and} \quad (V_j^G)_H = V_j \oplus V'_j$$

for some FH-module V'_i and V'_j. By the Krull-Schmidt theorem, V_i is isomorphic to an FH-direct summand of V'_j. Hence $(V_i)_N$ is isomorphic to an FN-direct summand of $(V'_j)_N$. Invoking (10) we infer that V is isomorphic to an FN-direct summand of $(V'_j)_N$ and that

$$(V_j^G)_N = (V_j)_N \oplus (V'_j)_N \cong m_j V \oplus (V'_j)_N$$

But, by (11), the multiplicity of V as an indecomposable direct summand of $(V_j^G)_N$ is exactly m_j. This contradiction proves the desired assertion.

(ii) Assume that there is an $i \in \{1,\dots,s\}$ such that V_i is irreducible. Then, by Theorem 4.1, $(V_i)_N$ is completely reducible. Thus, by (10), V is irreducible. Let W be an irreducible factor module of V_i^G. Owing to Theorem 2.4.10, V_i is isomorphic to an FH-submodule of W_H. Hence $(V_i)_N$ is isomorphic to an FN-submodule of W_N and so, by (10), W_N contains a submodule $X \cong m_i V$. It follows that for all $g \in G, W_N$ contains the FN-submodule $gX \cong m_i(g \otimes V)$.

Let T be a transversal for H in G. Then the FN-modules $t \otimes V, t \in T$, are irreducible and nonisomorphic and they all occur as direct summands of W_N with multiplicity at least m_i. Applying (11), we conclude that

$$\dim_F W \geqslant \dim_F V_i^G$$

and hence that $V_i^G = W$ is irreducible. ∎

Next we exhibit a relationship between idempotents in a ring R, and idempotents in a factor ring R/I, where I is a nil ideal. In what follows $\bar{r}$ denotes the image of $r \in R$ in $\bar{R} = R/I$.

4.15. LEMMA. Let I be a nil ideal of a ring R.

(i) Each idempotent $\varepsilon \in \bar{R}$ can be lifted to an idempotent $e \in R$, that is, $\bar{e} = \varepsilon$.

(ii) If e and f are idempotents of R, then $eR \cong fR$ as right R-modules if and only if $\bar{e}\bar{R} \cong \bar{f}\bar{R}$ as right $\bar{R}$-modules.

(iii) Let $\bar{1} = \varepsilon_1 + \dots + \varepsilon_n$ be a decomposition of $\bar{1}$ into orthogonal idempotents in $\bar{R}$. Then there exist orthogonal idempotents $e_1, e_2, \dots, e_n \in R$ such that

$$1 = e_1 + \dots + e_n \quad \text{and} \quad \overline{e_i} = \varepsilon_i$$

Moreover, ε_i is primitive if and only if e_i is also.

Proof. (i) Given an idempotent ε of $\bar{R}$, choose $u \in R$ with $\bar{u} = \varepsilon$. Then $u-u^2 \in I$ and hence $(u-u^2)^m = 0$ for some $m \geqslant 1$. We have

$$1 = [u+(1-u)]^{2m} = \sum_{i=0}^{2m} \binom{2m}{i} u^{2m-i}(1-u)^i$$

Note that on the right each term after the first m is divisible by $(1-u)^m$, while the first m terms are divisible by u^m. Hence, if e denotes the sum of the first m terms, then $1 = e+(1-u)^m g$, where g is a polynomial in u. Now $u(1-u) \in I$, so

$$e \equiv u^{2m} + 2mu^{2m-1}(1-u) + \dots \equiv u(\text{mod } I),$$

that is, $\bar{e} = \bar{u} = \varepsilon$. Because $e(1-e) = e(1-u)^m g = 0, e$ is a required idempotent.

(ii) Assume that e and f are idempotents of R. We first show that $eR \cong fR$ if and only if there exist $a,b \in R$ such that $a = eaf, b = fbe, ab = e$, and $ba = f$. Assume first that $\theta : eR \longrightarrow fR$ is an isomorphism that maps e to b and a to f. Then $fb = b$ and $\theta(e)e = \theta(e)$, hence $be = b$, so $fbe = b$, and similarly $eaf = a$. Thus $e = (\theta^{-1}\theta)e = \theta^{-1}(b) = \theta^{-1}(fb) = \theta^{-1}(f)b = ab$, and similarly $f = ba$. Conversely, if a and b satisfy the condition above, then $x \mapsto bx$ is a homomorphism from eR to fR with inverse $y \mapsto ay$.

Assume that $eR \cong fR$ and let $a,b \in R$ satisfy conditions equivalent to this isomorphism. Then $\bar{a},\bar{b}$ satisfy the same conditions for the isomorphism $\bar{e}\bar{R} \cong \bar{f}\bar{R}$, so $\overline{eR} \cong \overline{fR}$. Now suppose that $\overline{eR} \cong \overline{fR}$ and $a,b \in R$ are such that $a \equiv eaf$, $b \equiv fbe, ab \equiv e, ba \equiv f(\text{mod } I)$, and set $a_1 = eaf$, $b_1 = fbe$. Then $a_1b_1 = e-z$, where $z \in eIe$. Because z is nilpotent, $1-z$ has an inverse $1-z'$ so that $z+z' = zz' = z'z$. Setting $z'' = ez'e$, we obtain $z+z'' = zz'' = z''z$ and hence $a_1b_1(e-z'') = e$. Now put $a_2 = a_1$, $b_2 = b_1(e-z'')$, then

$$a_2 \equiv a \quad b_2 \equiv b(\text{mod } I) \qquad a_2b_2 = e$$

Next write $b_2a_2 = f-y$, then $y \in fIf$. Since $(b_2a_2)^2 = b_2ea_2 = b_2a_2$, we have

$$f-y = (f-y)^2 = f^2-fy-yf+y^2 = f-2y+y^2$$

Hence $y^2 = y$ and since y is nilpotent, we find that $y = 0$ and so $b_2a_2 = f$. Because $a_2b_2 = e$ and $b_2a_2 = f$ imply $a_2 = ea_2f, b_2 = fb_2e$, the desired assertion is proved.

(iii) Assume that e and f are idempotents of R such that $\bar{e}\bar{f} = \bar{f}\bar{e} = 0$. We first show that there exists an idempotent g such that $\bar{g} = \bar{f}$ and $eg = ge = 0$.

To this end, note that fe is nilpotent and so $1-fe$ is a unit. Put

$$h = (1-fe)^{-1}f(1-fe)$$

Then $h^2 = h, he = 0$, and $\bar{h} = \bar{f}$. Now put $g = (1-e)h$. Then

$$\bar{g} = \bar{h} - \bar{e}\bar{h} = \bar{h} - \bar{e}\bar{f} = \bar{h} = \bar{f}\ , \ ge = eg = 0 \quad \text{and} \quad g^2 = g$$

proving the auxiliary assertion.

We now show that if $e_1,\ldots,e_n$ are idempotents such that $\overline{e_i e_j} = 0$ for all $i \neq j$, then there exist idempotents e_i' such that $\overline{e_i'} = \overline{e_i}$ and $e_i'e_j' = 0$ for all $i \neq j$. For $n = 1$ there is nothing to prove, so we may suppose that $n > 1$ and use induction on n. In other words, we may suppose that $e_i e_j = 0$ for $i \neq j, i,j > 1$. Put $e = e_2 + \ldots + e_n$, then e is an idempotent such that $\bar{e}_1\bar{e} = \bar{e}\bar{e}_1 = 0$. Invoking the auxiliary assertion, there exists an idempotent e_1' such that $\bar{e}_1' = \bar{e}_1$ and $ee_1' = e_1'e = 0$. Thus $e_1', e_2, \ldots, e_n$ are pairwise orthogonal idempotents. Hence, by (i) and the above, there exist pairwise orthogonal idempotents $e_1,\ldots,e_n$ such that $\bar{e}_i = \varepsilon_i$, $1 \leqslant i \leqslant n$. Therefore for $u=e_1+\ldots+e_n$ we have $\bar{u} = \bar{1}$, so $1-u$ is a nilpotent idempotent, and thus $u = 1$.

To prove the final assertion, assume that $e \in R$ is a primitive idempotent. Because $e \neq 0$, we have $\bar{e} \neq 0$. If $\bar{e} = x+y$ for some orthogonal idempotents $x,y \in \bar{R}$, then there exist orthogonal idempotents $e_1, e_2 \in R$ with $\bar{e}_1 = x, \bar{e}_2 = y$ and hence with $\bar{e} = \overline{e_1+e_2}$. Owing to (ii), $eR \cong (e_1+e_2)R$, which is possible only in the case $e_1 = 0$ or $e_2 = 0$. Thus either $x = 0$ or $y = 0$ and so $\bar{e}$ is primitive.

Conversely, suppose that $\bar{e}$ is primitive. Then $\bar{e} \neq 0$ and so $e \neq 0$. If $e = u+v$, where u,v are orthogonal idempotents, then $\bar{e} = \bar{u}+\bar{v}, \bar{u}\bar{v} = \bar{v}\bar{u}$ and so either $\bar{u} = 0$ or $\bar{v} = 0$. Hence either $u = 0$ or $v = 0$, proving that e is primitive. So the lemma is verified. ■

It is now an easy matter to prove the following useful observation.

4.16. PROPOSITION. Let I be a nil ideal of a ring R and let the ring $\bar{R} = R/I$ admit a decomposition

$$R = L_1 \oplus L_2 \oplus \cdots \oplus L_n$$

into finitely many indecomposable right ideals. Then there exists a decomposition of R,

$$R = I_1 \oplus I_2 \oplus \cdots \oplus I_n$$

into indecomposable right ideals such that

(i) L_i is the image of I_i under the natural map $R \longrightarrow \bar{R}$

(ii) $L_i \cong L_j$ as right $\bar{R}$-modules if and only if $I_i \cong I_j$ as right R-modules.

Proof. By assumption, there exist orthogonal primitive idempotents $\varepsilon_i \in \bar{R}$ with $1 = \varepsilon_1 + \cdots + \varepsilon_n$ and $L_i = \varepsilon_i\bar{R}$. Thanks to Lemma 4.15(iii), there exist orthogonal primitive idempotents $e_1,\ldots,e_n \in R$ with $\bar{e}_i = \varepsilon_i$ and

$$1 = e_1 + \cdots + e_n.$$

Setting $I_i = e_iR$, we have the decomposition

$$R = I_1 \oplus I_2 \oplus \cdots \oplus I_n$$

into indecomposable right ideals of R. Since $\bar{I}_i = \bar{e}_i\bar{R} = \varepsilon_i\bar{R} = L_i$, property (i) holds. Property (ii) being a consequence of Lemma 4.15(iii), the result is established. ∎

We next exhibit a general fact which provides a link between a module and its ring of endomorphisms. Let R be a ring and let V be a left R-module. For any subset S of $\operatorname{End}_R(V)$, let SV be the set of all finite sums $\sum s_i(v_i), s_i \in S$, $v_i \in V$. Then, clearly, SV is an R-submodule of V. Observe also that if I is a principal right ideal of $\operatorname{End}_R(V)$ generated by ϕ, then $IV = \phi(V)$.

4.17. LEMMA. Let R be a ring, let V be a left R-module, and let $\operatorname{End}_R(V)$ admit a decomposition

$$\operatorname{End}_R(V) = I_1 \oplus I_2 \oplus \cdots \oplus I_n$$

into finitely many indecomposable right ideals. Then

(i) $V = I_1V \oplus I_2V \oplus \ldots \oplus I_nV$ is a decomposition into indecomposable submodules.

(ii) $I_iV \cong I_jV$ as left R-modules if and only if $I_i \cong I_j$ as right $\mathrm{End}_R(V)$-modules.

Proof. (i) Put $E = \mathrm{End}_R(V)$ and write

$$1 = \phi_1 + \phi_2 + \ldots + \phi_n \qquad (\phi_i \in I_i)$$

Then $\phi_i^2 = \phi_i$, $\phi_i\phi_j = 0$, $i \neq j$, $I_i = \phi_iE$, and since I_i is indecomposable, each of the idempotents ϕ_i is primitive. If $v \in V$, then $v = \phi_1(v) + \ldots + \phi_n(v)$, so

$$V = I_1V + I_2V + \ldots + I_nV$$

Assume that $v_1, v_2, \ldots, v_n \in V$ satisfy $\phi_1(v_1) + \phi_2(v_2) + \ldots + \phi_n(v_n) = 0$. Then, taking the images of both sides under ϕ_i, we infer that each $\phi_i(v_i)$ is 0. Hence $V = \oplus I_iV$. Next put $\phi = \phi_i, I = I_i$, and assume that $IV = X_1 \oplus X_2$ for some R-modules X_1 and X_2. Then X_1 and X_2 are direct summands of V and we denote by $\pi_i \in E$ the canonical projection $V \longrightarrow X_i$. Then π_1 and π_2 are orthogonal idempotents such that for all $v \in IV$, $v = \pi_1(v)+\pi_2(v)$. Because $IV = \phi(V)$ we must also have $v = \phi(v)$ for all $v \in IV$. Thus $\phi = \pi_1+\pi_2$ and therefore either π_1 or π_2 must be 0. But $X_i = \pi_i(V)$, so $X_1 = 0$ or $X_2 = 0$, which proves that IV is indecomposable.

(ii) Assume that $I_i \cong I_j$ as right E-modules. We may view V as a left E-module via $\phi v = \phi(v), \phi \in E, v \in V$. Then $E \otimes_E V$ becomes a left E-module via

$$\psi(\phi \otimes v) = \psi\phi \otimes v \qquad (\psi, \phi \in E, v \in V)$$

Now the canonical map

$$\begin{cases} E \otimes_E V \longrightarrow V \\ \phi \otimes v \longmapsto \phi(v) \end{cases}$$

is an isomorphism of E-modules. Since I_i is a direct summand of E, we may identify $I_i \underset{E}{\otimes} V$ with its image in $E \underset{E}{\otimes} V$. But this image is I_iV, so $I_i \underset{E}{\otimes} V \cong I_iV$ which in turn implies that $I_iV \cong I_jV$.

Now assume that $\theta : I_iV \longrightarrow I_jV$ is an R-isomorphism and fix $v \in V$. Given $\phi \in I_i$, we have $\text{Im}\phi \subseteq I_iV$, so $\text{Im}(\theta\phi) \subseteq I_jV$. Because $I_jV = \phi_j(V)$, we deduce that $(\theta\phi)(v) = \phi_j(v_1)$ for some $v_1 \in V$. Thus

$$\phi_j(\theta\phi)(v) = \phi_j(v_1) = (\theta\phi)(v)$$

and so

$$\theta\phi = \phi_j(\theta\phi) \in I_j$$

If we now define $\lambda : I_i \longrightarrow I_j$ by $\lambda(\phi) = \theta\phi$, then λ is obviously a homomorphism of right E-modules. A similar argument shows that the mapping $\mu : I_j \longrightarrow I_i$ defined by $\mu(\psi) = \theta^{-1}\psi$ is also an E-homomorphism. Because μ is the inverse of λ, the result follows. ∎

4.18. LEMMA. Let R be a local ring. Then all (finitely generated) projective R-modules are free.

Proof. Let V be a finitely generated projective R-module, and let $v_1,\ldots,v_n$ be a minimal generating set for V. Suppose that we are given a relation

$$r_1v_1 + \ldots + r_nv_n = 0 \qquad (r_i \in R)$$

We claim that all such $r_i \in J(R)$. Indeed, if $r_i \notin J(R)$, then it is a unit in R, and we can solve for the corresponding v_i as an R-linear sum of the v_j's. This, however, is impossible since otherwise the generator v_i can be deleted. The claim is therefore substantiated.

Let W be a free R-module freely generated by $w_1,\ldots,w_n$ and map W onto V by sending w_i to v_i. By the foregoing, the kernel of this map, say S is contained in $J(R)W$. By hypothesis, V is projective so that we have $W = S \oplus V'$ with $V' \cong V$. Since $S \cong W/V'$ it must be finitely generated. Moreover,

$$J(R)W = J(R)S \oplus J(R)V'$$

and hence the inclusion $S \subseteq J(R)W$ forces $S = J(R)S$. Invoking Nakayama's lemma (Proposition 1.6.15), we conclude that $S = 0$ and hence that $V \cong W$. So the lemma is proved. ■

We now analyze the indecomposable components of the induced module V^G, where V is an indecomposable FN-module. The following preliminary observation will clear our path

4.19. **LEMMA.** Let H be a subgroup of G and let V be an FH-module.

(i) The map

$$\begin{cases} \text{End}_{FH}(V) \longrightarrow \text{End}_{FG}(V^G) \\ \phi \mapsto 1 \otimes \phi \end{cases}$$

is an injective homomorphism of F-algebras whose image consists of all $\psi \in \text{End}_{FG}(V^G)$ such that $\psi(1 \otimes V) \subseteq 1 \otimes V$

(ii) If $g \in N_G(H)$ and if $1 \otimes V \xrightarrow{\phi} g \otimes V$ is an FH-isomorphism, then ϕ can be uniquely extended to an FG-isomorphism $V^G \longrightarrow V^G$.

Proof. (i) It is clear that the given map is an injective homomorphism of F-algebras and that for all $\phi \in \text{End}_{FH}(V)$, $1 \otimes \phi$ sends $1 \otimes V$ into $1 \otimes V$. Suppose that $\psi \in \text{End}_{FG}(V^G)$ is such that $\psi(1 \otimes V) \subseteq 1 \otimes V$. Then there exists $\phi \in \text{End}_{FH}(V)$ such that $\psi(1 \otimes v) = 1 \otimes \phi(v)$ for all $v \in V$. Since ψ is uniquely determined by its restriction to $1 \otimes V$, we have $\psi = 1 \otimes \phi$, as required.

(ii) By a universal characterization of induced modules (Proposition 2.4.8), ϕ can be uniquely extended to an FG-homomorphism $\psi : V^G \longrightarrow V^G$. For an arbitrary $x \in G$, we have

$$\psi(x \otimes V) = x\psi(1 \otimes V) = x\phi(1 \otimes V) = x(g \otimes V) = xg \otimes V$$

Because V^G is the sum of all vector spaces on the right, ψ is surjective. Taking into account that V^G is finite-dimensional over F, we deduce that ψ

is an isomorphism. ■

With the aid of the lemma above we shall identify $\operatorname{End}_{FN}(V)$ with its image in $\operatorname{End}_{FG}(V^G)$ and any FN-isomorphism $1 \otimes V \longrightarrow g \otimes V$ with its extension to an FG-isomorphism $V^G \longrightarrow V^G$. We now fix the following assumptions and notation:

V is a G-invariant FN-module

$E = \operatorname{End}_{FG}(V^G)$, $A = \operatorname{End}_{FN}(V)$

$Q = G/N$

$\{t_x | x \in Q\}$ is a transversal for N in G with $t_1 = 1$ and $x = t_x^{-1}N$.

For each $x \in Q, \phi_x : 1 \otimes V \longrightarrow t_x \otimes V$ is an FN-isomorphism with $\phi_1 = 1$ (thus by our convention and by Lemma 4.19(ii), each ϕ_x is a unit of E)

For each $x \in Q, E_x$ is the F-subspace of E consisting of all homomorphisms that send $1 \otimes V$ into $t_x \otimes V$ (hence $E_1 = A$).

4.20. THEOREM. (Conlon (1964)). With the notation above, the ring E satsifies the following properties.

(i) $E = \bigoplus_{x \in Q} E_x$ and $E_x = A\phi_x = \phi_x A$ for all $x \in Q$

(ii) For all $x,y \in Q, \phi_x\phi_y\phi_{xy}^{-1} \in A$

(iii) $J(A) \cdot E = E \cdot J(A) \subseteq J(E)$ and, in particular, $J(A)E$ is a nilpotent ideal of E.

(iv) Assume that V is indecomposable, F algebraically closed (hence $A/J(A) \cong F$, by Lemma 1.7.4(ii)), and let $\beta(x,y)$ denote the image of $\phi_x\phi_y\phi_{xy}^{-1}$ in F^*. Then $\beta \in Z^2(Q,F^*)$ and $E/J(A)E \cong F^{\beta}Q$.

Proof. (i) First assume that $\sum_{x \in Q} \gamma_x = 0$ with $\gamma_x \in E_x$. Then, for all $v \in V, g \in G$, we have

$$\sum_{x \in Q} \gamma_x(g \otimes v) = 0$$

But $\gamma_x(g \otimes v) = g\gamma_x(1 \otimes v) \in gt_xV$ and

$$V^G = \bigoplus_{x \in Q} gt_x \otimes V$$

Thus $\gamma_x(g \otimes v) = 0$ for all $v \in V, g \in G$, and so $\gamma_x = 0$. Next assume that

$f \in E$. Let π_x be the projection of V^G onto $t_x \otimes V$. Then π_x is an FN-homomorphism such that

$$\sum_{x \in Q} \pi_x = 1 \qquad (12)$$

Let $g_x : 1 \otimes V \longrightarrow t_x \otimes V$ be the FN-homomorphism that is the restriction of $\pi_x \circ f$ to $1 \otimes V$. By a universal characterization of unduced modules (Proposition 2.4.8), there exists $f_x \in E$ such that

$$f_x(1 \otimes v) = g_x(1 \otimes v) \qquad \text{for all} \quad v \in V$$

Then, for all $x \in Q$, we have $f_x \in E_x$ and by (12) we also have

$$\begin{aligned}(\sum_x f_x)(1 \otimes v) &= \sum_x g_x(1 \otimes v) = \sum_x (\pi_x f)(1 \otimes v)\\ &= (\sum_x \pi_x) f(1 \otimes v) = f(1 \otimes v)\end{aligned}$$

proving that $f = \sum_x f_x$ and hence that $E = \bigoplus_x E_x$.

By the definition of $A, \phi_x A \subseteq E_x$. If $f \in E_x$, then $\phi_x^{-1} f \in A$ and so $f \in \phi_x A$. This proves that $E_x = \phi_x A$. Because for all $\psi \in A, \psi(t_x \otimes V) \subseteq t_x \otimes V$, we have $A\phi_x \subseteq E_x$. Conversely, assume that $f \in E_x$, say $f = \phi_x \psi$ for some $\psi \in A$. Then $f\phi_x^{-1}$ sends $t_x \otimes V$ to $t_x \otimes V$ and hence $1 \otimes V$ to $1 \otimes V$. Therefore $f\phi_x^{-1} \in A$, so $f \in A\phi_x$ and (i) is established.

(ii) Fix $x, y \in Q$. Our choice of the transversal implies that $t_y t_x t_{xy}^{-1} \in N$ and hence

$$t_y t_x \otimes V = t_{xy} \otimes V$$

It therefore follows that

$$\begin{aligned}(\phi_x \phi_y)(1 \otimes V) &\subseteq \phi_x(t_y \otimes V) \subseteq t_y \phi_x(1 \otimes V)\\ &= t_y t_x \otimes V\\ &= t_{xy} \otimes V\end{aligned}$$

so $\phi_x \phi_y \in E_{xy} = A\phi_{xy}$ and hence $\phi_x \phi_y \phi_{xy}^{-1} \in A$.

(iii) By (i), $\phi_x^{-1} A \phi_x = A$, so $\phi_x^{-1} J(A) \phi_x = J(A)$ and hence

$$J(A)E = \sum_{x\in Q} J(A)\phi_x A = \sum_{x\in Q} \phi_x J(A)A$$
$$= \sum_{x\in Q} \phi_x AJ(A)$$
$$= EJ(A)$$

Since A is a finite-dimensional algebra over F, we have $J(A)^n = 0$ for some $n \geqslant 1$. Hence $[J(A)E]^n = J(A)^n E = 0$, proving that $J(A)E \subseteq J(E)$.

(iv) Given a subset X of E, let $\tilde{X}$ denote its image in $E/J(A)E$. If $W_x = J(A)\phi_x (x \in Q)$, then $W_x \subseteq E_x$ and

$$J(A)E = \bigoplus_{x\in Q} J(A)A\phi_x = \bigoplus_{x\in Q} W_x$$

Because $E = \bigoplus_x E_x$ it follows that

$$\tilde{E} = \bigoplus_{x\in Q} \tilde{E}_x = \bigoplus_{x\in Q} \tilde{A}\tilde{\phi}_x$$

Now $A = F \oplus J(A)$, so $\tilde{A} = A + J(A)E = F + J(A)E$ is a central subring of $\tilde{E}$ isomorphic to F. By hypothesis, given $x,y \in Q$ there exists $z \in J(A)$ such that $\phi_x\phi_y\phi_{xy}^{-1} = \beta(x,y)+z$. Thus

$$\tilde{\phi}_x\tilde{\phi}_y\tilde{\phi}_{xy}^{-1} = \phi_x\phi_y\phi_{xy}^{-1} + J(A)E = \beta(x,y) + J(A)E$$

Identifying $\tilde{A}$ with its image in F, we therefore derive

$$\tilde{E} = \bigoplus_{x\in Q} F\tilde{\phi}_x \qquad \tilde{\phi}_x\tilde{\phi}_y = \beta(x,y)\tilde{\phi}_{xy} \qquad \lambda\tilde{\phi}_x = \tilde{\phi}_x\lambda$$

for all $\lambda \in F$ and $x,y \in Q$. The associativity of multiplication in $\tilde{E}$ now implies that $\beta \in Z^2(Q,F^*)$ and hence that $\tilde{E} \cong F^{\beta}Q$, thus completing the proof.■

As an application of the theorem above, we shall establish two results. The first, pertaining to arbitrary fields, is of a general nature. The second, valid for algebraically closed fields, gives a complete description of the indecomposable components of V^G exclusively in terms of the twisted group algebra $F^{\beta}Q$ constructed above.

4.21. THEOREM. Let F be an artibrary field, let V be an indecomposable G-invariant FN-module and let $E = \mathrm{End}_{FG}(V^G)$, $A = \mathrm{End}_{FN}(V)$. Let $E = \bigoplus_{i=1}^{n} I_i$ be a decomposition into indecomposable right ideals.

(i) $V^G = \bigoplus_{i=1}^{n} I_i V$ is a decomposition into indecomposable FG-modules

(ii) $I_i \cong I_j$ as right E-modules if and only if $I_i V \cong I_j V$ as FG-modules

(iii) Each I_i is a free A-module and $\dim_E I_i V = (A\text{-rank} I_i)(\dim_F V)$

Proof. (i) Owing to Lemma 4.17(i),

$$V^G = \bigoplus_{i=1}^{n} I_i V^G$$

is a decomposition into indecomposable submodules. Since $V \subseteq V^G$, we obviously have $I_i V \subseteq I_i V^G$. To prove the opposite inclusion, note that $I_i = \phi E$ for some idempotent $\phi \in E$. Since $I_i V^G = \phi(V^G)$ and since V^G is a sum of subspaces of the form $g \otimes V$, it suffices to verify that $\phi(g \otimes V) \subseteq I_i V$ for all $g \in G$. Let $x_g \in E$ be defined by $x_g(v) = gv, v \in V$. Then $\phi x_g \in I_i$ and for all $v \in V$, we have

$$\phi(g \otimes v) = \phi g(v) = (\phi x_g)(v) \in I_i V$$

and hence $I_i V = I_i V^G$, proving (i).

(ii) Direct consequence of Lemma 4.17(ii)

(iii) Since V is indecomposable, A is a local ring (Lemma 1.7.4). Thus, by Lemma 4.18, any projective A-module is free. By Theorem 4.20(i), E is a free A-module. Hence I_i is a projective A-module and therefore is free.

We now claim that the map

$$\begin{cases} E \otimes_A V \longrightarrow V^G \\ \phi \otimes v \longmapsto \phi(v) \end{cases}$$

is an F-isomorphism. Indeed, by Theorem 4.20(i), E is a free A-module of rank $(G:N)$, so $\dim_F(E \otimes_A V) = (\dim_F V)(G:N) = \dim_F V^G$. The map above is surjective, since

$$\phi_x(V) = t_x \otimes V$$

and since V^G is a direct sum of all such $t_x \otimes V$. This substantiates our claim.

Because I_i is a direct summand of E, $I_i \otimes_A V$ is identifiable with its image in $E \otimes_A V$ and so the induced map

$$I_i \otimes_A V \longrightarrow V^G$$

is injective. Since $\dim_F(I_i \otimes_A V) = (A\text{-rank} I_i)(\dim_F V)$ and since $I_i V$ is the image of $I_i \otimes_A V$, the result follows. ■

4.22. THEOREM (Conlon (1964)). Let F be an algebraically closed field, let V be an indecomposable G-invariant FN-module and let $\beta \in Z^2(G/N, F^*)$ be such that $E/J(A)E \cong F^{\beta}(G/N)$, where $E = \operatorname{End}_{FG}(V^G)$ and $A = \operatorname{End}_{FN}(V)$ (by Theorem 4.20(iv), such β always exists). Let

$$F^{\beta}(G/N) = L_1 \oplus \dots \oplus L_n$$

be a decomposition into indecomposable right ideals and let

$$E = I_1 \oplus I_2 \oplus \dots \oplus I_n$$

be the corresponding lifted decomposition of E (see Proposition 4.16(i)). Then the following properties hold:

(i) $V^G = \bigoplus_{i=1}^{n} I_i V$ is a decomposition into indecomposable FG-modules

(ii) $L_i \cong L_j$ as $F^{\beta}(G/N)$-modules if and only if $I_i V \cong I_j V$ as FG-modules

(iii) $\dim_F I_i V = (\dim_F L_i)(\dim_F V)$

Proof. Properties (i) and (ii) follow from Theorem 4.21(i) and (ii) and Proposition 4.16(ii). Owing to Theorem 4.21(iii), to prove (iii) it suffices to verify that for $I = I_i$

$$A\text{-rank } I = \dim_F L_i$$

The latter will follow provided we show that for $n = A\text{-rank } I$,

$$\dim_F[(I+J(A)E)/J(A)E] = n$$

To this end, choose an A-basis $\{d_1,\dots,d_n\}$ of I and extend it to an A-basis $\{d_1,\dots,d_n,d_{n+1},\dots,d_m\}$ of E (this is always possible since all direct summands of E are free A-modules). Then

$$I = \bigoplus_{i=1}^{n} d_i A \qquad E = \bigoplus_{i=1}^{m} d_i A \qquad J(A)E = EJ(A) = \bigoplus_{i=1}^{m} d_i J(A)$$

and hence $I \cap J(A)E = \bigoplus_{i=1}^{n} d_i J(A)$. Because

$$[I + J(A)E]/J(A)E \cong I/(I \cap J(A)E)$$

and since $A = J(A) \oplus F$, we conclude that each element in $[I + J(A)E]/J(A)E$ is a unique F-linear combination of the images $d_1,\dots,d_n$ in $[I + J(A)E]/J(A)E$. This proves (13) and hence the result. ■

We now apply the foregoing to prove the following fundamental result.

4.23. THEOREM. (Green(1959), Conlon(1964), Willems(1976)). Let F be an algebraically closed field, let V be an indecomposable FN-module, where N is a proper normal subgroup of G and let H be the inertia group of V. Then the following properties are equivalent:

(i) V^G is indecomposable

(ii) V^H is indecomposable

(iii) char$F = p > 0$ and H/N is a p-group

In particular, if char$F = p > 0$ and G/N is a p-group, then V^G is indecomposable.

Proof. That (i) is equivalent to (ii) is a consequence of Theorem 4.14. It therefore suffices to verify that (i) is equivalent to (iii) for the case $H = G$. By Theorem 4.22, V^G is indecomposable if and only if $F^{\beta}(G/N)$ has no nontrivial idempotents. The latter is equivalent to the requirement that $F^{\beta}(G/N)$ is local. Since $G/N \neq 1$, by hypothesis, the desired conclusion follows by virtue of Lemma 4.11. ■

We now turn our attention to projective covers of irreducible FG-modules.

4.24. LEMMA. Let H be a subgroup of G and let V and W be FH and FG-modules respectivley. Then

(i) $P(V^G)$ is isomorphic to a direct summand of $P(V)^G$

(ii) $P(W_H)$ is isomorphic to a direct summand of $P(W)_H$.

Proof. (i) By the definition of $P(V)$, there is an FH-epimorphism $P(V) \longrightarrow V$. Invoking Proposition 2.4.4(i), we infer that there is an epimorphism $P(V)^G \longrightarrow V^G$ of FG-modules. Hence, by Lemma 1.9.2(ii), $P(V^G)$ is isomorphic to a direct summand of $P(P(V)^G)$. But by Theorem 2.5.9(ii) $P(V)^G$ is projective. Hence $P(P(V)^G) = P(V)^G$ and the required assertion follows.

(ii) Since there is a homomorphism of $P(W)$ onto W and hence of $P(W)_H$ onto W_H, $P(W_H)$ is isomorphic to a direct summand of $P(P(W)_H)$. Since $P(W)$ is projective, so is $P(W)_H$, by virtue of Theorem 2.5.9(i). Hence $P(W_H)$ is isomorphic to a direct summand of $P(W)_H$. ■

4.25. LEMMA. Let N be a normal subgroup of G and let V and W be irreducible FN and FG-modules, respectively.

(i) If $(G:N) = p^n$, p a prime, and F is an algebraically closed field of characteristic p, then

$$P(V^G) \cong P(V)^G \cong P(U)$$

where U is an irreducible FG-module such that V is a constituent of U_N.

(ii) If $(G:N) \neq 0$ in F, then $P(W_N) \cong P(W)_N$.

Proof. (i) Owing to Lemma 1.9.1(iv), $P(V)$ is an indecomposable module. Hence, by Theorem 4.23, $P(V)^G$ is also indecomposable. Applying Lemma 4.24, we infer that $P(V^G) \cong P(V)^G$. Since V is isomorphic to a submodule of U_N, Theorem 2.4.10 implies that there is an FG-epimorphism $V^G \longrightarrow U$. Thus, by Lemma 1.9.2(ii), $P(U)$ is isomorphic to a direct summand of $P(V^G)$. Since by the above $P(V^G)$ is indecomposable, (i) is established.

(ii) Consider FG as an FN-module. Then, by Proposition 1.8(ii), we have

$$(FG)_N/J(FN)(FG)_N = (FG/J(FG))_N \tag{14}$$

Write

$$FG/J(FG) = \bigoplus_i d_i V_i \tag{15}$$

with nonisomorphic irreducible FG-modules V_i. Then, by Theorem 1.9.3(iii),

$$FG = \bigoplus_i d_i P(V_i)$$

and thus

$$(FG)_N = \bigoplus_i d_i P(V_i)_N \tag{16}$$

Bearing in mind that FG is a free (hence, projective) FN-module, we also have

$$\begin{aligned}
(FG)_N &= P((FG)_N) \\
&= P((FG)_N/J(FN)(FG)_N) \\
&= P((FG/J(FG))_N) && \text{(by (14))} \\
&= \oplus\, d_i(P((V_i)_N)) && \text{(by (15) and Lemma 1.9.1(iii))}
\end{aligned}$$

Thus the desired conclusion follows from (16) and Lemma 4.24(ii). ■

We have now come to the final result of this section.

4.26. THEOREM. (Fong's dimension formula, Fong(1961)). Let G be a p-solvable group, let F be an algebraically closed field of characteristic p, and let V be an irreducible FG-module. Then

$$\dim_F P(V) = |G|_p\,(\dim_F V)_{p'}$$

Proof. We argue by induction on $|G|$. Let N denote a maximal normal subgroup of G. Since G is p-solvable, either $|G/N| = p^n$ for some $n \geqslant 1$ or G/N is a p'-group. By Clifford's theorem, there exists a decomposition

$$V_N = e(V_1 \oplus \ldots \oplus V_s) \tag{17}$$

with nonisomorphic irreducible FN-modules V_i of equal dimension. Moreover, by Theorem 4.12, es divides $|G/N|$. Assume first that $|G/N| = p^n$. Because $\dim_F V = es \dim_F V_1$, we have

$$(\dim_F V)_{p'} = (\dim_F V_1)_{p'}$$

On the other hand, by Lemma 4.25(i), $P(V) \cong P(V_1^G) \cong P(V_1)^G$. Hence

$$\dim_F P(V) = |G/N| \dim_F P(V_1)$$

$$= |G/N|\,|N|_p (\dim_F V_1)_{p'} \quad \text{(by induction hypothesis)}$$

$$= |G|_p (\dim_F V)_{p'} \quad \text{(by (18)}$$

as required.

Now suppose that G/N is a p'-group. By Lemma 4.25(ii), $P(V_N) \cong P(V)_N$ and thus

$$\dim_F P(V) = \dim_F P(V_N)$$

On the other hand, by (17),

$$P(V_N) = e(P(V_1) \oplus \dots \oplus P(V_s))$$

so

$$\begin{aligned} \dim_F P(V) &= e \sum_{i=1}^{s} \dim_F P(V_i) \\ &= es|N|_p (\dim_F V_i)_{p'} && \text{(by induction hypothesis)} \\ &= |G|_p (\dim_F V)_{p'} && \text{(since } es \mid |G/N| \text{ and since } |N|_p = |G|_p) \end{aligned}$$

as required. ■

5. A CHARACTERIZATION OF ELEMENTS OF $J(FG)$: G IS p- SOLVABLE

Throughout this section, F denotes a field of characteristic $p > 0$, G a finite group, and c the sum of all p-elements of G (including 1). Our aim is to discover necessary and sufficient conditions under which

$$J(FG) = \ell(c)$$

In particular, it will be shown that $J(FG) = \ell(c)$ in case G is p-solvable. Given an FG-module V, we write χ_V for the character of G afforded by V and set

$$\text{Ker}\chi_V = \{x \in FG \mid \chi_V(gx) = 0 \quad \text{for all} \quad g \in G\}$$

If H is a subgroup of G, then 1_H denotes the trivial FH-module. Define the

F-linear map

$$\begin{cases} FG \xrightarrow{T} F \\ \Sigma x_g g \longmapsto \sum_{g\in G_p} x_g \end{cases}$$

and put

$$\mathrm{Ker}T = \{x \in FG \,|\, T(gx) = 0 \quad \text{for all} \quad g \in G\}$$

Finally, if V,W are FG-modules and V is irreducible, we write $n(V,W)$ for the multiplicity of V as a composition factor of W. The following simple observation illustrates how the map T can be brought into argument.

5.1. LEMMA. (i) $\mathrm{Ker}T = \ell(c)$ and, in particular, $\mathrm{Ker}T$ is an ideal of FG.
(ii) If $S \in \mathrm{Syl}_p(G)$ and $V = (1_S)^G$, then $\chi_V(x) = |N_G(S) : S|T(x)$ for all $x \in FG$. In particular $\mathrm{Ker}T = \mathrm{Ker}\chi_V$.

Proof. (i) If $x = \sum_{g\in G} x_g g \in FG$, then $tr(g^{-1}x) = x_g$ for all $g \in G$ and hence $x = 0$ if and only if $tr(gx) = 0$ for all $g \in G$. Since $G_p^{-1} = G_p$, we have $T(x) = tr(xc)$ (Lemma 3.12(i)). Hence $T(gx) = tr(g(xc))$ and so $T(gx) = 0$ for all $g \in G$ if and only if $xc = 0$.

(ii) Let X and Y be left transversals for $N_G(S)$ in G and S in $N_G(S)$, respectively, and for a given $g \in G$, set

$$M = \{P \in \mathrm{Syl}_p(G) \,|\, g \in P\}$$

Then $Z = XY$ is a left transversal for S in G and

$$\begin{aligned} \chi_V(g) &= |\{z \in Z \,|\, z^{-1}gz \in S\}|\cdot 1 \\ &= |\{z \in Z \,|\, g \in zSz^{-1}\}|\cdot 1 \\ &= |Y|\,|M|\cdot 1 \end{aligned} \tag{1}$$

Hence, if g is not a p-element, then by (1) $\chi_V(g) = T(g) = 0$.

Let g be a p-element. Then $g \in P$ if and only if $g \in N_G(P)$, for all $P \in \mathrm{Syl}_p(G)$, and therefore $M = \{P \in \mathrm{Syl}_p(G) \,|\, P^g = P\}$. We claim that $|M| \equiv 1(\mathrm{mod}\, p)$; if sustained, it will follow from (1) that $\chi_V(g) = |N_G(S) : S|T(g)$, since $T(g) = 1$, and hence the result. The p-group $\langle g \rangle$ acts on $\mathrm{Syl}_p(G)$ by conjugation and $|M|$ is exactly the number of orbits of length 1. Hence $1 \equiv |\mathrm{Syl}_p(G)| \equiv |M| \pmod{p}$, as required. ■

5.2. LEMMA. Assume that F is a splitting field for FG and let V be an irreducible FG-module. Then

$$\mathrm{Ker}\chi_V = \mathrm{ann}(V)$$

Proof. Let $n = \dim_F V$ and for each $x \in FG$, let $\Gamma_x \in \mathrm{End}_F(V)$ be defined by $\Gamma_x(v) = xv, v \in V$. Identifying $\mathrm{End}_F(V)$ with $M_n(F)$, we obtain a surjective homomorphism $FG \longrightarrow M_n(F)$ of F-algebras given by $x \longmapsto \Gamma_x$. Hence we need only verify that $Tr(\Gamma_{gx}) = 0$ for all $g \in G$ if and only if $\Gamma_x = 0$. If $\Gamma_x = 0$, then $\Gamma_{gx} = \Gamma_g\Gamma_x = 0$ and hence $Tr(\Gamma_{gx}) = 0$. Conversely, assume that $Tr(\Gamma_{gx}) = 0$ for all $g \in G$. Then $Tr(\Gamma_g\Gamma_x) = 0$ for all $g \in G$ and hence $Tr(A\cdot\Gamma_x) = 0$ for all $A \in M_n(F)$. Since the latter is possible only in the case $\Gamma_x = 0$, the result follows. ■

5.3. LEMMA. Assume that F is a splitting field for FG, let V be an FG-module and let $V_1,\ldots,V_s$ be all nonisomorphic irreducible FG-modules. Then $J(FG) = \mathrm{Ker}\chi_V$ if and only if $p \nmid n(V_i,V)$ for all $i \in \{1,\ldots,s\}$.

Proof. Put $r_i = n(V_i,V)$ and denote by χ_i the character of G afforded by V_i, $1 \leqslant i \leqslant s$. Then, by the definition of χ_V,

$$\chi_V = r_1\chi_1 + \ldots + r_s\chi_s$$

and we claim that

$$\mathrm{Ker}\chi_V = \bigcap_{p\nmid r_i} \mathrm{Ker}\chi_i \tag{2}$$

If $x \in \bigcap_{p\nmid r_i} \mathrm{Ker}\chi_i$, then for all $g \in G$,

$$\chi_V(gx) = \sum_{p\nmid r_i} r_i\chi_i(gx) = 0$$

proving that $\mathrm{Ker}\chi_V \supseteq \bigcap_{p\nmid r_i} \mathrm{Ker}\chi_i$.

Conversely, assume that $x \in \mathrm{Ker}\chi_V$. Then

$$\sum_{p \nmid r_i} r_i \chi_i(gx) = 0 \qquad \text{for all } g \in G \tag{3}$$

Let $\varepsilon_1 = e_1 + J(FG), \ldots, \varepsilon_s = e_s + J(FG)$ be all block idempotents of $FG/J(FG)$. By renumbering the V_i, if necessary, we may assume that ε_i acts as the identity transformation on V_i and annihilated all V_j with $j \neq i$. Hence, for any $x \in FG$,

$$\chi_i(e_j x) = \begin{cases} \chi_i(x) & \text{if } i = j \\ 0 & \text{if } i \neq j \end{cases} \tag{4}$$

Now (3) implies that $\sum_{p \nmid r_i} r_i \chi_i(e_j(gx)) = 0$, for all $g \in G$. Invoking (4), we deduce that $\chi_i(gx) = 0$ for all $g \in G$ and all i such that $p \nmid r_i$. Hence $x \in \bigcap_{p \nmid r_i} \mathrm{Ker}\chi_i$, proving (2).

Applying (2) and Lemma 5.2, we derive

$$\mathrm{Ker}\chi_V = \bigcap_{p \nmid r_i} \mathrm{ann}(V_i)$$

Since $J(FG) = \bigcap_{i=1}^{s} \mathrm{ann}(V_i)$ and since for any proper subset X of $\{V_1, \ldots, V_s\}$, $J(FG) \subset \bigcap_{V \in X} \mathrm{ann}(V)$, the result follows. ■

5.4. LEMMA. Let F be a splitting field for FG, let V be an irreducible FG-module and let $S \in \mathrm{Syl}_p(G)$.

(i) $n(V,FG) = \dim_F P(V)$

(ii) $\dim_F P(V) = |S| n(V, (1_S)^G)$

Proof. (i) Let e be a primitive idempotent of FG such that $V \cong FGe/J(FG)e$. Then, by Proposition 1.10.16 (iii), $n(V,FG) = \dim_F eFG$. But, by Theorem 3.6(ii), $\dim_F eFG = \dim_F FGe$. Since $P(V) \cong FGe$, (i) is established.

(ii) Since 1_S is the only irreducible FS-module, the composition length of FS is $|S|$. Let

$$0 = X_0 \subset X_1 \subset \ldots \subset X_{|S|} = FS$$

be a composition series of FS. Then, by Lemma 2.4.2(i) and Proposition 2.4.3(i), we have

$$0 = X_0^G \subset X_1^G \subset \dots \subset X_{|S|}^G = (FS)^G \cong FG,$$

where, by Proposition 2.4.4(iii),

$$X_i^G/X_{i-1}^G \cong (X_i/X_{i-1})^G \cong (1_S)^G \qquad \text{for all } i \in \{1,\dots,|S|\}$$

Invoking (i), we therefore derive

$$|S|n(V,(1_S)^G) = n(V,FG) = \dim_F P(V),$$

as asserted. ∎

As a final step in the preparation for the proof of the main result, we establish the following property due to Okuyama (1980).

5.5. LEMMA. Let F be a splitting field for FG, let e be a primitive idempotent of FG and let $Z_{p'} = \sum_{i=1}^{r} FS_i^+$, where $S_1,\dots,S_r$ are all p-regular sections of G. Then, for any $z \in Z_{p'}$, $ez = 0$ if and only if $tr(ez) = 0$.

Proof. If $ez = 0$, then clearly $tr(ez) = 0$. Conversely, assume that $tr(ez) = 0$. Given $y \in FG$, we have $ey = eye + (eey - eye)$ and so

$$ezFG \subseteq ezFGe + [FG,FG]$$

Because F is a splitting field for G, $eFGe/eJ(FG)e \cong F$ and so

$$eFGe = Fe + eJ(FG)e$$

Invoking Theorem 3.14, it follows that

$$ezFG \subseteq Fez + eJ(FG)ze + [FG,FG] = Fez + [FG,FG]$$

and hence that $tr(ezFG) = 0$. Thus $ez = 0$ as asserted. ∎

We have now come to the demonstration for which this section has been developed.

5.6. THEOREM (Brockhaus (1985)). Let F be a splitting field for FG, let $V_1,\dots,V_r$ be all nonisomorphic irreducible FG-modules and let $S \in \mathrm{Syl}_p(G)$. Then the following conditions are equivalent:

(i) $J(FG) = \ell(c)$

(ii) $\mathrm{Soc}FG = FGc$

(iii) $p \nmid \dfrac{\dim P(V_i)}{|S|}$ for all $i \in \{1,\ldots,r\}$

(iv) $cP(V_i) \neq 0$ for all $i \in \{1,\ldots,r\}$

(v) $cP(V_i) \cong V_i$ for all $i \in \{1,\ldots,r\}$

(vi) $tr(ec) \neq 0$ for every primitive idempotent e of FG

Proof. (i) ⇔ (ii): Since $c \in Z(FG)$, we have $\ell(c) = \ell(FGc)$. If $J(FG) = \ell(c)$, then by Proposition 1.6.29, $\mathrm{Soc}FG = r(\ell(FGc)$ and hence, by Theorem 3.6(iii), $\mathrm{Soc}FG = FGc$. Conversely, assume that $\mathrm{Soc}FG = FGc$. Then, by Proposition 1.6.30 and Theorem 3.8(iii), $J(FG) = \ell(FGc) = \ell(c)$. Thus (i) is equivalent to (ii).

(i) ⇔ (iii): It is a consequence of Lemmas 5.1(ii) and 5.3 that

$$J(FG) = \mathrm{Ker}T \quad \text{if and only if} \quad p \nmid n(V_i,(1_S)^G) \quad \text{for all} \quad i \in \{1,\ldots,r\}$$

Hence, by Lemma 5.1(i),

$$J(FG) = \ell(c) \quad \text{if and only if} \quad p \nmid n(V_i,(1_S)^G) \quad \text{for all} \quad i \in \{1,\ldots,r\}$$

and so the desired conclusion follows by virtue of Lemma5.4.(ii).

(ii) ⇔ (iv) and (ii) ⇔ (v): Let $FG = FGe_1 \oplus \ldots \oplus FGe_t$ where each FGe_i is an indecomposable left ideal and $V_i \cong FGe_i/J(FG)e_i$, $1 \leqslant i \leqslant r$. Then

$$\begin{aligned} FGce_i &\subseteq FGe_i \cap FGc \\ &\subseteq FGe_i \cap \mathrm{Soc}FG \qquad \text{(by Theorem 3.14)} \\ &= \mathrm{Soc}(FGe_i) \end{aligned}$$

Since $\mathrm{Soc}(FG) = \bigoplus_{i=1}^{t} \mathrm{Soc}(FGe_i)$ and $FGc = \bigoplus_{i=1}^{t} FGce_i$, it follows that (ii) is equivalent to

$$\mathrm{Soc}(FGe_i) = FGce_i \qquad \text{for all} \quad i \in \{1,\ldots,t\}$$

or to

$$\mathrm{Soc}(P(V_i)) = cP(V_i) \qquad \text{for all} \quad i \in \{1,\ldots,r\}$$

But, by Theorem 3.8(iii), $\mathrm{Soc}(P(V_i)) \cong V_i$ for all $i \in \{1,\ldots,r\}$, hence the assertion.

(iv) ⇔ (vi): By Lemma 5.5, (vi) is equivalent to $ce \neq 0$ for any primitive idempotent e of FG. Since the latter is obviously equivalent to (iv), the result follows. ∎

5.7. COROLLARY (Formanek and Snider (1977), Tsushima (1971a)). Let F be an arbitrary field of characteristic p and let G be a p-solvable group. Then

$$\mathrm{Soc}FG = FGc \quad \text{and} \quad J(FG) = \ell(c)$$

Proof. By Proposition 1.6.29 and Theorem 3.6(iii), it suffices to verify that $J(FG) = \ell(c)$. By Corollary 1.18, we may harmlessly assume that F is algebraically closed. Now apply Theorems 5.6 and 4.26. ∎

5.8. COROLLARY. Let F be an arbitrary field of characteristic p, let G be a p-solvable group and let G_p be the set of all p-elements of G including 1. Then

$$J(FG) = \{\sum_{g\in G} x_g g \in FG \mid \sum_{s\in G_p} x_{gs} = 0 \quad \text{for all} \quad g \in G\}$$

Proof. Direct consequence of Lemma 3.12(ii) and Corollary 5.7. ∎

6. A CHARACTERIZATION OF ELEMENTS OF $J(Z(FG))$.

Throughout this section, F denotes an arbitrary field of characteristic $p > 0$, G a finite group and c the sum of all p-elements of G (including 1). It is a consequence of Theorem 3.14 that $J(Z(FG)) = Z(FG) \bigcap_{i=1}^{r} \ell(S_i^+)$, where $S_1,\ldots,S_r$ are all p-regular sections of G. Furthermore, by Corollary 5.7,

$$J(Z(FG)) = Z(FG) \cap \ell(c)$$

if G is p-solvable. The aim of this section is to prove the remarkable fact that

$$J(Z(FG)) = Z(FG) \cap \ell(c)$$

for an arbitrary finite group G.

6.1. LEMMA. Let e be any idempotent in an ideal I of a finite-dimensional algebra A over a field. Then the decompositions of e into primitive idempotents of A and of I are the same.

Proof. Let $e = \sum_{i=1}^{n} e_i$, where the e_i are orthogonal primitive idempotents of A. Then, for all $j \in \{1,\dots,n\}$,

$$ee_j = \sum_{i=1}^{n} e_i e_j = e_j^2 = e_j$$

so that $e_j \in I$, as required. ∎

6.2. THEOREM (Iizuka-Watanabe (1973)). $J(Z(FG)) = Z(FG) \cap \ell(c)$

Proof. (Külshammer (1981a)). Put

$$R(FG) = \{a \in FG \mid tr(a^{p^n}) = 0 \text{ for all sufficiently large } n \in \mathbb{N}\}$$

It is obvious that $R(FG)$ is an F-subspace of FG. Denote by X and Y the sets of all p-regular elements (excluding 1) and all p-elements (including 1), respectively. Since for all $y \in Y, y-1$ is nilpotent, we have

$$R(FG) \subseteq \sum_{x \in X} Fx + \sum_{y \in Y} F(y-1)$$

To prove the reverse containment, we need only verify that if $a \in R(FG)$ with $\mathrm{Supp}\, a \subseteq Y$, then $a \in \sum_{y \in Y} F(y-1)$. But if $\mathrm{Supp}\, a \subseteq Y$ and $a \in R(FG)$, then there exist $a_y \in F$ such that

$$a = \sum_{y \in Y} a_y(y-1) + \sum_{y \in Y} a_y$$

and $\sum_{y \in Y} a_y \in R(FG)$, which is possible only in the case $a = \sum_{y \in Y} a_y(y-1)$. This proves that

$$R(FG) = \sum_{x \in X} Fx + \sum_{y \in Y} F(y-1) \tag{1}$$

It follows from (1) that $Fc \subseteq R(FG)^{\perp}$. To prove the reverse containment, assume that $tr(a\,R(FG)) = 0$ for some $a = \sum_{g \in G} a_g g \in FG$. Then, by (1), for all $x \in X, y \in Y$, we have $tr(ax) = 0$ and $tr(ay) = tr\,a$. Thus, for all

$x \in X, y \in Y, a_{x^{-1}} = 0$ and $a_1 = a_{y^{-1}}$. This proves that $a = a_1 c$ and hence that

$$Fc = R(FG)^{\perp} \tag{2}$$

We now claim that $R(FG)$ does not contain blocks of FG. Let $e = \Sigma e_g g$ be a block idempotent of FG and assume, by way of contradiction, that $FGe \subseteq R(FG)$. By Theorem 2.2.4, there exists a p-regular element $g \in G$ such that $e_g \neq 0$. By hypothesis, $tr(eg^{-p^n}) = 0$ for all sufficiently large n. We may choose n such that $g^{-p^n} = g^{-1}$. But then

$$0 = tr(eg^{-p^n}) = tr(eg^{-1}) = e_g ,$$

a contradiction.

It is now an easy matter to complete the proof. Indeed, by Proposition 1.10.6, $J(Z(FG)) = Z(FG) \cap J(FG)$ and hence, by Theorem 3.14,

$$J(Z(FG)) \subseteq \ell(c) \cap Z(FG)$$

To prove the reverse containment, it suffices to show that $\ell(c) \cap Z(FG)$ is a nilpotent ideal of $Z(FG)$. Assume the contrary. Then $\ell(c) \cap Z(FG)$ contains a primitive idempotent, say e, by virtue of Corollary 1.6.38. By Lemma 6.1, e is a block idempotent of FG and so $FGe \subseteq \ell(c)$. However, by (2) and Theorem 3.6(i),

$$\ell(c) = \ell(R(FG)^{\perp}) \subseteq {}^{\perp}(R(FG)^{\perp}) = R(FG)$$

which implies that $R(FG)$ contains the block FGe, a contradiction. ■

6.3. COROLLARY. $J(Z(FG)) = \{\Sigma x_g g \in Z(FG) \mid \sum_{s \in G_p} x_{gs} = 0 \text{ for all } g \in G\}$

where G_p is the set of all p-elements of G (including 1).

Proof. Direct consequence of Theorem 6.2 and Lemma 3.12(ii). ■

For later use, we next establish the following result due to Reynolds (1972).

6.4. THEOREM. Let $S_1, S_2, \ldots, S_r$ be all p-regular sections of G and let $Z_{p'} = \sum_{i=1}^{r} FS_i^+$. Then $Z_{p'}$ is an ideal of $Z(FG)$ which satisfies the following properties:

(i) The primitive idempotents of $Z_{p'}$ are the same as block idempotents of FG of defect zero. In particular, a block B of FG is of zero defect if and only if B is a simple F-algebra.

(ii) $Z^2_{p'} = Z(B_1) \oplus \ldots \oplus Z(B_r)$

where $B_1,\ldots,B_r$ are all blocks of FG of defect zero. In particular, if F is a splitting field for FG, then $\dim_F Z^2_{p'}$ is equal to the number of blocks of FG of defect zero.

Proof. That $Z_{p'}$ is an ideal of $Z(FG)$ is a consequence of Theorem 3.10(iii) and Lemma 3.13(ii).

(i) Assume that e is a primitive idempotent of $Z_{p'}$. Then, by Lemma 6.1, e is a block idempotent of FG. Because Supp$e$ consists of p'-elements of G, it follows that each p-regular section in Supp$e$ is a conjugacy class of G. The latter implies that Supp$e$ is a union of conjugacy classes of p-defect zero. Thus e is a block idempotent of defect zero.

Conversely, suppose that e is a block idempotent of defect zero. Then Supp$e$ is a union of conjugacy classes of p-defect zero. Because these classes are also p-regular sections, we deduce that $e \in Z_{p'}$.

To prove the final statement, assume that $B = B(e)$ is a block of FG. Since $J(B) = J(FG)e$, we have $J(B) = 0$ if and only if $e \in \mathrm{Soc}(FG) \cap Z(FG)$. By Theorem 3.10(iii), $\mathrm{Soc}(FG) \cap Z(FG) = Z_{p'}$, and thus $J(B) = 0$ if and only if $e \in Z_{p'}$. Hence, by the foregoing, $J(B) = 0$ if and only if B is of defect zero. But $J(B) = 0$ if and only if B is a simple F-algebra, hence the assertion.

(ii) By Theorem 3.10(iii) and Lemma 3.13(ii),

$$Z^2_{p'} = Z(B_1) \oplus \ldots \oplus Z(B_r)$$

where $B_1,\ldots,B_r$ are all blocks of FG which are simple F-algebras. By (i), $B_1,\ldots,B_r$ are all blocks of FG of defect zero.

Assume that F is a splitting field for FG. Since each B_i is a block of $FG/J(FG)$, it follows that $B_i \cong M_{n_i}(F)$ for some $n_i \geqslant 1$, $1 \leqslant i \leqslant r$. Thus

$\dim_F Z(B_i) = 1$, $1 \leqslant i \leqslant r$, and the required assertion follows. ■

We close by providing circumstances under which $Z_{p'}^2 = 0$ and hence $Z_{p'} \subseteq J(Z(FG))$. For this we need the following two simple observations.

6.5. LEMMA. Let P be a normal p-subgroup of G. Then the kernel of the Brauer homomorphism

$$\pi : Z(FG) \longrightarrow Z(FG)$$

is a nilpotent ideal.

Proof. Let $C \in C\ell(G)$ be such that $C \cap C_G(P) = \emptyset$. Invoking Theorem 2.2.3, it suffices to verify that C^+ is nilpotent. Since $FG \cdot I(P)$ is a nilpotent ideal of FG (Proposition 1.1(i)), the latter will follow provided we show that $C^+ \in FG \cdot I(P)$.

The group P acts by conjugation as a permutation group of the set C. Since $C \cap C_G(P) = \emptyset$, the size of each orbit under this action is divisible by p. But for all $x \in P$ and $g \in C$, we have

$$x^{-1}gx - g = x^{-1}g(x-1) + (x^{-1}-1)g \in FG \cdot I(P)$$

Hence $C^+ \in FG \cdot I(P)$ and the result follows. ■

In what follows, given $C \in C\ell(G)$, we write $I[C]$ for the ideal of $Z(FG)$ which is the F-linear span of X^+, where X ranges all elements in $C\ell(G)$ with $\delta_G(X) \leqslant \delta_G(C)$ (see Lemma 2.2.5).

6.6. LEMMA. Let P be a normal p-subgroup of G and let $C_1, \ldots, C_k$ (respectively, $C_{k+1}, \ldots, C_t$) be all conjugacy classes of G with defect group P (respectively, all conjugacy classes of G whose defect groups are conjugate to proper subgroups of P). Denote by A the F-linear span of $C_1^+, \ldots, C_k^+$ and by B the F-linear span of $C_{k+1}^+, \ldots, C_t^+$. Then

(i) A is a subalgebra of $Z(FG)$, B is a nilpotent ideal of $Z(FG)$ and for any $C \in \{C_1, \ldots, C_k\}$, $I[C] = A \oplus B$ (direct sum of F-spaces)

(ii) The defect groups of any block idempotent of FG contain P. If the order of P is p^d, then any block idempotent of FG of defect d lies in A.

Proof. (i) Fix $i,j \in \{1,\dots,k\}$. By Lemma 2.2.5, we have

$$C_i^+C_j^+ = \lambda_1 C_1^+ + \dots + \lambda_k C_k^+ + \lambda_{k+1}C_{k+1}^+ + \dots + \lambda_t C_t^+$$

where $\lambda_r \in F$, $1 \leqslant r \leqslant t$. Let

$$\pi : Z(FG) \longrightarrow Z(FG)$$

be the Brauer homomorphism. Since $C_1, C_2, \dots, C_k$ have P as their defect group, $C_r \subseteq C_G(P)$, $1 \leqslant r \leqslant k$, and so $\pi(C_r^+) = C_r^+$ for all $r \in \{1,\dots,k\}$. On the other hand, since the defect groups of $C_{k+1},\dots,C_t$ are conjugate to proper subgroups of P, we have

$$C_{k+1} \cap C_G(P) = \emptyset \quad \text{for all} \quad r \in \{1,\dots,t-k\}$$

Thus $\pi(C_{k+1}^+) = \dots = \pi(C_t^+) = 0$ and therefore

$$C_i^+C_j^+ = \pi(C_i^+)\pi(C_j^+) = \pi(C_i^+C_j^+) = \sum_{r=1}^{k} \lambda_r C_r^+ \in A ,$$

proving that A is a subalgebra.

By the definition of $I[C]$, $I[C] = A \oplus B$ as F-spaces. By Theorem 2.2.3, we also have $B = \text{Ker}\pi \cap I[C]$ and so B is an ideal. Finally, since $\text{Ker}\pi$ is nilpotent (Lemma 6.5), B is a nilpotent ideal.

(ii) Let e be a block idempotent of FG with D as a defect group. By Proposition 2.2.6, $\text{Supp}e$ is a union of conjugacy classes whose defect groups are conjugate to subgroups of D. Let $C \in C\ell(G)$ with $C \subseteq \text{Supp}e$. If $P \nsubseteq D$ then $C \cap C_G(P) = \emptyset$ and so, by Lemma 6.5, e is nilpotent which is impossible. Thus $P \subseteq D$.

Now assume that e is of defect d. Then, by the above, P is a defect group of e. Hence, by (i) and Proposition 2.2.6(i), $e = a+b$ with $a \in A, b \in B$. But there is an integer n with $b^{p^n} = 0$. Hence

$$e = e^{p^n} = (a+b)^{p^n} = a^{p^n} + b^{p^n} = a^{p^n} \in A$$

and the result follows. ■

6.7. COROLLARY. Let $S_1, S_2, \ldots, S_r$ be all p-regular sections of G and let $Z_{p'} = \sum_{i=1}^{r} FS_i^+$. If G has a nontrivial normal p-subgroup, then

$$Z_{p'}^2 = 0$$

and, in particular, $Z_{p'} \subseteq J(Z(FG))$.

Proof. Apply Lemma 6.6(ii) and Theorem 6.4(ii). ■

7. FROBENIUS GROUPS.

A transitive permutation group G in which only the identity fixes more than one letter, but the subgroup fixing a letter is nontrivial is called a *Frobenius group*. In what follows we shall use the following two standard facts for the proof of which we refer to Gorenstein (1968).

(1) (Frobenius). Let G be a Frobenius group and let H be the subgroup fixing a letter. Then the subset of G consisting of the identity together with those elements which fix no letters forms a normal subgroup N of G of order $|G:H|$

(2) (Thompson). Let a group G admit an automorphism of prime order which fixes only the identity element of G. Then G is nilpotent

We shall refer to the normal subgroup N and the subgroup H of G in (1) as the *Frobenius kernel* and *Frobenius complement*, respectively. Our aim in this section is to reduce the calculation of $J(FG)$ to that of $J(FH)$. We start by recording some group-theoretic results.

7.1. LEMMA. Let G be a Frobenius group with complement H and kernel N. Then

(i) $G = NH$ with $N \cap H = 1$

(ii) Every $h \neq 1$ in H induces by conjugation an automorphism of N which fixes only the identity element of N

(iii) $|H|$ divides $|N| - 1$

(iv) N is nilpotent

(v) If $g \in G - N$, then g lies in a conjugate of H

(vi) For every $x \neq 1$ in N, $C_G(x) \subseteq N$

Proof. (i) By the definition of N, we have $N \cap H = 1$. But $|N| = |G:H|$ by (1) and hence $|G| = |NH|$. Thus $G = NH$ as required.

(ii) Since G is transitive and H is the subgroup fixing a letter, our representation is equivalent to that on the cosets of H. By (i), we can take the elements of N themselves as a transversal for H in G. Suppose that $n^h = n$ for some $h \neq 1$ in H and $n \neq 1$ in N. Then it is immediate that h fixes the coset Hn as well as the coset H. But, by definition of a Frobenius group, only the identity fixes more than one letter. Hence $n^h \neq n$ for any $h \neq 1$ in H and $n \neq 1$ in N, proving (ii).

(iii) For a fixed $n \neq 1$ in N, the set $\Gamma_n = \{n^h | H \in H\}$ must consist of $m = |H|$ distinct elements of N. But clearly for x, y in N with $x \neq 1$ and $y \neq 1$, we have either $\Gamma_x = \Gamma_y$ or $\Gamma_x \cap \Gamma_y = \emptyset$. Hence $|N-\{1\}|$ is a multiple of m and (iii) follows.

(iv) Since $H \neq 1$, we may choose an element $h \in H$ of prime order. By (ii), h induces an automorphism of N which fixes only the identity element of N. Hence, by (2), N is nilpotent.

(v) If $g \in G - N$, then g fixes a letter. But the conjugates of H are simply the subgroups fixing a letter. Hence g lies in a conjugate of H.

(vi) By (v), N consists precisely of the elements of G which lie in no conjugate of $H - \{1\}$. We conclude that $C_G(x) \subseteq N$ for $x \neq 1$ in N, as required. ■

7.2. LEMMA. Let H be a nontrivial subgroup of G. Then G is a Frobenius group with complement H if and only if $H \cap H^g = 1$ for all $g \in G - H$.

Proof. Let $Hg_i, g_i \in G, g_1 = 1, 1 \leqslant i \leqslant n$, be a complete set of cosets of H in G. If an element of G fixes two cosets of H, then by transitivity some element of $H - \{1\}$ fixes one of the cosets Hg_i with $i > 1$. Thus G is a Frobenius group with complement H if and only if no element of $H - \{1\}$ fixes any $Hg_i, i > 1$. However for $h \neq 1$ in $H, Hg_i = Hg_i h$ if and only if $g_i h g_i^{-1} \in H$ or equivalently, $h \in H \cap H^{g_i}$. Thus G and H have the given properties if and only if $H \cap H^{g_i} = 1$ for $i \geqslant 2$. Since the latter is equivalent to $H \cap H^g = 1$ for all $g \in G - H$, the result follows. ■

Turning our attention to $J(FG)$, we next establish the following elementary result.

7.3. LEMMA. Let F be an arbitrary field of characteristic $p > 0$, let N be a normal p'-subgroup of a group G and let $e = |N|^{-1} \sum_{n \in N} n$. If G has a subgroup H such that $G = NH$ and $N \cap H = 1$, then

(i) $J(FG)e = J(FH)e$

(ii) $\dim_F J(FH)e = \dim_F J(FH)$

Proof. The element e is obviously a central idempotent of FG. Hence $J(FGe) = J(FG)e$ by Proposition 1.6.36. On the other hand, by Proposition 2.1.9, $FGe \cong F(G/N) \equiv FH$. Thus

$$\begin{cases} FH \longrightarrow FGe \\ x \longmapsto xe \end{cases}$$

is an isomorphism of F-algebras. Hence $J(FGe) = J(FH)e$ and

$$J(FG)e = J(FGe) = J(FH)e\ , \quad J(FH) \cong J(FH)e$$

as required. ■

7.4. LEMMA. Let F be a field such that $\mathrm{char}F \nmid |G|$ and such that F is a splitting field for FG. Denote by $V_1, V_2, \ldots, V_r$ all nonisomorphic irreducible FG-modules and by χ_i the character of FG afforded by V_i, $1 \leqslant i \leqslant r$. Then

$$e_i = \frac{\chi_i(1)}{|G|} \sum_{g \in G} \chi_i(g^{-1})g \qquad (1 \leqslant i \leqslant r)$$

are all distinct block idempotents of FG. In particular, $\{e_1, \ldots, e_r\}$ is an F-basis for $Z(FG)$.

Proof. Because FG is semisimple (Maschke's theorem) and F is a splitting field for FG, there exists exactly r block idempotents, say $v_1, \ldots, v_r$, of FG. Thus we have:

(a) $FG = FGv_1 \oplus \ldots \oplus FGv_r \qquad (v_i v_j = 0$ for $i \neq j)$

(b) $FGv_i \cong M_{n_i}(F)$, where $n_i = \dim_F V$

(c) The irreducible constituents of the FG-module FGv_i are isomorphic to V_i

and V_i occurs with multiplicity n_i.

Let ρ be the regular character of FG, i.e. the character afforded by FG. It follows from (a) - (c) that

(d) $\rho = \chi_1(1)\chi_1 + \ldots + \chi_r(1)\chi_r$

(e) $\chi_i(FGv_j) = 0 \ (1 \leqslant i,j \leqslant r, i \neq j)$

Let x be an arbitrary element of FG. By (a) there is a unique element $x_i \in FGv_i$ such that $x = x_1 + \ldots + x_r$. Because $v_i v_j = 0$ for $i \neq j$, we see that $x_i = xv_i$ $(1 \leqslant i \leqslant r)$. Hence (e) implies

$$\chi_j(xv_i) = \chi_j(x_i) = \begin{cases} 0, & \text{if } j = 1,2,\ldots,r \text{ and } j \neq i \\ \chi_i(x) & \text{if } j = i \end{cases}$$

In view of (d), this gives

(f) $\rho(xv_i) = \chi_i(1)\chi_i(x)$ for all $x \in FG$, $1 \leqslant i \leqslant r$

We now fix $i \in \{1,\ldots,r\}$ and write $v_i = \sum_{g\in G} x_g g$ for some $x_g \in F$. Evidently, x_g is the coefficient of 1 when $g^{-1}v_i$ is written as a linear combination of elements of G:

$$g^{-1}v_i = x_g \cdot 1 + \ldots$$

Applying ρ and using the equality $\rho(\Sigma x_g g) = x_1|G|$, we obtain

$$\rho(g^{-1}v_i) = |G|x_g \qquad \text{for all } g \in G$$

Hence, by (f), we derive

$$|G|x_g = \chi_i(1)\chi_i(g^{-1})$$

and therefore

$$x_g = \frac{\chi_i(1)}{|G|}\chi_i(g^{-1})$$

Thus $v_i = e_i$ for all $i \in \{1,\ldots,r\}$ and the result follows. ■

Let G be a group of linear transformations of a vector space V. By the *fixed-point space* of V (relative to G) we understand the set of those $v \in V$

for which $g(v) = v$ for all $g \in G$. A weaker version of the following result would be sufficient for the proof of the main theorem of this section. However, the more general result will be needed later.

7.5. LEMMA. Let G be a group of linear transformations of a vector space V over a field F. Assume that X and Y are two bases for V such that each element of G induces a permutation of the sets X and Y. Then X and Y have the same number of G-orbits. Furthermore, if $\text{char}F = 0$, then for each $g \in G$, the number of fixed elements of X is equal to that of Y.

Proof. Let $X_1, X_2, \ldots, X_n$ and $Y_1, \ldots, Y_m$ be all G-orbits of X and Y, respectively, and let W be the fixed-point space of V. Since G acts transitively on each X_i, it is immediate that the elements

$$\sum_{x \in X_i} x \qquad (1 \leqslant i \leqslant n)$$

constitute a basis for W. Thus $n = \dim W$ and, by the same argument, $m = \dim W$ proving that $n = m$.

Fix $g \in G$ and denote by A and B the matrices of the linear transformation g with respect to X and Y, respectively. Then A and B are similar and so $tr\,A = tr\,B$. Now assume that $\text{char}F = 0$. Since g permutes the elements of X and Y, $tr\,A$ and $tr\,B$ are, respectively, the numbers of fixed elements of X and Y. So the lemma is true. ∎

The next observation is crucial.

7.6. LEMMA. Let N be a normal subgroup of G, let F be an algebraically closed field such that $\text{char}F \nmid |N|$ and let H be a subgroup of G with $G = NH$.

(i) The number of conjugacy classes of G contained in N is equal to the number of H-conjugacy classes of block idempotents of FN.

(ii) For each $h \in H$, the number of irreducible FN-modules V with $V \cong {}^hV$ equals the number of conjugacy classes C of N with $C = hCh^{-1}$.

(iii) If G is a Frobenius group with complement H and kernel N, then V^G is irreducible for any irreducible FN-module V with $V \not\cong 1_N$.

Proof. (i) Let χ be the character of N afforded by an irreducible FN-

module V, and let

$$e_\chi = |N|^{-1} \sum_{n\in N} \chi(n^{-1})n$$

If ${}^g\chi$ is the character of N afforded by ${}^gV$, then ${}^g\chi(n) = \chi(g^{-1}ng)$ for all $n \in N$. Hence

$$ge_\chi g^{-1} = |N|^{-1} \sum_{n\in N} \chi(n^{-1})gng^{-1} = \sum_{n\in N} {}^g\chi(n^{-1})n = e_{{}^g\chi}$$

Let $C_1 = \{1\}, C_2, \ldots, C_r$ be all conjugacy classes of N. By Lemma 7.4, there exists exactly r nonisomorphic irreducible FN-modules, say $V_1 = 1_N, V_2, \ldots, V_r$ and if χ_i is the character of FN afforded by V_i, then the elements e_{χ_i}, $1 \leqslant i \leqslant r$, form an F-basis for $Z(FN)$. Now H acts by conjugation as a group of linear transformations of $Z(FN)$. Put

$$X = \{C_1^+, \ldots, C_r^+\} \quad \text{and} \quad Y = \{e_{\chi_1}, \ldots, e_{\chi_r}\}$$

Then X and Y are F-bases of $Z(FN)$ and each element of H induces a permutation of X and Y. Since the H-orbits of X and Y are conjugacy classes of G contained in N and H-conjugacy classes of block idempotents of FN, respectively, the assertion follows by appealing to Lemma 7.5.

(ii) Let χ be the character of N afforded by an irreducible FN-module V. Then, for any $h \in H$, $V \cong {}^hV$ if and only if $\chi = {}^h\chi$ or if and only if $he_\chi h^{-1} = e_\chi$. Since $\mathrm{char}F \nmid |N|$, the number of χ with $\chi = {}^h\chi$ is independent of the characteristic of F (see Curtis and Reiner (1981,p.420)). Hence we may assume that $\mathrm{char}F = 0$, in which case (ii) follows by virtue of Lemma 7.5.

(iii) Let V be a nontrivial irreducible FN-module and let W be an irreducible factor module of V^G. Then, by Theorem 2.4.10, V is isomorphic to a submodule of W_N. Hence if the inertia group of V is N, then by Theorem 4.1(ii), $W \cong V^G$ and V^G is irreducible. We are therefore left to verify that ${}^hV \not\cong V$ for all $h \neq 1$ in H.

Let $C \in C\ell(N)$ be such that $C = hCh^{-1}$ and $C \neq 1$. Then there exists $x \neq 1, n \in N$ with $hxh^{-1} = nxn^{-1}$ and hence with $n^{-1}h \in C_G(x)$. But by Lemma

7.1(vi), $C_G(x) \subseteq N$, so $n^{-1}h \in N$ and thus $h = 1$, a contradiction. Thus h fixes no nonidentity conjugacy class of H and hence, by (ii), ${}^hV \not\cong V$, as required. ■

We have now accumulated all the information necessary to prove the following result.

7.7. THEOREM. (Motose (1974a)). Let G be a Frobenius group with complement H and kernel N and let F be an arbitrary field of characteristic $p > 0$.

(i) If p divides $|N|$ and P is a Sylow p-subgroup of N (and hence of G, by Lemma 7.1(iii)), then $J(FG) = FG \cdot I(P)$

(ii) If p does not divide $|N|$, then

$$J(FG) = J(FH)e \quad \text{and} \quad \dim_F J(FG) = \dim_F J(FH)$$

where

$$e = |N|^{-1} \sum_{n \in N} n$$

Proof. (i) Assume that p divides $|N|$ and that P is a Sylow p-subgroup of N. By Lemma 7.1 (iv), N is nilpotent and hence $P \lhd N$. Thus $P \lhd G$ and the result follows by virtue of Proposition 1.1(iii).

(ii) Suppose that p does not divide $|N|$. By Corollary 1.18, we may assume that F is algebraically closed. Note that $FN = FNe \oplus FN(1-e)$ and $FNe \cong 1_N$. Hence $FN(1-e)$ is a direct sum of nontrivial irreducible FN-modules. Invoking Lemma 7.6 and Proposition 2.4.3(i), we deduce that

$$FG(1-e) \cong (FN(1-e))^G \qquad \text{is completely reducible}$$

Thus $J(FG)(1-e) = 0$ and therefore $J(FG) = J(FGe)$. Now apply Lemma 7.3. ■

7.8. COROLLARY (Wallace (1958)). Let P be a Sylow p-subgroup of G, let F be a field of characteristic p and let G be a Frobenius group with complement P and kernel N. Then

$$J(FG) = I(P)e$$

where

$$e = |N|^{-1} \sum_{n\in N} n$$

In particular,

$$\dim_F J(FG) = |P| - 1$$

Proof. Apply Theorem 7.7 for $H = P$ and Corollary 1.2. ■

We close this section by considering the following two examples.

7.9. EXAMPLE. Let $G = S_3$, i.e. $G = \langle a,b \mid a^3 = 1, b^2 = 1, bab^{-1} = a^2\rangle$, and let $p = 2$. Then G is a Frobenius group with complement $P = \langle b\rangle$ and kernel $N = \langle a\rangle$. The element e in this case is given by

$$e = \frac{1}{3}(1+a+a^2) = (1+a+a^2).$$

Since $I(P) = \{\lambda(b+1) \mid \lambda \in F\}$, Corollary 7.8 implies that

$$\begin{aligned} J(FG) &= \{\lambda(1+a+a^2)(1+b) \mid \lambda \in F\} \\ &= \{\lambda(1+a+a^2+b+ab+a^2b) \mid \lambda \in F\} \end{aligned}$$

7.10. EXAMPLE. Let $G = A_4$, i.e. $G = \langle a,b,c \mid a^2 = b^2 = c^3 = 1, ab=ba, cac^{-1}=b, cbc^{-1} = ab\rangle$ and let $p = 3$. Then G is a Frobenius group with complement $P = \langle c\rangle$ and kernel $N = \langle a,b\rangle$. The element e in this case is given by

$$e = \frac{1}{4}(1+a+b+ab) = 1+a+b+ab$$

Hence, by Corollary 7.8,

$$J(FG) = \{\lambda(1+a+b+ab)(c-1)+\mu(1+a+b+ab)(c^2-1) \mid \lambda,\mu \in F\}$$

7.11. REMARK. Let G be a Frobenius group with complement $P \in \mathrm{Syl}_p(G)$. According to Burnside (1955,p.335), P has no noncyclic subgroup of order p^2. Hence, by Zassenhaus (1949,p.119), we have

(i) If $p \neq 2$, then P is cyclic

(ii) If $p = 2$, then P is either cyclic or a generalized quaternion group. ■

8. UPPER AND LOWER BOUNDS FOR $\dim_F J(FG)$

Throughout this section, G denotes a finite group, F an arbitrary field of characteristic $p > 0$, 1_G or 1_{FG} the trivial FG-module and

$$u = \dim_F P(1_G)$$

Our aim is to provide an upper and lower bounds for $\dim_F J(FG)$. It turns out that $J(FG)$ is of maximal F-dimension if and only if the Sylow p-subgroup P of G is normal, while $J(FG)$ is of minimal F-dimension if and only if P is disjoint from its conjugates. Some of the results presented here were originally stated for splitting fields. The following simple observation will eliminate that assumption.

8.1. LEMMA. Let L be a field extension of F. Then

$$\dim_F P(1_{FG}) = \dim_L P(1_{LG})$$

Proof. Put $V = 1_{FG}$. Then $V_L = 1_{FL}$ and the result follows by virtue of Proposition 1.19 and Corollary 1.18. ∎

8.2. LEMMA. Let P be a Sylow p-subgroup of G and let V be a projective FG-module. Then $|P|$ divides $\dim_F V$.

Proof. By Corollary 1.2, FP is a local ring. Since FG is a free FP-module, V_P is a projective module. Now apply Lemma 4.18. ∎

8.3. LEMMA. Let V be an irreducible FG-module. Then $P(V)$ is isomorphic to a direct summand of $P(1_G) \otimes_F V$. In particular,

$$\dim_F P(V) \leqslant u \dim_F V$$

Proof. By Proposition 2.5.7, $P(1_G) \otimes_F V$ is a projective FG-module. On the other hand, by Theorem 3.8(iii), $1_G \subseteq P(1_G)$. Hence

$$V \cong 1_G \otimes_F V \subseteq P(1_G) \otimes_F V = \bigoplus_{i=1}^{n} P(V_i)$$

for some irreducible FG-modules $V_1,\ldots,V_n$. It follows that V is isomorphic to a submodule of

$$\mathrm{Soc}\left(\bigoplus_{i=1}^{n} P(V_i)\right) = \bigoplus_{i=1}^{n} \mathrm{Soc}(P(V_i))$$

But, Theorem 3.8(iii), $\mathrm{Soc}P(V_i) \cong V_i$ for all i. Hence $V \cong V_i$ for some $i \in \{1,\ldots,n\}$, and thus $P(V) \cong P(V_i)$ is a direct summand of $P(1_G) \otimes_F V$. ■

We are now ready to prove the following important result.

8.4. THEOREM. Let G be a group of order $p^a m$ where $(p,m) = 1$. Then

(i) (Wallace (1958)) $\dim_F J(FG) \geqslant p^a - 1$

(ii) (Brauer and Nesbitt (1941)) $\dim_F J(FG) \leqslant |G| - \frac{|G|}{u}$

Proof. (i) Write $FG = FGe_1 \oplus \ldots \oplus FGe_r$ for some primitive idempotents $e_1,\ldots,e_r$ of FG with $FGe_1/J(FG)e_1 \cong 1_G$. Then, by Lemma 8.2,

$$\dim_F J(FG)e_1 = \dim_F FGe_1 - 1 \geqslant p^a - 1 \tag{1}$$

Since $J(FG)e_1 \subseteq J(FG)$, the required assertion follows.

(ii) By Lemma 8.1 and Corollary 1.18, we may assume that F is algebraically closed. Let $V_1, V_2, \ldots, V_n$ be all nonisomorphic irreducible FG-modules. Then

$$\begin{aligned}
\dim_F FG/J(FG) &= \sum_{i=1}^{n} (\dim_F V_i)^2 && \text{(by Corollary 1.10.17)}\\
&= \frac{1}{u}\sum_{i=1}^{n} (\dim_F V_i)(u \dim_F V_i)\\
&\geqslant \frac{1}{u}\sum_{i=1}^{n} (\dim_F V_i)(\dim_F P(V_i)) && \text{(by Lemma 8.2)}\\
&= \frac{|G|}{u} && \text{(by Corollary 1.10.17)}
\end{aligned}$$

and the result follows. ■

We next provide necessary and sufficient conditions for $J(FG)$ to be of maximal F-dimension. The lemma which follows disposes of some of the technicalities.

8.5. LEMMA. Let P be a p-group acting on a p'-group Q such that P stabilizes each conjugacy class of Q. Then P acts trivially.

Proof. Let C be a conjugacy class of Q. By assumption P acts on C. Since $p \nmid |C|$, there exists $c \in C$ such that $c \in C_Q(P)$. Hence

$$C = c^Q \subseteq \bigcup_{g\in Q} [C_Q(P)]^g$$

and therefore

$$Q = \bigcup_{g\in Q} [C_Q(P)]^g$$

The desired assertion is now a consequence of the following general observation: If H is a subgroup of G with $G = \bigcup_{g\in G} H^g$, then $G = H$.

Indeed, assume that $G \neq H$. Then $H^g \neq H$ for some $g \in G$ and hence $|H \cap H^g| < |H|$. But then

$$|G| < |H||G:N_G(H)| \leqslant |G|$$

a contradiction. So the lemma is true. ■

8.6. THEOREM (Wallace (1961)). Let G be a p-solvable group. Then

$$\dim_F J(FG) = |G| - \frac{|G|}{u}$$

if and only if the Sylow p-subgroup of G is normal.

Proof. (Brockhaus (1982)). By Lemma 8.1, we may harmlessly assume that F is algebraically closed. Hence, by Theorem 4.26, $u = |P|$ where P is a Sylow p-subgroup of G. Thus we are left to verify that

$$\dim_F J(FG) = |G| - |G:P| \quad \text{if and only if} \quad P \lhd G$$

If $P \lhd G$, then $\dim_F J(FG) = |G| - |G:P|$, by virtue of Corollary 1.4.

Assume that $\dim_F J(FG) = |G| - |G:P|$. We argue by induction on $|G|$. The case $|G| = 1$ being obvious, assume that $|G| > 1$ and that the result holds for p-solvable groups of smaller order. Because G is p-solvable, there is a normal subgroup N of G such that either G/N is a p'-group or G/N is of order p.

Case 1: G/N is a p'-group. Then $P \in \mathrm{Syl}_p(N)$ and

$$\begin{aligned}\dim_F J(FN) &= |G:N|^{-1}\dim_F J(FG) \qquad \text{(by Proposition 1.8(ii))}\\ &= |G:N|^{-1}(|G| - |G:P|)\\ &= |N| - |N:P|\end{aligned}$$

Thus, by induction, $P \lhd N$ and therefore $P \lhd G$.

Case 2: $|G/N| = p$ and $p \nmid |N|$. Then $N = O_{p'}(G)$ and $G = NP$. Since $FN \cap J(FG) \subseteq J(FN) = 0$ (by Maschke's theorem) the map

$$\begin{cases} FN \xrightarrow{\phi} FG/J(FG) \\ x \longmapsto x + J(FG) \end{cases}$$

is an injective homomorphism of F-algebras. Furthermore, ϕ is an isomorphism since

$$\dim_F(FG/J(FG)) = |G| - \dim_F J(FG) = |N|$$

Define $\psi : FG \longrightarrow FN$ by $\psi(x) = \phi^{-1}(x+J(FG))$. Then ψ is an F-algebra homomorphism such that ψ is the identity map on FN. Let C be a conjugacy class of N and $g \in P$. Then

$$g^{-1}C^+g = \psi(g^{-1}C^+g) = \psi(g)^{-1}\psi(C^+)\psi(g) = \psi(g)^{-1}C^+\psi(g) = C^+$$

and hence $g^{-1}Cg = C$. Invoking Lemma 8.5, we deduce that $N = C_N(P)$ and hence that $P \lhd G$.

Case 3: $|G:N| = p$ and $p \mid |N|$. Let $Q \in Syl_p(N)$. Then

$$\begin{aligned} \dim_F J(FN) &\geqslant \dim_F J(FG) + |N| - |G| \quad \text{(by Proposition 1.6(ii))} \\ &= |G| - |G:P| + |N| - |G| = |N| - |G:P| \\ &= |N| - |N:Q| \end{aligned}$$

and therefore, by Theorem 8.4(ii), $\dim_F J(FN) = |N| - |N:Q|$. It follows, by induction, that $Q \lhd N$ and hence that $Q \lhd G$. We now apply Corollary 1.5 to deduce that

$$\begin{aligned} \dim_F J(F(G/Q)) &= \dim_F J(FG) + |G:Q| - |G| \\ &= |G| - |G:P| + |G:Q| - |G| \\ &= |G/Q| - |G/Q:P/Q| \end{aligned}$$

Since $P/Q \in Syl_p(G/Q)$ it follows by induction that $P/Q \lhd G/Q$. Thus $P \lhd G$

and the result is established. ■

Remark. The result above holds for an arbitrary group G. This was proved by Brockhaus (1985) whose argument relies on classification theorem of finite simple groups.

We now head towards providing circumstances under which $J(FG)$ is of minimal F-dimention.

By the *principal block* of FG we understand the block which contains the trivial FG-module. Let V be an FG-module. By the *kernel* of V, written $\mathrm{Ker}V$, we understand the kernel of the representation of G afforded by V. Assume that B is a block of FG. Following Brauer (1971), we define the *kernel* of B, written $\mathrm{Ker}B$, by

$$\mathrm{Ker}B = \bigcap_{V \in X} \mathrm{Ker}V$$

where X is the set of all irreducible FG-modules in B. The following results are extracted from a work of Michler (1973).

8.7. LEMMA. Let $B = B(e)$ be a block of FG. Then

$$\mathrm{Ker}B = \{g \in G \mid (1-g)e \in J(FG)e\}$$

Proof. We may write e in the form $e = e_1 + \dots + e_s$, where $e_1, \dots, e_s$ are orthogonal primitive idempotents of FG, and where

$$V_i = FGe_i/J(FG)e_i \qquad (1 \leqslant i \leqslant r, r \leqslant s)$$

are all nonisomorphic irreducible FG-modules in B.

If $g \in \mathrm{Ker}B$, then G acts as the identity operator on each V_i and so

$$(1-g)e_i \in J(FG)e_i$$

for all $i \in \{1, \dots, s\}$. Thus $(1-g)e \in J(FG)e$.

Conversely, assume that $g \in G$ satisfies $(1-g)e \in J(FG)e$. Then

$$gv_i = gev_i = v_i \text{ for all } v_i \in V_i \qquad 1 \leqslant i \leqslant r$$

since $J(FG)V_i = 0$ and $v_i = ev_i$ for all $v_i \in V_i$, $1 \leqslant i \leqslant r$. This proves the lemma. ■

8.8. LEMMA. Let $B = B(e)$ be a block of FG. Then the following properties of $g \in G$ are equivalent:

(i) $ge = e$

(ii) g is a p-regular element of $\mathrm{Ker}B$

Proof. Suppose first that $ge = e$. Assume, by way of contradiction, that g is not p-regular. Replacing g by its suitable power, if necessary we may assume that g has order p. Let D denote the cyclic group generated by g and write $e = \sum_{i=1}^{t} \lambda_i g_i$ with $\lambda_i \in F$ and $t = |G|$. Then $e = ge$ implies that D acts fixed-point free on $\{g_i | 1 \leqslant i \leqslant t\}$ by left multiplication. Hence we may assume that

$$e = \sum_{j=1}^{k} \lambda_j (g+g^2+ \ldots + g^{p-1}+g^p)g_j ,$$

where $k = t/p$. Because $e = e^2$ and $e = g^s e$ for all $s \geqslant 1$, it follows that

$$\begin{aligned} e = e^2 &= \sum_{j=1}^{k} \lambda_j (ge+g^2e + \ldots + g^{p-1}e+g^pe)g_j \\ &= \sum_{j=1}^{k} p\lambda_j g_j e \\ &= 0, \end{aligned}$$

a contradiction. Consequently (i) implies (ii).

Now assume that g is a p-regular element of $\mathrm{Ker}B$. Since $J(FG)$ is nilpotent, it follows from Lemma 8.7 that

$$0 = (1-g)^{p^k} e = e - g^{p^k} e \tag{2}$$

for some integer $k \geqslant 0$. If r is the order of g, then $(p^k, r) = 1$ and hence $g = g^{p^k u + rv}$, for some integers u and v. But then (2) ensures that

$$ge = g^{p^k u+rv} e = g^{p^k u} e = e^u = e ,$$

proving (i). So the lemma is proved. ■

A group G is said to be *p-nilpotent* if G has a normal p'-subgroup N, called the *normal p-complement*, such that G/N is a p-group. Expressed otherwise, G is p-nilpotent if $G = 0_{p',p}(G)$. We have now accumulated all the information necessary to establish the following two useful results.

8.9. PROPOSITION. Let $B = B(e)$ be a block of FG and let $N = \mathrm{Ker}B$.

(i) N is a p-nilpotent normal subgroup of G

(ii) $N/0_{p'}(N) = 0_p(G/0_{p'}(N))$

Proof. (i) Let X be the set of all p-regular elements of N. Then, by Lemma 8.8, X is multiplicatively closed. Hence N must be p-nilpotent.

(ii) By (i), $X = 0_{p'}(N)$ and N/X is a normal p-subgroup of G/X. Set $f = |X|^{-1} \sum_{x\in X} x$. Then, by Proposition 2.1.9, f is a central idempotent of FG such that $FGf \cong F(G/X)$. Applying Lemma 8.8, we also have $e = fe = ef$. Assume that M/X is a normal p-subgroup of G/X. Then, by Proposition 1.1(i),

$$(1-m)f \in J(FG)f = J(FGf) \qquad \text{for all } m \in M$$

and therefore

$$(1-m)e = (1-m)fe \in J(FG)fe = J(FG)e, \qquad \text{for all } m \in M$$

proving that $M \subseteq N$, by appealing to Lemma 8.7. So the proposition is true. ∎

8.10. PROPOSITION. Let $B = B(e)$ be the principal block of FG.

(i) $\{g \in G \mid ge = e\} = 0_{p'}(G)$

(ii) $\mathrm{Ker}B = 0_{p',p}(G)$

Proof. (i) Assume that $g \in 0_{p'}(G)$ and put $f = |0_{p'}(G)|^{-1} \sum_{x\in 0_{p'}(G)} x$ Then f is a central idempotent of FG, so ef and $e(1-f)$ are also central idempotents of FG. Since FGe is indecomposable, either $ef = 0$ or $e(1-f) = 0$. But if V is the trivial FG-module, then $fv = v$ for all $v \in V$, so $(ef)v = e(fv) = ev = v$ and thus $ef \neq 0$. Hence $e(1-f) = 0$ and $e = ef$. Since $fg = f$, we conclude that

$$eg = efg = ef = e,$$

proving that $O_{p'}(G) \subseteq \{g \in G \mid ge = e\}$

Conversely, assume that $ge = e$ for some $g \in G$. Then, by Lemma 8.8, g is a p-regular element of $\mathrm{Ker}B$. Applying Proposition 8.10(i), we therefore derive

$$g \in O_{p'}(\mathrm{ker}B) \subseteq O_{p'}(G)$$

as required.

(ii) Owing to Proposition 8.10(ii), it suffices to verify that $O_{p'}(\mathrm{Ker}B) = O_{p'}(G)$. The latter will follow provided we show that $O_{p'}(G) \subseteq O_{p'}(\mathrm{Ker}B)$. So assume that $g \in O_{p'}(G)$. Then, by (i), $ge = e$ and thus, by Lemma 8.8, $g \in O_{p'}(\mathrm{Ker}B)$, as asserted. ∎

We are now ready to prove the following result.

8.11. THEOREM (Wallace (1958)). Let P be a Sylow p-subgroup of G. Then

$$\dim_F J(FG) = |P| - 1$$

if and only if either $G = P$ or G is a Frobenius group with complement P.

Proof. Owing to Corollaries 7.8 and 1.2, we need only verify the "only if" part. So assume that $G \neq P$ and $\dim_F J(FG) = |P| - 1$. By Corollary 1.18, there is no loss of generality in assuming that F is algebraically closed. Write

$$FG = FGe_1 \oplus \ldots \oplus FGe_n$$

where $e_1,\ldots,e_r$ are mutually orthogonal primitive idempotents of FG and $FGe_1/J(FG)e_1 \cong 1_G$. By (1) $\dim_F J(FG)e_1 \geqslant |P| - 1$. Since $J(FG)e_1 \subseteq J(FG)$, our assumption on $\dim_F J(FG)$ forces

$$J(FG)e_2 = J(FG)e_3 = \ldots = J(FG)e_n = 0 \tag{3}$$

It follows from (3) that each FGe_i, $i \in \{2,\ldots,n\}$, is an irreducible module. Furthermore, by Proposition 1.9.4, $FGe_i \not\cong 1_G$ for all $i \in \{2,\ldots,n\}$. Note also that by Proposition 1.10.16(iii) and Theorem 3.6(ii), the multiplicity of 1_G as a composition factor of FG is $\dim_F FGe_1$. Hence all composition factors of FGe_1 are isomorphic to 1_G. Thus e_1 is linked to none of the e_i, $2 \leqslant i \leqslant n$, which implies, by Proposition 1.10.14, that FGe_1 is the principal block. Since 1_G is the only irreducible module belonging to FGe_1, we conclude from Proposition 8.10(ii) that G is p-nilpotent.

Setting $N = O_{p'}(G)$ we have $G = NP$. Write

$$FN = FNv_1 \oplus \ldots \oplus FNv_s$$

where $v_1, \ldots, v_s$ are mutually orthogonal primitive idempotents of FN. Then

$$FG = FGv_1 \oplus \ldots \oplus FGv_s$$

and, by Proposition 2.4.3(i) and Theorem 4.23, each FGv_i is indecomposable. Hence, by the Krull-Scmidt theorem, $s = n$ and $FGv_i \cong FGe_i$, $1 \leqslant i \leqslant n$, after possible renumbering of the FGv_i. Since $FGv_i \cong (FNv_i)^G$ is irreducible for all $i \in \{2,\ldots,n\}$, by virtue of (3), it follows that

$$V^G \text{ is irreducible for any irreducible } FN\text{-module } V \not\cong 1_N$$

Hence, by Theorem 2.4.10, $(V^G)_N$ contains V with multiplicity 1. Thus, by Clifford theorem, for each $h \neq 1$ in P, $V \not\cong {}^hV$. By Lemma 7.6, we conclude that each $h \neq 1$ in P fixes no conjugacy class of N distinct from 1. Hence for any $1 \neq x \in N$, $C_G(x) \subseteq N$.

Suppose that $g \in G - N$. Then $g^{p^\lambda} \in N$ for some positive integer λ; but $gg^{p^\lambda}g^{-1} = g^{p^\lambda}$ and therefore g is in the centralizer of the p-regular element g^{p^λ}, which is impossible unless $g^{p^\lambda} = 1$. Hence g is an element of some Sylow p-subgroup. Therefore an element either lies in N or in a Sylow p-subgroup of G. Hence the number of elements, excluding 1, in Sylow p-subgroups is exactly $m(p^a-1)$, where $|G| = p^a m$, $(p,m) = 1$. The latter implies that $P \cap P^g = 1$ for all $g \in G - P$. Hence, by Lemma 7.2, G is a Frobenius group with complement P. ■

As a preparation for the proof of the final result, we record the following fact.

8.12. LEMMA. Let K be a finite field. Then each element k of K can be written in the form $k = x^2+y^2$ for some $x,y \in K$.

Proof. The result is trivial if $\mathrm{char}K = 2$, since in this case each element of K is a square. Now assume that $\mathrm{char}K \neq 2$. Set $L = \{a^2 | a \in K^*\}$. Then L is a proper subgroup of K^* and since K^* is cyclic, $K^* = L \cup kL$ for some $k \in K$. If $k\ell_o = x^2+y^2$ for some $\ell_o \in L$ and some $x,y \in K$, then writing any $\ell \in L$ in the form $\ell = c^2\ell_o$, $c \in K^*$, we obtain

$$k\ell = k\ell_o c^2 = (cx)^2 + (cy)^2$$

Hence it suffices to prove that at least one element of the form x^2+y^2 is a nonsquare. Assume, by way of contradiction, that for all $x,y \in K$, x^2+y^2 is a square. Then $K_1 = \{0\} \cup L$ is a subfield of K. If $|K| = q$, then we have $|K_1| = \frac{q+1}{2}$. But $\frac{q+1}{2} \nmid q$, a contridiction. ■

8.13. THEOREM (Brauer and Nesbitt (1941)). Let $G = SL(2,p)$ and let F be an arbitrary field of characteristic p. Then

$$\dim_F J(FG) = \frac{p(p+1)(4p-7)}{6}$$

Proof. By Corollary 1.18, we may assume that F is algebraically closed. For the sake of clarity, we divide the proof into three steps.

Step 1. Here we prove that the number of p-regular classes of G is equal to p.

We denote by $\mathbb{F}_p$ the field of p elements. It suffices to show that every element of $\mathbb{F}_p$ is the trace of some p-regular element in G and that all p-regular elements of G with the same trace are conjugate. This is so since conjugate elements of G have the same trace.

It is clear that there exist p-regular elements in G with trace 2 and -2, namely I and $-I$, where I is the identity matrix. If $\lambda \in \mathbb{F}_p$ and $\lambda \neq \pm 2$, put

$$\Gamma(\lambda) = \begin{pmatrix} 0 & 1 \\ -1 & \lambda \end{pmatrix}$$

Then $\Gamma(\lambda) \in G$ and $tr(\Gamma(\lambda)) = \lambda$. We show that $\Gamma(\lambda)$ is a p-regular element. Assume the contrary. Then the p-component $\Gamma(\lambda)_p$ of $\Gamma(\lambda)$ is conjugate to an element

$$A = \begin{pmatrix} 1 & a \\ 0 & 1 \end{pmatrix}$$

for some $a \neq 0$, and $\Gamma(\lambda)$ is conjugate to an element in the centralizer of A. However the centralizer of A in G consists of all elements of the form

$$\pm \begin{pmatrix} 1 & \alpha \\ 0 & 1 \end{pmatrix}$$

and all these elements have trace ± 2, contrary to the assumption $\lambda \neq \pm 2$. Thus $\Gamma(\lambda)$ is a p-regular element.

Let C be a p-regular element of G with trace λ. Then

$$f(X) = X^2 - \lambda X + 1$$

is the characteristic polynomial of C. If $\lambda = \pm 2$, $f(X) = (X \pm 1)^2$ and C has the eigenvalues 1,1 or -1,-1. Thus C is conjugate in $GL(2,p)$ to a matrix of the form $\pm A$ (A as above), and $C^{2p} = I$. Since C is p-regular, it follows easily that $C = \pm I$. If $\lambda \neq \pm 2$, then C is not a scalar multiple of I, so there is a vector v_1 such that v_1 and $v_2 = v_1 C$ are linearly independent. Invoking the Cayley-Hamilton theorem, we derive

$$0 = f(C) = C^2 - \lambda C + I$$

Hence

$$v_2 C = v_1 C^2 = -v_1 + \lambda v_2$$

and therefore C is conjugate in $GL(2,p)$ to $\Gamma(\lambda)$.

We are now left to verify that C is conjugate to $\Gamma(\lambda)$ in G. To this end, write

$$D^{-1} C D = \Gamma(\lambda)$$

for some $D \in GL(2,p)$. We wish to find $E \in C_{GL(2,p)}(\Gamma(\lambda))$ such that $\det DE = 1$. if accomplished, then

$$(DE)^{-1}C(DE) = E^{-1}\Gamma(\lambda)E = \Gamma(\lambda)$$

and $DE \in G$. Thus we need only verify that the determinants of $C_{GL(2,p)}(\Gamma(\lambda))$ take all values in $\mathbb{F}_p^*$.

An elementary calculation shows that

$$C_{GL(2,p)}(\Gamma(\lambda)) = \left\{ \begin{pmatrix} x & y \\ -y & x+\lambda y \end{pmatrix} \middle| \; x,y \in \mathbb{F}_p, \; x^2+\lambda xy+y^2 \neq 0 \right\}$$

Consequently we have to prove that the quadratic form $x^2+\lambda xy+y^2$ represents all values in $\mathbb{F}_p$. This is obvious for $p = 2$. If $p \neq 2$, then

$$x^2 + \lambda xy + y^2 = (x + \tfrac{1}{2}\lambda y)^2 + (1 - \tfrac{1}{4}\lambda^2)y^2$$

If $1 - \frac{1}{4}\lambda^2$ is not a square, it is trivial that this form represents all values in $\mathbb{F}_p$. Otherwise, $1 - \frac{1}{4}\lambda^2$ is a nonzero square since $\lambda \neq \pm 2$. Hence the form is equivalent to x^2+y^2 and represents all values, by virtue of Lemma 8.12.

Step 2. Here we construct irreducible FG-modules V_m $(0 \leqslant m \leqslant p-1)$ *for which*

$$\dim_F V_m = m + 1$$

It will then follow, from Step 1 and Theorem 2.3.2, that these V_m are all non-isomorphic irreducible FG-modules.

Let V_m denote the F-space of homogeneous polynomials of degree m in the independent variables x and y. If $A = (a_{ij}) \in G$, we put

$$(x^i y^{m-i})A = (a_{11}x + a_{12}y)^i (a_{21}x + a_{22}y)^{m-i}$$

Then V_m becomes a (right) FG-module of F-dimension $m+1$. We next prove that V_m is irreducible for $0 \leqslant m \leqslant p-1$. So assume that W is nonzero submodule of V_m and choose

$$0 \neq f = \sum_{j=0}^{n} a_j x^j y^{m-j} \in W$$

where $a_n \neq 0$ and $n \leqslant m$. Given $\mu \in \mathbb{F}_p$, we put

$$S(\mu) = \begin{pmatrix} 1 & \mu \\ 0 & 1 \end{pmatrix} \text{ and } T(\mu) = \begin{pmatrix} 1 & 0 \\ \mu & 1 \end{pmatrix}$$

Then there exist polynomials f_j, $0 \leq j \leq n$, with $f_0 = f$ and $f_n = a_n y^m$ such that

$$fS(\mu) = \sum_{j=0}^{n} a_j (x+\mu y)^j y^{m-j} = \sum_{j=0}^{n} f_j(x,y)\mu^j \in W$$

Note that

$$\sum_{\mu=1}^{p-1} \mu^{-i}(f(S(\mu))) \in W$$

Bearing in mind that in $\mathbb{F}_p$

$$\sum_{\mu=1}^{p-1} \mu^i = \begin{cases} -1 \text{ if } p-1 \text{ divides } i \\ 0 \quad \text{otherwise} , \end{cases}$$

we derive

$$\begin{aligned}\sum_{\mu=1}^{p-1} \mu^{-i}(f(S(\mu))) &= \sum_{\mu=1}^{p-1} \mu^{-i} \sum_{j=0}^{n} f_j \mu^j \\ &= \sum_{j=0}^{n} f_j \sum_{\mu=1}^{p-1} \mu^{j-i} \\ &= - \sum_{p-1 | j-i} f_j \\ &= \begin{cases} -f_i & \text{for } 1 \leq i \leq p-2,\ 1 \leq i \leq n \\ -f_0 - f_{p-1} & \text{for } i = 0, p-1 = n = m \end{cases}\end{aligned}$$

Because $f_0 = f \in W$, we infer that $f_i \in W$ for all $i \in \{1,\dots,n\}$. In particular, since $f_n = a_n y^m$, we have $y^m \in W$. A similar argument shows that W contains the element

$$\begin{aligned}\sum_{\mu=1}^{p-1} \mu^{-j}(y^m T(\mu)) &= \sum_{\mu=1}^{p-1} \mu^{-j}(\mu x+y)^m \\ &= \begin{cases} \binom{m}{j} & \text{for } 0 \leq j \leq m < p-1 \\ & \text{and } 0 < j < p-1 = m \\ x^{p-1}+y^{p-1} & \text{for } m = p-1, j=0 \end{cases}\end{aligned}$$

But $y^{p-1} \in W$ has already been shown in the case $m = p-1$. Hence $x^j y^{m-j} \in W$ for all $j \in \{0,1,\dots,m\}$. The conclusion is that $W = V_m$ and therefore V_m is irreducible for all $m \in \{0,1,\dots,p-1\}$.

Step 3. Completion of the proof.

By Step 2, the dimensions of irreducible FG-modules are $1,2,\dots,p$. Invoking

Corollary 1.10.17, we deduce that

$$\dim_F J(FG) = \dim_F FG - \sum_{i=1}^{p} i^2 = p(p^2-1) - \frac{p}{6}(p+1)(2p+1)$$
$$= \frac{p(p+1)(4p-7)}{6}$$

thus completing the proof. ■

9. A CHARACTERIZATION OF $\dim_F J(Z(FG))$ AND ITS APPLICATIONS.

Throughout this section, G denotes a finite group and F an arbitrary field of characteristic $p > 0$. As usual, we write $C\ell(G)$ and $\exp(G)$ for the set of conjugacy classes of G and the exponent of G, respectively. Let $\mathbb{N}_0$ be the set of all nonnegative integers. Given a subset X of G and $n \in \mathbb{N}_0$ we put

FX = the F-linear span of X

$$T_n(FG) = \{x \in FG \mid x^{p^n} \in [FG,FG]\}$$

$$X^+ = \sum_{x\in X} x \quad \text{(if } X \text{ is empty, then by definition } X^+ = 0)$$

$$X^{p^n} = \{x^{p^n} \mid x \in X\}$$

$$X^{1/p^n} = \{g \in G \mid g^{p^n} \in X\}$$

We denote by c the sum of all p-elements of G (including 1).

Recall that a defect group of $C \in C\ell(G)$ is a Sylow p-subgroup of $C_G(g)$ with $g \in C$. We write $\mathrm{Def}_p(C)$ for the set of all defect groups of C. Finally, $G_{p'}$ denotes the set of all p-regular elements of G. Our aim is to characterize the number $\dim_F J(Z(FG))$ in terms of the rank of a matrix associated with G. A number of important applications is also provided. As a first step, we describe $T_n(FG)$ and $T_n(FG)^{\perp}$ explicitely.

9.1. LEMMA. For any $n \in \mathbb{N}_0$, the following properties hold

(i) $T_n(FG)$ is a $Z(FG)$-module and

$$T_n(FG) = \{\sum x_g g \in FG \mid \sum_{g\in C^{1/p^n}} x_g = 0 \text{ for all } C \in C\ell(G)\}$$

(ii) $T_n(FG)^{\perp}$ is an ideal of $Z(FG)$ and

$$T_n(FG)^{\perp} = \sum_{C\in C\ell(G)} F(C^{1/p^n})^+$$

Proof. (i) Assume that $y \in T_n(FG)$ and $z \in Z(FG)$. Then

$$(yz)^{p^n} = y^{p^n} z^{p^n} \in [FG,FG]$$

and so $yz \in T_n(FG)$, proving that $T_n(FG)$ is a $Z(FG)$-module.

Now fix $x = \Sigma x_g g \in FG$. Because

$$x^{p^n} \equiv \Sigma x_g^{p^n} g^{p^n} \qquad (\text{mod}[FG,FG])$$

it follows that $x \in T_n(FG)$ if and only if $\Sigma x_g^{p^n} g^{p^n} \in [FG,FG]$. The required assertion is now a consequence of the equality $\Sigma x_g^{p^n} = (\Sigma x_g)^{p^n}$ and Lemma 2.2.2.

(ii) Owing to Lemma 3.9, $Z(FG) = [FG,FG]^{\perp}$. Since $T_n(FG) \supseteq [FG,FG]$, we have

$$T_n(FG)^{\perp} \subseteq [FG,FG]^{\perp} = Z(FG)$$

But, by (i), $T_n(FG)$ is a $Z(FG)$-module, so $T_n(FG)^{\perp}$ is an ideal of $Z(FG)$.

Let $C \in C\ell(G)$ and put $X = C^{1/p^n}$. Then $X^{-1} = (C^{-1})^{1/p^n}$ and, by (i), for $x \in T_n(FG)$, we have

$$tr(xX^+) = \sum_{g\in X^{-1}} x_g = 0$$

proving that $X^+ \subseteq T_n(FG)^{\perp}$.

To prove the opposite containment, suppose that $x = \Sigma x_g g \in T_n(FG)^{\perp}$. Then $tr(xy) = 0$ for all $y \in T_n(FG)$. In particular, if $a,b \in C^{1/p^n}$ then $a^{-1},b^{-1} \in (C^{-1})^{1/p^n}$ and thus, by (i), $tr(x(a^{-1}-b^{-1})) = 0$. Hence

$$x_a = tr(xa^{-1}) = tr(xb^{-1}) = x_b$$

Finally assume that $g \notin C^{1/p^n}$ for all $C \in C\ell(G)$. Then $g^{-1} \notin C^{1/p^n}$ for all $C \in C\ell(G)$ and hence, by (i), $g^{-1} \in T_n(FG)$. Thus

$$x_g = tr(xg^{-1}) = 0$$

as asserted. ∎

9.2. LEMMA. Let D be a defect group of $C \in C\ell(G)$, let $x \in FG$ and let

$q = p^n$ for some $n \in \mathbb{N}_0$.

(i) $tr(x^q C^+) = [tr(x(C^{1/q})^+)]^q$

(ii) Under the assumption that $q \geqslant \exp(D)$, the following hold:

(a) If C is p-singular, then $C^{1/q} = \emptyset$ and $tr(x^q C^+) = 0$

(b) If C is p-regular, then $C^{1/q}$ is a p-regular section containing C^i, where i is the inverse of q modulo the order of the elements in C. Moreover,

$$tr(x^q C^+) = tr(x^q((C^q)^{1/q})^+) = [tr(x(C^{1/q})^+)]^q$$

Proof. (i) Write $x = \sum_{g\in G} x_g g$, $x_g \in F$, so that

$$x^q \equiv \sum_{g\in G} x_g^q g^q \pmod{[FG,FG]}$$

Since $[FG,FG]$ is a $Z(FG)$-module, we have

$$x^q C^+ \equiv \sum_{g\in G} x_g^q g^q C^+ \pmod{[FG\ FG]}$$

Taking into account that $tr([FG,FG]) = 0$, it follows that

$$\begin{aligned} tr(x^q C^+) &= tr(\sum_{g\in G} x_g^q g^q C^+) \\ &= \sum_{g^{-q}\in C} x_g^q = (\sum_{g^{-1}\in C^{1/q}} x_g)^q \\ &= [tr(x(C^{1/q})^+)]^q, \end{aligned}$$

as desired.

(ii) (a) Suppose that $C^{1/q} \neq \emptyset$. Then there exists $g \in C^{1/q}$ so that $g^q \in C$ and $g_p \in C_G(g^q)$. Since $q \geqslant \exp(D)$, we must have $(g^q)_p = (g_p)^q = 1$, proving that C is p-regular. The desired conclusion now follows by virtue of (i).

(b) Suppose that C is p-regular. Then $C^i \subseteq C^{1/q}$ since for any $x \in C$, $(x^i)^q = x$. Assume that $g \in C^{1/q}$. Then $g^q = (g_p)^q (g_{p'})^q = (g_{p'})^q \in C$ and hence for any $x,y \in C^{1/q}$, $(x_{p'})^q$ and $(y_{p'})^q$ are conjugate. Thus $x_{p'}$ and $y_{p'}$ are also conjugate. Conversely, suppose that $x \in G$ is such that $x_{p'}$ is conjugate to $g_{p'}$, say $x_{p'} = y^{-1} g_{p'} y$. Then

$$(x_{p'})^q = y^{-1}(g_{p'})^q y = y^{-1} g^q y \in C$$

and D is a Sylow p-subgroup of $C_G(g_{p'})$. Thus $(x_p)^q = 1$ and so $x^q \in C$, proving that $C^{1/q}$ is a p-regular section.

Since C is p-regular, C^q is a p-regular conjugacy class of G and so, by the foregoing, $(C^q)^{1/q}$ is a p-regular section. Suppose that $X \in C\ell(G), X \neq C$ and $X \in (C^q)^{1/q}$. Then X is a p-singular class and $\delta(X)$ is conjugate to a subgroup of D. Thus, by (a), $tr(x^q X) = 0$ and therefore

$$tr(x^q((C^q)^{1/q})^+) = tr(x^q C^+)$$

The required assertion now follows by appealing to (i). ■

For any $C \in C\ell(G)$, define $C(p) \in C\ell(G)$ by

$$C(p) = \{g_p | g \in C\}$$

We are now ready to prove the following important result.

9.3. THEOREM (Külshammer (1982)). Let C be a conjugacy class of G with D as a defect group and let $n \in \mathbb{N}_0$ be such that $q = p^n \geqslant \exp(D)$. Denote by $C_1, \ldots, C_m$ all p-regular conjugacy classes of G.

(i) If $z \in Z(FG)$ then $C(p) = K(p)$ for any $K \in C\ell(G)$ with $K \subseteq \text{Supp} z^q C^+$

(ii) If C is p-regular, then for any $z \in Z(FG)$ there exist $\lambda_1, \ldots, \lambda_m \in F$ such that

(a) $z(C^{1/q})^+ = \sum_{i=1}^{m} \lambda_i (C^{1/q})^+$

(b) $z^q((C^q)^{1/q})^+ = \sum_{i=1}^{m} \lambda_i^q ((C_i^q)^{1/q})^+$

(c) $z^q C^+ = \sum_{i=1}^{m} \lambda_i^q C_i^+$

Proof. (i) Let G be a counter-example of minimal order and write

$$z^q C^+ = \sum_{g \in G} \lambda_g g \qquad (\lambda_g \in F)$$

Then there exists $g \in G$ with $\lambda_g \neq 0$ and such that for $h = g_{p'}$, $h \notin C(p)$. Let $\pi : Z(FG) \longrightarrow Z(FC_G(h))$ be the Brauer homomorphism. Then

$$\lambda_g = tr(g^{-1}\pi(z^g C^+)) = tr(g^{-1}\pi(z)^q \pi(C^+)) \neq 0 \qquad (1)$$

and hence $C \cap C_G(h) \neq \emptyset$. Let $C_1,\ldots,C_r$ be all conjugacy classes of $C_G(h)$ contained in C and for each $i \in \{1,\ldots,r\}$, let D_i be a defect group of C_i. Since each D_i is G-conjugate to a subgroup of $D, q \geqslant \exp(D_i)$ for all $i \in \{1,\ldots,r\}$. By (1), we may write

$$\lambda_g = \sum_{i=1}^{r} tr(g^{-1}\pi(z)^q C_i^+)$$

so that $tr(g^{-1}\pi(z)^q C_i^+) = 0$ for a suitable $i \in \{1,\ldots,r\}$. Because $h \notin C_i(p)$, the group $C_G(h)$ is also a counter example to (i). Hence $h \in Z(G)$ and so for all $K \in C\ell(G), Kh^{-1} \in C\ell(G)$, Ch^{-1} is p-singular with D as a defect group and gh^{-1} is p-regular. Since hg^{-1} is p-regular, we have $y^q = hg^{-1}$ for some $y \in G$. Applying Lemma 9.1(ii), we therefore deduce that

$$\lambda_g = tr(y^q z^q C^+ h^{-1}) = 0,$$

a contradiction.

(ii) Denote by $C_{m+1}, C_{m+2},\ldots,C_r$ all p-singular classes of G and let $z \in Z(FG)$. Owing to Lemma 9.1(ii), there exist $\lambda_1,\ldots,\lambda_r \in F$ such that

$$z(C^{1/q})^+ = \sum_{i=1}^{r} \lambda_i (C_i^{1/q})^+$$

We may harmlessly assume that each $C_i^{1/q}$ is nonempty. Fix $i \in \{1,\ldots,r\}$ and $g \in C_i^{1/q}$. Then $\lambda_i = tr(g^{-1}z(C^{1/q})^+)$ and hence, by Lemma 9.2(ii)b,

$$\lambda_i^q = tr(g^{-g} z^q C^+).$$

But, by (i) $\mathrm{Supp} z^q C^+$ is a union of some p-regular classes. Thus $\lambda_i = 0$ for all $i \in \{m+1,\ldots,r\}$, proving (a) and (c).

Since C^q is a p-regular class of G with defect group D, it follows from (c) that

$$z^q((C^q)^{1/q})^+ = \sum_{i=1}^{m} \mu_i[(C_i^q)^{1/q}]^+ \tag{2}$$

with $\mu_i \in F$. Fix $i \in \{1,\ldots,m\}$ and $g \in C_i^{1/q}$. Then $g^q \in C_i \subseteq (C_i^q)^{1/q}$ and so, by (2), $\mu_i = tr(g^{-q}z^q((C^q)^{1/q})^+)$. Applying Lemma 9.2(ii)(b), we derive

$$\lambda_i^q = [tr(g^{-1}z(C^{1/q})^+)]^q = tr(g^{-q}z^q((C^q)^{1/q})^+) = \mu_i$$

as asserted. ∎

For future use we next quote the following standard result.

9.4. THEOREM (Wedderburn-Malcev). Let A be a finite-dimensional algebra over a field K. If $A/J(A)$ is separable, then

(i) There exists a semisimple subalgebra $\Delta(A)$ of A such that

$$A = \Delta(A) \oplus J(A) \qquad \text{(direct sum of } K\text{-spaces)}$$

(ii) If $\Delta_1(A)$ and $\Delta_2(A)$ are subalgebras of A with $A = \Delta_i(A) \oplus J(A), i = 1,2$, then

$$(1-x)\Delta_2(A)(1-x)^{-1} = \Delta_1(A) .$$

for some $x \in J(A)$.

Proof. See Curtis and Reiner (1962). ∎

The following useful observation will be derived with the aid of the result above.

9.5. LEMMA. (i) $\dim_F Z(FG)/J(Z(FG)) = \dim_F Z(FG)c$. In particular, if F is a splitting field for $Z(FG)$, then $\dim_F Z(FG)c$ is equal to the number of blocks of FG.

(ii) There exists a unique F-subalgebra $\Delta(Z(FG))$ of $Z(FG)$ such that

$$Z(FG) = \Delta(Z(FG)) \oplus J(Z(FG)) \qquad \text{(direct sum of } F\text{-spaces)}$$

Proof. (i) Let $\phi : Z(FG) \longrightarrow Z(FG)c$ be defined by $\phi(z) = zc$ for all $z \in Z(FG)$. Then ϕ is a surjective F-linear map with $\text{Ker}\phi = \ell(c) \cap Z(FG)$. Now apply Theorem 6.2.

(ii) Owing to Corollary 1.18, $Z(FG)/J(Z(FG))$ is a separable F-algebra. The desired conclusion now follows by appealing to Theorem 9.4. ∎

From now on, the maps:

$$\pi : FG \rightarrow FG_p, \quad \text{and} \quad \delta : Z(FG) \longrightarrow \Delta(Z(FG))$$

designate the natural projections.

The following result illustrates how the algebra $\Delta(Z(FG))$ can be brought into argument.

9.6. LEMMA. (i) $\delta(z) = \pi(zc)$ for all $z \in Z(FG)$

(ii) $\Delta(Z(FG))$ contains all central idempotents of FG.

Proof. (i) Invoking Theorem 9.3(ii), (b), (c) for a sufficiently large n and $C = \{1\}$, we obtain

$$z^{p^n} = \pi(z^{p^n}c) \quad \text{for all} \quad z \in Z(FG) \tag{3}$$

Since $\Delta(Z(FG))$ is a commutative separable F-algebra, it must be a direct sum of separable field extensions of F. Assume that E/F is one of them. Then we may write $E = F(y)$ for some $y \in \Delta(Z(FG))$. Moreover, the separability of E/F ensures that $E = F(y^{p^n})$ and so, by (3), that $\pi(zc) = z$ for all $z \in E$. Hence, $\pi(zc) = z$ for all $z \in \Delta(Z(FG))$. Applying Theorem 6.2, we infer that $\pi(zc) = \delta(z)$ for all $z \in Z(FG)$, as asserted.

(ii) Let e denote a central idempotent of FG. Invoking (3) together with (i), we derive

$$e = \pi(ec) = \delta(e)$$

as required. ∎

For any p-subgroup D of G, let $I_D(G)$ be the F-linear span of all C^+, $C \in C\ell(G)$, with $\delta(C) \underset{G}{\subseteq} D$. Note that by Lemma 2.2.5, $I_D(G)$ is an ideal of $Z(FG)$.

9.7. LEMMA. Let D be a p-subgroup of G and let $e_1,\ldots,e_n$ be all block idempotents of FG with a subgroup of D as a defect group.

(i) $\{e_1,\ldots,e_n\}$ is the set of all primitive idempotents of the ideal

$$I_D(G) \cap \Delta(Z(FG)) \quad \text{of} \quad \Delta(Z(FG))$$

(ii) $\delta(I_D(G)) = I_D(G) \cap \Delta(Z(FG)) = \sum_{i=1}^{n} \Delta(Z(FG))e_i$

Proof. (i) Assume that e is a primitive idempotent of $I_D(G) \cap \Delta(Z(FG))$.

By Lemma 6.1, e is a primitive idempotent of $\Delta(Z(FG))$. Therefore, by Lemma 9.6(ii), e is a block idempotent of FG. Because $e \in I_D(G)$, we also have $e \in \{e_1,\ldots,e_n\}$.

Conversely, by hypothesis, each $e_i \in I_D(G)$ and so, by Lemma 9.6(ii), each e_i lies in $I_D(G) \cap \Delta(Z(FG))$. Again, applying Lemma 6.1, we conclude that each e_i is in fact a primitive idempotent of $I_D(G) \cap \Delta(Z(FG))$.

(ii) Let $I = I_D(G) \cap \Delta(Z(FG))$. Bearing in mind that $\Delta(Z(FG))$ is a direct sum of fields and that I is an ideal of $\Delta(Z(FG))$, the second equality is true by virtue of (i). Since $\pi(I_D(G)) \subseteq I_D(G)$, Lemma 6.1(i) implies that

$$\begin{aligned}\delta(I_D(G)) &= \pi(I_D(G)c)\\ &\subseteq \pi(I_D(G)) \cap \Delta(Z(FG))\\ &\subseteq I\end{aligned}$$

On the other hand, we have

$$\Delta(Z(FG))e_i = \delta(\Delta(Z(FG)e_i) \subseteq \delta(I_D(G)) \qquad (1 \leqslant i \leqslant n)$$

proving that $I \subseteq \delta(I_D(G))$. ∎

Let D denote a p-subgroup of G. We write $\hat{I}_D(G)$ for the F-linear span of all C^+ with $C \in C\ell(G)$ such that D contains properly a defect group of C. Because $\hat{I}_D(G)$ is the sum of all $I_L(G)$ with $L \subset D$, we see that $\hat{I}_D(G)$ is an ideal of $Z(FG)$. Denote by $V_D(G)$ the F-linear span of all C^+, where C is a p-regular class of G with defect group D.

9.8. LEMMA. Let D be a p-subgroup of G and let e_D be the sum of all block idempotents of FG with defect group D. Then

$$\dim_F[Z(FG)e_D/J(Z(FG)e_D] = \dim_F\{[\delta(V_D(G)) + \hat{I}_D(G)]/\hat{I}_D(G)\}$$

(By convention, $e_D = 0$ if there are no blocks of FG with defect group D).

Proof. Let X and Y be the left and right hand side spaces, respectively. By Lemma 9.6(ii), $e_D \in \Delta(Z(FG))$ and thus

$$Z(FG)e_D = \Delta(Z(FG))e_D \oplus J(Z(FG))e_D$$

proving that $\dim_F X = \dim_F \Delta(Z(FG))e_D$. On the other hand, by Lemma 9.7,

$$\delta(I_D(G)) = \Delta(Z(FG))e_D \oplus [\delta(I_D(G)) \cap \hat{I}_D(G)]$$

Bearing in mind that

$$[\delta(I_D(G)) + \hat{I}_D(G)]/\hat{I}_D(G) \cong \delta(I_D(G))/[\delta(I_D(G)) \cap \hat{I}_D(G)],$$

we infer that

$$\dim_F X = \dim_F\{[\delta(I_D(G)) + \hat{I}_D(G)]/\hat{I}_D(G)\} \tag{4}$$

Invoking the argument employed in the proof of Lemma 9.7(ii), we have

$$\delta(\hat{I}_D(G)) \subseteq \hat{I}_D(G).$$

Thus the right hand side of (4) is the F-linear span of the elements $\delta(C^+) + \hat{I}_D(G)$, where $C \in C\ell(G)$ is such that D is a defect group of C. Owing to Lemmas 9.6(i) and 9.7(ii), $\delta(C^+) \in FG_{p'} \cap I_D(G)$ and hence

$$\delta(C^+) \equiv \sum_i \lambda_i \delta(C_i^+) \pmod{\hat{I}_D(G))}$$

for some p-regular classes C_i with defect group D and some $\lambda_i \in F$. This implies that

$$[\delta(I_D(G) + \hat{I}_D(G)]/\hat{I}_D(G) = Y$$

and thus

$$\dim_F X = \dim_F Y$$

Finally note that if there are no blocks of FG with defect group D, then either $\delta(I_D(G)) = 0$ or $\delta(I_D(G)) = \delta(I_D(G))e$ for some block idempotent e of FG with $e \in \hat{I}_D(G)$. In both cases we have $\dim_F Y = 0$, as required. ■

9.9. LEMMA. Let Q be a p-subgroup of G, let S be a p-regular section of G and let C be a unique p-regular class of G contained in S.

(i) The number r of Sylow p-subgroups of G containing Q is congruent to 1 modulo p

(ii) $S^+ = \sum_{g\in C} \sum_{P\in \mathrm{Syl}_p(C_G(g))} gP^+$

Proof. (i) The group Q acts on the set $\mathrm{Syl}_p(G)$ by conjugation. Given $P \in \mathrm{Syl}_p(G), P \supseteq Q$ if and only if $\{P\}$ is a Q-orbit. Since the size of each nontrivial Q-orbit is divisible by p, we have

$$|\mathrm{Syl}_p(G)| \equiv r(\mathrm{mod}\ p)$$

On the other hand, by Sylow theorem $|\mathrm{Syl}_p(G)| \equiv 1(\mathrm{mod}\ p)$. Thus $r \equiv 1\ (\mathrm{mod}\ p)$, as asserted.

(ii) Fix $x \in S$ and denote by r_x the number of Sylow p-subgroups of $C_G(x_{p'})$ containing $<x_p>$. Then r_x is the coefficient of x in the right hand side of (ii). By (i), $r_x \equiv 1(\mathrm{mod}\ p)$ and so the above coefficient is 1. Since for any $g \in C$ and $y \in P \in \mathrm{Syl}_p(C_G(g)), gy \in S$ the required assertion follows. ■

9.10. LEMMA. Let S be a p-regular section of G and let C_0 be a unique p-regular class of G contained in S. Fix $s \in C_0$, $P \in \mathrm{Syl}_p(C_G(s))$ and, for given $C,K \in C\ell(G)$, put

$$\lambda(C,K) = |C|_p^{-1}|C_0|\cdot|\{(c,k) \in C\times K | c^{-1}ks \in P\}|$$

and write

$$K^+S^+ = \sum_{C\in C\ell(G)} \alpha_{KS}^C C^+ \qquad (\alpha_{KS}^C \in F)$$

Then $\lambda(C,K)$ is an integer such that

$$|C|_{p'}\alpha_{KS}^C = \lambda(C,K)\cdot 1_F$$

Proof. Let $c_0 \in C \in C\ell(G)$. Invoking Lemma 9.9(ii), we derive

$$\alpha_{KS}^C = |\{(k,s',Q,x) | k \in K, s' \in C_0, x \in Q \in \mathrm{Syl}_p(C_G(s')), ks'x = c_0\}|\cdot 1_F$$

On the other hand, an easy calculation shows that

$$|\{(k,s',Q,x)\,|\,k \in K, s' \in C_o, x \in Q \in \mathrm{Syl}_p(C_G(s')), ks'x = c_o\}|$$

$$= |\{(k,s',Q,x)\,|\,k \in K, s' \in C_o, x \in Q \in \mathrm{Syl}_p(C_G(s')), ks'x \in C\}|/|C|$$

$$= |\{(c,k,s',Q)\,|\,c \in C, k \in K, s' \in C_o, Q \in \mathrm{Syl}_p(C_G(s'), c^{-1}ks' \in Q\}|/|C|$$

$$= |C_o||\mathrm{Syl}_p(C_G(s))||\{(c,k) \in C \times K | c^{-1}ks \in P\}|/|C|$$

Because the number above is an integer and $|\mathrm{Syl}_p(C_G(s))| \equiv 1 \pmod{p}$, we conclude that $\lambda(C,K)$ is an integer such that

$$|C|_{p'}\alpha^{C}_{ks} = \lambda(C,K)\cdot 1_F$$

as required. ■

We have now come to the demonstration for which this section has been developed. The following result for the case $D \triangleleft G$ is due to Robinson (1983).

9.11. THEOREM (Kulshammer (1984b)). Let $C_1, C_2, \ldots, C_m$ be all p-regular classes of G having the same defect group D, let e_D be the sum of all block idempotents of FG with defect group D and let P be a Sylow p-subgroup of G. Define the matrix $M_D = (m_{ij}) \in M_m(\mathbb{F}_p)$ as follows:

$$m_{ij} = \{|\{(x,y) \in C_i \times C_j | x^{-1}y \in P\}|/p^{a-d}\}\cdot 1_F \qquad 1 \leq i,j \leq m$$

where p^a and p^d are the orders of P and D, respectively (by Lemma 9.10 applied for $S^+ = c, K = C_j$ and $C = C_i$, the number in parenthesis is an integer). Then

$$\dim_F Z(FG)e_D/J(Z(FG))e_D = \mathrm{rank}\, M_D$$

(By convention, $e_D = 0$ if there are no blocks of FG with defect group D)

Proof. Let $g_1, g_2, \ldots, g_r$ be a full set of (P,P)-double coset representatives in G and put

$$n_{ij} = |\{x \in g_iP \cap C_j | P \cap g_iPg_i^{-1} \in \mathrm{Syl}_p(C_G(x))\}|\cdot 1_F \qquad 1 \leq i \leq r, 1 \leq j \leq m$$

Then

$$N_D = (n_{ij}) \in M_{r\times m}(\mathbb{F}_p)$$

and we denote by N_D^t the transpose of N_D.

Fix $j \in \{1,\ldots,m\}$ and denote by α_{ij} the coefficient of C_i^+ in $C_j^+ c$. By Lemma 9.7(ii), $\delta(C_j^+) \in I_D(G)$ and, by Lemma 9.6(i), $\delta(C_j^+) = \pi(C_j^+ c)$. Thus

$$\delta(C_j^+) \equiv \sum_{i=1}^{m} \alpha_{ij} C_i^+ \pmod{I_D(G)} \tag{5}$$

Applying Lemma 9.10 for $S^+ = c, K = C_j$ and $C = C_i$, we also have

$$\begin{aligned} |C_i|_{p'} \alpha_{ij} &= \{|C_i|_p^{-1} |\{(x,y) \in C_i \times C_j \mid x^{-1}y \in P\}|\} \cdot 1_F \\ &= m_{ij} \\ &= \{|C_i|_p^{-1} |\bigcup_{gP \in G/P} (gP \cap C_i) \times (gP \cap C_j)|\} \cdot 1_F \end{aligned} \tag{6}$$

We now claim that the result will follow provided we show that

$$|C_i|_{p'} \alpha_{ij} = \sum_{k=1}^{r} n_{ki} n_{kj} \tag{7}$$

Indeed, put $M' = (\alpha_{ij}), v_j = \sum_{i=1}^{m} \alpha_{ij} C_i^+$ and $W = [\delta(V_D(G) + \hat{I}_D(G)]/\hat{I}_D(G)$. By (5), W is the F-linear span of $\{v_1 + \hat{I}_D(G), \ldots, v_m + \hat{I}_D(G)\}$. Denote by V the F-linear span of $\{v_1, \ldots, v_m\}$. Because $V \subseteq V_D(G)$ and $V_D(G) \cap \hat{I}_D(G) = 0$, we see that $\dim_F V = \dim_F W = \operatorname{rank} M'$. By (6) and (7), we also have

$$\operatorname{rank} M_D = \operatorname{rank} M' = \operatorname{rank} N_D^t N_D$$

Applying Lemma 9.8, we conclude that

$$\dim Z(FG)e_D / J(Z(FG)e_D) = \operatorname{rank} N_D^t N_D = \operatorname{rank} M_D \tag{8}$$

as claimed.

By the foregoing, we are left to verify (7). To this end, assume that $xP \subseteq PgP$ so that $xP = ygP$ for some $y \in P$. Then

$$\begin{aligned} |(gP \cap C_i) \times (gP \cap C_j)| &= |y(gP \cap C_i)y^{-1} \times y(gP \cap C_j)y^{-1}| \\ &= |(ygP \cap C_i) \times (ygP \cap C_j)| \end{aligned}$$

Applying (7) and the fact that PgP contains exactly $|P:P \cap gPg^{-1}|$ left cosets, we obtain

$$|C_i|_p\alpha_{ij} = \{|C_i|_p^{-1}\sum_{k=1}^{r}|P:P\cap g_kPg_k^{-1}||(g_kP\cap C_i)\times(g_kP\cap C_j)|\}\cdot 1_F \qquad (9)$$

Fix $k\in\{1,\dots,r\}$. Then the group $(P\cap g_kPg_k^{-1})\times(P\cap g_kPg_k^{-1})$ acts on

$$(g_kP\cap C_i)\times(g_kP\cap C_j)$$

by the rule

$$(a,b)^{(x,y)} = (x^{-1}ax, y^{-1}by) \qquad (a\in g_kP\cap C_i, b\in g_kP\cap C_j, x,y\in P\cap g_kPg_k^{-1})$$

since

$$x^{-1}ax\in g_kPg_k^{-1}(g_kP)P = g_kP \quad\text{and}\quad y^{-1}by\in g_kPg_k^{-1}(g_kP)P = g_kP$$

Note also that $(a,b)^{(x,y)} = (a,b)$ if and only if

$$x\in C_G(a)\cap P\cap g_kPg_k^{-1} = P\cap aPa^{-1}\cap C_G(a) = C_P(a)$$

and

$$y\in C_G(b)\cap P\cap g_kPg_k^{-1} = C_P(b)$$

Thus the size ℓ of the orbit of (a,b) is given by

$$\ell = |(P\cap g_kPg_k^{-1}):C_P(a)||(P\cap g_kPg_k^{-1}):C_P(b)|$$

But, if either $C_P(a)\notin \mathrm{Syl}_p(C_G(a))$ or $C_P(b)\neq P\cap g_kPg_k^{-1}$, we have

$$|P:(P\cap g_kPg_k^{-1})|\ell \equiv 0 \pmod{p|C_i|_p}$$

On the other hand, if $C_P(a)\in \mathrm{Syl}_p(C_G(a))$ and $C_P(b) = P\cap g_kPg_k^{-1}$, then

$$\begin{aligned}|P\cap g_kPg_k^{-1}| &= |C_P(b)| \leqslant |C_G(b)|_p\\ &= |C_G(a)|_p = |C_P(a)|\\ &\leqslant |P\cap g_kPg_k^{-1}|\end{aligned}$$

and hence

$$P \cap g_k P g_k^{-1} = C_P(a) \in \mathrm{Syl}_p(C_G(a))$$

$$P \cap g_k P g_k^{-1} = C_P(b) \in \mathrm{Syl}_p(C_G(b))$$

Thus in this case we have $\ell = 1$ and

$$|P:(P \cap g_k P g_k^{-1})|\ell = |C_i|_p$$

Applying (9), we therefore conclude that

$$|C_i|_p \alpha_{ij} = \sum_{k=1}^{r} n_{ki} n_{kj}$$

proving (7) and hence the result. ■

It is now an easy matter to provide a characterization of $\dim_F J(Z(FG))$.

9.12. COROLLARY. Let r be the number of conjugacy classes of G and let X be the set of all nonisomorphic defect groups of p-regular classes of G. For each $D \in X$, let the matrix $M_D \in M_m(\mathbb{F}_p)$ be defined as in Theorem 9.11. Then

$$\dim_F J(Z(FG)) = r - \sum_{D \in X} \mathrm{rank} M_D$$

Proof. Our choice of X ensures that it contains all nonisomorphic defect groups of blocks of FG. Hence

$$Z(FG) = \bigoplus_{D \in X} Z(FG)e_D$$

where e_D is the sum of all block idempotents with defect group D and $e_D = 0$ if there are no such idempotents. Accordingly,

$$Z(FG)/J(Z(FG)) \cong \bigoplus_{D \in X} Z(FG)e_D/J(Z(FG))e_D$$

Taking dimensions of both sides and applying Theorem 9.11, we derive

$$r - \dim_F J(Z(FG)) = \sum_{D \in X} \mathrm{rank} M_D ,$$

whence the result. ■

We close this section by providing some applications of Theorem 9.11.

9.13. COROLLARY (Green (1968)). Let D be a defect group of a block of FG. Then there exists a p-regular element $t \in C_G(D)$ and a Sylow p-subgroup Q of G such that

$$D = Q \cap tQt^{-1}$$

Proof. We keep the notation of Theorem 9.11. Since D is a defect group of a block of FG, it follows from (8) that $\text{rank}N_D^tN_D \neq 0$. Hence there exists a p-regular class C of G with defect group D, an element $g \in G$ and $d \in gP \cap C$ such that $P \cap gPg^{-1} \in \text{Syl}_p(C_G(d))$. Since $d \in gP$, we have $gPg^{-1} = dPd^{-1}$. Taking into account that D is a defect group of C, we conclude that

$$P \cap dPd^{-1} = xDx^{-1} \qquad \text{for some } x \in G$$

Setting $t = x^{-1}dx$ and $Q = x^{-1}Px$, it follows that $t \in C_G(D)$ and $D = Q \cap tQt^{-1}$, as asserted. ■

In order to provide further information about $\dim_F J(Z(FG)$ it is desirable to investigate the rank of the matrix M_D defined in Theorem 9.11. In general, this is a formidable task and no precise formula for the rank of M_D in terms of the group structure of G has been discovered so far. However, it is quite possible to provide some important useful information, which we present below.

A conjugacy class C of G is said to be *nilpotent* if C^+ is a nilpotent element of $Z(FG)$.

9.14. LEMMA. Let m_D be the number of nonnilpotent p-regular classes of G with defect group D and let ℓ_D be the number of conjugacy classes of G which are contained in $O_{p'}(G)$ and have defect group D. Then

$$\ell_D \leqslant \text{rank}M_D \leqslant m_D$$

Proof. Let $C_1, C_2, \ldots, C_m$ be all p-regular classes of G with defect group D. Then $[\delta(V_D(G)) + \hat{I}_D(G)]/\hat{I}_D(G)$ is the F-linear span of $\delta(C_i^+) + \hat{I}_D(G), 1 \leqslant i \leqslant m$. If C_i^+ is nilpotent, then $\delta(C_i^+)$ is a nilpotent element of $\Delta(Z(FG))$. But $\Delta(Z(FG))$ is semisimple, so $\delta(C_i^+) = 0$. Hence

$$\dim_F\{[\delta(V_D(G)) + \hat{I}_D(G)]/\hat{I}_D(G)\} \leqslant m_D$$

and therefore, by Lemma 9.8 and Theorem 9.11, $\text{rank}M_D \leqslant m_D$.

Let $C_1, C_2, \ldots, C_\ell, \ell = \ell_D$, be all conjugacy classes of G contained in $O_{p'}(G)$ and having defect group D. For each $i \in \{1,\ldots,\ell\}$, fix $y_i \in C_i$. Assume that x is conjugate to y_i, y is conjugate to y_j and $x^{-1}y \in P \in \text{Syl}_p(G)$. Since $x, y \in O_{p'}(G)$, $x^{-1}y$ is p-regular. Thus $x^{-1}y = 1$, so $x = y$. This proves that

$$|\{(x,y) \in C_i \times C_j \mid x^{-1}y \in P\}| = \begin{cases} 0 & \text{if } i \neq j \\ |G:C_G(y_i)| & \text{if } i=j \end{cases}$$

Hence $m_{ij} = 0$ for $i \neq j$. Now $m_{ii}, 1 \leqslant i \leqslant \ell$, is the residue (mod$p$) of

$$|G:C_G(y_i)|/|P:D|$$

so $m_{ii} \neq 0$ as $|D| = |C_G(y_i)|_p$. Hence $\text{rank}M_D \geqslant \ell$, as asserted. ∎

We are now ready to provide the following application of Corollary 9.12.

9.15. THEOREM. Define the numbers r, r_1, and r_2 as follows:

r = the number of conjugacy classes of G

r_1 = the number of nonnilpotent p-regular classes of G

r_2 = the number of conjugacy classes of G contained in $O_{p'}(G)$

Then $r - r_1 \leqslant \dim_F J(Z(FG)) \leqslant r - r_2$.

Proof. We use the notation of Corollary 9.12 and Lemma 9.14. It is a consequence of Lemma 9.14 that

$$\sum_{D\in X} \ell_D \leqslant \sum_{D\in X} \text{rank}M_D \leqslant \sum_{D\in X} m_D$$

Applying Corollary 9.12, we therefore deduce that

$$r - \sum_{D\in X} m_D \leqslant \dim_F J(Z(FG)) \leqslant r - \sum_{D\in X} \ell_D$$

But $\sum_{D\in X} m_D$ is the number of nonnilpotent p-regular classes of G, while $\sum_{D\in X} \ell_D$ is the number of conjugacy classes of G contained in $O_{p'}(G)$. Thus

$$r - r_1 \leq \dim_F J(Z(FG)) \leq r - r_2$$

as required. ∎

9.16. COROLLARY. Let G be p-nilpotent. Then $\dim_F J(Z(FG))$ is equal to the number of p-singular classes of G.

Proof. Since G is p-nilpotent, all p-regular classes of G are contained in $O_{p'}(G)$ and therefore are nonnilpotent. Hence $r_1 = r_2$ and $r - r_1$ is the number of p-singular classes of G. Now apply Theorem 9.15. ∎

To provide further applications of Theorem 9.11, we need to introduce the Brauer correspondence. The following preliminary observations will clear our path.

9.17. LEMMA. Let P be a p-subgroup of G and let $H = N_G(P)$

(i) If $C \in C\ell(G)$ with $\delta(C) = P$ then $C \cap C_G(P)$ is a conjugacy class of H with defect group

(ii) If L is a conjugacy class of H with defect group P, then there exists $C \in C\ell(G)$ such that $L = C \cap C_G(P)$ and $\delta(C) = P$.

Proof. (i) There is an element $g \in C$ such that P is a Sylow p-subgroup of $C_G(g)$. Hence $g \in C \cap C_G(P)$. If $g' \in C \cap C_G(P)$, then there is an $x \in G$ such that $xgx^{-1} = g'$. Because $g' \in C_G(P)$, P is a Sylow p-subgroup of $C_G(g') = xC_G(g)x^{-1}$. But xPx^{-1} is one also and so there is a $y \in C_G(g')$ with $yPy^{-1} = xPx^{-1}$; hence $y^{-1}x \in H$. Since y commutes with g', we have

$$g' = y^{-1}xgx^{-1}y$$

so that g' is conjugate to g in H. Finally, because $C_G(P) \lhd N_G(P)$ any element of H which is H-conjugate to g, lies in $C \cap C_G(P)$.

(ii) By assumption, there is an element h in L such that P is a Sylow p-subgroup of the centralizer of h in H. Thus $h \in C_G(P)$ and so $L \subseteq C \cap C_G(P)$, where C is the conjugacy class of G containing h. We next show that P is a defect group of C. It will then follow, by (i), that $L = C \cap C_G(P)$ and hence the proof will be complete. Let Q denote a Sylow p-subgroup of $C_G(h)$ with $P \subseteq Q$; then Q is a defect group of C. If $Q \neq P$, choose a subgroup P_1 of

Q such that $(P_1:P) = p$. Then $P \subseteq N_G(P) = H$ and $P_1 \subseteq C_H(h)$, contrary to the fact that P is a defect group of L. Hence $Q = P$ and the result follows. ■

9.18. LEMMA. Let P be a p-subgroup of G, let $H = N_G(P)$ and let

$$\pi : Z(FG) \longrightarrow Z(FH)$$

be the Brauer homomorphism. Then π induces an injection of the block idempotents of FG having P as a defect group into nonzero idempotents of $Z(FH)$. Moreover, for any block idempotent e of FG with defect group P, $\text{Supp}\pi(e)$ is a union of H-conjugacy classes having defect group P.

Proof. Let e be a block idempotent of FG with defect group P. Owing to Proposition 2.2.6, there exist $C_1,\ldots,C_r$ in $C\ell(G)$ which satisfy conditions (a) and (b) of Proposition 2.2.6 and such that

$$e = \lambda_1 C_1^+ + \ldots + \lambda_r C_r^+$$

with some nonzero λ_i in F. Hence, in the notation of Proposition 2.2.6, the classes $C_1,\ldots,C_t$ have defect group P, while defect groups of $C_{t+1},\ldots,C_r$ have order less than $|P|$. Thus

$$C_k \cap C_G(P) = \emptyset \quad \text{for all} \quad k \in \{t+1,\ldots,r\}$$

It follows that $\pi(C_k^+) = 0, t \leqslant k \leqslant r$, and therefore $\pi(e) = \pi(C_1^+) + \ldots + \pi(C_t^+)$. Now

$$\text{Supp}\pi(C_i^+) = C_i \cap C_G(P) \qquad (1 \leqslant i \leqslant t)$$

and, by Lemma 9.17(i), the $C_i \cap C_G(P), 1 \leqslant i \leqslant t$, are distinct conjugacy classes of H with defect group P. Hence $\pi(e)$ satisfies the required property.

Assume that e' is another block idempotent of FG with defect group P, and let $\pi(e) = \pi(e')$. Then $\text{Supp}(e-e')$ is a union of conjugacy classes whose defect groups are conjugate to proper subgroups of P. Let γ be the irreducible representation of $Z(FG)$ associated with e. By Proposition 2.2.6(ii), $\gamma(e-e') = 0$ and so $\gamma(e) = \gamma(e')$. But if e and e' really were distinct, then $\gamma(e) = 1$ and $\gamma(e') = 0$. Hence $e = e'$ and the result follows. ∎

9.19. LEMMA. Let A be a commutative algebra over a field K and let $\Gamma_i : A \longrightarrow M_{n_i}(K), i = 1,2,$ be two irreducible representations of A. Then Γ_1 is equivalent to Γ_2 if and only if there exists a K-algebra isomorphism $f : \Gamma_1(A) \longrightarrow \Gamma_2(A)$ which renders commutative the following diagram

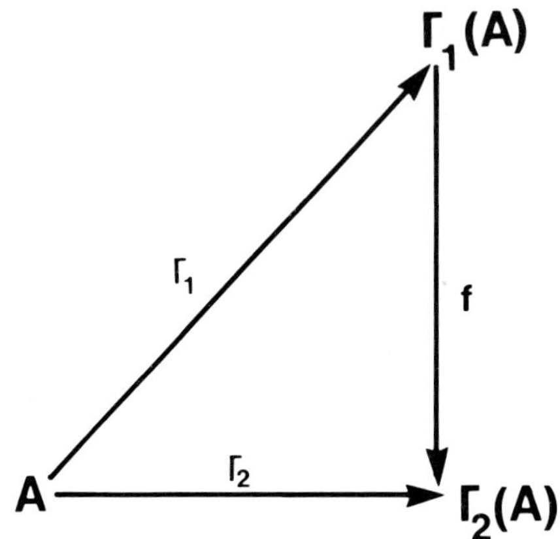

Proof. Assume that Γ_1 is equivalent to Γ_2. Then $n_1 = n_2$ and there exists a nonsingular $n_1 \times n_1$-matrix C such that $\Gamma_2(a) = C^{-1}\Gamma_1(a)C$ for all $a \in A$. Hence the map $f : \Gamma_1(A) \longrightarrow \Gamma_2(A)$ defined by $f(x) = C^{-1}xC$ is an isomorphism of K-algebras such that $\Gamma_2 = f\Gamma_1$. To prove the converse, denote by e the block idempotent of A such that $\Gamma_1(e) = 1$. Then

$$\Gamma_2(e) = (f\Gamma_1)(e) = f(1) = 1$$

and so, by Proposition 1.10.8(v), Γ_1 is equivalent to Γ_2. So the lemma is proved. ∎

9.20. LEMMA. Let $\rho : A \longrightarrow B$ be a homomorphism of commutative algebras over a field K, let e be a block idempotent of A and let

$$\rho(e) = u_1 + u_2 + \ldots + u_n$$

where the u_i are block idempotents of B. Denote by λ_i the irreducible representation of B associated with u_i and set $K_i = (\lambda_i\rho)(A)$, $1 \leqslant i \leqslant n$. Then, for all $i,j \in \{1,\ldots,n\}$, there exists a K-algebra isomorphism $Q_{ij} : K_j \longrightarrow K_i$

which renders commutative the following diagram

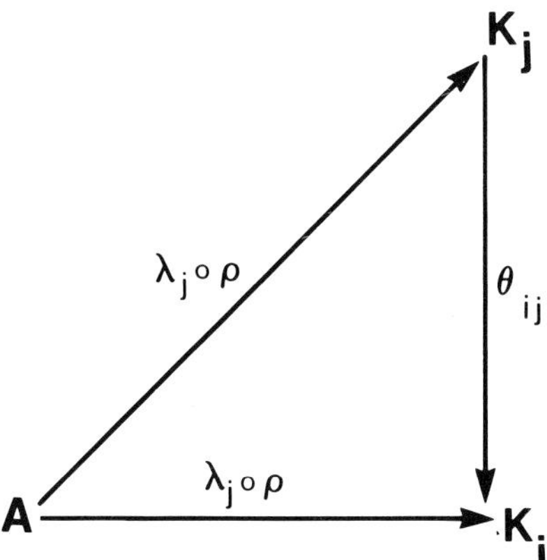

Proof. Fix $i,j \in \{1,\dots,n\}$. Owing to Proposition 1.10.8(v),

$$\lambda_i\rho(e) = 1 = \lambda_j\rho(e) \tag{10}$$

Because $\lambda_i(B)$ is a finite field extension of K and since $K_i \subseteq \lambda_i(B)$, we conclude that K_i is also a finite field extension of K. Thus, replacing K_i by its image in the regular representation of K_i, we may assume that $\lambda_i\rho$ is an irreducible representation of A, $1 \leqslant i \leqslant n$. By (10) and Proposition 1.10.8(v), it follows that $\lambda_i\rho$ and $\lambda_j\rho$ are equivalent. The desired conclusion is now a consequence of Lemma 9.19. ∎

We are now ready to prove the following important result.

9.21. THEOREM (Brauer correspondence). Let P be a p-subgroup of G and let H be the normalizer of P in G. Then the Brauer homomorphism

$$\pi : Z(FG) \longrightarrow Z(FH)$$

determines a bijective correspondence between the block idempotents of FG which have P as one of their defect groups and the block idempotents of FH which have P as their unique defect group.

Proof. For the sake of clarity, we shall divide the proof into two steps.

Step 1. Here we prove that $\pi(e)$ is a block idempotent of FH with defect group P.

Owing to Lemma 9.18, it suffices to show that $\pi(e)$ is a primitive idempotent of $Z(FG)$. Assume the contrary, say,

$$\pi(e) = e_1 + e_2 + \dots + e_t ,$$

$t > 1$, where the e_i are orthogonal primitive idempotents of $Z(FH)$. Denote by γ_i the irreducible representation of $Z(FH)$ associated with e_i and by F_i the image of $Z(FG)$ under the homomorphism $\gamma_i \circ \pi$, $1 \leqslant i \leqslant t$. By Lemma 9.20, there exist F-isomorphisms $Q_{ij} : F_j \longrightarrow F_i$ which render commutative the following diagram

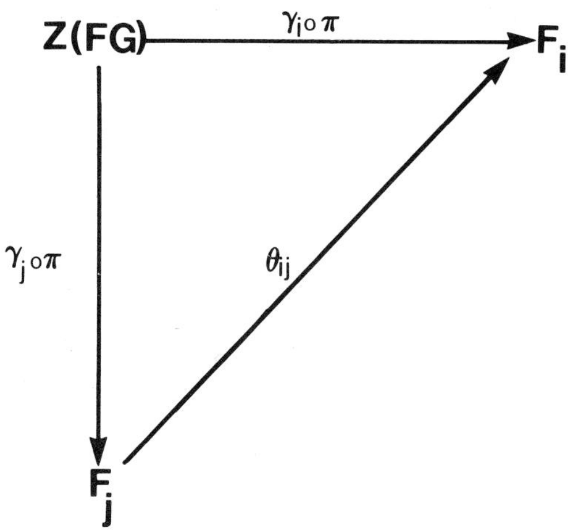

Owing to Lemma 6.6(ii), the defect groups of each e_i contain P. Assume that $k \in \{1,2,\dots,t\}$ is such that the defect groups of e_k contain P properly. Then, by Proposition 2.2.6(ii), $\gamma_k(A) = 0$ where A is as in Lemma 6.6 with respect to FH. It follows that $(\gamma_k \circ \pi)(e) = 0$, which is impossible. Hence all the e_i are in A.

Let $L_1,\dots,L_r$ denote all conjugacy classes of H with defect group P. Then, by Lemma 9.17(ii), there exist conjugacy classes $C_1,\dots,C_r$ of G with defect group P and with $L_i = C_i \cap C_G(P)$, $1 \leqslant i \leqslant r$. Thus

$$e_i = \sum_{j=1}^{r} \lambda_{ij}\pi(C_j^+)$$

for some $\lambda_{ij} \in F$. Because $t > 1$, the commutativity of the diagram above yields

$$\begin{aligned} 0 = \gamma_2(e_1) &= \gamma_2 \pi(\sum_{j=1}^{r} \lambda_{ij} C_j^+) \\ &= \theta_{21} \gamma_1 \pi(\sum_{j=1}^{r} \lambda_j C_j) = \theta_{21} \gamma_1(e_1) \\ &= 1, \end{aligned}$$

a contradiction. Thus $\pi(e)$ is primitive.

Step 2. Let f be a block idempotent of FH with defect group P. By Lemma 9.18, we are left to verify that $\pi(e) = f$ for some block idempotent e of FG with defect group P.

Owing to Lemma 6.6(ii), f lies in A. Let γ denote the irreducible representation of $Z(FH)$ associated with f. Since by Lemma 9.17 every conjugacy class L of H with defect group P is of the form $C \cap C_G(P)$ for some $C \in C\ell(G)$ with $\delta(C) = P$, A is contained in $\pi(Z(FG))$. Hence $\gamma\pi \neq 0$, and therefore $\gamma\pi$ is an irreducible representation of $Z(FG)$.

Let e denote the block idempotent of FG associated with $\gamma \circ \pi$. Write

$$e = \lambda_1 C_1^+ + \dots + \lambda_t C_t^+ + \lambda_{t+1} C_{t+1}^+ + \dots + \lambda_r C_r^+ \qquad (\lambda_i \in F)$$

with the C_i as in Proposition 2.2.6. Then $(\gamma\pi)(C_{t+k}^+) = 0$ for all $k \in \{1,\dots,r-t\}$, while $(\gamma\pi)(C_s^+) = 0$ for some $s \in \{1,\dots,t\}$. Because $\pi(C_s^+) \neq 0$, we have $C_s \cap C_G(P) \neq \emptyset$. Hence the defect groups of C_s, and so of e, contain subgroups conjugate to P. But if the defect groups of e were to contain P properly, then by Proposition 2.2.6 and Lemma 6.5, $\gamma(A) = 0$ which contradicts $\gamma(f) = 1$. Thus the defect groups of e are conjugate to P. Therefore, by Step 1, $\pi(e)$ is primitive. But then since $(\gamma\pi)(e) = \gamma(f)$, we have $\pi(e) = f$, thus completing the proof. ■

9.22. LEMMA. Let D be a subgroup of G of order p^d and let p^a be the order of Sylow p-subgroups of G. Then $N_G(D)/D$ has a normal p-complement under either of the following hypotheses

(i) $d = a-1$ and is the smallest prime divisor of $|G|$

(ii) $d = a-2$, G is of odd order and p is the smallest prime divisor of G.

Proof. Let P denote a Sylow p-subgroup of G. A classical theorem of Burnside (Burnside (1955)) asserts that if P lies in the centre of its normalizer in G, then G has a normal p-complement. We claim that

If P is abelian and $q \nmid |\mathrm{Aut}P|$ for any prime $q \neq p$ dividing $|G|$, then G has a normal p-complement. (11)

Indeed, assume that $g \in N_G(P)$ and let $\langle g\rangle = \langle g_1\rangle \times \ldots \times \langle g_n\rangle$ be the decomposition of $\langle g\rangle$ into direct product of primary components. Owing to Burnside's theorem mentioned above, it suffices to verify that $g \in C_G(P)$. We may clearly assume that $p \nmid |\langle g\rangle|$. Since conjugation by g_i induces a homomorphism $\langle g_i\rangle \longrightarrow \mathrm{Aut}P$, the assumption that $q \nmid |\mathrm{Aut}P|$ ensures that the image of g_i is the identity automorphism, as claimed.

Suppose that (i) holds. Then $N_G(D)/D$ has a Sylow p-subgroup S of order p. Because $|\mathrm{Aut}S| = p-1$ and p is the smallest prime divisor of $|G|$, it follows from (11) that $N_G(D)/D$ has a normal p-complement.

Now assume that (ii) holds. Then $N_G(D)/D$ has a Sylow p-subgroup S of order p or p^2. By the foregoing, we may assume that $|S| = p^2$ in which case

$$|\mathrm{Aut}S| = \begin{cases} p(p-1) & \text{if } S \cong \mathbb{Z}_{p^2} \\ (p-1)^2 p(p+1) & \text{if } S \cong \mathbb{Z}_p \times \mathbb{Z}_p \end{cases}$$

Let q denote a prime divisor of $|G|$ distinct from p. Since G is of odd order, $q > p+1$ and so in both cases $q \nmid |\mathrm{Aut}S|$. Hence, by (ii), $N_G(D)/D$ has a normal p-complement, as required. ∎

We are now ready to provide an important application of Theorem 9.11. Part (ii)d of the following result is due to Brauer and Nesbitt (1941), while the rest generalizes theorems due to Gow (1978), Kawada (1966) and Michler (1972).

9.23. THEOREM. Let F be a field of characteristic $p > 0$ which is a splitting field for $Z(FG)$, let D be a p-subgroup of G and let p^d and p^a denote the orders of D and Sylow p-subgroups of G, respectively. Define the numbers n_D, m_D, s_D, and ℓ_D by:

n_D = the number of blocks of G with defect group D

m_D = the number of nonnilpotent p-regular classes of G with defect group D

s_D = the number of p-regular classes of G with defect group D

ℓ_D = the number of conjugacy classes of G contained in $O_{p'}(G)$ and having defect group D. Then

(i) $\ell_d \leqslant n_D \leqslant m_D$

(ii) The equality $n_D = s_D$ is true under either of the following hypotheses:

(a) G is p-nilpotent

(b) $DC_G(D)/D$ is p-nilpotent

(c) $N_G(D)/D$ is p-nilpotent

(d) D is a Sylow p-subgroup of G

(e) The Sylow p-subgroups of G are metacyclic and $(p^2-1,|G|) = 1$

(f) $d = a-1$ and p is the smallest prime divisor of $|G|$

(g) $d = a-2$, G is of odd order and p is the smallest prime divisor of $|G|$.

Proof. (i) We keep the notation of Theorem 9.11. If e is a block idempotent of FG, then

$$Z(FG)e/J(Z(FG)\ e \cong F$$

since F is a splitting field for $Z(FG)$. Hence

$$\dim_F Z(FG)e_D\ /J(Z(FG))e_D = n_D \qquad (12)$$

and so, by Theorem 9.11, $n_D = \text{rank}M_D$. The desired assertion is now a consequence of Lemma 9.14.

(ii) If G is p-nilpotent, then $s_D = \ell_D = m_D$ and so, by (i), $n_D = s_D$. By Lemma 9.22, each of the conditions (f), (g) implies (c). Since $DC_G(D)$ is a subgroup of $N_G(D)/D$, (c) implies (b). Moreover, (e) implies (a) (see Huppert (1967),p.437). Since (d) obviously implies (c), we are left to verify (b).

Suppose that $DC_G(D)/D$ is p-nilpotent. If the result is true for $N_G(D)$, then by Lemma 9.17 and Theorem 9.21, the corresponding result is true for G. Thus we may harmlessly assume that $D \lhd G$.

Let C_i, C_j be p-regular classes of G with defect group D and let $x \in C_i, y \in C_j$ with $x^{-1}y \in P \in \text{Syl}_p(G)$. Because $D \lhd G$ we have $D \in \text{Syl}_p(C_G(x))$ and $D \in \text{Syl}_p(C_G(y))$ so that $x,y \in C_G(D)$. Then $x^{-1}D$ and yD are p-regular elements of the p-nilpotent group $DC_G(D)/D$. Thus $xy^{-1}D$ is a p-regular element. However, $x^{-1}yD$ is also a p-element, so $x^{-1}y \in D$, say $x^{-1}y = d$ with $d \in D$. It follows that $y = xd$ with $xd = dx$ which is only possible in the case $d = 1$. Hence $x^{-1}y = 1$ and so $x = y$. By repeating the argument of Lemma 9.14, we conclude that $\mathrm{rank}M_D \geq s_D$. But $\mathrm{rank}M_D = n_D$, so by (i), $n_D = s_D$, as asserted. ■

10. MORITA'S THEOREM.

Throughout this section, N denotes a normal subgroup of a finite group G, F a field of characteristic $p > 0$ and

$$Z^*(FN) = Z(FG) \cap Z(FN)$$

Our aim is to provide an important decomposition theorem due to Morita which will be frequently applied in the investigation of $J(FG)$.

We start by exhibiting some connections between the blocks of FG and FN.

10.1. LEMMA. Assume that f is a block idempotent of FN.

(i) For all $g \in G$, gfg^{-1} is a block idempotent of FN. If γ is the irreducible representation of $Z(FN)$ associated with f, then γ^g defined by

$$\gamma^g(x) = \gamma(g^{-1}xg) \qquad (x \in Z(FN))$$

is the irreducible representation of $Z(FN)$ associated with gfg^{-1}.

(ii) The set

$$G(f) = \{g \in G \mid gfg^{-1} = f\}$$

is a subgroup of G containing N.

(iii) If T is a left transversal for $G(f)$ in G, then the $tft^{-1}, t \in T$, are all distinct conjugates of f and $f^* = \sum_{t\in T} tft^{-1}$ is an idempotent of $Z^*(FN)$ which is independent of the choice of T.

(iv) $\mathrm{Supp}f^*$ is a union of some p-regular classes of G contained in N.

Proof. (i) Follows from the fact that conjugation by g is an automorphism of FN.

(ii) Straightforward

(iii) It is obvious that the $tft^{-1}, t \in T$, are all distinct conjugates of f. Hence f^* is an idempotent of FN which is independent of the choice of T. If $g \in G$ and $T_1 = gT$, then T_1 is also a left transversal for $G(f)$ in G and

$$gf^*g^{-1} = \sum_{t \in T} (gt)f(gt)^{-1} = \sum_{x \in T_1} xfx^{-1} = f^*$$

This proves that $f^* \in Z(FG)$ and hence that $f^* \in Z^*(FN)$.

(iv) Apply Theorem 2.3.4. ■

Let f be a block idempotent of FN. In what follows we shall refer to the group $G(f)$ as the *inertia group* of f. We say that a block idempotent $e \in FG$ *covers* f (or $B = B(e)$ *covers* $b = b(f)$) if e occurs in the decomposition of f^* into the sum of block idempotents of FG.

Observe that conjugation by $g \in G$ induces an action of G on the sets of all block idempotents of FN and all blocks of FN, and that f^* is just the sum of the elements in the G-orbit of f. We shall refer to the G-orbit of f as the *G-conjugacy class* of f. Similarly, the G-orbit of a block $b = b(f)$ will be called the *G-conjugacy class* of b.

Assume that $b = b(f)$ is a block of FN and put

$$G(b) = \{g \in G | gbg^{-1} = b\}$$

Then $G(b) = G(f)$ and we refer to $G(b)$ as the *inertia group of the block* b.

10.2. LEMMA. (i) The mapping $f \mapsto f^*$ induces a bijective correspondence between the G-conjugacy classes of block idempotents of $Z(FN)$ and all block idempotents of $Z^*(FN)$.

(ii) Every block idempotent of FG covers exactly one G-conjugacy class of block idempotents of FN.

(iii) A block $B = B(e)$ of FG covers the block $b = b(f)$ of FN if and only if $ef \neq 0$.

Proof. (i) Let f be a block idempotent of $Z(FN)$. By Lemma 10.1(iii), f^* is an idempotent of $Z^*(FN)$. Assume that $f^* = u+v$, where u and v are nonzero orthogonal idempotents of $Z^*(FN)$. Since $u,v \in Z(FN)$, there exist block idempotents u_i, v_j in FN, $1 \leqslant i \leqslant n$, $1 \leqslant j \leqslant m$, such that

$$u = \sum_{i=1}^{n} u_i \quad \text{and} \quad v = \sum_{j=1}^{m} v_j$$

Thus

$$f^* = \sum_{i=1}^{n} u_i + \sum_{j=1}^{m} v_j$$

Since f^* is a unique sum of block idempotents of FN and f is one of them, we may assume that $u_1 = f$. Because $u_1 u = u_1$, we have $f = u_1 u \in Z(FN)u$. Bearing in mind that $u \in Z(FG)$, we therefore deduce that

$$gfg^{-1} \in gZ(FN)ug^{-1} = gZ(FN)g^{-1}u = Z(FN)u$$

for all $g \in G$. Hence $f^* = zu$ for a suitable $z \in Z(FN)$. But $f^* = u+v$, so

$$v = vf^* = v(zu) = z(vu) = 0,$$

a contradiction. Thus f^* is a block idempotent of $Z^*(FN)$.

Now assume that x is an arbitrary block idempotent of $Z^*(FN)$. Then

$$x = f_1 + f_2 + \dots + f_s \,, \qquad (*)$$

where the f_i are uniquely determined block idempotents of FN. Put $f = f_1$ and denote by $g_1,\dots,g_t$ a left transversal for $G(f)$ in G. Then

$$x = g_j \; xg_j^{-1} = g_jfg_j^{-1} + \sum_{k=2}^{s} g_jf_kg_j^{-1}$$

for all $j \in \{1,2,\dots,t\}$. Because every $g_jfg_j^{-1}$ is a block idempotent of FN, the uniqueness of (*) implies that

$$g_jfg_j^{-1} \in \{f_1,\dots,f_s\}$$

for all $j \in \{1,2,\dots,t\}$. Hence

$$x = f^*+y$$

for some idempotent y in $Z(FN)$ such that $yf^* = f^*y = 0$. Since x and f^*

are block idempotents of $Z^*(FN)$, we infer that $y = 0$ and hence that $x = f^*$. Finally, since for any two block idempotents f_1 and f_2 of $Z(FN)$, $f_1^* = f_2^*$ if and only if f_1 and f_2 are G-conjugate, (i) follows.

(ii) Fix a block idempotent e of FG. Owing to (i), we can find block idempotents $f_1, f_2, \ldots, f_n$ of FN such that $f_1^*, f_2^*, \ldots, f_n^*$ are all block idempotents of $Z^*(FN)$. Thus there exists a unique $k \in \{1,2,\ldots,n\}$ such that e occurs in the decomposition of f_k^* as the sum of block idempotents of FG. This proves that e covers f_k and that, for any block idempotent f of FN, e covers f if and only if $f^* = f_k^*$. Because the latter is equivalent to f and f_k being G-conjugate, the required assertion follows.

(iii) It is clear that $ef \neq 0$ if and only if $ef^* \neq 0$. Since the latter holds if and only if e covers f, the result follows. ∎

10.3. LEMMA. Let A be a finite-dimensional local commutative algebra over F. Then there exists a subfield $E \supseteq F$ of A such that

$$A = E \oplus J(A) \qquad \text{(direct sum of } F\text{-spaces).}$$

Proof. Observe that if $A = E \oplus J(A)$, then $E \cong A/J(A)$ is an F-algebra and hence $E \supseteq F$. We divide the rest of the proof into two steps.

Step 1. Here we establish the case where $J(A)^2 = 0$.

Put $A^p = \{a^p | a \in A\}$ and note that if $a^p \neq 0$ then $a \notin J(A)^2$, since $J(A)^2 = 0$. Therefore, for all $a \in A$, a and a^p are units of A whenever $a^p \neq 0$. If $a^p b = 1$ for some $b \in A$, then $b = (a^{-1})^p \in A^p$ and hence A^p is a subfield of A. Let E denote a maximal subfield of A containing A^p. Clearly $E \cap J(A) = 0$. We denote by $\bar{E}$ the image of E in $\bar{A} = A/J(A)$. To prove that $A = E \oplus J(A)$, assume by way of contradiction that $\bar{a} = a + J(A) \notin \bar{E}$. Since $A^p \subseteq E$, $\bar{a}^p \in \bar{E}$ and so $a^p \in E$. Consider the polynomial $f(X) = X^p - a^p \in E[X]$. If $f(X)$ is not irreducible, then it has an irreducible factor $(X-a)^r \in E[X]$ with $1 < r < p$, because $a \notin E$. Now ra is a coefficient of $(X-a)^r$ and so $ra \in E$. Since $(r,p) = 1$, $1 = sr+tp$ for some $t,s \in \mathbb{Z}$. Bearing in mind that $\text{char}E = p$, we obtain

$$a = sra + tpa = s(ra) \in E,$$

contrary to our choice of a. Thus $f(X)$ is irreducible, in which case $E[a]$ is a subfield of A properly containing E, a contradiction.

Step 2. Completion of the proof.

Assume the contrary and denote by A a counter-example of minimal F-dimension. By Step 1, $J(A)^2 \neq 0$ and so $A_1 = A/J(A)^2$ has smaller dimension than A. Clearly, $J(A_1) = J(A)/J(A)^2$ and so $J(A_1)^2 = 0$. Because A_1 is also local, it follows from Step 1 that

$$A_1 = E_1 \oplus J(A_1) \tag{1}$$

for a suitable subfield E_1 of A_1 containing F. Write $E_1 = B/J(A)^2$ for some F-subalgebra B of A. Then, by (1), we have

$$A = B + J(A) \quad \text{and} \quad B \cap J(A) = J(A)^2 \tag{2}$$

Since B is a local proper F-subalgebra of A with $J(B) = J(A)^2$, there is a subfield E of B for which

$$B = E \oplus J(A)^2 \tag{3}$$

Applying (2) and (3), we therefore derive

$$A = E + J(A)$$

and

$$E \cap J(A) = E \cap B \cap J(A) = E \cap J(A)^2 = 0,$$

a contradiction. ∎

10.4. LEMMA. Let e and f be block idempotents of FG and FN, respectively, and let γ_e and γ_f denote the corresponding irreducible representations of $Z(FG)$ and $Z(FN)$. Then e covers f if and only if there exists an F-algebra isomorphism

$$\pi : \gamma_f(Z^*(FN)) \longrightarrow \gamma_e(Z^*(FN))$$

such that

$$\pi\gamma_f(x) = \gamma_e(x) \qquad \text{for all} \quad x \in Z^*(FN)$$

In particular, if F is algebraically closed, then e covers f if and only if

$$\gamma_f(x) = \gamma_e(x) \qquad \text{for all} \quad x \in Z^*(FN)$$

Proof. Assume that e covers f. Then there are block idempotents $e = e_1, e_2, \ldots, e_r$ in FG such that

$$f^* = e_1 + e_2 + \ldots + e_r$$

and, by Lemma 10.2(i), f^* is a block idempotent of $Z^*(FN)$. Denote by J the Jacobson radical of $Z^*(FN)$. Then $Z^*(FN)f^*$ is a finite-dimensional commutative local algebra over F with Jacobson radical $J_1 = Jf^*$. Applying Lemma 10.3, we have

$$Z^*(FN)f^* = E \oplus J_1$$

for some subfield E of $Z^*(FN)f^*$ containing F. Thus, given $x \in Z^*(FN)$, there is a unique $y \in E$ and a unique $z \in J_1$ with

$$xf^* = y + z \tag{4}$$

Because f^* is the identity element of $Z^*(FN)f^*$ and since $f^*e = e$, it follows from (4) that

$$xe + ye + ze$$

Since $z \in J_1 \subseteq J(Z(FG))$, we have

$$0 = \gamma_e(z) = \gamma_e(z)\gamma_e(e) = \gamma_e(ze)$$

and thus

$$\gamma_e(x) = \gamma_e(xe) = \gamma_e(ye) + \gamma_e(ze) = \gamma_e(y) \tag{5}$$

Now the image of the restriction γ' of γ_e to $Z^*(FN)$ is a finite field extension of F (Proposition 1.10.8(iv)). Hence the restriction γ of γ_e to F is an F-algebra isomorphism of E onto $\gamma_e(Z^*(FN))$, by virtue of (5). Thus

$$\gamma(E) = \gamma'(Z^*(FN)) = \gamma_e(Z^*(FN))$$

Because f occurs in the decomposition of f^* into the sum of block idempotents of FN, it follows similarly from (4) that the restriction δ of γ_f to E is an F-algebra isomorphism from E onto $\gamma_f(Z^*(FN))$ such that

$$\gamma_f(x) = \gamma_f(y) = \delta(y) \tag{6}$$

where x and y are chosen as in (4). Invoking (5) and (6), we therefore deduce that $\pi = \gamma\delta^{-1}$ is a required isomorphism.

Conversely, suppose that π satisfies the hypothesis of the theorem. Owing to Lemma 10.1(iii), f^* is a central idempotent of FG. Hence

$$f^* = e_1 + e_2 + \ldots + e_k$$

where the e_i are suitable block idempotents of FG. If $e \neq e_i$ for all $i \in \{1,\ldots,k\}$, then $\gamma_e(e_i) = 0$ for all i by Proposition 1.10.8(v). Thus

$$0 = \gamma_e(f^*) = \pi\gamma_f(f^*) = \pi\gamma_f(f) = \pi(1) = 1,$$

a contradiction. Hence e covers f and the result follows. ■

10.5. LEMMA. Let L be a conjugacy class of N containing x. Then

$$N_G(L) = C_G(x)N$$

Proof. Assume that $g \in N_G(L)$. Then, for any $n \in N$, there exists $n_1 \in N$ with $g^{-1}n^{-1}xng = n_1^{-1}xn_1$. Hence $ngn_1^{-1} \in C_G(x)$ and $g \in C_G(x)N$ since N is normal. Conversely, suppose that $g \in C_G(x)N$, say $g = yn$ with $y \in C_G(x)$ and $n \in N$. Then $y^{-1}xy = ng^{-1}xgn^{-1} = x$ and so $g^{-1}xg = n^{-1}xn$. Hence, for all $h \in N$,

$$g^{-1}(h^{-1}xh)g = (g^{-1}h^{-1}g)(g^{-1}xg)(g^{-1}hg)$$
$$= (g^{-1}hg)^{-1}(n^{-1}xn)(g^{-1}hg) \in L,$$

proving that $g \in N_G(L)$. So the lemma is true. ∎

The following result is extracted from a work of Michler(1972).

10.6. PROPOSITION. Let f be a block idempotent of FN. Among the block idempotents $e_1,\ldots,e_s$ of FG covering f, let e_1 be of maximal defect and let D_i be a defect group of e_i, $1 \leqslant i \leqslant s$. Then

(i) For all $i \in \{1,\ldots,s\}$, D_i is conjugate to a subgroup of D_1

(ii) For all $i \in \{1,\ldots,s\}$, D_i is conjugate to a subgroup of $G(f)$

(iii) $D_1 \cap N$ is a defect group of f

(iv) Assume that F is a splitting field for $Z(FN)$ and choose D_1 to be a subgroup of $G(f)$. Then D_1N is a subgroup of $G(f)$ of p'-index.

Proof. (i) Denote by γ_i the irreducible representation of $Z(FG)$ associated with e_i, $1 \leqslant i \leqslant s$. Write $f^* = \lambda_1 C_1^+ + \ldots + \lambda_r C_r^+$ where $0 \neq \lambda_i \in F$ and $C_1,\ldots,C_r$ are suitable p-regular classes of G contained in N (see Lemma 10.1 (iv)). Because $\gamma_1(f^*) = \gamma_1(e_1) = 1$, it follows that $\gamma_1(C_k^+) \neq 0$ for some $k \in \{1,\ldots,r\}$. Let μ be the irreducible representation of $Z(FN)$ associated with f. Then, by Lemma 10.4, there exist F-algebra isomorphisms π_i from $\mu(Z^*(FN))$ onto $\gamma_i(Z^*(FN))$ such that

$$\mu(x) = \pi_i^{-1}\gamma_i(x) \quad \text{for all } x \in Z^*(FN), i \in \{1,\ldots,s\}$$

Because $C_k^+ \in Z^*(FN)$, it follows that

$$0 \neq \gamma_1(C_k^+) = \pi_1\mu(C_k^+) = \pi_1\pi_i^{-1}\gamma_i(C_k^+)$$

and thus $\gamma_i(C_k^+) \neq 0$ for all $i \in \{1,\ldots,s\}$. Thus, by Proposition 2.2.6(iv), each D_i is conjugate to a subgroup of a defect group, say D, of C_k. Taking into account that

$$f^* = \lambda_1 C_1^+ + \ldots + \lambda_r C_r^+ = e_1 + \ldots + e_s$$

and that $C_k \subseteq \text{Supp} f^*$, we must have $C_k \subseteq \text{Supp} e_j$ for some $j \in \{1,\ldots,s\}$.

Therefore D is conjugate to a subgroup of D_j and hence, by the above, D is conjugate to D_j. Since D_1 is of maximal order among the $D_1,\dots,D_s$, we infer that D_1 is conjugate to D, proving (i).

(ii) Fix $x \in C_k$ together with a complete system $g_1,\dots,g_n$ of representatives of the double cosets $G(f)yC_G(x)$ in G and choose $x_{ij} \in C_G(x)$ such that

$$\{g_i x_{ij} \mid 1 \leqslant i \leqslant n,\ 1 \leqslant j \leqslant r_i\}$$

is a right transversal for $G(f)$ in $G(f)g_iC_G(x)$.

Assume that $g \in N$ is such that $g_i^{-1}gg_i = x$ for some $i \in \{1,\dots,n\}$, and write $f = \sum_{g\in N} \lambda_g g$ with $\lambda_g \in F$. Then

$$\begin{aligned} f^* &= \sum_{i=1}^{n} \sum_{j=1}^{r_i} (g_i x_{ij})^{-1} f (g_i x_{ij}) \\ &= \sum_{i=1}^{n} \sum_{j=1}^{r_i} \sum_{g\in N} \lambda_g (g_i x_{ij})^{-1} g (g_i x_{ij}), \end{aligned}$$

and so $\sum_{i=1}^{n} r_i \lambda_{g_i x g_i^{-1}}$ is the coefficient of x in f^*. Because $x \in \mathrm{Supp} f^*$, there is at least one integer $t \in \{1,\dots,n\}$ such that $p \nmid r_t$. Since

$$\begin{aligned} r_t &= \frac{|G(f)g_tC_G(x)|}{|G(f)|} = \frac{|G(f)g_tC_G(x)g_t^{-1}|}{|G(f)|} \\ &= \frac{|C_G(g_t^{-1}xg_t)|}{|G(f) \cap C_G(g_r^{-1}xg_t)|} \end{aligned}$$

it follows that $G(f)$ contains a Sylow p-subgroup of $C_G(g_t^{-1}xg_t)$. Taking into account that $g_t^{-1}xg_t \in C_k$, we infer that D is conjugate to a subgroup of $G(f)$. But, by (i), each D_i is conjugate to a subgroup of D, proving (ii).

(iii) Let x be as in (ii) and let L be the N-conjugacy class containing x. Then $C_k = \bigcup_{i=1}^{m} t_i^{-1}Lt_i$ where $\{t_i \mid 1 \leqslant i \leqslant m\}$ is a right transversal for $N_G(L)$ in G. Denote by μ the irreducible representation of $Z(FN)$ associated with f. Owing to Lemma 10.4, there exists an F-algebra isomorphism π from $\mu(T)$ onto $\gamma_1(T)$, where $T = Z^*(FN)$, such that $\pi\mu(y) = \gamma_1(y)$ for all $y \in T$.

Because $\gamma_1(C_k^+) \neq 0$, it follows that

$$0 \neq \gamma_1(C_k^+) = \pi\mu(C_k^+) = \pi(\sum_{i=1}^{m} \mu(t_i^{-1}L^+t_i)) \tag{7}$$

and hence that

$$\mu(t_q^{-1}L^+t_q) \neq 0 \quad \text{for some} \quad q \in \{1,\ldots,m\} \tag{8}$$

It is obvious that $t_q^{-1}L^+t_q$ is a central element of FN whose support is the N-conjugacy class containing $t_q^{-1}xt_q$. Since $t_q^{-1}xt_q \in C_k$, we may assume that D_1 is a Sylow p-subgroup of $C_G(t_q^{-1}xt_q)$. Now $N \cap C_G(t_q^{-1}xt_q) = C_N(t_q^{-1}xt_q)$ is a normal subgroup of $C_G(t_q^{-1}xt_q)$, so by Sylow theorem $D_1 \cap N$ is a Sylow p-subgroup of $C_N(t_q^{-1}xt_q)$. Thus $D_1 \cap N$ is a defect group of the N-conjugacy class L_1 containing $t_q^{-1}xt_q$.

It now follows, from (8) and Proposition 2.2.6(iv), that a defect group $\delta(f)$ of f is N-conjugate to a subgroup of $D_1 \cap N$. Now $L_1 \subseteq C_k \subseteq \text{Supp}f^*$ and f^* is the sum of the conjugates of f. Thus $L_1 \subseteq \text{Supp}gfg^{-1}$ for some $g \in G$. Because the defect groups of f and gfg^{-1} are of the same order, we have $|D_1 \cap N| \leqslant |\delta(f)|$. Hence $D_1 \cap N$ is N-conjugate to $\delta(f)$, proving (iii).

(iv) Since $f \in Z(FN)$, we have $N \subseteq G(f)$ and so D_1N is a subgroup of $G(f)$. For a given $g \in G$, gfg^{-1} is a block idempotent of FN whose associated irreducible representation is given by.

$$\mu^g(z) = \mu(g^{-1}zg) \qquad \text{for all} \quad z \in Z(FN)$$

Invoking (7), we derive

$$\sum_{i=1}^{m} \mu^{t_i}(L^+) \neq 0 \tag{9}$$

Assume that $g \in G(f)$. Then μ^g and μ are equivalent one dimensional repressentations of $Z(FN)$ and hence $\mu^g = \mu$.

Let $\{w_j \in G | 1 \leqslant j \leqslant s\}$ be a full system of representatives of the double cosets $G(f)wN_G(L)$ in G. Choose $t_{ij} \in G(f)$ such that $\{t_{ij}w_i | j = 1,\ldots,n_i\}$ is a left transversal for $N_G(L)$ in $G(f)w_iN_G(L)$. Then

$$\sum_{i=1}^{m} \mu^{t_i}(L^+) = \sum_{i=1}^{s} \sum_{j=1}^{n_i} \mu^{t_{ij}w_i}(L^+) = \sum_{i=1}^{s} n_i\mu^{w_i}(L^+) ,$$

$\mu^{t_{ij}w_i} = (\mu^{t_{ij}})^{w_i} = \mu^{w_i}$ for all $j \in \{1,\dots,n_i\}$. Invoking (9), we infer that

$$\sum_{i=1}^{s} n_i \mu^{w_i}(L^+) \neq 0$$

and hence that there is an integer $d \in \{1,\dots,s\}$ such that $p \nmid n_d$. Because

$$n_d = \frac{|G(f)w_d N_G(L)|}{|N_G(L)|} = \frac{|G(f)N_G(w_d L w_d^{-1})|}{|N_G(w_d L w_d^{-1})|}$$

$$= \frac{|G(f)|}{|G(f) \cap N_G(w_d L w_d^{-1})|}$$

it follows that

$$p \nmid |G(f) : (G(f) \cap N_G(w_d L w_d^{-1}))| \tag{10}$$

By Lemma 10.5, $N_G(w_d L w_d^{-1}) = C_G(w_d x w_d^{-1})N$, so by (10) $C_G(w_d x w_d^{-1})N$ contains a Sylow p-subgroup of $G(f)$. Thus

$$\nu(|G(f)|) \leqslant \nu(|C_G(w_d x w_d^{-1})N|) \tag{11}$$

where $\nu(n)$ denotes the highest power of p dividing n. Because $x \in C_k$ and, by (i), D and D_1 are conjugate, it follows that D_1 is G-conjugate to a Sylow p-subgroup of $C_G(w_d x w_d^{-1})$. The conclusion is that

$$\nu(|C_G(w_d x w_d^{-1})N|) = \nu(|D_1 N|) \tag{12}$$

proving (iv), by applying (11) and (12). ∎

At this point we begin to attack Morita's theorem directly.

10.7. LEMMA. Let S be a ring, let $1 = e_1 + e_2 + \dots + e_n$ be a decomposition of 1 into a sum of orthogonal idempotents, and let $R = e_1 S e_1$. Assume that G is a subgroup of the unit group of S which permutes the set $\{e_1,\dots,e_n\}$ transitively by conjugation. Then $S \cong M_n(R)$ as rings and R-modules.

Proof. Since G acts transitively, we may write $e_i = g_i^{-1} e_1 g_i$ for a suitable $g_i \in G$, $1 \leqslant i \leqslant n$, with $g_1 = 1$. Given $i,j \in \{1,\dots,n\}$, set

$$e_{ij} = g_i^{-1} e_1 g_j$$

and fix $k \neq j$. Then $g_j^{-1}e_1g_j \neq g_k^{-1}e_1g_k$ so that

$$(g_j^{-1}e_1g_j)(g_k^{-1}e_1g_k) = 0$$

Thus

$$\begin{aligned} e_{ij}e_{kt} &= (g_i^{-1}e_1g_j)(g_k^{-1}e_1g_t) \\ &= (g_i^{-1}g_j)(g_j^{-1}e_1g_j)(g_k^{-1}e_1g_k) \\ &= 0 \end{aligned}$$

and

$$\begin{aligned} e_{ij}e_{jt} &= (g_i^{-1}e_1g_j)(g_j^{-1}e_1g_t) \\ &= g_i^{-1}e_1g_t \\ &= e_{it} \end{aligned}$$

By definition, $e_{ii} = e_i$ and therefore $1 = e_{11} + e_{22} + \ldots + e_{nn}$. The desired assertion is now a consequence of Proposition 1.5.1. ∎

10.8. LEMMA. Let $\{e_1, \ldots, e_s\}$ be a full set of representatives of the G-conjugacy classes of block idempotents of FN. For each $i \in \{1, \ldots, s\}$, let e_i^* be the sum of all G-conjugates of e_i, let $G_i = G(e_i)$ and let $n_i = (G : G_i)$. Then

(i) $FGe_i^* \cong M_{n_i}(FG_ie_i)$ $\quad (1 \leqslant i \leqslant s)$

(ii) $FG \cong \prod_{i=1}^{s} M_{n_i}(FG_ie_i)$

Proof. By Lemma 10.1, each e_i^* is a central idempotent of FG. Taking into account that $1 = e_1^* + \ldots + e_s^*$ and that $e_i^*e_j^* = 0$ for $i \neq j$, we see that (ii) is a consequence of (i).

To prove (i), fix a block idempotent e of FN and put $G^* = G(e)$. Let $\{e = v_1, \ldots, v_n\}$ be the G-orbit of e. It is obvious that $n = (G : G^*)$. Note also that $S = FGe^*$ is a ring with identity element e^*, and S contains the group of units $\bar{G} = Ge^*$. By hypothesis, $\bar{G}$ acts transitively by conjugation on the set $\{v_1, \ldots, v_n\}$. Because $e^* = v_1 + \ldots + v_n$ is a decomposition of the identity element of S into a sum of orthogonal idempotents of S, Lemma 10.7 tells us that

$$FGe^* = S \cong M_n(eSe)$$

Furthermore, since $e^*e = e$, we have

$$eSe = e(FGe^*)e = eFGe$$

and it remains to prove that $eFGe = FG^*e$.

To this end, note that $eFGe$ is the F-linear span of the set $\{ege \mid g \in G\}$. If $g \in G^*$, then certainly $ege = ge$. On the other hand, if $g \notin G^*$, then $ege = g(g^{-1}eg)e = 0$. Hence $eFGe = FG^*e$ and the result follows. ∎

We have now come to the demonstration for which this section has been developed. The following result is essentially due to Morita (1951) (see also Nakayama and Shoda (1936), Clifford (1937), and Koshitani (1982c)).

10.9. THEOREM. Let F be an algebraically closed field of characteristic $p > 0$, let $N = O_{p'}(G)$ and let $\{e_1, \ldots, e_s\}$ be a full set of representatives of G-conjugacy classes of block idempotents of FN, where by convention

$$e_1 = |N|^{-1} \sum_{x \in N} x$$

For each $i \in \{1, \ldots, s\}$, let e_i^* be the sum of all G-conjugates of e_i (so that $e_1 = e_1^*$), let $G_i = G(e_i)$ and let $m_i = (G : G_i)$. Finally, write

$$FNe_i \cong M_{n_i}(F) \qquad (1 \leqslant i \leqslant s)$$

for suitable positive integer n_i.

(i) $FGe_i^* \cong M_{n_i m_i}(F^{\alpha_i}(G_i/N))$ and $FG \cong \prod_{i=1}^{s} M_{n_i m_i}(F^{\alpha_i}(G_i/N))$ for suitable $\alpha_i \in Z^2(G_i/N, F^*)$, $1 \leqslant i \leqslant s$. Furthermore, s coincides with the number of conjugacy classes of G contained in N and $FGe_1^* \cong F(G/N)$.

(ii) If FGe_i^* is a block of FG for some $i \in \{1, \ldots, s\}$, then a Sylow p-subgroup of G_i is a defect group of FGe_i^*.

(iii) If $P \in \mathrm{Syl}_p(G)$ is normal in G, then each FGe_i^* is a block of FG.

Proof. That s is equal to the number of conjugacy classes of G contained in N is a consequence of Lemma 7.6(i). For the sake of clarity, we divide the rest of the proof into two steps.

Step 1. Here we prove that if H is a normal subgroup of G and if $e \in FH$ is a central idempotent of FG with $FHe \cong M_n(F)$, then

$$FGe \cong M_n(F^{\alpha}(G/H)) \qquad \text{for some } \alpha \in Z^2(G/H,F^*)$$

To begin with, put $S = FGe$ and note that e is the identity element of the ring S and that $S \supseteq FHe$. Let $\{e_{ij} | 1 \leqslant i,j \leqslant n\}$ be a set of matrix units in FHe. Then $e = e_{11} + e_{22} + \ldots + e_{nn}$, so by Proposition 1.5.1

$$FGe = S \cong M_n(e_{11} S e_{11})$$

Since $ee_{11} = e_{11}$, we have $e_{11} S e_{11} = e_{11}(FGe)e_{11} = e_{11} FGe_{11}$ and thus

$$FGe \cong M_n(e_{11} FGe_{11})$$

It therefore suffices to verify that

$$e_{11} FGe_{11} \cong F^{\alpha}(G/H) \tag{13}$$

for some $\alpha \in Z^2(G/H,F^*)$. Owing to Noether-Skolem Theorem (see Reiner (1975)), given $g \in G, ge_{11}g^{-1}$ is conjugate to e_{11} by an invertible matrix $u' \in FHe$. Setting $u = u' + (1-e)$, we see that $u^{-1}e_{11}u = ge_{11}g^{-1}$. Hence, for each $g \in G$, there exists a unit $u \in FH$ such that ug commutes with e_{11}.

Let T denote a transversal for H in G, and for each $t \in T$ choose a unit $u_t \in FH$ such that $u_t t$ commutes with e_{11}. If y is any element of FG, then y can be written in the form

$$y = \sum_{t\in T} y_t t = \sum_{t\in T} (y_t u_t^{-1}) u_t t$$

with $y_t \in FH$ and hence

$$\begin{aligned} e_{11} y e_{11} &= \sum_{t\in T} e_{11} (y_t u_t^{-1}) u_t t e_{11} \\ &= \sum_{t\in T} e_{11} (y_t u_t^{-1}) e_{11} u_t t \end{aligned}$$

Since $e_{11}(y_t u_t^{-1})e_{11} \in e_{11}FHe_{11}$ and

$$e_{11}FHe_{11} = e_{11}(FHe)e_{11} = Fe_{11}, \tag{14}$$

it follows that $e_{11}FGe_{11}$ is the F-linear span of the set $\{e_{11}u_t t \mid t \in T\}$. Note that $e_{11}u_t t \neq 0$ (since $u_t t$ is a unit) and that the elements $e_{11}u_t t$ have disjoint supports (lying in the cosets of Ht). Hence $\{e_{11}u_t t \mid t \in T\}$ is an F-basis for $e_{11}FGe_{11}$.

For each $t \in T$, set $\overline{TH} = e_{11}u_t t$. Then, by the foregoing, $e_{11}FGe_{11}$ is an associative F-algebra with basis $\{\overline{tH} \mid t \in T\}$. Let $x,y \in T$ and write $xy = ht$ with $h \in H, t \in T$. Then $(u_t t)^{-1}(u_x x)(u_y y)$ is a nonzero element in FH and so, by (14), there exists a nonzero $\alpha(x,y)$ in F such that

$$e_{11}((u_t t)^{-1}(u_x x)(u_y y))e_{11} = \alpha(x,y)e_{11}$$

But then

$$\overline{xyH}^{-1}\,\overline{xH}\,\overline{yH} = e_{11}((u_t t)^{-1}(u_x x)(u_y y))e_{11} = \alpha(x,y)e_{11}$$

proving (13) and hence the required assertion.

Step 2. Completion of the proof.

Observe that the second isomorphism of (i) is a consequence of the first. Applying Lemma 10.8(i) and Step 1, we have

$$\begin{aligned} FGe_i^* &\cong M_{m_i}(FG_ie_i) \cong M_{m_i}(M_{n_i}(F^{\alpha_i}(G_i/N)) \\ &\cong M_{m_in_i}(F^{\alpha_i}(G_i/N)), \end{aligned}$$

proving (i).

Assume that $i \in \{1,\dots,s\}$ is such that FGe_i^* is a block of FG. Then, by Proposition 10.6(ii), there is a p-subgroup D_i of $G(e_i)$ which is a defect group of FGe_i^*. Invoking Proposition 10.6(iv), we see that D_i is in fact a Sylow p-subgroup of $G(e_i)$, proving (ii).

Finally, assume that $P \in \mathrm{Syl}_p(G)$ is normal in G. Then, by Corollary 9.13, P is a defect group of any block of FG. Hence, by Theorem 9.23(ii), the number of blocks of FG is equal to the number of p-regular classes of G with defect group P. But a p-regular C of G has defect group P if and only if $C \subseteq N$. The desired conclusion is therefore a consequence of (i). ∎

10.10. COROLLARY. Let G be a p-nilpotent group and let F be an algebraically closed field of characteristic p. If B is a block of FG with defect group D, then $B \cong M_n(FD)$ for some $n \geqslant 1$.

Proof. We keep the notation of Theorem 10.9. Let D_i be a Sylow p-subgroup of G_i, $1 \leqslant i \leqslant s$. Since G is p-nilpotent $G_i/N \cong D_i$ and hence, by Theorem 10.9(i) and Lemma 4.9, $FGe_i^* \cong M_{n_i m_i}(FD_i)$. Since FD_i is indecomposable, FGe_i^* is a block of FG. Hence, by Theorem 10.9(i),(iv), D_i is a defect group of FGe_i^* and the FGe_i^*, $1 \leqslant i \leqslant s$, are all blocks of FG. So the corollary is true. ∎

10.11. COROLLARY. Further to the assumptions and notation of Theorem 10.9, assume that G is p-nilpotent and let D_i be a Sylow p-subgroup of G_i, $1 \leqslant i \leqslant s$.

(i) For any block idempotent e_i of FN, e_i^* is the unique block idempotent of FG which covers e_i.

(ii) $\dim_F J(FGe_i^*) = n_i^2 m_i^2(|D_i|-1)$

(iii) $\dim_F J(FG) = \sum_{i=1}^{s} n_i^2 m_i^2(|D_i|-1)$

Proof. As we have seen in the proof of Corollary 10.10, $FGe_1^*,\ldots,FG^*e_s$ are all blocks of FG and $FGe_i^* \cong M_{n_i m_i}(FD_i)$. This proves (i) and also that (iii) is a consequence of (ii). Since $J(FD_i) = I(D_i)$, we have

$$J(FGe_i^*) \cong M_{n_i m_i}(I(D_i))$$

and hence

$$\dim_F J(FGe_i^*) = n_i^2 m_i^2 \dim_F I(D_i) = n_i^2 m_i^2(|D_i|-1),$$

as required. ∎

11. AN APPLICATION : CRITERIA FOR $J(FG) = FG \cdot J(Z(FG))$

Throughout this section, G denotes a finite group and F an algebraically closed field of characteristic $p > 0$. It is clear that

$$FG\cdot J(Z(FG)) \subseteq J(FG)$$

The problem that motivates this section is to discover necessary and sufficient conditions under which the equality $FG\cdot J(Z(FG)) = J(FG)$ holds. Our main tool will be a consequence of Morita's theorem (Corollary 10.10).

11.1. LEMMA. Let H be a subgroup of G, let R be an arbitrary ring and let

$$\ell(I(H)) = \{x \in RG \mid x\, I(H) = 0\}$$

Then $\ell(I(H)) = RG(\sum_{h\in H} h)$ and, in particular, $\ell(I(G)) = R(\sum_{g\in G} g)$.

Proof. Suppose that $x = \sum_{g\in G} x_g g \in \ell(I(H))$, $x_g \in R$. Then $x(h-1) = 0$ or $x = xh$, for all $h \in H$. Hence $x_g = x_{gh}$ for all $g \in G, h \in H$. Therefore x_g is constant on the cosets gH of H. This proves that

$$\ell(I(H)) \subseteq RG(\sum_{h\in H} h)$$

The reverse inclusion being obvious, the result follows. ■

11.2. LEMMA. Let P be a p-group and let R be a field of characteristic p. If t is the nilpotency index of $J(RP)$, then $J(RP)^{t-1} = R(\sum_{g\in P} g$

Proof. By Corollary 1.2, $J(RP) = I(P)$ and so

$$0 = J(RP)^t = J(RP)^{t-1}I(P)$$

It follows, from Lemma 11.1, that $J(RP)^{t-1} \subseteq R(\sum_{g\in P} g)$. But $J(RP)^{t-1} \neq 0$ and $R(\sum_{g\in P} g)$ is one-dimensional subspace of RG. Hence

$$J(RP)^{t-1} = R(\sum_{g\in P} g)$$

as asserted. ■

11.3. LEMMA. Assume that there exists a primitive idempotent u of FG such that all composition factors of FGu are isomorphic to the trivial FG-module. Then G is p-nilpotent.

Proof. Denote by 1_G the trivial FG-module. Then, by Proposition 1.10.16 (ii) and Theorem 3.6(ii), the multiplicity of 1_G as a composition factor of FG is $\dim_F FGu$. On the other hand, by hypothesis, $\dim_F FGu$ is equal to the multiplicity of 1_G as a composition factor of FGu. It follows that if

$$FG = FGe_1 \oplus \dots \oplus FGe_n$$

is a decomposition of FG into principal indecomposable modules with $u = e_1$, then none of the $FGe_i, 2 \leqslant i \leqslant n$, has 1_G as a composition factor. Thus e_1 is linked to none of the e_i, $2 \leqslant i \leqslant n$, which implies, by Proposition 1.10.14, that FGe_1 is the principal block of FG. Since 1_G is the only irreducible module belonging to FGe_1, we conclude, from Proposition 8.10(ii), that G is p-nilpotent. ■

11.4. LEMMA. Let $B = B(e)$ be the principal block of FG. If

$$I(G)J(FG)e = J(FG)I(G)e ,$$

then G is p-nilpotent.

Proof. Let u be a primitive idempotent of FG such that $FGu/J(FG)u$ is the trivial FG-module. Then, for a given $g \in G, gu \equiv u(\text{mod } J(FG)u)$, so $(g-1)u \in J(FG)u$ and therefore $I(G)u \subseteq J(FG)u$. Since the opposite containment is trivial, we conclude that $I(G)u = J(FG)u$. Hence we have

$$\begin{aligned} I(G)J(FG)u &= I(G)J(FG)eu \\ &= J(FG)I(G)eu \\ &= J(FG)I(G)u \\ &= J(FG)^2u \end{aligned}$$

Recurring this, we get

$$I(G)J(FG)^n u = J(FG)^{n+1}u \qquad \text{for all } n \geqslant 1$$

which shows that G acts trivially on each factor of the series

$$FGu \supset J(FG)u \supset \dots \supset J(FG)^s u = 0$$

This implies that each composition factor of FGu is isomorphic to the trivial FG-module. Now apply Lemma 11.3. ∎

We are now ready to prove

11.5. THEOREM (Tshushima (1978b), Külshammer (1979)). The following conditions are equivalent.

(i) $J(FG) = FG\cdot J(Z(FG))$

(ii) $J(B) = B\cdot J(Z(B))$ for each block B of FG

(iii) $J(B) = B\cdot J(Z(B))$ for the principal block B of FG

(iv) G is p-nilpotent with abelian Sylow p-subgroups.

Proof. (i) ⇒ (ii): Write $FG = B_1 \oplus \ldots \oplus B_n$, where $B_1,\ldots,B_n$ are all blocks of FG. Then

$$J(FG) = J(B_1) \oplus \ldots \oplus J(B_n)$$

and

$$J(Z(FG)) = J(Z(B_1)) \oplus \ldots \oplus J(Z(B_n))$$

Hence

$$\begin{aligned} FG\cdot J(Z(FG)) &= B_1J(Z(B_1)) \oplus \ldots \oplus B_nJ(Z(B_n)) \\ &= J(B_1) \oplus \ldots \oplus J(B_n). \end{aligned}$$

Thus $J(B_i) = B_iJ(Z(B_i))$ for all $i \in \{1,\ldots,n\}$, as required.

(ii) ⇒ (iii): Obvious

(iii) ⇒ (iv): Write $B = B(e)$ so that $J(B) = J(FG)e$. Since

$$B\cdot J(Z(B)) = FG\cdot J(Z(B)),$$

we have $J(FG)e = FG\cdot J(Z(B))$. Hence

$$\begin{aligned} I(G)J(FG)e &= I(G)FG\cdot J(Z(B)) = I(G)\cdot J(Z(B)) \\ &= J(Z(B))I(G) = (J(Z(B))FG)I(G) \\ &= J(FG)e\ I(G) = J(FG)I(G)e, \end{aligned}$$

proving that G is p-nilpotent, by applying Lemma 11.4.

Let P be a Sylow p-subgroup of G. Then, by Proposition 1.20, $B \cong FP$ and thus

$$J(FP) = FP\cdot J(Z(FP)) \tag{1}$$

To prove that P is abelian, we use induction on the order of P. It is clear that for any normal subgroup H of P.

$$J(F(P/H)) = F(P/H)J(Z(F(P/H)))$$

Let Z be the centre of P, let z be an element of order p in Z and put

$$\sigma = \sum_{g\in P'} g \quad \text{and} \quad \tau = \sum_{y\in Z} y$$

We may assume that $P/\langle z\rangle$ is abelian, by the induction hypothesis.

Assume by way of contradiction that P is not abelian. Then we have $P' = \langle z\rangle$. Given $x \in P$, we have $|xP'| = p$ and xP' is the conjugacy class of x unless $x \in Z$. Hence $J(Z(FP))$ is the F-linear span of the set

$$\{g-1, x\sigma \mid g \in Z, x \in P-Z\} \tag{2}$$

Let t be the nilpotency index of $J(FG)$. By Lemma 11.2, we have

$$J(FZ)^{t-1} = F\tau \tag{3}$$

We now claim that it suffices to prove that

$$J(Z(FP))^{t} = 0 \tag{4}$$

Indeed, if (4) is true, then $J(FP)^{t} = 0$ by virtue of (1). On the other hand, taking $y \in P-Z$, we have $(y-1)\tau \neq 0$ and, by (3), $\tau \in J(FZ)^{t-1}$, so

$$(y-1)\tau \in (y-1)J(FZ)^{t-1} \subseteq J(FP)^{t} = 0,$$

a contradiction.

To prove (4), we note that

$$(x\sigma)(y\sigma) = xy\sigma^2 = 0 \tag{5}$$

Since $G' \subseteq Z$ we also have $\sigma\tau = p\tau = 0$. Hence, applying (3), we have

$$(x\sigma)\prod_{i=1}^{t-1}(z_i-1) \in (x\sigma)J(FZ)^{t-1} = (x\sigma)F\tau = 0 \tag{6}$$

for all $z_1,\ldots,z_{t-1} \in Z$, $x \in P-Z$. Finally, by definition of t,

$$\prod_{i=1}^{t}(z_i-1) = 0 \tag{7}$$

where $z_1,\ldots,z_t$ are arbitrary elements of Z. Taking into account (2) and (5)-(7), we immediately deduce that (4) holds, as required.

(iv) $\Rightarrow$ (i): By Corollary 10.10, for each $i \in \{1,\ldots,n\}$, there exists an abelian p-group D_i such that $B_i \cong M_{n_i}(FD_i)$ for some $n_i \geqslant 1$. Hence $J(B_i) \cong M_{n_i}(J(FD_i))$ and $Z(B_i) \cong FD_i$. It follows that $J(B_i) \cong M_{n_i}(J(Z(B_i)))$ and hence $J(B_i) = B_iJ(Z(B_i))$, $1 \leqslant i \leqslant n$. Thus

$$J(FG) = B_1J(Z(B_1)) \oplus \ldots \oplus B_nJ(Z(B_n)) = FG\cdot J(Z(FG))$$

and the result follows. ■

12. GROUP ALGEBRAS WITH RADICALS OF SQUARE ZERO

Throughout this section, G denotes a finite group and F an arbitrary field of characteristic $p > 0$. Our aim is to discover the structure of G in case $J(FG)^2 = 0$.

12.1. LEMMA. Assume that the Sylow 2-subgroups of G are cyclic. Then G is 2-nilpotent.

Proof. Let P be a Sylow 2-subgroup of G. Since P is cyclic, Aut(P) is a 2-group. Hence the required assertion follows by virtue of property (ii) in the proof of Lemma 9.22. ■

12.2. LEMMA. Let B be the principal block of FG. Then the Sylow p-subgroups of G are defect groups of B. In particular, by Theorem 6.4(i), if $J(B) = 0$ then $J(FG) = 0$.

Proof. Let e be the block idempotent of FG belonging to B. We may write e in the form

$$e = \lambda_1 C_1^+ + \ldots + \lambda_r C_r^+ \qquad (0 \neq \lambda_i \in F)$$

for some p-regular classes $C_1,\ldots,C_r$ of G. Denote by V the trivial FG-module. Since V belongs to B, we have $ev = v$ for all $v \in V$. Hence

$$v = (\lambda_1|C_1| + \ldots + \lambda_r|C_r|)v \qquad \text{for all } v \in V$$

which implies that $p \nmid |C|$ for some p-regular class C in Supp$e$. Thus a Sylow p-subgroup of G is a defect group of C, as required. ∎

12.3. LEMMA. Let B be a block of FG with $J(B) \neq 0$. Then $J(B)^2 = 0$ if and only if all irreducible FG-modules in B are isomorphic and all projective indecomposable FG-modules in B have composition length 2.

Proof. Assume that $J(B)^2 = 0$ and write

$$B = Be_1 \oplus \ldots \oplus Be_n$$

where $e_1,\ldots,e_n$ are mutually orthogonal primitive idempotents of B. As $J(B) \neq 0$, there exists $i \in \{1,\ldots,n\}$ such that Be_i has composition length > 1. Since $J(B)(J(B)e_i) = 0$, it follows from Theorem 3.8 that $J(B)e_i$ is irreducible and $Be_i/J(B)e_i \cong J(B)e_i$. This shows that Be_i has composition length 2 and that the composition factors of Be_i are isomorphic. Since the idempotents $e_1,\ldots,e_n$ are linked (Proposition 1.10.14) and $Be_i/J(B)e_i \cong Be_j/J(B)e_j$ implies $Be_i \cong Be_j$ (Proposition 1.9.4), the required assertion follows. Conversely, assume that $J(B)^2 \neq 0$. Then $J(B)^2 e_i \neq 0$ for some $i \in \{1,\ldots,n\}$ and so

$$0 \subset (J)B^2 e_i \subset J(B)e_i \subset Be_i$$

Thus the composition length of Be_i is greater than 2, as required. ∎

We are now ready to prove the main result of this section. The equivalence of (i) and (iii) in the theorem below was established by Wallace (1962a).

12.4. THEOREM. Let F be an arbitrary field of characteristic $p > 0$ and let G be a group of order $p^a m$ with $a \geqslant 1$ and $(p,m) = 1$. Then the following conditions are equivalent:

(i) $J(FG)^2 = 0$

(ii) $J(B)^2 = 0$, where B is the principal block of FG

(iii) $p^a = 2$

(iv) $p = 2$ and G has a normal subgroup of odd order and of index 2

Proof. (i) ⇒ (ii): Obvious

(ii) ⇒ (iii): By hypothesis, p divides the order of G and so $J(FG) \neq 0$. Hence by Lemma 12.2, $J(B) \neq 0$. Let V be a projective indecomposible module in B. Since the trivial FG-module belongs to B, Lemma 12.3 implies that $\dim_F V = 2$. But $p^a \mid \dim_F V$ by Lemma 8.2. Thus $p^a = 2$ as required.

(iii) ⇒ (iv) By hypothesis, the Sylow 2-subgroup of G is cyclic of order 2. Now apply Lemma 12.1.

(iv) ⇒ (i): By hypothesis, G is 2-solvable and the Sylow 2-subgroup of G is of order 2. Hence the required assertion follows by virtue of Proposition 1.9.■

13. GROUP ALGEBRAS WITH CENTRAL RADICALS

Throughout this section, G denotes a finite group and F an arbitrary field of characteristic $p > 0$. Our aim is to present Spiegel's ring-theoretic proof (1970) of Wallace's theorem (1962b) which determines the structure of G in case

$$J(FG) \subseteq Z(FG)$$

13.1. LEMMA. If $J(FG) \subseteq Z(FG)$, then G is p-nilpotent with abelian Sylow p-subgroups. In particular, $p \nmid |G'|$.

Proof. Let K be the algebraic closure of F. Then, by Corollary 1.18,

$$J(KG) = K \otimes_F J(FG) \subseteq K \otimes_F Z(FG) = Z(KG)$$

and so, by Theorem 11.5, G is p-nilpotent with abelian Sylow p-subgroups. Since the latter implies that $G' \subseteq O_{p'}(G)$, the result follows. ■

13.2. LEMMA. Let X be an ideal of FG. Then the following conditions are equivalent:

(i) $X \subseteq Z(FG)$

(ii) $X \cdot I(G') = 0$

(iii) $X \subseteq FGs$, where $s = \sum_{g \in G'} g$

Proof. (i) ⇒ (ii): Let $x \in X$ and $a,b \in G$. Then

$$x(ab) = (xa)b = b(xa) = (bx)a = x(ba)$$

and $x(ab-ba) = 0$. Hence $x(aba^{-1}b^{-1}-1) = 0$. Since G' is generated by the $aba^{-1}b^{-1}, a,b \in G$, the required assertion is a consequence of Proposition 2.1.10.

(ii) ⇒ (iii): Direct consequence of Lemma 11.1

(iii) ⇒ (i): Clearly $s \in Z(FG)$. Since FGs is the F-linear span of the set $\{gs \mid g \in G\}$ and since for every $t \in G$, we have

$$\begin{aligned} t(gs) &= (tg)s = s(tg) \\ &= s(tg)(t^{-1}g^{-1}gt) \\ &= s(tgt^{-1}g^{-1})gt \\ &= sgt = (gs)t \end{aligned}$$

it follows that $FGs \subseteq Z(FG)$. ∎

13.3. LEMMA. If $J(FG) \subseteq Z(FG)$, then the following are equivalent:

(i) $J(FG) \neq FGs$, where $s = \sum_{g \in G'} g$

(ii) $p \nmid |G'|$

Proof. (i) ⇒ (ii): By Lemma 13.3, $J(FG) \subseteq FGs$. If $p \mid |G'|$, then $s^2 = |G'|s = 0$ and $FGs \subseteq J(FG)$. Hence $J(FG) = FGs$ as required

(ii) ⇒ (i): If $p \nmid |G'|$, then $e = |G'|^{-1}s$ is a nonzero idempotent in FGs. Thus $J(FG) \neq FGs$ and the result follows. ∎

The following theorem was proved by Wallace (1962b) for group algebras over algebraically closed fields. Its generalization and the proof given here are due to Spiegel (1970).

13.4. THEOREM. Let F be an arbitrary field of characteristic p and let G be a finite group. Then $J(FG) \subseteq Z(FG)$ if and only if G is one of the following three types:

(i) G has order prime to p

(ii) G is abelain

(iii) If P is a Sylow p-subgroup of G, then $G'P$ is a Frobenius group with complement P and kernel G'.

Proof. Let G be a nonabelian group whose order is divisible by p. It clearly suffices to prove that

$$J(FG) \subseteq Z(FG) \quad \text{if and only if (iii) holds}$$

By Theorem 8.11, the latter is equivalent to

$$J(FG) \subseteq Z(FG) \quad \text{if and only if} \quad \dim_F[J(F(G'P)] = |P|-1 \quad \text{and} \quad p \nmid |G'| \tag{1}$$

Assume that $J(FG) \subseteq Z(FG)$ and put $s = \sum_{g\in G'} g$. By Lemmas 13.1, 13.2 and 13.3, $J(FG) \subset FGs$ and $p \nmid |G'|$. Clearly $N = G'P$ is a normal subgroup of G whose index $(G:N)$ is not divisible by p. Hence, by Proposition 1.8(ii),

$$\dim_F J(FG) = (G:N)\dim_F J(FN) \tag{2}$$

Since $J(FG) \subset FGs$, we have

$$\dim_F J(FG) < \dim_F FGs = (G:G') \tag{3}$$

Invoking (2) and (3) we therefore derive

$$\dim_F J(FN) < |P|$$

On the other hand, by Theorem 8.4(i), $\dim_F J(FN) \geqslant |P|-1$. Thus $\dim_F J(FN) = |P|-1$, as required. Conversely assume that $\dim_F (FN) = |P|-1$ and that $p \nmid |G'|$. Let

$$s = \sum_{g\in G'} g \quad \text{and} \quad e = |G'|^{-1}s$$

Then $FNe \cong FP$ so $\dim_F J(FNe) = |P|-1 = \dim_F J(FN)$ and thus $J(FN) = J(FNe)$. Hence

$$J(FN) = J(FNe) = J(FN)e \subseteq FNs \tag{4}$$

Since N is a normal subgroup of G of p'-index. Proposition 1.8(ii) ensures that $J(FG) = FG\cdot J(FN)$. Invoking (4) we therefore conclude that

$$J(FG) \subseteq FGs$$

Hence $J(FG) \subseteq Z(FG)$ by Lemma 13.2. Thus completes the proof of the theorem. ∎

13.5. REMARK. Let G be a group such that $G'P$ is a Frobenius group with complement P and kernel G'. Then $P \cong G'P/G'$ is abelian and hence, by Remark 7.11, is in fact cyclic. Furthermore, by Lemma 7.1, G' is nilpotent. ∎

14. COMMUTATIVITY OF THE RADICAL OF THE PRINCIPAL BLOCK

Throughout this section, G denotes a finite group, F an arbitrary field of characteristic $p > 0$, 1_G the trivial FG-module and B_0 the principal block of FG. Our aim is to discover necessary and sufficient conditions for the commutativity of $J(B_0)$.

Let A be a finite-dimensional algebra over a field K and let B be a block of A. We say that B is *quasi-primary* if in decomposition

$$B = Be_1 \oplus \dots \oplus Be_n$$

as a direct sum of principal indecomposable modules, each Be_i appears with the same multiplicity. If K is a splitting field for B, then the following conditions are equivalent:

(i) B is quasi-primary

(ii) All irreducible B-modules have the same K-dimension

(iii) $B/J(B)$ is a direct product of isomorphic simple algebras

Indeed, the equivalence of (ii) and (iii) is trivial, while the equivalence of (i) and (ii) is a consequence of Proposition 1.10.16(ii).

Recall that if V and W are FG-modules, then the tensor product $V \otimes_F W$ is an FG-module via

$$g(v \otimes w) = gv \otimes gw \qquad (g \in G, v \in V, w \in W)$$

14.1. LEMMA. Let V be an FG-module of dimension 1 and let W be an irreducible FG-module. Then $V \otimes_F W$ is an irreducible FG-module

Proof. By hypothesis, $V = Fv$ for some nonzero v in V. Let X be a nonzero submodule of $V \otimes_F W$. Then every $x \in X$ can be written as $x = v \otimes w$ for a $w \in W$. It is a consequence of the definition of the action of G on $V \otimes_F W$ that the set

$$X_1 = \{w \in W | v \otimes w \in X\} \neq 0$$

is a submodule of W and hence $X_1 = W$. But V has dimension 1, so

$$X = \{v \otimes w | w \in X_1\} = V \otimes_F W,$$

which proves the lemma. ∎

14.2. LEMMA. Let U, V and W be FG-modules with $U \subseteq V$. Then

$$(V \otimes_F W)/(U \otimes_F W) \cong (V/U) \otimes_F W \qquad \text{as } FG\text{-modules}$$

Proof. The canonical map

$$\begin{cases} V \otimes_F W \longrightarrow (V/U) \otimes_F W \\ v \otimes w \longmapsto (v+U) \otimes w \end{cases}$$

is surjective F-homomorphism whose kernel is $U \otimes_F W$. Since the map obviously preserves the action of G, the result follows. ∎

14.3. LEMMA. Let U be any indecomposable projective FG-module. If the composition factors of $P(1_G)$ have dimension 1, then the dimensions of all composition factors of U are equal.

Proof. Put $V = U/J(FG)U$. We know, from Lemma 8.3, that U is a direct summand of $P(1_G) \otimes_F V$. Let

$$P(1_G) = X_1 \supset X_2 \supset \ldots \supset X_n = 0$$

be a composition series of $P(1_G)$. Owing to Lemmas 12.1 and 12.2,

$$P(1_G) \otimes_F V = X_1 \otimes_F V \supset X_2 \otimes_F V \supset \ldots \supset X_n \otimes_F V = 0$$

is a composition series of $P(1_G) \otimes_F V$ and all composition factors have the same dimension, namely $\dim_F V$. This proves the required assertion. ∎

14.4. LEMMA. The following conditions are equivalent:

(i) All irreducible FG-modules belonging to the same block have the same F-dimension

(ii) All irreducible FG-modules belonging to the principal block have F-dimension 1.

(iii) All composition factors of $P(1_G)$ have F-dimension 1.

Proof. (i) ⇒ (ii): Obvious

(ii) ⇒ (iii): Direct consequence of the fact that all composition factors of $P(1_G)$ belong to B_0.

(iii) ⇒ (i): Apply Lemma 14.3 and the fact that the indecomposable projective modules in a block are linked. ■

The proof of the implications (i) ⇒ (ii) and (ii) ⇒ (iii) below is extracted from a work of Hamernik (1975b).

14.5. PROPOSITION (Morita (1951)). Let F be an algebraically closed field of characteristic $p > 0$ and let P be a Sylow p-subgroup of G. Then the following conditions are equivalent:

(i) $PO_{p'}(G) \lhd G$ and $G/PO_{p'}(G)$ is abelian

(ii) B_0 is quasi-primary

(iii) Each block B of FG is quasi-primary

Proof. (i) ⇒ (ii): Set $\bar{G} = G/O_{p'}(G)$ and $\bar{P} = PO_{p'}(G)/O_{p'}(G)$. Since G is p-solvable (and hence p-constrained), it follows from Proposition 1.20 that $B_0 \cong F\bar{G}$. By hypothesis, $\bar{G}$ is a group with normal Sylow p-subgroup $\bar{P}$ and abelian factor group. Hence $F\bar{G}$ has only irreducible representations of dimension 1. Since the same is true for B_0, the assertion follows.

(ii) ⇒ (iii): Follows from Lemma 14.4.

(iii) ⇒ (i): Write $B_0 = B_0(e)$ and note that, by Proposition 8.10(i),

$$O_{p'}(G) = \{g \in G \mid ge = e\}$$

Let

$$\Gamma : g \mapsto (\Gamma_{ij}(g))$$

be the matrix representation of G afforded by the FG-module FGe. Since $g \in \mathrm{Ker}\Gamma$ if and only if $ge = e$, we have $\mathrm{Ker}\Gamma = O_{p'}(G)$. We may therefore regard Γ as a faithful representation of $G/O_{p'}(G)$. Since the composition factors of FGe belong to B_0, the assumption that B_0 is quasi-primary implies that all these factors are one-dimensional. Thus we may assume that the entries $\Gamma_{ij}(g)$ of the matrix $(\Gamma_{ij}(g))$ are zero if $i < j$. Then it is easily verified that the set of all elements $g \in G$ such that $\Gamma_{jj}(g) = 1$ for all j forms a Sylow p-subgroup, say Q of $G/O_{p'}(G)$. Moreover, Q is normal in $G/O_{p'}(G)$ and the corresponding factor group is abelian. This shows that $PO_{p'}(G) \lhd G$ and that $G/PO_{p'}(G)$ is abelian. So the proposition is true. ■

14.6. LEMMA. Let F be a splitting field for FG and let B be a block of FG such that $J(B)$ is commutative.

(i) If e is a primitive idempotent of B with $J(B)^2e \neq 0$, then all composition factors of Be are one-dimensional

(ii) If B is the principal block, then B is quasi-primary

Proof. (i) We claim that it suffices to show that $\dim_F \mathrm{Soc}(Be) = 1$. Indeed, assume that $\dim_F \mathrm{Soc}(Be) = 1$ and let V be a composition factor of Be. Since the Cartan matrix of FG is symmetric (Theorem 3.6(vi)), $Be/J(B)e$ is a composition factor of $P(V)$. But, by Theorem 3.8,

$$Be/J(B)e \cong \mathrm{Soc}(Be)$$

and so $P(V)$ has a composition factor of F-dimension 1. If $J(B)^2P(V) = 0$, then the composition factors of $P(V)$ have the same dimension equal to $\dim_F V$. Hence $\dim_F V = 1$. If $J(B)^2P(V) \neq 0$ then $\mathrm{Soc}P(V) \cong V$ is one-dimensional, by hypothesis. This substantiates our claim.

Since $J(B)^2e \neq 0$, we have $\mathrm{Soc}(Be) \subseteq J(B)^2e \subseteq J(B)^2$. Thus we need only show that all composition factors of $J(B)^2$ are one-dimensional. The latter will follow provided we prove that for all $x,y \in J(B)$ and $a,b \in G$, $(ab)(xy) = (ba)(xy)$. This is so since then G' will be in the kernel of the representation of G afforded by $J(B)^2$.

Using the assumption that $J(B)$ is commutative, we have

$$\begin{aligned}(ab)(xy) &= a(bx)y = a(y(bx)) = (ay)(bx)\\ &= (bx)(ay) = b(x(ay)) = b(ay)x = (ba)(yx)\\ &= (ba)(xy),\end{aligned}$$

as required.

(ii) By Lemma 14.4, it suffices to verify that all composition factors of $P(1_G)$ are one-dimensional. By (i), we may assume that $J(B)^2P(1_G) = 0$. But then all composition factors of $P(1_G)$ are isomorphic to 1_G. So the lemma is true. ∎

We have now come to the demonstration for which this section has been developed.

14.7. THEOREM (Hamernik (1975)). Let F be an arbitrary field of characteristic $p > 0$, let G be a finite group and let B_0 be the principal block of FG. Then the following conditions are equivalent:

(i) B_0 is commutative

(ii) $J(B_0) \subseteq Z(B_0)$

(iii) $J(B_0)$ is commutative

(iv) G is p-nilpotent with abelian Sylow p-subgroups

Proof. (i) ⇒ (ii): Obvious

(ii) ⇒ (iii): Obvious

(iii) ⇒ (iv): Let K be the algebraic closure of F and write $B_0 = B_0(e)$. Then

$$e = e_1 + e_2 + \dots + e_s$$

for some block idempotents $e_1,\dots,e_s$ of KG. Since e acts as identity on 1_{FG}, one of the e_i, say e_1, acts as identity on 1_{KG}. Setting $B_i^* = KGe_i$, $1 \leqslant i \leqslant s$, it follows that

$$K \otimes_F B_0 = KGe = B_1^* \oplus \dots \oplus B_s^*$$

and that B_1^* is the principal block of KG. Invoking Corollary 1.18, we derive

$$J(K \otimes_F B_0) = K \otimes_F J(B_0) = J(B_1^*) \oplus \ldots \oplus J(B_s^*)$$

which shows that $J(B_1^*)$ is commutative. Hence we may harmlessly assume that F is algebraically closed.

By Lemma 14.6, B_0 is quasi-primary and so, by Proposition 14.5,

$$PO_{p'}(G) \lhd G \quad \text{and} \quad G/PO_{p'}(G) \quad \text{is abelian}$$

where $P \in \mathrm{Syl}_p(G)$. Because G is p-solvable, Proposition 1.20 implies that

$$e = |O_{p'}(G)|^{-1} \sum_{x \in O_{p'}(G)} x$$

is the block idempotent in B_0. Setting $H = PO_{p'}(G)$, it follows from Lemma7.3. that

$$J(FHe) = I(P)e$$

On the other hand, by Proposition 1.8(ii), $J(FG) = FG \cdot J(FH)$. Hence

$$J(B_0) = J(FG)e = FG \cdot J(FHe) = FG \cdot I(P)e$$

and therefore $FG \cdot I(P)e$ is commutative.

Choose an arbitrary $s \in P$ and $u \in N_G(P)$. By Burnside's criterion, it suffices to check that $usu^{-1} = s$. To this end, we put $usu^{-1} = s_1 \in P$. Then we have

$$\begin{aligned} e(1-s)(1-s_1) &= e(1-s)e\; u(1-s)u^{-1} \\ &= [e(1-s)u\; e(1-s)]u^{-1} \\ &= [ue(1-s)e(1-s)]u^{-1} && \text{(since } FG \cdot I(P)e \text{ is commutative)} \\ &= u\; e(1-s)^2 u^{-1} \\ &= [u\; e(1-s)][(1-s)eu^{-1}] \\ &= [(1-s)eu^{-1}][ue(1-s)] && \text{(since } FG \cdot I(P) \text{ is commutative)} \\ &= e(1-s)^2 \end{aligned}$$

Since for any $x \in FP$, $ex = 0$ implies $x = 0$, we deduce that

$$(1-s)(1-s_1) = (1-s)^2$$

and hence that

$$s_1 - ss_1 = s - s^2$$

If $p \neq 2$ this yields $s_1 = s$. If $p = 2$, then we have $s_1 + ss_1 = s + s^2$, in which case either $s_1 = s$ or $s_1 = s^2$. But if $s_1 = s^2$, then $s = s^3$ and so $s_1 = s^2 = 1$. Hence $s = 1$ and therefore in all cases $s_1 = s$. This completes the proof of the theorem. ■

15. CRITERIA FOR THE COMMUTATIVITY OF $J(FG)$

Throuthout this section, G denotes a finite group and F an arbitrary field of characteristic $p > 0$. Our aim is to determine the structure of G in case $J(FG)$ is commutative. The following lemma is stated in a stronger form than required.

15.1. LEMMA. Let N be a normal p'-subgroup of G. Then the following conditions are equivalent:

(i) $J(FG) \cong J(F(G/N))$

(ii) $\dim_F J(FG) = \dim_F J(F(G/N))$

(iii) Every block B of FG with $J(B) \neq 0$ contains an irreducible FG-module V such that $N \subseteq \text{Ker}V$.

Furthermore, if for $P \in \text{Syl}_p(G)$, $PN \lhd G$ then (i), (ii) and (iii) are equivalent to each of the following conditions:

(iv) $J(FG) = FG \cdot I(P)e$, where $e = |N|^{-1} \sum_{g \in G} g$

(v) $\dim_F J(FG) = \dfrac{|G|(|P|-1)}{|P||N|}$

Proof. Owing to Proposition 2.1.9, e is a central idempotent of FG such that $FGe \cong F(G/N)$.

(i) $\Rightarrow$ (ii): Obvious

(ii) $\Rightarrow$ (iii): Since $J(FG) = J(FGe) \oplus J(FG(1-e))$ and, by hypothesis, $\dim_F J(FGe) = \dim_F J(FG)$, we have $J(FG(1-e)) = 0$. Let B be a block of FG with $J(B) \neq 0$ and let V be an irreducible FG-module in B. Since $J(FG(1-e)) = 0$, we have $B \subseteq FGe$ and hence $ev = v$ for all $v \in V$. Thus, if $n \in N$, then since

$ne = e$, we have

$$nv = (ne)v = ev = v \qquad \text{for all } v \in V$$

This proves that $N \subseteq \mathrm{Ker}V$, as required.

(iii) $\Rightarrow$ (i): Let B be a block of FG with $J(B) \neq 0$. By hypothesis, B contains an irreducible FG-module V with $N \subseteq \mathrm{Ker}V$. Then N acts as identity on V and so $ev = v$ for all $v \in V$. Thus $B \subseteq FGe$ and therefore

$$J(FG) = J(FGe) \cong J(F(G/N))$$

as required.

Now assume that $PN \triangleleft G$. Then PN/N is a normal Sylow p-subgroup of G/N and hence, by Proposition 1.1(iii) and Corollary 1.4,

$$J(F(G/N)) = F(G/N) \cdot I(PN/N) \tag{1}$$

and

$$\dim_F J(F(G/N)) = \frac{|G|(|P|-1)}{|P||N|} \tag{2}$$

By Proposition 2.1.9, the map $F(G/N) \longrightarrow FGe$, $gN \longmapsto ge$ determines an isomorphism of F-algebras. Since $FG \cdot I(P)e$ is the image of $F(G/N)I(PN/N)$ under this map, we deduce from (1) that

$$J(FGe) = FG \cdot I(P)e \cong J(F(G/N)) \tag{3}$$

Thus, by (3), (iv) is equivalent to (i), while by (2), (v) is equivalent to (ii). So the lemma is true. ∎

In the result below, the equivalence of (i), (ii) and (v) was established by Wallace (1965).

15.2. THEOREM. Let F be an arbitrary field of characteristic $p \neq 2$ and let $P \in \mathrm{Syl}_p(G)$. Then the following conditions are equivalent:

(i) $J(FG) \subseteq Z(FG)$

(ii) $J(FG)$ is commutative

(iii) $p \nmid |G'|$ and $\dim_F J(FG) = \dfrac{|G|(|P|-1)}{|P||G'|}$

(iv) $p \nmid |G'|$ and $J(FG) = FG \cdot I(P)e$, where $e = |G'|^{-1} \sum_{g \in G'} g$

(v) The group G is one of the following three types

(a) G has order prime to p

(b) G is abelian

(c) $G'P$ is a Frobenius group with complement P and kernel G'.

Proof. (i) $\Rightarrow$ (ii): Obvious

(ii) $\Rightarrow$ (iii): Owing to Corollary 1.18, we may harmlessly assume that F is algebraically closed. Since $J(FG)$ is commutative, so is $J(B_0)$ where B_0 is the principal block of FG. Hence, by Theorem 14.7, G is p-nilpotent and P is abelian. Thus $G' \subseteq O_{p'}(G)$ and therefore $p \nmid |G'|$.

We now claim that

$$\text{if } J(B) \neq 0, \text{ then } J(B)^2 \neq 0 \tag{4}$$

Assume by way of contradiction that $J(B) \neq 0$ but $J(B)^2 = 0$. If V is an irreducible FG-module in B, then by Lemma 12.3, $\dim_F P(V) = 2\dim_F V$. On the other hand, by Theorem 4.27, $\dim_F P(V) = |P|(\dim_F V)_{p'}$. Hence $|P| = 2(\dim_F V)_{p'}$, contrary to the assumption that $p \neq 2$.

With $N = G'$ we certainly have $PN \lhd G$. Let B be a block of FG with $J(B) \neq 0$. Then, by (4), $J(B)^2 \neq 0$ and so, by Lemma 14.6, B has an irreducible module V of dimension 1. But then $N \subseteq \mathrm{Ker}V$ and so the desired conclusion follows by virtue of Lemma 15.1.

(iii) $\Leftrightarrow$ (iv): Direct consequence of Lemma 15.1

(iii) $\Rightarrow$ (v): Let G be a nonabelian group whose order is divisible by p and let $M = G'P$. By the property (1) in the proof of Theorem 13.4, it suffices to show that

$$\dim_F J(FM) = |P|-1.$$

But M is a normal subgroup of G of p'-index and so, by Proposition 1.8,

$$\dim_F J(FM) = \frac{\dim_F J(FG)}{|G:M|} = \frac{|G|(|P|-1)}{|G:G'P||G'||P|}$$
$$= |P|-1$$

as required.

(v) ⇒ (i): Direct consequence of Theorem 13.4. ■

The next example illustrates that the hypothesis $p \neq 2$ in Theorem 15.2 cannot be ommited without sacrifising the conclusion.

15.3. EXAMPLE (Koshitani (1979)). Assume that F is algebraically closed and $p = 2$. Let the group G be defined by

$$G = \langle a,b \mid a^3 = 1, b^4 = 1;\ b^{-1}ab = a^2 \rangle$$

Then G is 2-nilpotent with $O_{p'}(G) = \langle a \rangle$ and $P = \langle b \rangle \in \mathrm{Syl}_p(G)$. The block idempotents of $FO_{p'}(G)$ are given by

$$e_o = (1+a+a^2),\ e_1 = (1+\varepsilon a+\varepsilon^2 a), \text{ and } e_2 = (1+\varepsilon^2 a+\varepsilon a)$$

Furthermore $b^{-1}e_1 b = e_2$ and so $G(e_1) = \langle a,b^2 \rangle$ is of index 2 in G. Hence, by Theorem 10.9,

$$FG \cong F\mathbb{Z}_4 \times M_2(F\mathbb{Z}_2)$$

and therefore

$$J(FG) \cong I(\mathbb{Z}_4) \times M_2(I(\mathbb{Z}_2))$$

The first factor is commutative and the second is also commutative, since $I(\mathbb{Z}_2)^2 = 0$. Thus $J(FG)$ is commutative and, by looking at the second factor, $J(FG) \not\subseteq Z(FG)$. ■

In order to examine the case $p = 2$, we need the following preliminary observation.

15.4. LEMMA. Let I be an ideal of FG. If I is commutative, then $I^2 \subseteq Z(FG)$.

Proof. Let $x,y \in I$ and $z \in FG$. Then

$$y(xz) = (xz)y, \quad x(zy) = (zy)x$$

so that z commutes with yx and hence $I^2 \subseteq Z(FG)$. ∎

Turning our attention to the case $p = 2$, we now prove

15.5. LEMMA. Let F be an algebraically closed field of characteristic 2, let G be a nonabelian group of order $2^a m$, where m is odd and $a > 1$, and let $P \in Syl_2(G)$ and $N = G'P$. Then the following conditions are equivalent:

(i) $J(FG)$ is commutative

(ii) G is 2-nilpotent, G' is of odd order and for each $x \in G'-1, 4 \nmid |C_N(x)|$

(iii) G is 2-nilpotent, G' is of odd order and each nonprincipal block of FN is of defect 1 or 0.

Proof (i) ⇒ (ii): Since $J(FG)$ is commutative, so is $J(B_o)$ where B_o is the principal block of FG. Hence, by Theorem 14.7, G is 2-nilpotent and P is abelian. In particular, $G' \subseteq O_{2'}(G)$ and so G' is of odd order. Now $N \lhd G$ and therefore, by Proposition 1.8(i), $J(FN) \subseteq J(FG)$. Hence $J(FN)$ is commutative. It will next be shown that $G' = N'$ and hence that FN has a unique (up to isomorphism) irreducible module of dimension 1.

Put $x = \sum_{g \in N'} g$, $y = \sum_{g \in P} g$, and $z = \sum_{g \in G'} g$. Since N' is of odd order, Lemma 7.3 implies that

$$J(F(N'P)) \supseteq I(P)x$$

whence

$$J(FG) \supseteq J(FN) \supseteq J(F(N'P)) \supseteq I(P)x$$

On the other hand, by Lemmas 15.4 and 13.2, $J(FG)^2 \subseteq FGz$. It therefore follows that

$$FGz \supseteq J(FG)^2 \supseteq I(P)^2 x^2 = I(P)^2 x$$

But $I(P)^2 \neq 0$, since otherwise $a = 1$ by Theorem 12.4. Hence, by Lemma 11.2, $y \in I(P)^2$ and therefore $yx \in FGz$. It follows, from Lemma 11.1, that $yx(1-g) = 0$ for all $g \in G'$. Since $1 \in \mathrm{Supp}\, yx$, $g \in \mathrm{Supp}\, yxg = \mathrm{Supp}\, yx \subseteq N'P$ and thus

$G' \subseteq N'P$. Because G' is of odd order, we conclude that $N' = G'$.

Assume by way of contradiction that there exists $x \in G'-1$ with $4 \mid |C_N(x)|$. Then N has at least two 2-regular class of defect $d \geqslant 2$. Hence, by Theorem 9.23, FN has at least 2 blocks with defect $d \geqslant 2$. Thus FN has a nonprincipal block B with defect group D such that $|D| \geqslant 4$. By Corollary 10.10, $B \cong M_n(FD)$ for some $n \geqslant 1$. If $n = 1$, then B has a one-dimensional irreducible module, which is impossible by the foregoing discussion. Thus $n > 1$ and $J(B) \cong M_n(I(D))$. But since $|D| \geqslant 4$, $I(D)^2 \neq 0$ and hence $J(B)$ is noncommutative, a contradiction.

(ii) $\Rightarrow$ (iii): By hypothesis, each nonidentity 2-regular class of G is of defect 1 or 0. The desired conclusion is therefore a consequence of Theorem 9.23.

(iii) $\Rightarrow$ (i): Let B be a block of FN with defect group D. Since N is 2-nilpotent, we have $B \cong M_n(FD)$ for some $n \geqslant 1$ (Corollary 10.10). If B is nonprincipal, then by hypothesis $|D| \leqslant 2$. Hence $J(B)^2 \cong M_n(I(D))^2 = 0$.

Since N is a normal subgroup of G of odd index, we have $J(FG) = FG \cdot J(FN)$. Put

$$e_1 = |G'|^{-1} \sum_{g \in G'} g \quad \text{and} \quad e_2 = 1 - e_1$$

Then we have

$$J(FG) = FGJ(FN)e_1 \oplus FGJ(FN)e_2$$

Since $FG \cdot J(FN)e_1 \subseteq FGe_1$, $FG \cdot J(FN)e_1$ is a central ideal of FG, by virtue of Lemma 13.2. Now every block of FN contained in FNe_2 is nonprincipal. Hence, by the above, $J(FNe_2)^2 = J(FN)^2 e_2 = 0$. Thus

$$[FG \cdot J(FN)e_2]^2 = FG \cdot J(FN)^2 e_2 = 0$$

and so $FG \cdot J(FN)e_2$ is commutative. This proves that $J(FG)$ is commutative, as required. ■

We have now accumulated all the information necessary to prove the following result.

15.6. THEOREM. (Motose and Ninomiya (1980)). Let F be an arbitrary field of characteristic 2, let G be a group of order $2^a m$, where m is odd, and let $P \in \mathrm{Syl}_2(G)$ and $N = G'P$. Then the following conditions are equivalent:

(i) $J(FG)$ is commutative

(ii) Either $a \leqslant 1$, or $G' = 1$, or $a > 1, G$ is 2-nilpotent, $G' \neq 1$ is of odd order and for each $x \in G'-1$, $4 \nmid |C_N(x)|$

(iii) Either $a \leqslant 1$, or $G' = 1$, or $a > 1, G$ is 2-nilpotent, $G' \neq 1$ is of odd order and $|P \cap P^x| \leqslant 2$ for all $x \in N-P$

(iv) Either $a \leqslant 1$, or $G' = 1$, or $a > 1, G$ is 2-nilpotent, $G' \neq 1$ is of odd order and, for every involution s in P, $C_N(s)/\langle s \rangle$ is either a 2-group or a Frobenius group with complement $P/\langle s \rangle$.

Proof. By Corollary 1.18, we may harmlessly assume that F is algebraically closed. If $a = 0$, then $J(FG) = 0$ and if $a = 1$, then by Theorem 12.4, $J(FG)^2 = 0$. Thus if $a \leqslant 1$, then $J(FG)^2 = 0$ and therefore $J(FG)$ is commutative. Since $J(FG)$ is obviously commutative in case $G' = 1$, we may also assume that $a > 1$ and $G' \neq 1$. Hence, by Lemma 15.5, (i) and (ii) are equivalent.

(i) $\Rightarrow$ (iii): Assume that $J(FG)$ is commutative. Since (i) and (ii) are equivalent, G' is of odd order. Let $x \in G'-1$ and $y \in P \cap P^x$. Then $xyx^{-1}y^{-1} \in P \cap G' = 1$ and so $y \in C_N(x)$. This shows that

$$P \cap P^x \subseteq C_N(x)$$

and hence, by Lemma 15.5, $|P \cap P^x| \leqslant 2$.

(iii) $\Rightarrow$ (iv): Let s be an arbitrary involution in P. If $C_N(s) = P$, then $C_N(s)/\langle s \rangle$ is a 2-group. Now assume that $C_N(s) \neq P$. Then, for all $x \in C_N(s)-P$,

$$P \cap P^x = \langle s \rangle$$

since $|P \cap P^x| \leqslant 2$. Hence, by Lemma 7.2, $C_N(s)/\langle s \rangle$ is a Frobenius group with complement $P/\langle s \rangle$.

(iv) $\Rightarrow$ (ii): Let $x \in G'-1$ and let $S \in Syl_2(C_N(x))$. It suffices to prove that $|S| \leqslant 2$. We may assume that $S \neq 1$, in which case there is an involution, say s, in S. Then

$$x \in C_N(S) \subseteq C_N(s)$$

which shows that $C_N(s)$, and hence $C_N(s)/\langle s\rangle$, is not a 2-group. Now $S \subseteq P^u$ for some $u \in N$ and therefore, by hypothesis, $C_N(s)/\langle s\rangle$ is a Frobenius group with complement $P^u/\langle s\rangle$. Thus we have $S \subseteq P^u \cap P^{ux} = \langle s\rangle$, as required. ∎

15.7. COROLLARY. (Motose and Ninomiya (1980)). Let F be an arbitrary field of characteristic 2, let G be a nonabelian group of order $2^a m$, $(m,2) = 1$, $a > 1$, and let $J(FG)$ be commutative. Then the Sylow 2-subgroups of G are either cyclic or isomorphic to $\mathbb{Z}_2 \times \mathbb{Z}_{2^{a-1}}$.

Proof. Let $P \in Syl_2(G)$ and let $N = G'P$. Then, by Theorem 15.6, G' is of odd order, P is abelian and

$$|P \cap P^x| \leqslant 2 \quad \text{for all} \quad x \in N-P$$

If $P \cap P^x = 1$ for all $x \in N-P$, then N is a Frobenius group with complement P and kernel G'. Hence P is cyclic, by virtue of Remark 13.5.

On the other hand, if $P \cap P^x = \langle s\rangle$ for some $x \in N-P$ and some involution s in P, then $xsx^{-1}s^{-1} \in P \cap G' = 1$. Hence $x \in C_N(s) - P$ and so $C_N(s) = P$. Thus, by the condition (iv) of Theorem 15.6, $C_N(s)/\langle s\rangle$ is a Frobenius group with complement $P/\langle s\rangle$. Since $P/\langle s\rangle$ is abelian and is a Sylow 2-subgroup of $C_N(s)/\langle s\rangle$, it follows from Remark 7.11 that $P/\langle s\rangle$ is cyclic. Thus P is either cyclic or $P \cong \mathbb{Z}_2 \times \mathbb{Z}_{2^{a-1}}$. ∎

16. THE RADICAL OF BLOCKS AND NORMAL SUBGROUPS

Throughout this section, G denotes a finite group and F an arbitrary field of characteristic $p > 0$. Our aim is to determine the Jacobson radical $J(B)$ of those blocks B of FG for which the defect group of B is contained in a given normal subgroup of G. As a preparation for the proof of the main result, we shall prove a number of important properties of independent interest.

It will be fundamental for future investigations to provide a useful character -rization of defect groups of blocks. The following preliminary results will clear our path.

16.1. LEMMA. Let W be an indecomposable FG-module with vertex Q and let H be any subgroup of G such that W is H-projective. Write

$$V_H = V_1 \oplus \ldots \oplus V_n$$

where the V_i are indecomposable FH-modules, and denote by Q_i a vertex of $V_i, 1 \leqslant i \leqslant n$. Then for each $i \in \{1,\ldots,n\}$, there exists $g_i \in G$ such that $g_iQg_i^{-1} \subseteq Q$.

Proof. By Lemma 2.6.2, each V_i is $(x_iQx_i^{-1} \cap H)$-projective for some $x_i \in G$. Owing to Lemma 2.6.3(ii), there exists $y_i \in H$ such that $x_iQx_i^{-1} \cap H \supseteq y_iQy_i^{-1}$. Setting $g_i = x_i^{-1}y_i$, we conclude that $g_iQg_i^{-1} \subseteq Q$, as required. ∎

16.2. LEMMA. Let W be an indecomposable FG-module with vertex $Q \subseteq P$, where P is a Sylow p-subgroup of G. Then $(P;Q)$ divides $\dim_F W$. In particular, if $\dim_F W$ is prime to p, then P is a vertex of W.

Proof. We first consider the case where F is algebraically closed. Let V be an indecomposable FP-module with vertex X and source S. Since X is a subnormal subgroup of P, it follows from Theorem 4.23 that S^P is indecomposable. Hence $V \cong S^P$ and so

$$\dim_F V = (P:X)\dim_F S \tag{1}$$

Now write

$$W_P = V_1 \oplus V_2 \oplus \ldots \oplus V_n$$

where the V_i are indecomposable FP-modules, and denote by Q_i a vertex of V_i. Then by (1), $(P:Q_i)$ divides $\dim_F V_i$ and, by Lemma 16.1, $(P:Q)$ divides $(P:Q_i)$, $1 \leqslant i \leqslant n$. Thus, for all i, $(P:Q)$ divides $\dim_F V_i$ and hence $(P:Q)$ divides $\dim_F W$. This establishes the case where F is algebraically closed.

In general, let E be the algebraic closure of F, and let U be a source of W corresponding to the vertex Q. Then

$$E \otimes_F W = W_1 \oplus \dots \oplus W_m$$

where the W_i are indecomposable EG-modules. Since for all $i \in \{1,\dots,m\}$, W_i is a component of $E \otimes_F W$ and $E \otimes_F W$ is a component of $E \otimes_F U^G = (E \otimes_F U)^G$, we infer that W_i is a component of $(E \otimes_F U)^G$ for the EQ-module $E \otimes_F U$. This proves that each W_i is Q-projective and so has vertex $Q_i \subseteq Q$. By the algebraically closed field case, $(P:Q_i)$ divides $\dim_E W_i$. Since $(P:Q)$ divides each $(P:Q_i)$ and $\dim_F W = \dim_E W_1 + \dots + \dim_E W_m$, we infer that $(P:Q)$ divides $\dim_F W$. So the lemma is true. ∎

Given subsets X,Y of FG, we denote by FX the F-linear span of X and by $C_X(Y)$ the centralizer of Y in X, i.e.

$$C_X(Y) = \{x \in X \mid xy = yx \text{ for all } y \in Y\}$$

In particular, $C_{FG}(G) = Z(FG)$. Let $S \subseteq H$ be subgroups of G. We shall refer to the map

$$T_{H,S} : C_{FG}(S) \longrightarrow C_{FG}(H)$$

introduced below, as the *relative trace map*.

16.3. LEMMA. Let $S \subseteq H$ be subgroups of G and let X be a left transversal for S in H. For each $z \in C_{FG}(S)$, put

$$T_{H,S}(z) = \sum_{x \in X} xzx^{-1}$$

(i) $C_{FG}(H) \subseteq C_{FG}(S)$, $T_{H,S}(z) \in C_{FG}(H)$ and $T_{H,S}$ is independent of the choice of X.

(ii) $T_{H,S} : C_{FG}(S) \longrightarrow C_{FG}(H)$ is a homomorphism of $C_{FG}(H)$-bimodules. In particular, $T_{H,S}(C_{FG}(S))$ is an ideal of $C_{FG}(H)$.

(iii) If D is a subgroup of S, then for all $z \in C_{FG}(D)$,

$$T_{H,D}(z) = T_{H,S}(T_{S,D}(z))$$

(iv) If D is a p-subgroup of G, then

$$I_D(G) = T_{G,D}(C_{FG}(D))$$

where $I_D(G)$ is defined in Lemma 2.2.5.

Proof. (i) Because any element of FG which centralizes H must also centralize S, we have $C_{FG}(H) \subseteq C_{FG}(S)$. If $y = xs$ for some $x \in X$ and $s \in S$, then

$$yzy^{-1} = (xs)z(xs)^{-1} = xszs^{-1}x^{-1} = xzx^{-1}$$

and so $T_{H,S}(z)$ is independent of the choice of X. Fix $h \in H$ and note that hX is another left transversal for S in H. Thus

$$hT_{H,S}(z)h^{-1} = \sum_{x\in X} (hx)z(hx)^{-1} = T_{H,S}(z)$$

as asserted.

(ii) The map $T_{H,S}$ obviously preserves addition. Fix $y \in C_{FG}(H)$ and $z \in C_{FG}(S)$. Then, for all $x \in X$, we have $yx = xy$ and thus

$$T_{H,S}(yz) = \sum_{x\in X} xyzx^{-1} = y\sum_{x\in X} xzx^{-1} = yT_{H,S}(z)$$

A similar argument proves that $T_{H,S}(zy) = T_H(z)y$ as required.

(iii) Denote by Y a left transversal for D in S. Then XY is a left transversal for D in H. Thus

$$\begin{aligned} T_{H,D}(z) &= \sum_{\substack{x\in X\\ y\in Y}} (xy)z(xy)^{-1} \\ &= \sum_{x\in x} \Big(\sum_{y\in Y} xyzy^{-1}x^{-1}\Big) \\ &= \sum_{x\in X} x\Big(\sum_{y\in Y} yzy^{-1}\Big)x^{-1} \\ &= T_{H,S}(T_{S,D}(z)), \end{aligned}$$

as asserted.

(iv) Let $z = C^{+} \in I_D(G)$, where C is a conjugacy class of G. Fix $g \in C$, put $L = C_G(g)$ and choose $Q \in \mathrm{Syl}_p(L)$. Because $z \in I_D(G)$, we may assume that $Q \subseteq D$. Moreover, by the definition of L, we must have $z = T_{G,L}(g)$. Applying (iii) to compute $T_{G,Q}(g)$ in two different ways, we find

$$T_{G,L}(T_{L,Q}(g)) = T_{G,D}(T_{D,Q}(g)) \tag{2}$$

Since $T_{L,Q}(g) = (L:Q)g$ and $(L:Q)$ is prime to p, we may define

$$w = (L:Q)^{-1}T_{D,Q}(g)$$

Then $w \in C_{FG}(D)$ and, by (2), $T_{G,D}(w) = z$, proving that

$$I_D(G) \subseteq T_{G,D}(C_{FG}(D))$$

Denote by L the D-conjugacy class containing $g \in G$ and by C the G-conjugacy class containing g. Since the elements L^+ form an F-basis for $C_{FG}(D)$, it suffices to prove that $T_{G,D}(L^+) \in I_D(G)$. But

$$T_{G,D}(L^+) = (C_G(g) : C_D(g))C^+$$

and so either $T_{G,D}(L^+) = 0$ or $C_D(g)$ is a defect group of C. Thus in both cases $T_{G,D}(L^+) \in I_D(G)$, as desired. ∎

16.4. LEMMA. Let K be a subgroup of G and let D,H be subgroups of K. Denote by X a set of representatives for the double cosets DxH in K and for each $k \in K$ put $H_k = kHk^{-1} \cap D$.

(i) $T_{K,H}(a) = \sum_{x\in X} T_{D,H_x}(xax^{-1})$ for all $a \in C_{FG}(H)$

(ii) $T_{K,H}(a)T_{K,D}(b) = \sum_{x\in X} T_{K,H_x}(xax^{-1}b)$ for all $a \in C_{FG}(H), b \in C_{FG}(D)$

(iii) $T_{K,H}(C_{FG}(H)) \subseteq \sum_{k\in K} T_{D,H_k}(C_{FG}(H_k))$

(iv) $T_{K,H}(C_{FG}(H))T_{K,D}(C_{FG}(D)) \subseteq \sum_{k\in K} T_{K,H_k}(C_{FG}(H_k))$

Proof. (i) Given $x \in X$, denote by $T(x)$ a left transversal for H_x in D. Then the set

$$Y = \bigcup_{x\in X} T(x)x$$

is obviously a left transversal for H in K. Thus, for all $a \in C_{FG}(H)$, we have

$$\sum_{x\in X} T_{D,H_x}(xax^{-1}) = \sum_{x\in X}\sum_{t\in T(x)} (tx)a(tx)^{-1}$$
$$= \sum_{y\in Y} yay^{-1}$$
$$= T_{K,H}(a)$$

as asserted.

(ii) We have

$$T_{K,H}(a)T_{K,D}(b) = T_{K,D}(T_{K,H}(a)b) \qquad \text{(by Lemma 16.3(iii))}$$
$$= T_{K,D}(\sum_{x\in X} T_{D,H_x}(xax^{-1}b)) \qquad \text{(by (i) and Lemma 16.3(ii))}$$
$$= \sum_{x\in X} T_{K,H_x}(xax^{-1}b) \qquad \text{(by Lemma 16.3(iii))}$$

as required.

(iii) Direct consequence of (i)

(iv) Follows from (ii). ■

16.5. LEMMA. Let A be a finite-dimensional algebra over F and let e be a primitive idempotent of A. If $I_1, I_2, \ldots, I_n$ are ideals of A such that $e \in I_1 + \ldots + I_n$, then $e \in I_i$ for some $i \in \{1,\ldots,n\}$.

Proof. Without loss of generality we may assume that $n = 2$. Because e is primitive, e is the identity element of the local ring eAe and $e \in I_1 + I_2$ implies $e \in eI_1e + eI_2e$. Thus we may assume that A is local and that $e = 1$. If $1 = a+b$ with $a \in I_1, b \in I_2$, then not both a and b lie in $J(A)$, say $a \notin J(A)$. Since A is local, a has inverse a', which gives $1 = aa' \in I_1$, as required. ■

It is now easy to provide a useful characterization of defect groups of blocks.

16.6. PROPOSITION. Let $B = B(e)$ be a block of FG. Then the following conditions are equivalent:

(i) D is a defect group of B

(ii) D is a minimal element in the set of all subgroups H of G such that

$$e \in T_{G,H}(C_{FG}(H))$$

Proof. Suppose that D is a defect group of B. Because $e \in I_D(G)$, Lemma 16.3(iv) implies that $e \in T_{G,D}(C_{FG}(D))$. Now assume that $H \subset D$ and that $e \in T_{G,H}(C_{FG}(H))$. Then, by Lemma 16.3(iv), $e \in I_H(G)$ contrary to the assumption that D is a defect group of e. Hence (i) implies (ii).

Assume that H is a subgroup of G such that $e \in T_{G,H}(C_{FG}(H))$ and let D be a minimal element in the set of all such subgroups H of G. Owing to Lemma 16.4(iv) (with $K = G$), we have

$$e = ee \in T_{G,H}(C_{FG}(H))T_{G,D}(C_{FG}(D))$$

$$\subseteq \sum_{g \in G} T_{G,H_g}(C_{FG}(H_g))$$

Applying Lemma 16.5, we infer that

$$e \in T_{G,H_g}(C_{FG}(H_g)) \qquad \text{for some } g \in G$$

Then the minimality of D ensures that $gHg^{-1} \supseteq D$. Now suppose that H (and hence gHg^{-1}) is a defect group of B. Then $gHg^{-1} \supseteq D$ and in view of the implication (i) $\Rightarrow$ (ii), we obtain $gHg^{-1} = D$, as required. ■

16.7. LEMMA. Let D be a p-subgroup of G and let e be an idempotent in $I_D(G)$ (e.g. e is a block idempotent of FG with D as a defect group)

(i) For any FG-module V, eV is D-projective

(ii) If V is an indecomposable FG-module in a block $B = B(e)$ with defect group D, then V is D-projective.

Proof. If V satisfies (ii), then $eV = V$ and so (ii) is a consequence of (i). To prove (i), we employ the relative trace map. Owing to Lemma 16.3(iv), we may choose $w \in C_{FG}(D)$ such that $e = T_{G,D}(w)$. Consider the mapping $\theta : eV \longrightarrow eV$ defined by $\theta(x) = wx$ for all $x \in eV$. Since w commutes with all elements g with $g \in D$, $\theta \in \mathrm{End}_{FD}(V_D)$. Let T denote a transversal for D in G and let $x \in eV$. Then

$$(\sum_{t\in T} t\,\theta\, t^{-1})x = \sum_{t\in T} twt^{-1}x$$
$$= (\sum_{t\in T} twt^{-1})x$$
$$= ex = x$$

and the result follows by virtue of Theorem 2.5.4. ■

16.8. LEMMA. Let $B = B(e)$ be a block of FG and let I be an ideal of $Z(FG)$. If $e \notin I$, then Ie is a nilpotent ideal of $Z(FG)$.

Proof. Let γ be the irreducible representation of $Z(FG)$ associated with e. By Proposition 2.10.8,

$$\mathrm{Ker}\gamma = Z(FG)(1-e) + J(Z(FG))e$$

We claim that $I \subseteq \mathrm{Ker}\gamma$; if sustained it will follow that

$$Ie \subseteq (\mathrm{Ker}\gamma)e = J(Z(FG))e$$

as required. Assume by way of contradiction that $\gamma(I) \neq 0$. Since $\gamma(Z(FG))$ is a field and $0 \neq \gamma(I)$ is an ideal of $\gamma(Z(FG))$, it follows that

$$\gamma(Z(FG)) = \gamma(I)$$

Thus $\gamma(e) = \gamma(x)$ for some $x \in I$ and therefore

$$e-x \in Z(FG)(1-e) + J(Z(FG))e$$

Hence $e-x = re + z(1-e)$ for some $r \in J(Z(FG))$ and some $z \in Z(FG)$. Consequently,

$$e = xe + re$$

and

$$e-xe = re \in J(Z(FG))$$

Now $J(Z(FG))$ is a nilpotent ideal of $Z(FG)$ and $Z(FG)$ is a commutative algebra over the field F of characteristic $p > 0$. Hence there exists a positive integer n such that

$$0 = (re)^{p^n} = (e-xe)^{p^n} = e^{p^n} - (xe)^{p^n} = e-(xe)^{p^n}$$

Therefore $e = (xe)^{p^n} \in I$, a contradiction. ∎

16.9. LEMMA. Let e be a block idempotent of FG with defect group D and let $\hat{I}_D(G)$ be the F-linear span of all C^+, $C \in C\ell(G)$, with $\delta(C) \underset{G}{<} D$. Then $\hat{I}_D(G)$ is an ideal of $Z(FG)$ and $\hat{I}_D(G)e$ is a nilpotent ideal of $Z(FG)$.

Proof. By Lemma 2.2.5, $I_D(G)$ is an ideal of $Z(FG)$. Since $\hat{I}_D(G)$ is the sum of all $I_H(G)$ with $H \subset D$, we see that $\hat{I}_D(G)$ is an ideal of $Z(FG)$. By the definition of the defect group of e, $e \notin \hat{I}_D(G)$. Hence $\hat{I}_D(G)e$ is a nilpotent ideal of $Z(FG)$, by virtue of Lemma 16.8. ∎

16.10. LEMMA. Let e be a block idempotent of FG with defect group D, let E be any field extension of F and write

$$e = e_1 + \ldots + e_n$$

as a sum of block idempotents of EG. Then, for all $i \in \{1,\ldots,n\}$, e_i has D as a defect group.

Proof. Denote by $I^*_D(G)$ and $\hat{I}^*_D(G)$ the counterparts of $I_D(G)$ and $\hat{I}_D(G)$, respectively, in EG. Fix $i \in \{1,\ldots,n\}$ and note that, by the definition of the defect group of e_i, it suffices to verify that $e_i \in I^*_D(G)-\hat{I}^*_D(G)$. Since $e \in I_D(G) \subseteq I^*_D(G)$ and since $I^*_D(G)$ is an ideal of $Z(EG)$, we have

$$e_i = e_i e \in I^*_D(G)$$

On the other hand,

$$\hat{I}^*_D(G)e = E \otimes_F \hat{I}_D(G)e$$

and so, by Lemma 16.9, $\hat{I}^*_D(G)e$ is a nilpotent ideal of $Z(EG)$. Hence if $e_i \in \hat{I}^*_D(G)$, then $e_i = e_i e \in \hat{I}^*_D(G)e$, a contradiction. So the lemma is true. ∎

Let S be a subring of a ring R. An R-module M is said to be *S-projective* if any exact sequence

$$0 \longrightarrow Y \longrightarrow X \longrightarrow M \longrightarrow 0$$

of R-modules and R-homomorphisms which splits as an S-sequence, also splits as an R-sequence. If $L \subseteq S \subseteq R$ are subrings and M is L-projective, then obviously M is S-projective. We state without proof the following basic fact (see Higman (1955, Theorem 4) and Hochschild (1956, Proposition 2).

16.11. LEMMA. Let S be a subring of a ring R and let M be an R-module. Then the following conditions are equivalent:

(i) M is S-projective

(ii) The canonical map $\psi : R \underset{S}{\otimes} M \longrightarrow M$ given by $\psi(r \otimes m) = rm$ is a retraction, i.e. there exists an R-homomorphism $\phi : M \longrightarrow R \underset{S}{\otimes} M$ with $\psi\phi = 1$.

(iii) There exists an S-module V such that M is a direct summand of $R \underset{S}{\otimes} V$.

(iv) Consider any diagram of R-modules and R-homomorphisms

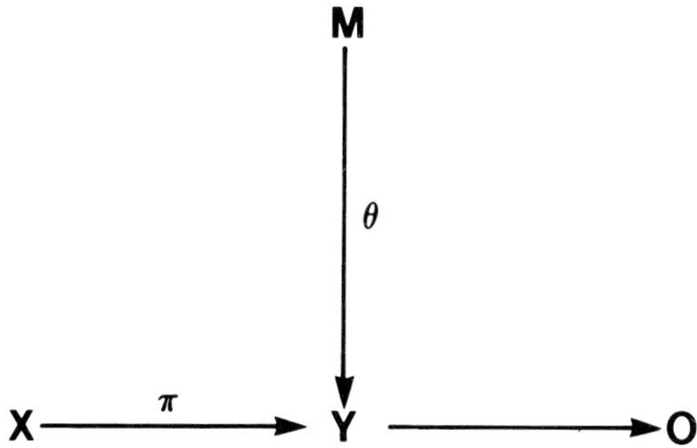

with exact row. If there is an S-homomorphism $\phi : M \longrightarrow X$ making the diagram commute, then there is an R-homomorphism $\psi : M \longrightarrow X$ making the daigram commute.■

In order to take advantage of Lemma 16.9, we make the following simple observation .

16.12. LEMMA. Assume that the following is a commutative diagram of R-modules and R-homomorphisms

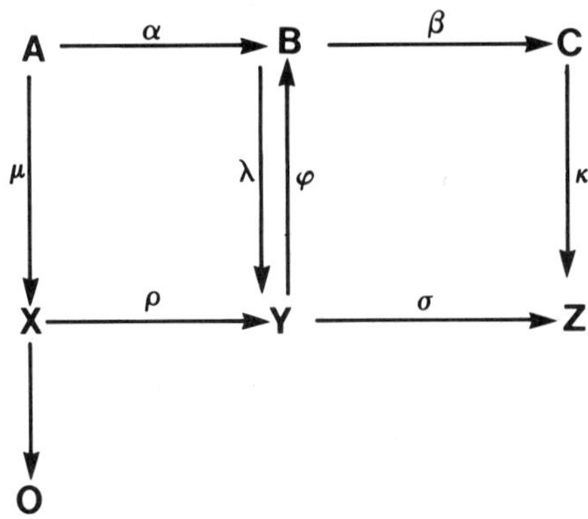

If $\lambda\phi = 1_Y$, then the following are equivalent:

(i) $\mathrm{Im}(\phi\lambda\alpha) \subseteq \mathrm{Im}\alpha$

(ii) There is a $\psi : Z \longrightarrow C$ such that $\psi\sigma = \beta\phi$ and $k\psi = 1_Z$ (in particular, k is a retraction).

Proof. Assume that $\mathrm{Im}(\phi\lambda\alpha) \subseteq \mathrm{Im}\alpha$. Because $\mathrm{Im}\alpha = \mathrm{Ker}\beta$, we have

$$\beta\phi\rho\mu = \beta\phi\lambda\alpha = 0$$

which shows that $\beta\phi\rho = 0$, by virtue of the fact that μ is an epimorphism. Hence

$$\beta\phi(\mathrm{Ker}\sigma) = \beta\phi(\mathrm{Im}\rho) = 0$$

and thus $\psi = \beta\phi\sigma^{-1}$ is a well-defined R-homomorphism from Z to C with the required properties.

Now assume that (ii) holds. Then

$$\beta\phi\lambda\alpha = \psi\sigma\lambda\alpha = \psi\sigma\rho\mu = 0$$

and therefore $\mathrm{Im}(\phi\lambda\alpha) \subseteq \mathrm{Ker}\beta = \mathrm{Im}\alpha$. ∎

The next observation will allow us to take full advantage of the results so far obtained.

16.13. LEMMA. Let S be a subring of a ring R, let $A \subseteq S$ be a subset such that $AR = RA$ and let U be an S-projective R-module. Then $M = U/AU$ is an

S-projective R-module (the fact that AU is a submodule of U is a consequence of the equality $R(AU) = ARU = AU$).

Proof. Consider the diagram below where all maps are canonical.

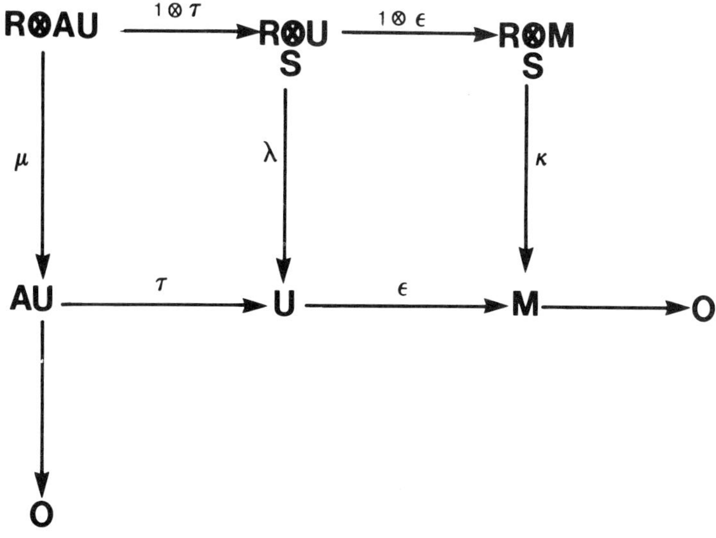

Bearing in mind that

$$\mathrm{Im}(1 \otimes \tau) = R \otimes_S AU = A(R \otimes_S U)$$

we obtain

$$\begin{aligned}\mathrm{Im}(\phi\lambda(1 \otimes \tau)) &= \phi\lambda A(R \otimes_S U)\\ &= A\phi\lambda(R \otimes_S U)\\ &\subseteq A(R \otimes_S U)\\ &= \mathrm{Im}(1 \otimes \tau)\end{aligned}$$

The desired assertion is now a consequence of Lemma 16.12. ■

We have now accumulated all the information necessary to prove the following fundamental result.

16.14. THEOREM. (Green (1959), Knorr (1976)). Let G be a finite group, let F be an arbitrary field of prime characteristic p and let B be a block of FG.

If P is a p-subgroup of G, then the following conditions are equivalent:

(i) P is a defect group of B

(ii) All indecomposable FG-modules in B are P-projective and there is an irreducible FG-module M in B such that P is a vertex of M.

Proof. Suppose that (i) $\Rightarrow$ (ii) is true and that (ii) holds. Denote by Q a defect group of B. Then all indecomposable FG-modules in B are Q-projective and there is an irreducible FG-module V in B such that Q is a vertex of V. Since V is P-projective, Q is conjugate to a subgroup of P. On the other hand, because M is Q-projective, P is conjugate to a subgroup of Q. Hence P and Q are conjugate. Owing to Lemma 16.7, if P is a defect group of B, then all indecomposable FG-modules in B are P-projective. Thus we need only verify the following assertion:

If P is a defect group of B, then there is an irreducible FG-module M in B such that P is a vertex of M. (3)

For the sake of clarity, we divide the proof into three steps.

Step 1. The case where F is algebraically closed.

Let d be the defect of B and let p^m be the order of Sylow p-subgroups of G. By Theorem 2.2.7, there exists an irreducible FG-module M in B such that

$$p^{m-d} \text{ is the highest power of } p \text{ dividing } \dim_F M$$

Denote by Q a vertex of M. Because M is indecomposable, M is P-projective and so $Q \subseteq_G P$. On the other hand, by Lemma 16.2, $(S:Q)$ divides $\dim_F V$, where S is a Sylow p-subgroup of G. Because $|P| = p^d$, this means that $Q =_G P$ and thus P is a vertex of M, proving (3).

Step 2. Let E be the algebraic closure of F. Our aim is to prove that every irreducible FG-module M is FG-projective.

Owing to Corollary 1.18, $FG/J(FG)$ is a separable F-algebra and

$$J(EG) = E \otimes_F J(FG) \tag{4}$$

Denote by U a projective EG-module such that $M \cong U/J(EG)U$. Then, by (4),

$J(EG)U = J(FG)U$ and so $M \cong U/J(FG)U$. Invoking Lemma 16.13 with $A = J(FG)$, we conclude that M is FG-projective.

Step 3. Here we complete the proof by establishing (3) for an arbitrary F. Let e denote the block idempotent of FG contained in B. Because e is a central idempotent of EG, we may write $e = e_1 + \dots + e_n$, where the e_i are block idempotents of EG. Put $B' = EGe_1$ and note that, by Lemma 16.10, P is a defect group of B'. Applying Step 1, we deduce that there is an irreducible EG-module V in B' such that P is a vertex of V. Now $J(EG)V = 0$ and so, by (4) $J(FG)V = 0$. Thus V_{FG} is completely reducible and so

$$V_{FG} = \bigoplus_{i=1}^{s} M_i$$

for some irreducible FG-modules M_i. Since multiplication by e induces the identity on V, all M_i are in B. By Step 2, V is FG-projective, so Lemma 16.11 ensures that V is a direct summand of

$$EG \otimes_{FG} V = \bigoplus_{i=1}^{s} (EG \otimes_{FG} M_i)$$

But V is irreducible, so V is a direct summand of $EG \otimes_{FG} M$ for some $M \in \{M_1, \dots, M_s\}$. Denote by Q a vertex of M. Then M is a direct summand of $FG \otimes_{FQ} M$ and thus V is a direct summand of

$$EG \otimes_{FG} (FG \otimes_{FQ} M) \cong EG \otimes_{FQ} M$$

Applying Lemma 16.11, we infer that V is FQ-projective. A fortiori, V is EQ-projective and thus P is conjugate to a subgroup of Q. Now M is an irreducible FG-module in B. Consequently M is P-projective and so Q is conjugate to a subgroup of P. This proves that P and Q are conjugate, as we wished to show. ■

As a final preliminary result, we prove

16.15. LEMMA. Let N be a normal subgroup of G and let V be an FG-module. Then the following properties hold:

(i) $J(FN)V$ is an FG-module

(ii) If V is N-projective, then $V/J(FG)V$ is N-projective if and only if

$$J(FG)V = J(FN)V$$

(iii) If V is projective and indecomposable, then N contains a vertex of $V/J(FG)V$ if and only if

$$J(FG)V = J(FN)V$$

Proof. (i) For a given $g \in G$, conjugation by g is an automorphism of FN and hence of $J(FN)$. Bearing in mind that

$$gx = (gxg^{-1})g \qquad x \in J(FN)$$

we infer that $FGJ(FN) = J(FN)FG$. Hence

$$FG[J(FN)V] = J(FN)FGV = J(FN)V$$

as required.

(ii) Put $M = V/J(FG)V$ and assume that M is N-projective. The inclusion $J(FN)V \subseteq J(FG)V$ induces an exact sequence

$$0 \longrightarrow J(FG)V/J(FN)V \longrightarrow V/J(FN)V \longrightarrow M \longrightarrow 0$$

Because $(V/J(FN)V)_N$ is completely reducible, the above sequence splits as an FN-sequence. But M is N-projective, so there exists an FG-homomorphism $f : M \longrightarrow V/J(FN)V$ such that

$$V/J(FN)V = f(M) \oplus J(FG)V/J(FN)V$$

Bearing in mind that M and $f(M)$ are completely reducible, we derive

$$\begin{aligned} J(FG)V/J(FN)V &= J(FG)[V/J(FN)V] \\ &= J(FG)f(M) \oplus J(FG)[J(FG)V/J(FN)V] \\ &= J(FG)[J(FG)V/J(FN)V] \end{aligned}$$

Hence $J(FG)V/J(FN)V = 0$ and $J(FG)V = J(FN)V$.

Conversely, suppose that $J(FG)V = J(FN)V$. Since $A = J(FN)$ satisfies the condition of Lemma 16.13, it follows that $M = V/J(FG)V = V/J(FN)V$ is N-projective.

(iii) Direct consequence of (ii). ∎

We are at last in a position to achieve our main objective, which is to prove the following result.

16.16. THEOREM (Knorr (1976)). Let N be a normal subgroup of G and let $B = B(e)$ be a block of FG with defect group D. Then the following are equivalent:

(i) $D \subseteq N$

(ii) $[J(FG)]^n V = [J(FN)]^n V$ for all indecomposable FG-modules V in B and all positive integers n

(iii) $J(B) = J(FN)B$

(iv) $J(FG)V = J(FN)V$ for all projective indecomposable FG-modules V in B.

Proof. (i) ⇒ (ii): Assume that V is an indecomposable FG-module in B. By Theorem 16.14, V is D-projective and so V is N-projective. The desired conclusion now follows from Lemma 16.5(ii) by induction on n.

(ii) ⇒ (iii): Write $e = e_1 + \dots + e_n$ as a sum of primitive idempotents of FG. Then each FGe_i is an indecomposable module in B and so

$$J(FG)FGe_i = J(FN)FGe_i$$

Bearing in mind that

$$J(FG)e = J(FG)e_1 + \dots + J(FG)e_n = J(FG)FGe_1 + \dots + J(FG)FGe_n$$

we derive

$$\begin{aligned} J(B) &= J(FG)e = J(FN)FGe_1 + \dots + J(FN)FGe_n \\ &= J(FN)FGe \\ &= J(FN)B \end{aligned}$$

(iii) ⇒ (iv): Write $B = \bigoplus_{i=1}^{n} U_i$, where the U_i are projective indecomposable modules in B. Then

$$\bigoplus_{i=1}^{n} J(FG)U_i = J(B) = J(FN)B = \bigoplus_{i=1}^{n} J(FN)U_i$$

Since $J(FN)U_i \subseteq J(FG)U_i$ for all i, we have

$$J(FG)U_i = J(FN)U_i \qquad \text{for all } i$$

Hence $J(FG)U = J(FN)U$ for all projective indecomposable modules V in B

(iv) ⇒ (i): Owing to Theorem 7.6.7, there exists an irreducible FG-module V in B such that D is a vertex of V. Now $V \cong U/J(FG)U$ for some projective indecomposable FG-module U in B. The desired conclusion is now a consequence of Lemma 16.15(iii). ■

16.17. COROLLARY. Let B be a block of FG with normal defect group D. Then

$$J(B) = I(D)B$$

Proof. By Theorem 16.16, applied for the case $N = D$, we have $J(B) = J(FD)B$. Since $J(FD) = I(D)$, the result follows. ■

16.18. COROLLARY. Define $O_{p,p'}(G)$ by $O_{p,p'}(G)/O_p(G) = O_{p'}(G/O_p(G))$. If B is a block of FG with defect group D contained in $O_{p,p'}(G)$, then

$$J(B) = F(O_{p,p'}(G))\cdot I(O_p(G))B$$

Proof. Put $N = O_{p,p'}(G)$ and $P = O_p(G)$. Then P is a normal Sylow p-subgroup of N. Hence, by Proposition 1.1, $J(FN) = FN\cdot I(P)$. Since $D \subseteq N$, the result follows by virtue of Theorem 16.16. ■

Recall that G is called p-solvable if each of its composition factors is either a p-group or a p'-group. Thus G is solvable in the ordinary sense if and only if it is p-solvable for all primes p. Clearly G is p-solvable if and only if it has a series of normal subgroups

$$1 = V_0 \subseteq V_1 \subseteq \ldots \subseteq V_n = G \tag{4}$$

for which each factor group V_{i-1}/V_i is either a p-group or a p'-group. For each such group we can therefore define the *upper p-series*

$$1 = P_0 \subseteq N_0 \subset P_1 \subset P_2 \subset \ldots \subset P_\ell \subseteq N_\ell = G \tag{5}$$

inductively by the rule that

$$N_k/P_k = O_{p'}(G/P_k) \quad \text{and} \quad P_{k+1}/N_k = O_p(G/N_k)$$

Thus, in particular,

$$N_0 = O_{p'}(G)\ ,\ P_1 = O_{p',p}(G) \quad \text{and} \quad N_1 = O_{p',p,p'}(G)$$

The number ℓ which is the least integer such that $N_\ell = G$ is called the p-length of G and is denoted by $\ell_p(G)$. Observe that $\ell_p(G)$ may also be defined as the smallest number of p-factors occuring in any series such as (4), the minimum being attained for the upper p-series (5). Thus G is of p-length 1 if and only if

$$G/O_{p',p}(G) \text{ is a } p'\text{-group}$$

With this information at our disposal, we now prove

16.19. COROLLARY. Let G be a p-solvable group of p-length 1 and let $P \in \mathrm{Syl}_p(G)$. Then, for any block B of FG,

$$J(B) = J(FO_{p',p}(G))B$$

Proof. Put $N = O_{p',p}(G)$. Since G is of p-length 1, N contains all p-subgroups of G. Now apply Theorem 16.16. ■

To present our final application of Theorem 16.16, we quote the following standard group-theoretic fact.

16.20. PROPOSITION. Let G be a p-solvable group with abelian Sylow p-subgroups. Then G is of p-length 1.

Proof. See Gorenstein (1968,p.228). ■

16.21. COROLLARY. Let G be a p-solvable group with abelian Sylow p-subgroups, and let $P \in \mathrm{Syl}_p(G)$. Then, for any block B of FG,

$$J(B) = J(FO_{p',p}(G))B$$

Proof. Apply Proposition 16.20 and Corollary 16.19. ■

17. GROUP ALGEBRAS WITH RADICALS EXPRESSIBLE AS PRINCIPAL IDEALS

Let G be a finite group and let F be a field of characteristic $p > 0$. The problem that motivates this section is to discover conditions under which $J(FG) = xFG = FGx$ for some $x \in FG$. The main result, due to Morita (1951), solves this problem for the case where F is algebraically closed. We have chosen, for the sake of clarity and generality, to deal with the problem within the framework of a general theory, namely the theory of uniserial algebras, although a shorter proof can be obtained by means of ad hoc arguments.

Throughout this section, A denotes a finite-dimensional algebra over an arbitrary field F. Unless explicitely stated otherwise, all A-modules are assumed to be left and finitely generated.

Let V denote an A-module. Then the *dual module* $V^* = \mathrm{Hom}_F(V,F)$ is a right A-module with the action of A on V^* given by

$$(\psi a)(v) = \psi(av) \qquad (a \in A, v \in V, \psi \in V^*)$$

If V is a right A-module, then a similar definition gives a left A-module V^*. The following properties of left (or right) A-modules V and W are easily verified:

(a) V is indecomposable if and only if V^* is indecomposable

(b) V is projective if and only if V^* is injective

(c) $(V^*)^* \cong V$

(d) $(V \oplus W)^* \cong V^* \oplus W^*$

Let V be an A-module. Then the descending chain

$$V \supseteq J(A)V \supseteq J(A)^2V \supseteq \cdots$$

of submodules of V is called the (lower) *Loewy series* of V. Because $J(A)$ is nilpotent, there is an integer k, called the *Loewy length* of V such that

$$J(A)^{k-1}V \neq 0 \quad \text{but} \quad J(A)^kV = 0$$

Note that if $J(A)^iV = J(A)^{i+1}V$ for $i < k$, then

$$J(A)^{k-1}V = J(A)^{k-i-1}J(A)^{i}V = J(A)^{k-i-1}J(A)^{i+1}V = J(A)^{k}V = 0,$$

a contradiction. The conclusion is that, if k is the Loewy length of V, then

$$V \supset J(A)V \supset J(A)^2V \supset \ldots \supset J(A)^{k-1}V \supset 0$$

The module V is said to be *uniserial* if the above chain is a composition series of V. Thus V is uniserial if and only if it has a unique composition series. Because a uniserial module has a unique irreducible submodule, it must be *indecomposable*. We shall refer to the algebra A us being *uniserial* if every indecomposable A-module is uniserial. Note that the dual of a uniserial module is a uniserial right module. Hence, if A is uniserial, then every indecomposable right A-module is uniserial.

17.1. LEMMA. Let V be a (left) module over an arbitrary ring R. Then the following conditions are equivalent.

(i) V is injective

(ii) Every homomorphism $W \longrightarrow V$ of R-modules can be extended to a homomorphism from any module containing W to V.

(iii) If I is a left ideal of R and $\phi \in \mathrm{Hom}_R(I,V)$, then ϕ can be extended to an element of $\mathrm{Hom}_R(R,V)$.

Proof. (i) $\Rightarrow$ (ii): Assume that V is injective and that $\phi : W \longrightarrow V$ is a homomorphism of R-modules. If X is an R-module containing W and $\alpha : W \longrightarrow X$ is the natural injection, then by definition there exists an R-homomorphism $\phi^* : X \longrightarrow V$ which renders commutative the following diagram:

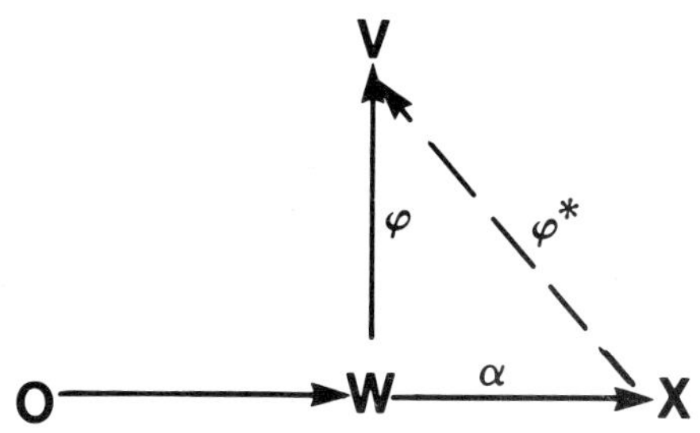

Hence ϕ^* extends ϕ.

(ii) ⇒ (iii): Obvious

(iii) ⇒ (i): Assume that V satisfies (iii) and let X,Y,α,β be given as in the following diagram

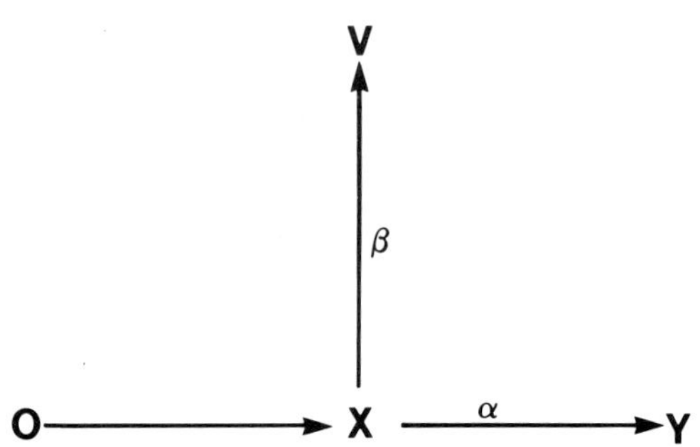

Put $\alpha(X) = X_o$ and observe that because α is injective we can define $\gamma_o : X_o \longrightarrow V$ by $\gamma_o = \beta\alpha^{-1}$. Now let Z be the set of all ordered pairs (U,γ) such that U is a submodule of $Y, U \supseteq X_o$, $\gamma : U \longrightarrow V$, and γ extends γ_o. If $(U_1,\gamma_1),(U_2,\gamma_2)$ are in Z, we write $(U_1,\gamma_1) \leqslant (U_2,\gamma_2)$ if and only if $U_1 \subseteq U_2$ and γ_2 extends γ_1. Since $(X_o,\gamma_o) \in Z$, it follows easily from Zorn's lemma that Z contains a maximal element, say, (U,γ). We show that $U = Y$.

Let $y \in Y$ and let $I = \{r \in R | ry \in U\}$. Then I is a left ideal of R and we have an R-homomorphism $\phi : I \longrightarrow V$ given by $\phi(r) = \gamma(ry)$. By hypothesis, ϕ extends to an R-homomorphism $\phi^* : R \longrightarrow V$. With this map we can now define $\gamma^* : U+Ry \longrightarrow V$ by $\gamma^*(u+ry) = \gamma(u) + \phi^*(r)$. This is easily seen to be well defined and $(U,\gamma) \leqslant (U+Ry,\gamma^*)$. Hence by the maximality of (U,γ) we have $y \in U$, and because y was arbitrary we have $U = Y$. So the lemma is true. ■

17.2. LEMMA. Let V be an A-module. Then there exists a (finitely generated) injective A-module that contains V as a submodule.

Proof. Set $M = \mathrm{Hom}_F(A,V)$ and note that M is an A-module via

$$(a\phi)(b) = \phi(ba) \qquad (a,b \in A, \phi \in M)$$

We claim that M is injective. Indeed, let I be a left ideal of A and let $\psi : I \longrightarrow M$ be an A-homomorphism. Define $\mu : I \longrightarrow V$ by

$$\mu(x) = \psi(x)(1) \qquad \text{for all } x \in I$$

Then μ is clearly an F-homomorphism and hence we may extend μ to an F-homomorphism $\lambda : A \longrightarrow V$. Now define $\theta : A \longrightarrow M$ by

$$\theta(a)(b) = \lambda(ba) \qquad \text{for all } a,b \in A$$

Then θ is an A-homomorphism extending ψ and so M is injective, by Lemma 17.1. Because the map $V \longrightarrow M$ given by $v \mapsto f_v$ with $f_v(a) = av, a \in A, v \in V$, is an injective A-homomorphism, the result follows. ∎

17.3. LEMMA. If A is a Frobenius algebra, then A is an injective left A-module.

Proof. Let I be a left ideal of A and let $f \in \mathrm{Hom}_A(I,A)$. By Lemma 17.1, it suffices to show that f can be extended to an element of $\mathrm{End}_A(A)$. By hypothesis, there exists $\psi \in A^*$ such that (A,ψ) is a Frobenius algebra. Given $a \in A$, define $\lambda_a : I \longrightarrow A$ by $\lambda_a(x) = xa$. Then each $\lambda_a \in \mathrm{Hom}_A(I,A)$ and clearly λ_a can be extended to an element λ'_a of $\mathrm{End}_A(A)$ by putting $\lambda'_a(x) = xa$. Thus we need only verify that for $M = \{\lambda_a | a \in A\}$

$$\dim_F(\mathrm{Hom}_A(I,A)) \leqslant \dim_F M$$

To this end, note that the mapping $a \mapsto \lambda_a$ is an F-homomorphism of A onto M with kernel $r(I)$. Therefore, by Theorem 3.6(iii),

$$\dim_F M = \dim_F A - \dim_F r(I) = \dim_F I = \dim_F(\mathrm{Hom}_F(I,F)).$$

and so prove (1) it suffices to show that the F-linear map

$$\begin{cases} \mathrm{Hom}_A(I,A) \longrightarrow \mathrm{Hom}_F(I,F) \\ \qquad f \mapsto \psi f \end{cases}$$

is injective. But if $(\psi f)(I) = 0$, then the left ideal $f(I)$ is in the kernel of ψ. Thus $f(I) = 0$ and therefore $f = 0$, as required. ∎

A submodule W of an A-module V is called *essential* if W has nonzero intersection with any nonzero submodule of V. An A-module M containing V is called an *injective hull* of V if M is injective and, for any injective

A-module M_1 with $V \subseteq M_1 \subseteq M$, we have $M_1 = M$. By Lemma 17.2, such M always exists.

17.4. LEMMA. Let V be a submodule of an injective A-module M and let W be a submodule of M maximal with respect to V being an essential submodule of W. Then $M = X \oplus W$ for a suitable A-submodule X of M.

Proof. Let X be an A-submodule of M such that $X \cap W = 0$ and $X' \cap W \neq 0$ if $X' \subseteq M$ and X is a proper submodule of X'. Let $\bar{W}$ be the image of W in M/X. Then $\bar{W}$ is an essential submodule of M/X. Since M is injective, the injective homomorphism $\bar{W} \longrightarrow M, w + X \longmapsto w, w \in W$, can be extended to an injective homomorphism $\psi : M/X \longrightarrow M$ (Lemma 17.1). Setting $K = \psi(M/X)$, we have $V \subseteq W \subseteq K$. Let Y be a nonzero submodule of K, say $Y = \psi(Z)$, where Z is a nonzero submodule of M/X. Then $Z \cap \bar{W} \neq 0$ and hence $0 \neq \psi(Z) \cap \psi(\bar{W}) = Y \cap W$. It follows that $(Y \cap W) \cap V \neq 0$, since V is an essential submodule of W. Hence $Y \cap V \neq 0$, so V is an essential submodule of K, therefore $K = W = \psi(\bar{W})$. It follows that $\bar{W} = M/X$ and hence that $M = X \oplus W$, as required. ∎

17.5. LEMMA. Let V be a submodule of an injective A-module M.

(i) M is an injective hull of V if and only if V is an essential submodule of M.

(ii) If M is indecomposable, then M is an injective hull of V. The converse is true if $\mathrm{Soc}V$ is irreducible.

(iii) Any two injective hulls of V are isomorphic.

Proof. (i) Suppose that M is an injective hull of V and let W be as in Lemma 17.4. Then W is an injective A-module with $V \subseteq W \subseteq M$. Hence $W = M$ and so V is an essential submodule of M. Conversely, assume that V is an essential submodule of M and denote by W an injective A-module with $V \subseteq W \subseteq M$. By Lemma 17.1, the inclusion map $V \longrightarrow W$ extends to an A-homomorphism $M \longrightarrow W$. Since V is an essential submodule of M, the kernel of this homomorphism is zero. Thus $M = W$ and hence M is an injective hull of V.

(ii) If M is indecomposable, then V is an essential submodule of M, by Lemma 17.4. Thus, by (i), M must be an injective hull of V. Conversely, assume that M is an injective hull of V and that $\mathrm{Soc}V$ is irreducible. Then, by (i), V is an essential submodule of M and, by hypothesis $\mathrm{Soc}V$ is contained in any nonzero submodule of V. Hence, if $M = X \oplus Y$ for some nonzero submodules X,Y of M, then $\mathrm{Soc}V \subseteq X \cap Y$, a contradiction. Thus M is indecomposable.

(iii) Let M_1 and M_2 be two injective hulls of V. Because V is an essential submodule of M_1, it follows from Lemma 17.1 that there is an injective A-homomorphism $M_2 \longrightarrow M_1$. A similar argument shows the existence of an injective A-homomorphism $M_1 \longrightarrow M_2$ and thus $M_1 \cong M_2$. ■

17.6. LEMMA. Let A be a Frobenius algebra and let V be an A-module. Then V is injective if and only if V is projective.

Proof. By Lemma 17.3 the regular module ${}_AA$ is injective. Hence every free A-module is injective. If V is projective, then V is a direct summand of a free A-module. Hence V is a direct summand of an injective module and so is injective.

Conversely, assume that V is injective. Since V is a direct sum of finitely many indecomposable modules, we may assume that V is indecomposable. Let V_1 be an irreducible submodule of V. Owing to Lemma 17.5(ii), V is an injective hull of V_1. By Theorem 3.8, there exists a primitive idempotent e of A such that $V_1 \cong \mathrm{Soc}(Ae)$. But, by the foregoing, Ae is an indecomposable injective A-module. Thus, by Lemma 17.5(ii), Ae is an injective hull of $\mathrm{Soc}(Ae)$. Hence $V \cong Ae$ by Lemma 17.5(iii), as required. ■

An algebra A is said to be of *finite representation type* if it has only finitely many isomorphism classes of indecomposable modules.

17.7. LEMMA. Let V be a uniserial A-module. Then V is isomorphic to a factor module of a projective indecomposable A-module. In particular, if A is uniserial, then A is of finite representation type.

Proof. There exists a projective A-module X and a surjective A-homomorphism $f : X \longrightarrow V$. Write $X = \bigoplus_{i=1}^{n} X_i$, where each X_i is indecomposable. If $f(X_i) \subset V$ for all I, then $f(X_i) \subseteq J(A)V$ for all i. Thus $f(X) \subseteq J(A)V$, a contradiction. Consequently, $f(X_i) = V$ for some i, as required. ■

We are now ready to provide an important characterization of uniserial algebras.

17.8. THEOREM (Nakayama (1939)). The following conditions are equivalent:

(i) The algebra A is uniserial

(ii) All indecomposable projective and injective A-modules are uniserial

(iii) For any primitive idempotent e of A, the modules Ae and eA are uniserial.

Proof. Properties (a) - (d) of dual modules ensure that (ii) is equivalent to (iii). Since (i) obviously implies (ii), it suffices to verify that (ii) implies (i).

Suppose that all indecomposable projective and injective A-modules are uniserial. Let V denote an A-module. It will be shown, by induction on $\dim_F V$, that V is a direct sum of uniserial modules. This will clearly prove the result. If $\dim_F V = 1$, then the assertion is obvious. Assume that $\dim_F V > 1$. If V is irreducible, then V is obviously uniserial. Thus we may assume that V has a proper irreducible submodule V'. By induction hypothesis, V' must be uniserial. Denote by W a uniserial submodule of V with $\dim_F W$ as large as possible. Then we must have $W \neq 0$. Let X be a submodule of V, maximal with respect to the property that $W \cap X = 0$. Assume that X_1/X and X_2/X are irreducible submodules of V/X. Then $X_1 \cap W \neq 0$ and $X_2 \cap W \neq 0$ and hence $X_1 \cap X_2 \cap W \neq 0$ since W is uniserial. Hence $X_1 \cap X_2 \neq X$ and so $X_1 = X_2$, proving that $\mathrm{Soc}(V/X)$ is irreducible. By Lemma 17.5(ii), V/X is a submodule of an indecomposable injective A-module, and so is uniserial. Invoking Lemma 17.7, we deduce that there is a surjective homomorphism $f : P \longrightarrow V/X$ for a suitable indecomposable projective A-module P. Since P is projective, there exists a homomorphism $g : P \longrightarrow V$ which renders commutative the following diagram:

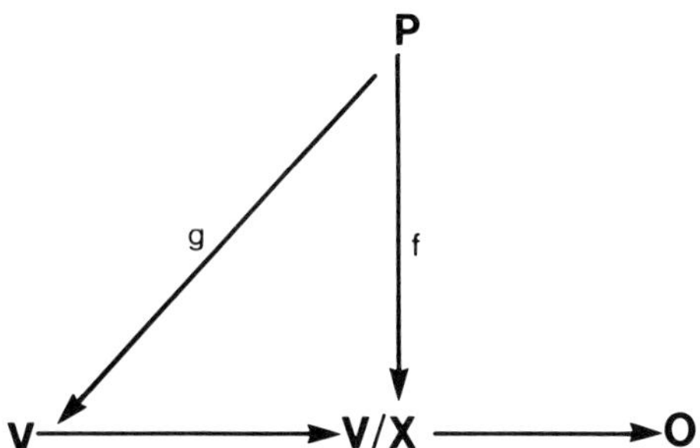

Because P is uniserial, so is $g(P)$. Now $W \cap X = 0$ implies that W is isomorphic to a submodule of V/X. Therefore, by the maximality of $\dim_F W$,

$$\dim_F g(P) \leqslant \dim_F W \leqslant \dim_F (V/X)$$

On the other hand, the commutativity of the diagram ensures that

$$\dim_F g(P) \geqslant \dim_F (V/X)$$

Hence $g(P) \cong V/X \cong W$. It follows that $V = g(P) + X$ and $g(P) \cap X = 0$. Thus $V = g(P) \oplus X$ and the desired assertion follows by induction hypothesis. ∎

17.9. COROLLARY. Let A be a Frobenius algebra. Then A is uniserial if and only if for each primitive idempotent e of A, the A-module Ae is uniserial.

Proof. Apply Theorem 17.8 and Lemma 17.6.

17.10. LEMMA. Let I be an ideal of A such that $I = Ax = yA$ for some x,y in A. Then $I = xA = Ay$.

Proof. Given an A-module M, let $c(M)$ denote the composition length of M. The map $A \longrightarrow I$, $a \mapsto ax$ is a surjective A-homomorphism with kernel $\ell(x) = \ell(xA)$. Hence $c(I) = c(A/\ell(xA))$. However, $xA \subseteq I$ and $\ell(xA) \supseteq \ell(I)$. Thus

$$c(I) \leqslant c(A/\ell(I))$$

On the other hand, $Ay \cong A/\ell(yA) = A/\ell(I)$ and $I \supseteq Ay$. Hence

$$c(I) \geqslant c(Ay) = c(A/\ell(I))$$

By combining these two inequalities, we deduce that $c(I) = c(Ay)$. Consequently, $I = Ay$ and, by a similar argument, $I = xA$. So the lemma is verified. ∎

We have now at our disposal all the information necessary to prove the following important result.

17.11. THEOREM. (Morita (1951), Azumaya and Nakayama (1948)). Let A be an algebra over an arbitrary field F. Then the following conditions are equivalent:

(i) There exist x,y in A such that $J(A) = xA = Ay$

(ii) There exists $x \in A$ such that $J(A) = xA = Ax$.

(iii) A is uniserial and each block of A is quasi-primary.

Proof. Clearly (ii) implies (i). By Lemma 17.10, (i) implies (ii) and hence we need only verify that (ii) is equivalent to (iii). For the sake of clarity, we divide the rest of the proof into three steps.

Step 1. Here we prove that (ii) implies that A *is uniserial.* Assume that $J(A) = xA = Ax$ for some x in A, and let e be any primitive idempotent of A. Let I be any proper nonzero submodule of Ae. Then $I \subseteq J(A)e$ since $J(A)e$ is a unique maximal submodule of Ae. Hence there exists $\mu \geqslant 1$ such that $I \subseteq J(A)^{\mu}e$ and $I \nsubseteq J(A)^{\mu+1}e$. Let a be an element of I with $a \notin J(A)^{\mu+1}e$. Because $J(A)^{\mu}e = x^{\mu}Ae$, there exists $a_o \in Ae$, $a_o \notin J(A)e$, such that $a = x^{\mu}a_o$. Then $Aa_o \subseteq Ae$ and $Aa_o \nsubseteq J(A)e$, so that $Aa_o = Ae$, since $J(A)e$ is a unique maximal submodule of Ae. Hence

$$I \supseteq Aa = Ax^{\mu}a_o = J(A)^{\mu}a_o = J(A)^{\mu}e$$

proving that $I = J(A)^{\mu}e$. This shows that $Ae, J(A)e, J(A)^2e, \ldots$ are the only submodules of Ae. Similarly $eA, eJ(A), eJ(A)^2, \ldots$ exhaust all right submodules of eA. Hence A is uniserial, by appealing to Theorem 17.8.

Step 2. We now prove that (ii) implies (iii). By Step 1, we need only show that each block of A is quasi-primary. To this end, write

$$A = Ae_{11} \oplus \ldots \oplus Ae_{1,f(1)} \oplus \ldots \oplus Ae_{n1} \oplus \ldots \oplus Ae_{n,f(n)}$$

where the e_{ij} are mutually orthogonal primitive idempotents, $Ae_{ki} \cong Ae_{k1}$ for $i \in \{1,\dots,f(k)\}$ and $Ae_{ki} \not\cong Ae_{tj}$ if $k \neq t$. Then it is easily seen that $J(A)^{\mu}(= Ax^{\mu} = x^{\mu}A)$ is expressible as a direct sum of $Ax^{\mu}e_{ki}$:

$$J(A)^{\mu} = \sum_{k=1}^{n} \sum_{i=1}^{f(k)} Ax^{\mu}e_{ki} \qquad (\mu = 1,2,\dots) \tag{1}$$

and similarly as a direct sum of $Ae_{ij}x^{\mu}$

$$J(A)^{\mu} = \sum_{i=1}^{n} \sum_{j=1}^{f(i)} Ae_{ij}x^{\mu} \qquad (\mu = 1,2,\dots) \tag{2}$$

Because $Ae_{ki} \cong Ae_{k1}$ for all $i \in \{1,2,\dots,f(k)\}$, we have

$$Ax^{\mu}\, e_{ki} \cong Ax^{\mu}e_{k1} \qquad (1 \leqslant i \leqslant f(k)) \tag{3}$$

and

$$Ae_{tj}x^{\mu} \cong Ae_{t1}x^{\mu} \qquad (1 \leqslant j \leqslant f(t)) \tag{4}$$

where $Ax^{\mu}e_{ki}$ and $Ae_{tj}x^{\mu}$ with $Ax^{\mu}e_{ki} = 0$ and $Ae_{tj}x^{\mu} = 0$ are excluded.

Next, we prove that

$$Ax^{\mu}e_{ki} \not\cong Ax^{\mu}e_{tj} \qquad \text{if} \quad k \neq t \tag{5}$$

$$Ae_{tj}x^{\mu} \not\cong Ae_{ki}x^{\mu} \qquad \text{if} \quad t \neq k \tag{6}$$

$$Ax^{\mu}e_{ki} \text{ and } Ae_{tj}x^{\mu} \text{ are indecomposable left ideals} \tag{7}$$

where $Ax^{\mu}e_{ki}$ and $Ae_{tj}x^{\mu}$ with $Ax^{\mu}e_{ki} = 0$ and $Ae_{tj}x^{\mu} = 0$ are excluded.

Suppose that $Ax^{\mu}e_{ki} \cong Ax^{\mu}e_{tj}$. Then, by Step 1, there exists e_{sm} such that

$$J(A)^{\mu}e_{ki}/J(A)^{\mu+1}e_{ki} \cong Ae_{sm}/J(A)e_{sm}$$

and

$$J(A)^{\mu}e_{tj}/J(A)^{\mu+1}e_{tj} \cong Ae_{sm}/J(A)e_{sm}$$

Choose $a \in e_{sm}J(A)^{\mu}e_{ki}, b \in e_{sm}J(A)^{\mu}e_{tj}$ with $a \notin J(A)^{\mu+1}, b \notin J(A)^{\mu+1}$. Then we have

$$e_{sm}J(A)^{\mu} = aA = bA$$

and therefore $e_{sm}J(A)^{\mu}$ is a homomorphic image of both $e_{ki}A$ and $e_{tj}A$. But then

$$e_{ki}A/e_{ki}J(A) \cong e_{tj}A/e_{tj}J(A)$$

and hence $k = t$, proving (5).

Property (6) is a direct consequence of the fact that $Ae_{tj}x^{\mu}$ is a homomorphic image of Ae_{tj}. To prove (7), note that $Ax^{\mu}e_{ki}$ is a homomorphic image of some Ae_{sm} and $Ae_{tj}x^{\mu}$ is a homomorphic image of Ae_{tj}. Thus both $Ax^{\mu}e_{ki}$ and $Ae_{tj}x^{\mu}$ are uniserial and therefore indecomposable, proving (7).

Invoking (3) - (7) and the Krull-Schmidt theorem, we deduce that there is a one-to-one correspondence ϕ between the subsets of $\{1,2,\ldots,n\}$ such that

$$Ax^{\mu}e_{k1} \cong Ae_{\phi(k),1}x^{\mu} \tag{8}$$

and

$$f(k) = f(\phi(k)) \tag{9}$$

where we exclude those $Ax^{\mu}e_{ki}$ and $Ae_{tj}x^{\mu}$ which reduce to zero.

From (8) it follows further that

$$\text{if } Ax^{\mu}e_{k1} \neq 0, \text{ then } J(A)^{\mu}e_{k1}/J(A)^{\mu+1}e_{k1} \cong Ae_{\phi(k),1}/J(A)e_{\phi(k),1} \tag{10}$$

Hence the number $f(k)$ is the same for any Ae_{ki} contained in a fixed block of A. This proves that each block of A is quasi-primary.

Step 3. Completion of the proof.

By Step 2, it suffices to show that (iii) implies (ii). To this end, we put

$$J(A)e_{k1}/J(A)^2e_{k1} \cong Ae_{\phi(k),1}/J(A)e_{\phi(k),1}$$

Then the proof of (5) shows that $k \neq t$ implies $\phi(k) \neq \phi(t)$. Since Ae_{k1} and $Ae_{\phi(k),1}$ are contained in the same block, we have $f(k) = f(\phi(k))$, since each block of A is quasi-primary. For an element c_{ki} such that

$$c_{ki} \in e_{\phi(k),i}J(A)e_{ki}\ ,\ c_{ki} \notin J(A)^2,$$

we have

$$Ac_{ki} = J(A)e_{ki} \quad \text{and} \quad c_{ki}A = e_{\phi(k),i}\, J(A) \tag{11}$$

Setting $x = \sum_{k=1}^{n} \sum_{i=1}^{f(k)} c_{ki}$, we obtain a direct decomposition

$$Ax = \sum_{k=1}^{n} \sum_{i=1}^{f(k)} Ac_{ki}$$

Applying (11), we infer that $J(A) = Ax$. Similarly xA is a direct sum of $c_{ki}A = e_{\phi(k),i}J(A)$ and $J(A) = xA$. Hence $J(A) = Ax = xA$ and the result follows. ∎

For the rest of this section, F *denotes a field characteristic* $p > 0$. Our next aim is to discover necessary and sufficient conditions under which FG is of finite representation type. Given a subset X of G, we write X^+ for the sum of elements in X and denote by FX^+ the F-linear span of X^+. The next three lemmas and Theorem 17.15 (below) will not be required in their full generality.

17.12. LEMMA. Let G be a p-group.

(i) If I is a proper left ideal of FG, then FG/I is an indecomposable FG-module.

(ii) FG has a unique minimal left ideal, namely FG^+

(iii) Each nonzero submodule of FG is indecomposable.

Proof. (i) Owing to Corollary 1.2, $J(FG) = I(G)$ and so $FG/J(FG) \cong F$. Hence $J(FG)/I$ is a unique maximal submodule of FG/I and thus FG/I is indecomposable.

(ii) It is obvious that FG^+ is a minimal left ideal of FG. Assume that X is an arbitrary minimal left ideal of FG. Since the representation of G afforded by X is irreducible the isomorphism $FG/J(FG) \cong F$ ensures that G acts trivially on X and $\dim_F X = 1$. Hence $X = Fx$ for some $x \in FG$ and $gx = x$ for all $g \in G$. But then $x = \lambda G^+$ for some $\lambda \neq 0$ in F. Thus $X = FG^+$ as asserted.

(iii) Direct consequence of (ii). ∎

17.13. LEMMA. Suppose that $G = \langle g \rangle$ is cyclic of order p^n.

(i) Each indecomposable FG-module is a homomorphic image of FG

(ii) For any $i \in \{1,2,\dots,p^n\}$, FG has exactly one submodule of codimension i, namely $J(FG)^i$

(iii) For any $i \in \{1,2,\dots,p^n\}$, put $V_i = FG/J(FG)^i$. Then

(a) V_i is an indecomposable FG-module of dimension i, and V_i possesses an F-basis $\{v_1,\dots,v_i\}$ such that

$$gv_1 = v_1 \;, \; gv_k = v_{k-1} + v_k \qquad (2 \leqslant k \leqslant i)$$

(b) Each indecomposable FG-module is isomorphic to V_i for some $i \in \{1,2,\dots,p^n\}$.

Proof. (i) We first note that $FG \cong F[X]/(X^{p^n} - 1)$. Because $F[X]$ is a principal ideal domain, each $F[X]$ (and hence FG)-module is the direct sum of cyclic $F[X]$-submodules. Hence, if V is an indecomposable FG-module, then V is cyclic. It follows that there is a surjective homomorphism $FG \longrightarrow V$ of FG-modules, proving (i).

(ii) The submodules of FG correspond to the ideals I of $F[X]$ for which

$$I \supseteq (X^{p^n} - 1)$$

Since $F[X]$ is a principal ideal domain, $I = (f)$ for some f dividing $X^{p^n} - 1$ But $X^{p^n} - 1 = (X-1)^{p^n}$, so $I = ((X-1)^i)$ for some i, $0 \leqslant i \leqslant p^n$, and the image I^* of I in FG is

$$I^* = FG(g-1)^i$$

Because $J(FG) = I(G)$ and G is generated by g, we also have

$$J(FG) = FG(g-1)$$

Hence the submodules of FG are the

$$J(FG)^i = FG(g-1)^i \qquad (0 \leqslant i \leqslant p^n)$$

and

$$\dim_F FG/J(FG)^i = \dim_F F[X]/((X-1)^i) = i$$

as asserted.

(iii) Owing to Lemma 17.12(i), each V_i is indecomposable. Setting

$$v_k = (g-1)^{i-k} + J(FG)^i \qquad (1 \leqslant k \leqslant i-1)$$
$$v_i = 1 + J(FG)^i$$

we obtain an F-basis $\{v_1,\ldots,v_i\}$ of V_i such that

$$gv_1 = v_1 \ , \ gv_k = v_{k-1} + v_k \qquad (2 \leqslant k \leqslant i)$$

proving (a). Assume that V is an indecomposable FG-module. By (i), $V \cong FG/I$ for some FG-submodule I. By (ii), $I = FG(g-1)^i$ for some i and hence

$$V \cong FG/FG(g-1)^i \cong V_i$$

proving (b). ∎

17.14. LEMMA. Let $G = \langle a \rangle \times \langle b \rangle$, where $\langle a \rangle$ and $\langle b \rangle$ are cyclic groups of order p. Then, for any positive integer n, there exists an indecomposable FG-module of dimension $2n+1$.

Proof. Let V be a vector space over F of dimension $2n+1$ and let

$$\{x_0, x_1, \ldots, x_n, y_1, \ldots, y_n\}$$

be a basis of V. It is immediate to verify that V becomes an FG-module by setting

$$\begin{aligned} ax_i = bx_i &= x_i && (0 \leqslant i \leqslant n) \\ (a-1)y_i &= x_i && (1 \leqslant i \leqslant n) \qquad (12) \\ (b-1)y_i &= x_{i-1} && (1 \leqslant i \leqslant n) \end{aligned}$$

Let X and Y denote, respectively, the F-linear span of $\{x_0,\ldots,x_n\}$ and $\{y_1,\ldots,y_n\}$, and let $\pi : V \longrightarrow Y$ be the natural projection. Owing to (12), left multiplications by $(a-1)$ and $(b-1)$ provide injections $Y \longrightarrow X$ and map X to 0

Assume by way of contradiction that $V = V_1 \oplus V_2$ for some nonzero FG-submodules V_1 and V_2 of V. Put $r = \dim_F \pi(V_1)$ so that $1 \leqslant r \leqslant n$. Then

$$(a-1)V_1 \subseteq V_1 \quad \text{and} \quad (a-1)V_1 = (a-1)\pi(V_1)$$

and therefore

$$\dim_F((a-1)V_1) = r$$

Bearing in mind that $(b-1)V_1 \nsubseteq (a-1)V_1$, we derive

$$\dim_F V_1 \geqslant 2r+1 \qquad (0 \leqslant r \leqslant n)$$

A similar inequality is also true for V_2. Setting $s = \dim_F \pi(V_2)$, we have $r + s \geqslant n$ and

$$\dim_F V_2 \geqslant 2s+1$$

Hence

$$\dim_F V \geqslant 2(r+s) + 2 \geqslant 2n+2$$

contrary to the assumption that $\dim_F V = 2n+1$. This completes the proof of the lemma. ■

It is now an easy matter to prove the following general result.

17.15. THEOREM. (D G Higman (1954), Kasch, Kneser and Kupisch (1957)). Let F be an arbitrary field of characteristic $p > 0$ and let P be a Sylow p-subgroup of G.

(i) If P is not cyclic, then there exist indecomposable FG-modules of arbitrary large F-dimension

(ii) If P is cyclic, then there exist at most $|G|$ nonisomorphic indecomposable FG-modules. Any indecomposable FG-module is a homomorphic image of FG.

Proof. (i) Assume that P is not cyclic. Then P has an elementary abelian factor group of order p^2. Hence, by Lemma 17.14, for each $n \geqslant 1$ there exists an indecomposable FP-module V_n with $\dim_F V_n = 2n+1$. But V_n is isomorphic to a direct summand of $(V_n^G)_P$. Hence, by the Krull-Schmidt theorem, V_n

is isomorphic to a direct summand of W_P for some indecomposable direct summand W of V_n^G. It follows that

$$\dim_F W \geq \dim_F V_n = 2n+1,$$

as required.

(ii) Since P is cyclic, it follows from Lemma 17.13 that there are $|P|$ indecomposable FP-modules $V_1, V_2, \ldots$, where $\dim_F V_i = i$ and every indecomposable FP-module is isomorphic to some V_i. Assume that V is an indecomposable FG-module. Then V is P-projective, so V is isomorphic to a direct summand of $(V_P)^G$. Write $V_P = \oplus X_i$ for some indecomposable FP-modules X_i. Then, by the Krull-Schmidt theorem, V is isomorphic to a direct summand of X_i^G for some i and $\dim_F V \geq \dim_F X_i$. Consequently, V is isomorphic to a direct summand of some V_j^G with $j \leq \dim_F V$. Since $\dim V_j^G = (G:P)j$, the number of nonisomorphic indecomposable direct summands of V_j^G of dimension at least j is at most $(G:P)$. Hence the number of nonisomorphic indecomposable FG-modules is at most $|G|$.

By Lemma 17.3(i), V_j is a homomorphic image of FP. Hence V_j^G is a homomorphic image of $(FP)^G \cong FG$. But V is isomorphic to a direct summand of V_j^G. Thus V is a homomorphic image of FG, as asserted. ∎

We have now come to the demonstration for which this section has been developed.

17.16. THEOREM (Morita (1951)). Let F be an algebraically closed field of characteristic $p > 0$, let G be a finite group and let P be a Sylow p-subgroup of G. Then the following conditions are equivalent:

(i) There exist $x, y \in FG$ such that $J(FG) = xFG = FGy$

(ii) There exists $x \in FG$ such that $J(FG) = xFG = FGx$

(iii) G is p-solvable and P is cyclic

(iv) $PO_{p'}(G) \lhd G$ and P is cyclic

(v) $PO_{p'}(G) \lhd G, P$ is cyclic and $G/PO_{p'}(G)$ is cyclic of order dividing $p-1$.

Proof. (i) ⇒ (ii): Direct consequence of Lemma 17.10.

(ii) ⇒ (iii): By Theorem 17.11, FG is uniserial and each block of FG is quasi-primary. Hence, by Lemma 17.7 and Theorem 17.15, P is cyclic and, by Proposition 14.5, $PO_{p'}(G) \lhd G$. Since the latter implies that G is p-solvable (iii) is established.

(iii) ⇒ (iv): By Proposition 16.20, G is of p-length 1 and so

$$PO_{p'}(G) = O_{p',p}(G) \lhd G$$

(iv) ⇒ (v): Set $\bar{G} = G/O_{p'}(G)$ and $\bar{P} = PO_{p'}(G)/O_{p'}(G)$. Then $\bar{P} \lhd \bar{G}$ and $C_{\bar{G}}(\bar{P}) = \bar{P}$ since $C_{\bar{G}}(\bar{P}) \lhd \bar{G}$ and $O_{p'}(\bar{G}) = 1$. Hence $\bar{G}/\bar{P}$ is isomorphic to a subgroup of $\mathrm{Aut}(\bar{P})$. But $\bar{P} \cong P$ and $\bar{G}/\bar{P} \cong G/PO_{p'}(G)$. Thus $G/PO_{p'}(G)$ is isomorphic to a subgroup of $\mathrm{Aut}(P)$. If $|P| = p^n$ and $p > 2$, then $\mathrm{Aut}(P)$ is a cyclic group of order $p^{n-1}(p-1)$. Therefore $G/PO_{p'}(G)$ is a cyclic group of order dividing $p-1$. In case $p=2$ we have $G = PO_{p'}(G)$, by virtue of Lemma 12.1. Thus the required assertion is also true for $p = 2$.

(v) ⇒ (ii): Put $N = PO_{p'}(G)$ and observe that, by Proposition 1.8(ii),

$$J(FG) = FGJ(FN) = J(FN)FG$$

Hence, if $x \in FN$ is such that $J(FN) = FNx = xFN$, then

$$\begin{aligned} J(FG) &= FG(FNx) = (xFN)FG \\ &= FGx = xFG \end{aligned}$$

Thus we may harmlessly assume that $G = N$ in which case G is p-nilpotent. Let B be an arbitrary block of FG. It suffices to verify that $J(B) = xB = Bx$ for some $x \in B$. Since P is cyclic, it follows from Corollary 10.10 that $B \cong M_n(FD)$ for some cyclic p-subgroup D of G. Since $J(B) \cong M_n(J(FD))$ and since

$$J(FD) = FD(g-1) = (g-1)FD$$

where g is a generator of D, the result follows. ■

4
Group algebras of p-groups over fields of characteristic p

In this chapter we confine our attention to the case where G is a finite p-group and F a field of characteristic p. After examining dimension subgroups, we provide Jennings' description of bases for $J(FG)^n$ and a formula for the nilpotency index $t(G)$ of $J(FG)$. As an application, we determine the right and left annihilators of powers of $I(G)$ and exhibit a remarkable symmetry of the numbers $c_n = \dim_F I(G)^n/I(G)^{n+1}$. We then compute $t(P)$ for some individual p-groups P, such as metacyclic p-groups, p-groups of order $\leqslant p^4$, etc. The results obtained are applied to investigate the structure of P in case $t(P)$ attains its highest, second highest, lowest and second lowest values. In the final section, all p-groups P with $t(P) \leqslant 7$ are characterized.

I. DIMENSION SUBGROUPS IN CHARACTERISTIC $p > 0$ AND RELATED RESULTS.

Throughout this section, F denotes an arbitrary field of characteristic $p > 0$ and G a finite group. Given subgroups G_1 and G_2 of G, we write $[G_1,G_2]$ for the subgroup of G generated by all commutators

$$[x,y] = x^{-1}y^{-1}xy \qquad \text{with } x \in G_1, y \in G_2$$

The *lower central series* $\{\gamma_n(G)\}$ of G is defined inductively by

$$\gamma_1(G) = G \quad \text{and} \quad \gamma_{n+1}(G) = [\gamma_n(G),G]$$

It is clear that these groups are, in fact, characteristic subgroups of G and that $\gamma_n(G)/\gamma_{n+1}(G)$ is central in $G/\gamma_{n+1}(G)$. Furthermore, G is called *nilpotent* if $\gamma_{c+1}(G) = 1$ for some integer $c \geqslant 0$. When this occurs, the smallest such integer c with $\gamma_{c+1}(G) = 1$ is, by definition, the *nilpotency class of* G. If $n \geqslant 1$ is an integer, we put

$$G^n = \langle g^n | g \in G\rangle$$

It is clear that all such subgroups G^n are characteristic in G.

The *Brauer–Jennings–Zassenhaus series* for G is defined inductively by

$$M_{1,p}(G) = G$$

and for $n \geqslant 2$

$$M_{n,p}(G) = [M_{n-1,p}(G), G]M_{i,p}(G)^p$$

where i is the smallest integer satisfying $ip \geqslant n$. One immediately verifies, by induction, that this is a decreasing sequence of characteristic subgroups of G.

The *Lazard series* for G is defined by

$$L_{n,p}(G) = \prod_{ip^j \geqslant n} \gamma_i(G)^{p^j}$$

The following observations are easy consequences of the definition:

(a) $L_{1,p}(G) = G$

(b) $L_{n,p}(G)$ is a product of finitely many of the characteristic subgroups $\gamma_i(G)^{p^j}$

(c) $\{L_{n,p}(G)\}$ is a decreasing sequence of characteristic subgroups of G.

Indeed, $L_{1,p}(G)$ contains the factor $\gamma_1(G)^{p^0} = G$ and so $L_{1,p}(G) = G$. The reason for (b) is that, since $L_{n,p}(G) \supseteq \gamma_n(G)^{p^0} = \gamma_n(G)$, all additional factors with $i \geqslant n$ are redundant. Also, for $i < n$ only the minimal j with $ip^j \geqslant n$ is really required. Finally, (c) is a consequence of the fact that increasing n decreases the number of defining factors for $L_{n,p}(G)$.

The *dimension subgroups* $D_n(FG)$ of G are defined by

$$D_n(FG) = G \cap (1 + I(G)^n)$$

Let $G = G_1 \supseteq G_2 \supseteq \cdots \supseteq G_n \supseteq \cdots$ be a decreasing sequence of normal subgroups of G. Then $\{G_i\}$ is said to be an N_p*-sequence* if the following two conditions hold:

$$[G_i, G_j] \subseteq G_{i+j}$$

$$g \in G_i \text{ implies } g^p \in G_{ip}$$

Of course, if G is a p-group, then the above implies that G_i/G_{2i} (and hence

G_i/G_{i+1}) is elementary abelian.

1.1. LEMMA. Let $I(G) = I_1 \supseteq I_2 \supseteq \dots \supseteq I_n \supseteq \dots$ be a decreasing sequence of ideals of FG satisfying $I_iI_j \subseteq I_{i+j}$ for all i,j, and let $G_i = G \cap (1+I_i)$ for all i. Then

$$G = G_1 \supseteq G_2 \supseteq \dots \supseteq G_n \supseteq \dots$$

is an N_p-sequence. In particular, the dimension subgroups $D_n(FG)$ form an N_p-sequence.

Proof. Since $I(G) = I_1$ and $I_i \supseteq I_{i+1}$, we have $G_i \supseteq G_{i+1}$ and $G_1 = G$. Since I_i is the kernel of the homomorphism $G \longrightarrow (G+I_i)/I_i, g \longmapsto g+I_i, I_i$ is a normal subgroup of G. Thus $G = G_1 \supseteq G_2 \supseteq \dots \supseteq G_n \supseteq \dots$ is a decreasing chain of normal subgroups of G.

Now let $x \in G_i, y \in G_j$, and note that

$$xy-yx = (x-1)(y-1) - (y-1)(x-1) \in I_{i+j}$$

since both I_iI_j and I_jI_i are contained in I_{i+j}. Furthermore,

$$x^{-1}y^{-1}xy - 1 = x^{-1}y^{-1}(xy-yx) \in I_{i+j}$$

which shows that $[x,y] \in G_{i+j}$ and hence that $[G_i,G_j] \subseteq G_{i+j}$.

Finally, fix $g \in G_i$. Then $g-1 \in I_i$ so that

$$g^p - 1 = (g-1)^p \in I_i^p \subseteq I_{ip}$$

Hence $g^p \in G_{ip}$ and the result follows. ∎

1.2. LEMMA. Let G be a group of order p^n and let

$$G = H_1 \supset H_2 \supset \dots \supset H_n \supset H_{n+1} = 1$$

be a composition series of G. For each $i \geqslant 1$ choose $g_i \in H_i - H_{i+1}$. Then the elements

$$\lambda(a_1,a_2,\dots,a_n) = \prod_{i=1}^{n} (g_i-1)^{a_i}$$

with $0 \leqslant a_i < p$ form an F-basis for FG. Moreover, these elements with $\lambda(0,0,\dots,0) = 1$ deleted form an F-basis for $I(G)$.

Proof. Since the number of the given elements $\lambda(a_1,\dots,a_n)$ is precisely the order of G, it suffices to prove that they span FG. The case $G = 1$ being trivial, we argue by induction on $|G|$. Since $|H_2| < |G|$, it follows from the induction hypothesis that the elements

$$\lambda(0,a_2,\ldots,a_n) = \prod_{i=2}^{n} (g_i-1)^{a_i}$$

span FH_2. Now G/H_2 is cyclic of order p and is generated by the image of g_1. Hence $\{1,g_1,\ldots,g_1^{p-1}\}$ is a transversal for H_2 in G, and each element of FG can be written as

$$x = \sum_{i=0}^{p-1} g_1^i x_i$$

with $x_i \in FH_2$. Finally, bearing in mind that

$$g_1^i = \{(g_1 - 1) + 1\}^i = \sum_{j=0}^{i} \binom{i}{j} (g -1)^j$$

the result follows. ■

Let G be a finite p-group and let

$$G = G_1 \supseteq G_2 \supseteq \ldots \supseteq G_d \supseteq G_{d+1} = 1$$

be an N_p-sequence of G. For each $g \neq 1$ in G, define the *height* of g, written $\nu(g)$, to be the largest m for which $g \in G_m$, and put $\nu(1) = \infty$. If $\lambda = \lambda(a_1,\ldots,a_n)$ is as in Lemma 1.2, we define the *weight* $w(\lambda)$ of λ by

$$w(\lambda) = \sum_{i=1}^{n} a_i \nu(g_i)$$

Let I_i be the F-linear span of all products of the form $\prod_{j=1}^{k} (x_j-1)$ for some k with $x_j \in G$ and $\sum_{j=1}^{k} \nu(x_j) \geqslant i$. Clearly, I_i is an ideal of FG because it is closed under multiplication by 1 and $g-1$, for any $g \in G$. From the definition of ν, it follows that

$$I(G) = I_1 \supseteq I_2 \supseteq \ldots \supseteq I_i \supseteq \ldots$$

is a decreasing sequence of ideals of FG with $I_i I_j \subseteq I_{i+j}$. We shall refer to this sequence as the *filtration of* $I(G)$ *determined by the* N_p*-sequence* $\{G_i\}$.

1.3. LEMMA. Let G be a finite p-group, let

$$G = H_1 \supset H_2 \supset \ldots \supset H_{n+1} = 1$$

be a composition series of G which refines a given N_p-sequence

$$G = G_1 \supseteq G_2 \supseteq \dots \supseteq G_d \supseteq G_{d+1} = 1$$

and let $\{I_i\}$ be the filtration of $I(G)$ determined by this sequence.

(i) If $\lambda = \lambda(a_1,\dots,a_n)$ is as in Lemma 1.2, then the elements λ with $w(\lambda) = t$ span I_t modulo I_{t+1}

(ii) The elements $\lambda = \lambda(a_1,\dots,a_n)$ with $w(\lambda) \geqslant t$ form an F-basis for I_t.

Proof. (i) Note first that, by definition of $w(\lambda)$ and I_t, if $w(\lambda) = t$, then $\lambda \in I_t$. We are therefore left to verify that any product

$$u = \prod_{j=1}^{k} (x_j - 1) \qquad (x_j \in G)$$

with $\sum_{j=1}^{k} \nu(x_j) \geqslant i$ can be written as an F-linear combination of λ's with $w(\lambda) = t$ plus a term in I_{t+1}. We argue by induction on k.

To begin with, we fix $i \in \{1,\dots,k\}$ and write $u = \alpha(x_i - 1)\beta$ for some $\alpha,\beta \in FG$. Put $\nu(x_i) = m$ so that $g_i \in G_m$. Since the sequence $\{H_s\}$ is a refinement of $\{G_s\}$, we have

$$G_m = H_r \supset H_{r+1} \supset \dots \supset H_{n+1} = 1$$

for some integer r. Invoking Lemma 1.2, we infer that $x_i - 1$ is an F-linear combination of terms of the form

$$\lambda' = (g_1 - 1)^{a_r}(g_{r+1} - 1)^{a_{r+1}} \dots (g_n - 1)^{a_n}$$

with the a_s not all zero. However, each g_s above satisfies $\nu(g_s) \geqslant m = \nu(x_i)$ and so

$$x_i - 1 = \sum_{\nu(g_s)=m} b_s(g_s - 1) + \gamma$$

for some $b_s \in F, \gamma \in I_{m+1}$. Hence

$$u = \alpha(x_i - 1)\beta \equiv \sum_{\nu(g_s)=\nu(g_i)} b_s \alpha(g_s - 1)\beta \pmod{I_{t+1}}$$

and therefore it suffices to verify that each product $\alpha(g_s - 1)\beta$ is an F-linear combination of the λ's modulo I_{t+1}. Moreover, by applying this argument to each $x_i - 1$ in turn, we may assume that each x_i is some g_i, with the same height. If $k = 1$, then $u = g_j - 1$ for some j with $\nu(g_j) \geqslant t$. Hence

u is a basis element λ of weight $\geqslant t$. But $w(\lambda) > t$ implies $\lambda \in I_{t+1}$ and thus the result is true for $k = 1$.

Assume that the result is true for all such products with less than k terms and write

$$u = \alpha(g_i-1)(g_j-1)\beta$$

where (g_i-1) and (g_j-1) are adjacent factors. Put $r = \nu(g_i), s = \nu(g_j)$ and consider the identity

$$\begin{aligned}(g_i-1)(g_j-1) &= (g_j-1)(g_i-1) + (g_i^{-1}g_j^{-1}g_ig_j-1)\\ &\quad + (g_jg_i-1)(g_i^{-1}g_j^{-1}g_ig_j-1)\end{aligned}$$

Setting $h = g_i^{-1}g_j^{-1}g_ig_j$ it follows that

$$h \in [G_r, G_s] \subseteq G_{r+s},$$

so $\nu(h) \geqslant r+s$ and thus $(g_jg_i-1)(h-1) \in I_{r+s+1}$. Therefore

$$\begin{aligned}u &= \alpha(g_i-1)(g_j-1)\beta\\ &\equiv \alpha(g_j-1)(g_i-1)\beta + \alpha(h-1)\beta \pmod{I_{t+1}}\end{aligned}$$

Observe that $\alpha(h-1)\beta$ is a product of at most $k-1$ factors but with total height at least t. Consequently, by induction, $\alpha(h-1)\beta$ is an F-linear combination modulo I_{t+1} of the λ's of weight t. Thus we are left to verify that $\alpha(g_j-1)(g_i-1)\beta$ is such an F-linear combination of the λ's.

By the foregoing, in proving the result we may harmlessly interchange adjacent factors. This implies that we can put all the factors in their natural order. Thus it suffices to consider u of the form

$$u = (g_1-1)^{b_1}(g_2-1)^{b_2}\cdots(g_n-1)^{b_n}$$

with $b_i \geqslant 0$, $\sum_{i=1}^{n} b_i = k$ and $\sum_{i=1}^{n} b_i\nu(g_i) \geqslant t$. If $b_i < p$ for all i, then $u = \lambda(b_1,\ldots,b_n)$ and $w(\lambda) \geqslant t$ so that the result follows. Finally, assume that $b_i \geqslant p$ for some i. Then $(g_i-1)^p = (g_i^p-1)$ occurs in u and

$$u = \alpha(g_i^p-1)\beta$$

is a product of fewer than k terms. Moreover, if $\nu(g_i) = s$, then $g_i \in G_s$ so that $g_i^p \in G_{sp}$ and $\nu(g_i^p) \geqslant sp$. The conclusion is that in this new representation of u, the sum of the heights is still $\geqslant t$. Hence the required assertion follows by induction.

(ii) We know that the λ's are linearly independent and that $w(\lambda) \geqslant t$ implies $\lambda \in I_t$. It therefore suffices to verify that these λ's span I_t. Invoking a repeated application of (i), we see that the set of λ's with $s \geqslant w(\lambda) \geqslant t$ span I_t modulo I_{s+1} for any $s \geqslant t$. Thus we need only show that $I_{s+1} = 0$ for all sufficiently large s.

Since $G_{d+1} = 1$, it follows that $\nu(g) \leqslant d$ for all $g \in G, g \neq 1$. Therefore in the product

$$u = (x_1-1)(x_2-1)\dots(x_k-1)$$

we have $\nu(x_1) + \nu(x_2) + \dots + \nu(x_k) \leqslant d_k$. Hence $I_{dk} \subseteq I(G)^k$ and since $I(G)$ is nilpotent by Corollary 3.1.2, the result follows. ■

We are now ready to prove the following result.

1.4. THEOREM. (Jennings(1941)). Let G be a finite p-group, let

$$G = G_1 \supseteq G_2 \supseteq \dots \supseteq G_d \supseteq G_{d+1} = 1$$

be a finite N_p-sequence with $|G_i/G_{i+1}| = p^{e_i}$, $1 \leqslant i \leqslant d$, and let $\{I_i\}$ be the filtration of $I(G)$ determined by this sequence. Put $I_o = FG$,

$$m = \sum_{n=1}^{d} (p-1)ne_n$$

and define the integers c_n by

$$\prod_{i=1}^{d} (1 + X^i + X^{2i} + \dots + X^{(p-1)i})^{e_i} = \sum_{n=o}^{m} c_n X^n$$

(i) $I_m \neq 0$ but $I_{m+1} = 0$

(ii) $c_n = \dim_F I_n/I_{n+1}$ for all $n \in \{0,1,\dots,m\}$

(iii) $G \cap (1 + I_i) = G_i$ for all $i \in \{1,2,\dots,d+1\}$

Proof. Note that the definition of N_p-sequence allows some of the $e_n's$ to be zero, but this does not effect our calculations. Observe also that if $\{H_s\}$ is any refinement of $\{G_s\}$ into a composition series and if $g_i \in H_i - H_{i+1}$ is chosen, then e_j is precisely the number of $g_i's$ of height j.

(i) Keep the notation of Lemma 1.2 so that the elements $\lambda(a_1,\dots,a_n) = \prod_{i=1}^{n} (g_i-1)^{a_i}$

form an F-basis for FG. Then the largest possible wieght of λ occurs when all a_i are equal to $p-1$. Since there are precisely e_j of the g_i's with height j, we have in this case that

$$w(\lambda) = (p-1)\sum_{i=1}^{n} \nu(g_i) = (p-1)\sum_{n=1}^{d} ne_n = m$$

Hence, by Lemma 1.3(i), $I_m \neq 0$ but $I_{m+1} = 0$.

(ii) Owing to Lemma 1.3(i), the number $f_n = \dim_F I_n/I_{n+1}$ coincides with the number of λ's of weight n. On the other hand, the weight of $\lambda = \lambda(a_1,\ldots,a_n)$ is

$$w(\lambda) = a_1\nu(g_1) + a_2\nu(g_2) + \ldots + a_n\nu(g_n)$$

which shows that f_n is equal to the number of ways of choosing $0 \leqslant a_i \leqslant p-1$ such that $w(\lambda) = n$. But the latter is the coefficient of X^n in the product

$$\prod_{i=1}^{n}(1 + X^{\nu(g_i)} + X^{2\nu(g_i)} + \ldots + X^{(p-1)\nu(g_i)})$$

Finally, since there are precisely e_j of the g_i's with $\nu(g_i) = j$ we get

$$\sum_{m=o}^{m} f_n X^n = \prod_{i=1}^{d}(1 + X^i + X^{2i} + \ldots + X^{(p-1)i})^{e_i}$$

This proves that $c_n = f_n$ for all $n \in \{0,1,\ldots,m\}$.

(iii) Fix $i \in \{1,2,\ldots,d+1\}$. If $g \in G_i$, then $\nu(g) \geqslant i$ so that $g-1 \in I_i$ by definition of I_i. Conversely, let $g \in G-G_i$ so that $\nu(g) < i$, and let $\{H_s\}$ be a refinement of $\{G_s\}$ into a composition series. Since $g \neq 1, g \in H_j - H_{j+1}$ for some $j \geqslant 1$ and in choosing our generator for H_j/H_{j+1} we can clearly take $g_j = g$. With this choice of generator, $g-1 = g_j-1$ is a basis element of weight $w(g-1) = \nu(g) < i$. Hence, by Lemmas 1.2 and 1.3(ii), $g-1 \notin I_i$ and the result follows. ■

In order to apply the previous result to the powers of the augmentation ideal of FG, we characterize the dimension subgroups in a purely group-theoretic manner. Assume that $\{G_i\}$ and $\{H_i\}$ are two N_p-sequences in G. Then we write $\{G_i\} \geqslant \{H_i\}$ if and only if $G_i \supseteq H_i$ for all i.

1.5. THEOREM. (Jennings (1941)). Let G be a finite p-group and let $\{I_i\}$ be the filtration of $I(G)$ determined by $\{D_i(FG)\}$.

(i) the sequence $\{D_i(FG)\}$ is the unique smallest N_p-sequence

(ii) $I_i = I(G)^i$ for all i

Proof. (i) By Lemma 1.1, the dimension subgroups $D_i(FG)$ form an N_p-sequence. Since $I(G)$ is nilpotent, $D_{d+1}(FG) = 1$ for some integer d. Assume that $\{G_i\}$ is an arbitrary N_p-sequence. Then obviously $\{G_i \cap D_i(FG)\}$ is an N_p-sequence $\leqslant \{D_i(FG)\}$. It therefore suffices to verify that there is no N_p-sequence smaller than $\{D_i(FG)\}$.

Fix an N_p-sequence $\{G_i\}$ with $\{G_i\} \leqslant \{D_i(FG)\}$, and let $\{J_i\}$ be the filtration of $I(G)$ determined by $\{G_i\}$. Since $I(G) = J_1$ we have

$$I(G)^i = J_1^i \subseteq J_i \tag{1}$$

On the other hand, let

$$u = (x_1-1)(x_2-1)\dots(x_k-1) \qquad (x_j \in G)$$

be any generating element of J_i so that

$$\nu(x_1) + \nu(x_2) + \dots + \nu(x_k) \geqslant i$$

Then $x_i \in G_{\nu(x_i)} \subseteq D_{\nu(x_i)}(FG)$ so that $x_i - 1 \in I(G)^{\nu(g_i)}$ and

$$u \in I(G)^{\nu(x_1)+\nu(x_2)+\dots+\nu(x_k)} \subseteq I(G)^i$$

This proves that $I_i \subseteq I(G)^i$ and hence, by (1), that $J_i = I(G)^i$. Applying Theorem 1.4(iii), we conclude that

$$G_i = G \cap (1+J_i) = G \cap (1+I(G)^i) = D_i(FG)$$

as required.

(ii) By (i), $\{I(G)^i\} = \{J_i\}$ is the filtration of $I(G)$ determined by $\{G_i\} = \{D_i(FG)\}$. Thus $J_i = I_i = I(G)^i$ and the result follows. ∎

The following result provides a more explicit characterization of dimension subgroups. Although it is valid for an arbitrary group G, the special case below will suffice for our subsequent investigations.

1.6. THEOREM. Let G be a finite p-group. Then, for all $n \geqslant 1$,

(i) (Jennings (1941)) $D_n(FG) = M_{n,p}(G)$

(ii) (Lazard (1953)) $D_n(FG) = L_{n,p}(G)$

Proof. We first show that $D_n(FG) \supseteq M_{n,p}(G)$. The case $n = 1$ being trivial, we argue by induction on n. Assume that $n \geqslant 2$ and that the assertion is true for all subscripts $< n$. We know, from Lemma 1.1, that $\{D_i(FG)\}$ is an N_p-sequence of G. Since

$$D_{n-1}(FG) \supseteq M_{n-1,p}(G)$$

this yields

$$\begin{aligned} D_n(FG) &\supseteq [D_{n-1}(FG), G] \\ &\supseteq [M_{n-1,p}(G), G] \end{aligned}$$

Let i be minimal with $ip \geqslant n$. Then $i \leqslant n-1$ so that $D_i(FG) \supseteq M_{i,p}(G)$ and we have

$$D_n(FG) \supseteq D_i(FG)^p \supseteq M_{i,p}(G)^p$$

Bearing in mind that $M_{n,p}(G)$ is generated by $[M_{n-1,p}(G), G]$ and $M_{i,p}(G)^p$, we derive $D_n(FG) \supseteq M_{n,p}(G)$.

We next show that $M_{n,p}(G) \supseteq L_{n,p}(G)$. Since

$$M_{i,p}(G) \supseteq [M_{i-1,p}(G), G] \qquad \text{for all } i$$

it follows by induction that $M_{i,p}(G) \supseteq \gamma_i(G)$. Furthermore, $M_{ip,p}(G) \supseteq M_{i,p}(G)^p$ which implies, by induction on j, that

$$M_{ip^j,p}(G) \supseteq M_{i,p}(G)^{p^j} \supseteq \gamma_i(G)^{p^j}$$

Taking into account that $\{M_{n,p}(G)\}$ is a decreasing sequence, we conclude that $M_{n,p}(G)$ contains all $\gamma_i(G)^{p^j}$ with $ip^j \geqslant n$ and hence $M_{n,p}(G) \supseteq L_{n,p}(G)$. We have therefore shown that

$$D_n(FG) \supseteq M_{n,p}(G) \supseteq L_{n,p}(G)$$

Thus it suffices to prove that $L_{n,p}(G) \supseteq D_n(FG)$. It can be shown (e.g. see Passman (1977)) that $\{L_{n,p}(G)\}$ is an N_p-sequence of G. But, by Theorem 1.5, $\{D_n(FG)\}$ is the unique minimal N_p-sequence of G. Hence $L_{n,p}(G) \supseteq D_n(FG)$ and the result follows. ■

We now turn our attention to the powers of the augmentation ideal of FG. Since the corresponding result entails a fair amount of notation, it will be especially useful to assemble most of it in one place.

G is a finite p-group

$G_i = M_{i,p}(G)$ and $|G_i/G_{i+1}| = p^{e_i}$

$G = G_1 \supseteq G_2 \supseteq \dots \supseteq G_d \supseteq G_{d+1} = 1$

$G = H_1 \supset H_2 \supset \dots \supset H_n \supset H_{n+1} = 1$ is a composition series of G which refines $\{G_i\}$.

$g_i \in H_i - H_{i+1}$, $1 \leqslant i \leqslant n$

$\nu(g_i)$ is the largest m for which $g_i \in G_m$

$t(G)$ is the nilpotency index of $J(FG) = I(G)$

1.7. THEOREM. (Jennings (1941)). With the notation above, the following properties hold:

(i) For all $k \geqslant 1$, the elements $\prod_{i=1}^{n}(g_i-1)^{a_i}$ with $0 \leqslant a_i < p$ and $\sum_{i=1}^{n} a_i\nu(g_i) \geqslant k$ form an F-basis for $I(G)^k$

(ii) For all $k \geqslant 1$, the elements $\prod_{i=1}^{n}(g_i-1)^{a_i}$ with $0 \leqslant a_i < p$ and $\sum_{i=1}^{n} a_i\nu(g_i) = k$ form an F-basis for $I(G)^k$ modulo $I(G)^{k+1}$

(iii) $t(G) = 1 + (p-1)\sum_{i=1}^{d} ie_i$

(iv) Let $m = \sum_{i=1}^{d}(p-1)ie_i$ and define the integers c_k by

$$\prod_{i=1}^{d}(1 + X^i + X^{2i} + \dots + X^{(p-1)i})^{e_i} = \sum_{k=o}^{m} c_k X^k$$

Then $c_k = \dim_F I(G)^k/I(G)^{k+1}$ for all $1 \leqslant k \leqslant m$.

Proof. Let $\{I_i\}$ be the filtration of $I(G)$ determined by $\{G_i\}$. By Theorem 1.6(i), $G_i = D_i(FG)$ and so, by Theorem 1.5(ii), $I_i = I(G)^i$ for all i. The required assertions now follow by appealing to Lemma 1.3 and Theorem 1.4. ■

As an application of Theorem 1.7(iv), we prove the following result.

1.8. THEOREM. Let G be a finite p-group and let $m = t(G) - 1$. Then

$$\dim_F I(G)^n/I(G)^{n+1} = \dim_F I(G)^{m-n}/I(G)^{m-n+1} \qquad (0 \leqslant n \leqslant m)$$

and

$$\dim_F I(G)^n + \dim_F I(G)^{m-n+1} = |G| \qquad (1 \leqslant n \leqslant m)$$

Proof. We keep the notation of Theorem 1.7 so that $m = \sum_{i=1}^{d} (p-1)ie_i$ and

$$\begin{aligned}
\sum_{n=0}^{m} c_n X^n &= \prod_{i=1}^{d} (1 + X^i + X^{2i} + \dots + X^{(p-1)i})^{e_i} \\
&= \prod_{i=1}^{d} X^{(p-1)ie_i} (X^{-(p-1)i} + X^{-(p-2)i} + \dots + 1)^{e_i} \\
&= X^m \prod_{i=1}^{d} (1 + X^{-i} + \dots + X^{-(p-1)i})^{e_i} \\
&= X^m \sum_{n=0}^{m} c_n X^{-n} \\
&= \sum_{n=0}^{m} c_{m-n} X^n
\end{aligned}$$

Comparing coefficients, we deduce that

$$c_m = c_{m-n} \qquad (0 \leqslant n \leqslant m)$$

Hence, if $1 \leqslant n \leqslant m-1$, then

$$\dim_F I(G)^n/I(G)^{n+1} = c_n = c_{m-n} = \dim_F I(G)^{m-n}/I(G)^{m-n+1}$$

On the other hand, by Lemma3.11.2, $\dim_F I(G)^m = 1$. Since $I(G)^{m+1} = 0$, we conclude that

$$\dim_F I(G)^0/I(G) = \dim_F FG/I(G) = 1 = \dim_F I(G)^m/I(G)^{m+1},$$

proving the case $n = 0$.

If $1 \leqslant n \leqslant m$, then by the foregoing

$$\begin{aligned}
\dim_F FG/I(G)^n &= \dim_F FG/I(G) + \sum_{k=1}^{n-1} \dim_F I(G)^k/I(G)^{k+1} \\
&= \dim_F I(G)^m + \sum_{k=1}^{n-1} \dim_F I(G)^{m-k}/I(G)^{m-k+1} \\
&= \dim_F I(G)^{m-n+1}
\end{aligned}$$

Therefore

$$|G| = \dim_F FG = \dim_F I(G)^n + \dim_F I(G)^{m-n+1}$$

as asserted. ■

Next we determine the left and right annihilators of powers of $I(G)$.

1.9. COROLLARY. (Hill (1970)). Let G be a finite p-group and let $m = t(G) - 1$. Then

$$\ell(I(G)^n) = r(I(G)^n) = I(G)^{m-n+1} \qquad \text{for all} \quad n \in \{1,2,\ldots,m\}$$

Proof. Since $I(G)^n I(G)^{m-n+1} = I(G)^{m-n+1} I(G)^n = I(G)^{m+1} = 0$, we have

$$I(G)^{m-n+1} \subseteq \ell(I(G)^n) \quad \text{and} \quad I(G)^{m-n+1} \subseteq r(I(G)^n)$$

On the other hand, by Theorem 3.3.6,

$$\dim_F \ell(I(G)^n) = |G| - \dim_F I(G)^n$$

and

$$\dim_F r(I(G)^n) = |G| - \dim_F I(G)^n$$

Applying Theorem 1.8, we therefore deduce that

$$\dim_F \ell(I(G)^n) = \dim_F r(I(G)^n) = \dim_F I(G)^{m-n+1}$$

Thus

$$\ell(I(G)^n) = r(I(G)^n) = I(G)^{m-n+1}$$

as asserted. ■

We close this section by providing some observations on the nilpotency index $t(G)$ of $J(FG)$.

1.10. PROPOSITION. (Wallace (1968)). Let G be a p-group of order p^n. Then

$$n(p-1)+1 \leqslant t(G) \leqslant p^n$$

Proof. By Proposition 3.1.8(i), $t(G) \leqslant |G|$ for an arbitrary finite group G. In particular, in our case we have $t(G) \leqslant p^n$. In the notation of Theorem 1. we have

$$t(G) = \sum_{i=1}^{d} (p-1)ie_i + 1$$

Since $\sum_{i=1}^{d} e_i = n$, we have

$$t(G) \geqslant \sum_{i=1}^{d} (p-1)e_i + 1 = n(p-1) + 1,$$

as required. ■

1.11. THEOREM. (Wallace (1968)). Let P be a normal p-subgroup of a finite group G. Then

$$t(P) + t(G/P) - 1 \leqslant t(G) \leqslant t(P)t(G/P)$$

Proof. By Proposition 1.1(ii), we have

$$J(F(G/P)) \cong J(FG)/FG\cdot I(P) \qquad (2)$$

Set $n = t(P)$ and $m = t(G/P)$. Since $J(FP) = I(P)$, we have

$$[FG\cdot I(P)]^n = FG\cdot I(P)^n = 0$$

Hence, by (2)

$$J(FG)^{mn} = (J(FG)^m)^n \subseteq [FG\cdot I(P)]^n = 0,$$

proving that $t(G) \leqslant t(P)t(G/P)$.

We now prove the second inequality. Since

$$[J(F(G/P))]^{m-1} \neq 0$$

it follows from (2) that there exists

$$w \in J(FG)^{m-1} \quad \text{and} \quad w \notin FG\cdot I(P)$$

Let T be a transversal for P in G and write $w = t_1s_1 + \dots + t_ns_n$ with $s_i \in FP$ and $t_i \in T$. Then there exists $j \in \{1,2,\dots,n\}$ such that $s_j \notin I(P)$, i.e.

$$s_j = \sum_{g\in P} \alpha_g g \quad \text{and} \quad \sum_{g\in P} \alpha_g \neq 0 \qquad (\alpha_g \in F)$$

Consider now $I(P)^{n-1}$ which, by Lemma3.11.2,is spanned by $\sum_{y\in P} y$. Then

$$(\sum_{y\in P} y)w \in I(P)^{n-1}J(FG)^{m-1}$$

But $(\sum_{y\in P} y)w \neq 0$, since

$$(\sum_{y\in P} y)w = w(\sum_{y\in P} y) = t_1s_1(\sum_{y\in P} y) + \dots + t_ns_n(\sum_{y\in P} y)$$

and in this expression of $(\sum_{y\in P} y)w$ as a sum of elements whose support lie in distinct cosets of P in G we have

$$t_js_j(\sum_{y\in P} y) = t_j(\sum_{g\in P} \alpha_g g)(\sum_{y\in P} y) \neq 0$$

upon noting that the coefficient of 1 in $(\sum_{g\in P} \alpha_g g)(\sum_{y\in P} y)$ is $\sum_{g\in P} \alpha_g \neq 0$.

It follows that

$$\begin{aligned} J(FG)^{n+m-2} &= J(FG)^{n-1} J(FG)^{m-1} \\ &\supseteq I(P)^{n-1} J(FG)^{m-1} \\ &\neq 0 \end{aligned}$$

Thus $t(G) \geqslant t(P) + t(G/P) - 1$, as required. ∎

Remark. The bounds given in the above theorem cannot be improved. Indeed, let $p = 2$ and $|G| = 4$. If P is a subgroup of order 2, then

$$t(P) = t(G/P) = 2$$

and so

$$4 = 2.2 \geqslant t(G) \geqslant 2 + 2 - 1 = 3$$

If G is cyclic then $t(G) = 4$ and if $G \cong \mathbb{Z}_2 \times \mathbb{Z}_2$ then $t(G) = 3$.

2. COMPUTATION OF $t(P)$ FOR SOME INDIVIDUAL p-GROUPS P

Throughout this section, P denotes a finite p-group and F an arbitrary field of characteristic p. Our aim is to compute the nilpotency index $t(P)$ of $J(FP) = I(P)$ for some individual p-groups P. For convenience, we put

$$M_n = M_{n,p}(P) \qquad \text{for all } n \geqslant 1$$

so that if

$$P = M_1 \supseteq M_2 \supseteq \cdots \supseteq M_d \supseteq M_{d+1} = 1$$

and

$$|M_i/M_{i+1}| = p^{e_i} \qquad (1 \leqslant i \leqslant d)$$

then by Jennings' formula (Theorem 1.7(iii))

$$t(P) = 1 + (p-1) \sum_{i=1}^{d} ie_i \tag{1}$$

Let A be an abelian subgroup of P, let $z \in Z(P)$ be of order p and assume that $\langle A,z \rangle$ is a normal subgroup of P. Then both A^p and $\langle A^p,z \rangle$ are normal in P. With this observation, we now prove

2.1. THEOREM. (Motose (1978)). Let A be an abelian subgroup of P, let $z \in Z(P)$ be of order p and assume that $\langle A,z \rangle \lhd P$ and that $P/\langle A^p,z \rangle$ is elementary abelian. Put

$$p^b = (P:A),\ p^{c_t} = (A^{p^t} : A^{p^{t+1}}) \text{ and } p^s = \text{the exponent of } A \qquad (t \geq 0)$$

(i) If P/A^p is elementary abelian, then

$$t(P) = (b + \sum_{t=o}^{s-1} c_t p^t)(p-1) + 1 \tag{2}$$

(ii) If P/A^p is not elementary abelian but of exponent p, then

$$t(P) = (b+1+\sum_{t=o}^{s-1} c_t p^t)(p-1) + 1 \tag{3}$$

(iii) If the exponent of P/A^p is larger than p, then

$$t(P) = (b+p-1 + \sum_{t=o}^{s-1} c_t p^t)(p-1) + 1 \tag{4}$$

Proof. By hypothesis, $P' \subseteq \langle A^p, z\rangle$ and so

$$[\langle A^{p^t}, z\rangle, P] = [A^{p^t}, P] \subseteq A^{p^{t+1}} \qquad (t \geq 1) \tag{5}$$

(i) Our hypotheses ensure that $P' \subseteq A^p$ and $P^p = A^p$. Invoking (5), we therefore deduce that

$$\begin{aligned} &M_2 = M_3 = \ldots = M_p = A^p \\ &M_{p+1} = M_{p+2} = \ldots = M_{p^2} = A^{p^2} \\ &- - - - - - - - - - - - - - - \\ &M_{p^{s-2}+1} = \ldots = M_{p^{s-1}} = A^{p^{s-1}} \\ &M_{p^{s-1}+1} = 1 \end{aligned}$$

Hence

$$\begin{aligned} &e_1 = b + c_0 \\ &e_2 = e_3 = \ldots = e_{p-1} = 0 \\ &e_p = c_1, e_{p+1} = \ldots = e_{p^2-1} = 0 \\ &- - - - - - - - - - - - - - - \\ &e_{p^{s-2}} = c_{s-2}, e_{p^{s-2}+1} = \ldots = e_{p^{s-1}-1} = 0 \\ &e_{p^{s-1}} = c_{s-1} \end{aligned}$$

and therefore, by (1), we have

$$t(P) = 1 + (p-1)\sum_{i=1}^{p^{s-1}} ie_i = 1 + (p-1)(b + \sum_{t=o}^{s-1} c_t p^t)$$

as required.

(ii) Since $P/\langle A^p,z\rangle$ is elementary abelian, we have $A^p \subseteq \langle P',A^p\rangle \subseteq \langle A^p,z\rangle$. By hypothesis, P/A^p is not elementary abelian but of exponent p. Hence it is easy to see that

$$\langle P',A^p\rangle = \langle A^p,z\rangle \quad \text{and} \quad A^p = P^p$$

Applying (5), we infer that if $p \neq 2$, then

$$\begin{aligned} M_2 &= \langle A^p,z\rangle \\ M_3 &= \dots = M_p = A^p \\ M_{p+1} &= \dots = M_{p^2} = A^{p^2} \\ &\text{- - - - - - - - - - - -} \\ M_{p^{s-2}+1} &= \dots = M_{p^{s-1}} = A^{p^{s-1}} \\ M_{p^{s-1}+1} &= 1 \end{aligned}$$

while if $p = 2$, then

$$\begin{aligned} M_2 &= \langle A^2,z\rangle \\ M_3 &= M_4 = A^4 \\ M_5 &= \dots = M_8 = A^8 \\ &\text{- - - - - - - - - -} \\ M_{2^{s-2}+1} &= \dots = M_{2^{s-1}} = A^{2^{s-1}} \\ M_{2^{s-1}+1} &= 1 \end{aligned}$$

Now $|P/\langle A^p,z\rangle| = p^{b+c_0-1}$ since

$$p^{b+c_0} = |P/A^p| = |P/\langle A^p,z\rangle| \; |\langle A^p,z\rangle/A^p|$$

Hence, in both cases $p \neq 2$ and $p = 2$, (3) follows from (1).

(iii) Since $P' \subseteq \langle A^p,z\rangle = P^p$, an application of (5) shows that

$$\begin{aligned} M_2 &= \dots = M_p = \langle A^p\; z\rangle \\ M_{p+1} &= \dots = M_{p^2} = A^{p^2} \end{aligned}$$

$$- - - - - - - - - - - - - -$$

$$M_{p^{s-2}+1} = \dots = M_{p^{s-1}} = A^{p^{s-1}}$$

$$M_{p^{s-1}+1} = 1$$

Thus (4) follows from (1). ■

2.2. COROLLARY. Let $|P| = p^r$ and assume that there exists a central element $z \in P$ of order p and an element $h \in P$ of order p^s such that

$$\langle h,z \rangle \lhd P \text{ and } P/\langle h^p,z \rangle \text{ is elementary abelian}$$

(i) If $P/\langle h^p \rangle$ is elementary abelian, then

$$t(P) = p^s + (r-s)(p-1)$$

(ii) $P/\langle h^p \rangle$ is not elementary abelian but of exponent p, then

$$t(P) = p^s + (r-s+1)(p-1)$$

(iii) If the exponent of $P/\langle h^p \rangle$ is larger than p, then

$$t(P) = p^s + (r-s+p-1)(p-1)$$

Proof. Setting $A = \langle h \rangle$, we have $A^p = \langle h^p \rangle, \langle A,z \rangle \lhd P$ and $P/\langle A^p,z \rangle$ is elementary abelian. Hence the hypotheses of Theorem 2.1 are satisfied. Furthermore, in the notation of Theorem 2.1, we have $c_t = 1$ for all t. Hence

$$\left(\sum_{t=o}^{s-1} c_t p^t\right)(p-1) = p^s - 1$$

and the result follows by appealing to Theorem 2.1. ■

Recall that a group G is called *metacyclic* if G has a cyclic normal subgroup whose factor group is also cyclic. Therefore, G can be generated by two elements a and b with defining relations

$$a^m = 1, b^s = a^t \text{ and } b^{-1}ab = a^r$$

where the positive integers $m,r,s,$ and t satisfy $r^s \equiv 1 (\text{mod } m)$ and $m|t(r-1)$. Set $d = (m,t)$ and $k = t/d$. Then, replacing a by a^k, if necessary, we may also assume that $t|m$.

2.3. THEOREM. (Koshitani (1977a), Motose (1978)). Let P be a metacyclic p-

group, say

$$P = \langle a,b \mid a^{p^n} = 1, b^{p^s} = a^{p^k}, b^{-1}ab = a^{\mu} \rangle \qquad (0 \leqslant k \leqslant n, \mu^{p^s} \equiv 1 (\mathrm{mod}\, p^n), p^{n-k} \mid (\mu-1))$$

Then

$$t(P) = \begin{cases} p^n + p^s - 1 & \text{if} \quad s \leqslant k \\ p^{n+s-k} + p^k - 1 & \text{if} \quad s > k \end{cases}$$

Proof. If P is cyclic, then the assertion is trivial. Hence we may assume that $s \geqslant 1$ and $k \geqslant 1$. Since $[b,a] \subseteq \langle a^p \rangle$, one easily verifies that

$$\langle b^{p^t}, a^{p^t} \rangle^p = \langle b^{p^{t+1}}, a^{p^{t+1}} \rangle$$

and

$$[b^{p^t}, P] \subseteq [\langle b^{p^t}, a^{p^t} \rangle, P] \subseteq \langle a^{p^{t+1}} \rangle$$

Hence if $s \leqslant k$, then

$$\begin{aligned}
M_2 &= \dots = M_p = \langle b^p, a^p \rangle \\
M_{p+1} &= \dots = M_{p^2} = \langle b^{p^2}\ a^{p^2} \rangle \\
&\text{- - - - - - - - - - - - - - - - - - -} \\
M_{p^{s-2}+1} &= \dots = M_{p^{s-1}} = \langle b^{p^{s-1}}, a^{p^{s-1}} \rangle \\
M_{p^{s-1}+1} &= \dots = M_{p^s} = \langle a^{p^s} \rangle \\
&\text{- - - - - - - - - - - - - - - - - - -} \\
M_{p^{n-2}+1} &= \dots = M_{p^{n-1}} = \langle a^{p^{n-1}} \rangle \\
M_{p^{n-1}+1} &= 1
\end{aligned}$$

while in case $s > k$, we have

$$\begin{aligned}
M_2 &= \dots = M_p = \langle b^p, a^p \rangle \\
M_{p+1} &= \dots = M_{p^2} = \langle b^{p^2}, a^{p^2} \rangle \\
&\text{- - - - - - - - - - - - - - - - - - -} \\
M_{p^{k-2}+1} &= \dots = M_{p^{k-1}} = \langle b^{p^{k-1}}, a^{p^{k-1}} \rangle \\
M_{p^{k-1}+1} &= \dots = M_{p^k} = \langle b^{p^{k-1}} \rangle
\end{aligned}$$

- -

$$M_{p^{n+s-k-2}+1} = \dots = M_{p^{n+s-k-1}} = \langle b^{p(n+s)-k-1}\rangle$$

$$M_{p^{n+s-k-1}+1} = 1$$

The desired conclusion now follows by applying (1). ■

We next record some group-theoretic information. A group D_n given by the defining relations:

$$D_n = \langle a,b \mid a^{2^{n-1}} = 1, b^2 = 1, b^{-1}ab = a^{-1}\rangle \qquad (n \geq 2)$$

is obviously of order 2^n and is called the *dihedral group* of order 2^n. The group Q_n given by

$$Q_n = \langle a,b \mid a^{2^{n-1}} = 1, b^2 = a^{2^{n-2}}, b^{-1}ab = a^{-1}\rangle \qquad (n \geq 3)$$

is again of order 2^n and is called the *generalized quaternion* group of order 2^n. For every prime $p \neq 2$, the group $M(p)$ is given by the following defining relations:

$$M(p) = \langle a,b,c \mid a^p = b^p = c^p = 1,\ a^{-1}ba = bc, a^{-1}ca = c, b^{-1}cb = c\rangle$$

The group $P = M_p(G)$ is obviously of order p^3 and has exponent p. Denote by $\Phi(P)$ the Frattini subgroup of P (i.e. the intersection of all maximal subgroups of P). Then

$$P' = \Phi(P) = Z(P) = \langle c\rangle \cong \mathbb{Z}_p \qquad (6)$$

Any p-group P satisfying (6) is called *extra-special*.

The *semidihedral group* $S(n)$ for $n > 3$ is a group of order 2^n defined by

$$S(n) = \langle a,b \mid a^{2^{n-1}} = 1, b^2 = 1, b^{-1}ab = a^{-1+2^{n-2}}\rangle$$

The following group $M_n(p)$ is defined for $n \geq 4$ if $p = 2$ and for $n \geq 3$ if $p \neq 2$.

$$M_n(p) = \langle a,b \mid a^{p^{n-1}} = 1, b^p = 1, b^{-1}ab = a^{1+p^{n-2}}\rangle$$

As an application of Corollary 2.2 and Theorem 2.3, we now prove

2.4. COROLLARY. The following properties hold:

(i) $t(D_n) = 2^{n-1} + 1$

(ii) $t(Q_n) = 2^{n-1} + 1$

(iii) $t(M(p)) = 4p-3$

(iv) $t(S(n)) = 2^{n-1} + 1$

(v) $t(M_n(p)) = p^{n-1} + p - 1$

Proof. Properties (i),(ii),(iv) and (v) follow from Theorem 2.3, while property (iii) is a consequence of Corollary 2.2(ii). ∎

We now determine $t(P)$ in the case $|P| \leqslant p^4$. If P is abelian then the formula for $t(P)$ can be obtained from Corollary 3.2.5. In case $|P| = p^3$ the corresponding formula follows from Corollary 2.4. Thus, we may harmlessly assume that P is nonabelian of order p^4. The defining relations for such P are well known (e.g. see Huppert (1967)). The results below can be easily verified by applying Theorem 2.1, Corollary 2.2, Theorem 2.3, Corollary 2.4, Corollary 3.2.4 and Jennings formula (1) (for direct calculations refer to Koshitani (1977a).

2.5. TABLE (Nonabelian groups of order p^4 where $p \geqslant 3$)

Defining relations of P	$t(P)$
$P = M_4(p)$	$p^3 + p - 1$
$P = M_3(p) \times \mathbb{Z}_p$	$p^2 + 2p - 2$
$P = M(p) \times \mathbb{Z}_p$	$5p - 4$
$P = \langle a,b \mid a^{p^2} = 1, b^{p^2} = 1, b^{-1}ab = a^{p+1} \rangle$	$2p^2 - 1$
$P = \langle a,b,c \mid a^p = b^p = c^{p^2} = 1, a^{-1}ba = bc^p, a^{-1}ca = c, b^{-1}cb = c \rangle$	$p^2 + 2p - 2$
$P = \langle a,b,c \mid a^p = b^p = c^{p^2} = 1, a^{-1}ba = b, a^{-1}ca = bc, b^{-1}cb = c \rangle$	$p^2 + 3p - 3$
$P = \langle a,b,c \mid a^p = b^p = c^{p^2} = 1, a^{-1}ba = bc^p, a^{-1}ca = bc, b^{-1}cb = c \rangle$	$p^2 + 3p - 3$
$P = \langle a,b,c \mid a^p = b^p = c^{p^2} = 1, a^{-1}ba = bc^{np}, a^{-1}ca = bc, b^{-1}cb = c \rangle$ where n is a quadratic nonresidue modulo p.	$p^2 + 3p - 3$
$P = \langle a,b,c,d \mid a^p = b^p = c^p = d^p = 1, b^{-1}cb = c, c^{-1}dc = d,$ $b^{-1}db = d, a^{-1}ba = b, a^{-1}ca = bc, a^{-1}da = cd \rangle$	$7p - 6$

$P=\langle a,b,c,d \mid a^p=b, b^p=c^p=d^p=1, b^{-1}cb=c, c^{-1}dc=d,$ $b^{-1}db=d, a^{-1}ca=bc, a^{-1}da=cd\rangle, p\geqslant 5$	p^2+3p-3
$P=\langle a,b,c \mid a^3=b^3, b^9=c^3=1, a^{-1}ba=bc, a^{-1}ca=b^3c, b^{-1}cb=c\rangle, p=3$	15

2.6. TABLE (Nonabelian groups of order 2^4)

Defining relations of P	$t(P)$
$P=D_4, Q_4, S(4)$, or $M_4(2)$	9
$P=D_3\times\mathbb{Z}_2$ or $Q_3\times\mathbb{Z}_2$	6
$P=\langle a,b \mid a^4=1, b^4=1, b^{-1}ab=a^3\rangle$	7
$P=\langle a,b,c \mid a^2=b^2=c^4=1, a^{-1}ba=bc^2, a^{-1}ca=c, b^{-1}cb=c\rangle$	6
$P=\langle a,b,c \mid a^2=b^2=c^4=1, a^{-1}ba=b, a^{-1}ca=bc, b^{-1}cb=c\rangle$	7

3. CHARACTERIZATION OF GROUPS P OF ORDER p^a WITH $t(P)=a(p-1)+1, t(P)=p^a$, $t(P)=p^{a-1}+p-1, t(P)=p^{a-1}$ AND $t(P)=(a+1)(p-1)+1$

Throughout this section, P denotes a group of order $p^a, a\geqslant 1$, and F an arbitrary field of characteristic $p>0$. By Proposition 1.8, the nilpotency index $t(P)$ of $J(FP)$ satisfies

$$a(p-1)+1\leqslant t(P)\leqslant p^a$$

It is therefore natural to investigate those groups P for which $t(P)=p^a$ or $t(P)=a(p-1)+1$. If P is an abelian group of order $p^a, a\geqslant 2$, then by Corollary 3.2.5, the second highest value of $t(P)$ is $p^{a-1}+p-1$ and in this case P is not cyclic but has a cyclic subgroup of index p. In what follows we generalize this fact by characterizing arbitrary groups P of order $p^a, a\geqslant 2$, for which $t(P)=p^{a-1}+p-1$. In the final part of the section we examine those groups P for which $t(P)=p^{a-1}$ or $t(P)=(a+1)(p-1)+1$, the latter being the second lowest value of $t(P)$, by virtue of Jennings' formula (Theorem 1.7(iii).

We start by recalling the following piece of information. The intersection $\Phi(G)$ of all maximal subgroups of G is called the Frattini subgroup of G. Suppose that $G = \langle X,g \rangle$ with $g \in \Phi(G)$ and let $H = \langle X \rangle$. If H is a proper subgroup of G, then $H \subseteq M$ for some maximal subgroup M of G. But then $g \in M$, so $G \subseteq M$, a contradiction. Thus $G = \langle X \rangle$, which means that *all elements of* $\Phi(G)$ *may be omitted from any generating set of* G. In particular, if $G/\Phi(G)$ is cyclic, then G is cyclic.

3.1. LEMMA. Let G be a finite p-group and let d be the minimal number of generators of G. Then $G/\Phi(G)$ is elementary abelian of order p^d.

Proof. If M is a maximal subgroup of G, then $M \lhd G$ and $(G:M) = p$. Thus $G' \subseteq M$ as G/M is abelian, and $x^p \in M$ for all $x \in G$ as $(G:M) = p$. Because the above holds for each maximal subgroup M of G, we have $G' \subseteq \Phi(G)$ and $x^p \in \Phi(G)$ for all $x \in G$. Hence $G/\Phi(G)$ is abelian and the image of each x in $G - \Phi(G)$ has order p. This proves that $G/\Phi(G)$ is elementary abelian. If $(G:\Phi(G)) = p^r$, then r is the minimal number of generators of $G/\Phi(G)$. But G has a generating set consisting of d elements, so $d \geqslant r$. On the other hand, if X is a generating set of $\Phi(G)$ and $\{g_1\Phi(G),\ldots,g_r\Phi(G)\}$ is a generating set of $G/\Phi(G)$, then $\{X,g_1,\ldots,g_r\}$ is a generating set of G. However the elements of $\Phi(G)$ may be omitted from any generating set of G; hence $G = \langle g_1,\ldots,g_r \rangle$. It follows that $d \leqslant r$ and so $(G:\Phi(G)) = p^d$, as asserted.■

We are now ready to prove

3.2. THEOREM. (Motose and Ninomiya (1975a)). Let P be a p-group of order $p^a, a \geqslant 1$.

(i) $t(P) = a(p-1) + 1$ if and only if P is elementary abelian

(ii) $t(P) = p^a$ if and only if P is cyclic

Proof. (i) For all $n \geqslant 1$, put $M_n = M_{n,p}(P)$ and write

$$P = M_1 \supseteq M_2 \supseteq \ldots \supseteq M_d \supseteq M_{d+1} = 1$$

If $|M_i/M_{i+1}| = p^{e_i}$, then by Jennings' formula (Theorem 1.7(iii))

$$t(P) = 1 + (p-1)\sum_{i=1}^{d} ie_i \tag{1}$$

Since $\{M_i\}$ is an N_p-sequence, each M_i/M_{i+1} is elementary abelian. Now assume that $t(P) = a(p-1) + 1$. Then, by (1) $\sum_{i=1}^{d} ie_i = a$. But $\sum_{i=1}^{d} e_i = a$ and thus

$$e_1 = a \quad \text{and} \quad e_i = 0 \quad \text{for} \quad i \neq 1$$

Hence P is elementary abelian.

Conversely, assume that P is elementary abelian. Then, by Corollary 3.2.5,

$$t(P) = \sum_{i=1}^{a} p - (a-1) = pa - (a-1) = a(p-1) + 1,$$

as required.

(ii) If P is cyclic, then $t(P) = p^a$, by virtue of Corollary 3.2.5. Conversely, assume that $t(P) = p^a$. Then, by Theorem 1.9,

$$\begin{aligned} |P| &= t(P) \leqslant t(\Phi(P))t(P/\Phi(P)) \\ &\leqslant |\Phi(P)||P/\Phi(P) \\ &= |P| \end{aligned}$$

whence

$$t(P/\Phi(P)) = |P/\Phi(P)| = p^b \qquad (b \leqslant a)$$

By Lemma 3.1, $P/\Phi(P)$ is elementary abelian and so by (i)

$$t(P/\Phi(P)) = b(p-1) + 1$$

It follows that

$$p^b = |P/\Phi(P)| = t(P/\Phi(P)) = b(p-1) + 1$$

which is possible only in the case $b = 1$. But then $P/\Phi(P)$ (and hence P) is cyclic. So the theorem is true. ■

The following simple observation will be required for the characterizations of those P for which $t(P) = p^{a-1} + p - 1$.

3.3. LEMMA. Let $q \geqslant 2$ and $n \geqslant 2$ be positive integers and let $r = \sum_{i=1}^{n} e_i$ where the e_i are positive integers with $e_1 \geqslant e_2 \geqslant \ldots \geqslant e_n$. Then the following conditions are equivalent:

(i) $\sum_{i=1}^{n} q^{e_i} - (n-1) \geqslant q^{r-1}$

(ii) $n=3, r=3, q=2$ or $n=2, e_2=1$

Proof. The implication (ii) $\Rightarrow$ (i) is obvious. So assume that (i) holds. Obviously, if $x \geqslant 2$ and $y \geqslant t \geqslant 2$, then

$$xy \geqslant x+y+t-2 \geqslant x+y \tag{2}$$

If $n \geqslant 3$ then by (2) we have

$$\begin{aligned} 0 &\geqslant q^{r-1} - \sum_{i=1}^{n} q^{e_i} + n-1 \\ &\geqslant q^{e_1+e_2-1} + q^{e_3+\dots+e_n} + q^{n-2} - 2 - \sum_{i=1}^{n} q^{e_i} + n-1 \\ &\geqslant q^{e_1+e_2-1} - q^{e_1} - q^{e_2} + (n-2)q + n-3 \\ &= (q^{e_1}-q)(q^{e_2-1}-1) + (n-3)(q+1) \geqslant 0 \end{aligned}$$

Thus $n=3, e_2=e_3=1$ and $(q^{r-2}-2)(q-1)=0$ which implies that $q=2$ and $r=3$. Finally, if $n=2$ then

$$q-1 \geqslant (q^{e_1}-q)(q^{e_2-1}-1)$$

and so $e_2=1$, as required. ∎

3.4. THEOREM. (Koshitani (1977a), Motose (1978)). Assume that P is noncyclic of order p^a. Then the following conditions are equivalent:

(i) $t(P) = p^{a-1}+p-1$

(ii) $t(P) > p^{a-1}$

(iii) P contains an element of order p^{a-1}.

Proof. (i) $\Rightarrow$ (ii): Trivial

(iii) $\Rightarrow$ (i): Let $h \in P$ be of order p^{a-1}. Then $\langle h\rangle$ is normal. Now $P/\langle h^p\rangle$ is not cyclic, for otherwise there exists $x \in P$ such that $x^p = h$ and P is cyclic. Hence $P/\langle h^p\rangle$ is elementary abelian and, by Corollary 2.2(i), $t(P) = p^{a-1}+p-1$.

(ii) $\Rightarrow$ (iii): First assume that P is abelian, say $P \cong \mathbb{Z}_{p^{e_1}} \times \dots \times \mathbb{Z}_{p^{e_n}}$. Then, by Corollary 3.2.5,

$$t(P) = \sum_{k=1}^{n} p^{e_k} - (n-1) > p^{a-1} \tag{3}$$

where $a = \sum_{k=1}^{n} e_k$ and $n \geqslant 2$. Owing to Lemma 3.3, (3) holds only for $n = 2, e_1 = a-1, e_2 = 1$. Thus $P \cong \mathbb{Z}_{p^{a-1}} \times \mathbb{Z}_p$ and so P contains an element of order p^{a-1}.

By the foregoing, we may assume that P is nonabelian so that $a \geqslant 3$. We argue by induction on a. If $a = 3$, then either P contains an element of order p^2 or $P \cong M(p)$. However, $t(M(p)) = 4p-3 \leqslant p^2$ by Corollary 2.4, and so the case $a = 3$ is established.

From now on, we assume that $a \geqslant 4$. Let z be a central element of order p, and put $\bar{P} = P/\langle z\rangle$. Then $\bar{P}$ is not cyclic and, by Theorem 1.9, $t(\bar{P}) > p^{a-2}$. Hence, by the induction hypothesis, there exists an element $\bar{h}$ of order p^{a-2} in $\bar{P}$. Because $\bar{P}$ is not cyclic, $\bar{P}/\langle \bar{h}^p\rangle$ is elementary abelian, by virtue of implication (iii) $\Rightarrow$ (i). Thus there exists an element h such that $\langle h,z\rangle$ is normal and $P/\langle h^p,z\rangle$ is elementary abelian. Now if $h^{p^{a-2}} = z^k (0 < k < p)$, then h is of order p^{a-1}. On the other hand, if h is of order p^{a-2} then, by Corollary 2.2,

$$t(P) = p^{a-2} + 2(p-1) \quad \text{or} \quad t(P) = p^{a-2} + 3(p-1) \quad \text{or} \quad t(P) = p^{a-2} + p^2 - 1$$

Since in all of these cases $t(P) \leqslant p^{a-1}$, we conclude that h is of order p^{a-1}, as required. ■

3.5. THEOREM. (Motose (1978)). Let P be a group of order $p^a, a \geqslant 1$. Then the following conditions are equivalent:

(i) $t(P) = p^{a-1}$

(ii) $P \cong M(3)$ or $P \cong \mathbb{Z}_2 \times \mathbb{Z}_2 \times \mathbb{Z}_2$

Proof. The implication (ii) $\Rightarrow$ (i) is a consequence of Corollary 2.4(iii) and Corollary 3.2.5. So assume that $t(P) = p^{a-1}$. If P is abelian, say $P \cong \mathbb{Z}_{p^{e_1}} \times \ldots \times \mathbb{Z}_{p^{e_n}}$, then $n \geqslant 2$ and

$$p^{a-1} = t(P) = \sum_{i=1}^{n} p^{e_i} - (n-1)$$

by Corollary 3.2.5. Hence, by Lemma 3.3, $P \cong \mathbb{Z}_2 \times \mathbb{Z}_2 \times \mathbb{Z}_2$.

Assume that P is nonabelian of order p^3. By Theorem 3.4, P is of exponent p and so $P \cong M(p)$. However, $p^2 = t(P) = 4p-3$, by Corollary 2.4(iii), and thus $p = 3$.

Finally, assume that P is nonabelian and that $a \geqslant 4$. We may choose P to be of minimal order with $t(P) = p^{a-1}$. Let $x \in P$ be a central element of order p and put $\bar{P} = P/\langle x\rangle$. Then $t(\bar{P}) \geqslant p^{a-2}$ by Theorem 1.9. If $t(\bar{P}) > p^{a-2}$ then, by Theorem 3.4, $\bar{P}$ contains an element $\bar{h}$ of order p^{a-2}. Hence there exists an element $h \in P$ of order p^{a-1} or p^{a-2} such that $P/\langle x,h^p\rangle$ is elementary abelian. Therefore, by Corollary 2.2,

$$p^{a-1} = t(P) = p^{a-1} + p - 1 \quad \text{or} \quad p^{a-1} = p^{a-2} + 2(p-1)$$
$$\text{or} \quad p^{a-1} = p^{a-2} + 3(p-1) \quad \text{or} \quad p^{a-1} = p^{a-2} + p^2 - 1$$

Since the above equalities are impossible, we infer that $t(\bar{P}) = p^{a-2}$. If $a \geqslant 5$, then $a-1 \geqslant 4$ and so our choice of P ensures that $\bar{P}$ is abelian. But then by the abelian case proved above, $|\bar{P}| = 2^3$ which contradicts $a \geqslant 5$. Thus

$$|\bar{P}| = p^3 \quad \text{and} \quad t(\bar{P}) = p^2$$

and hence $\bar{P} \cong M(3)$ or $\bar{P} \cong \mathbb{Z}_2 \times \mathbb{Z}_2 \times \mathbb{Z}_2$, by the foregoing discussion. If $\bar{P} \cong \mathbb{Z}_2 \times \mathbb{Z}_2 \times \mathbb{Z}_2$, then $t(P) = 5$ or 6, by Corollary 2.2. This contradicts the assumption that $t(P) = 2^3$. Thus $\bar{P} \cong M(3)$ and we can write

$$\bar{P} = \langle \bar{a},\bar{b},\bar{c} \mid \bar{a}^3 = \bar{b}^3 = \bar{c}^3 = 1, \bar{a}^{-1}\bar{b}\bar{a} = \bar{b}\bar{c}, \bar{a}^{-1}\bar{c}\bar{a} = \bar{c}, \bar{b}^{-1}\bar{c}\bar{b} = \bar{c}\rangle$$

Then we must have

$$P' \subseteq \langle x,c\rangle \ , \ P^3 \subseteq \langle x\rangle, \ [\langle x,c\rangle, P] \subseteq \langle x\rangle$$

Keeping the notation of Jennings' formula (1) in Sec. 2, we see that

$$1 \neq M_2 \subseteq \langle x,c\rangle \ , \ M_3 \subseteq \langle x\rangle$$
$$M_4 = M_5 = M_6 = M_2^3 \ , \ M_7 = 1$$

which implies that

$$2 \leqslant e_1 \leqslant 3, \ e_2 \leqslant 2, \ e_3 \leqslant 1, \ e_4 = e_5 = 0 \ \text{and} \ e_6 \leqslant 1$$

But then, by Jennings' formula,

$$3^3 = t(P) = (e_1 + 2e_2 + 3e_3 + 6e_6)(3-1) + 1 < 3^3$$

a contradiction. This completes the proof of the theorem. ■

3.6. THEOREM. (Motose (1978)). Assume that a group P of order p^a is not elementary abelian. Then the following conditions are equivalent:

(i) $t(P) = (a+1)(p-1) + 1$

(ii) $t(P) < (a+2)(p-1) + 1$

(iii) There exists a central element z of order p in P such that $P/\langle z\rangle$ is elementary abelian. When this is the case, $p = 2$ or P is of exponent p.

Proof. (i) $\Rightarrow$ (ii) Trivial

(iii) $\Rightarrow$ (i): Apply Corollary 2.2 (ii),(iii).

(ii) $\Rightarrow$ (iii): Put $M_n = M_{n,p}(P)$ and let $d \geq 1$ and $e_i \geq 0$ be such that

$$P = M_1 \supseteq M_2 \supseteq \dots \supseteq M_d \supseteq M_{d+1} = 1$$

and

$$|M_i/M_{i+1}| = p^{e_i} \qquad (1 \leq i \leq d)$$

Then, by Jennings' formula (Theorem 1.7(iii)), condition (ii) implies that

$$t(P) = 1 + (p-1) \sum_{i=1}^{d} ie_i < (a+2)(p-1) + 1 \tag{4}$$

Since $a = \sum_{i=1}^{d} e_i$, we conclude that

$$2 > \sum_{i=1}^{d} (i-1)e_i$$

Hence $e_3 = e_4 = \dots = e_d = 0$ and $e_2 \in \{0,1\}$. If $e_2 = 0$ then $M_2 = 1$ and so P is elementary abelian. Thus $e_2 = 1$ and hence $e_1 = a - 1$, proving that

$$t(P) = (a+1)(p-1) + 1 \tag{5}$$

by virtue of (4). Furthermore, M_2 is a central subgroup of P of order p. Let z be a generator of M_2. Then $P/\langle z\rangle$ is elementary abelian. Finally, assume that the exponent of P is greater than p. Invoking Corollary 2.2(iii) and (5), we derive

$$(a+p-1)(p-1) + 1 = t(P) = (a+1)(p-1) + 1$$

and so $p = 2$. This completes the proof of the theorem. ■

4. CHARACTERIZATIONS OF p-GROUPS P WITH $t(P) \leq 7$.

Throughout this section, F is a field of characteristic $p > 0$ and $t(P)$ is the nilpotency index of $J(FP)$, where P is a p-group of order $p^a, a \geqslant 1$. Our aim is to characterize those P for which $t(P) \leqslant 7$. By hypothesis, $t(P) > 1$. The case $t(P) = 2$ being a consequence of a general result, namely Theorem 3.12.4, we may assume that $t(P) > 2$. All the results presented below are extracted from Koshitani (1977a), Motose (1978) and Motose and Ninomiya (1975a)

4.1. PROPOSITION. The following conditions are equivalent:

(i) $t(P) = 3$

(ii) $P \cong \mathbb{Z}_3$ or $P \cong \mathbb{Z}_2 \times \mathbb{Z}_2$

Proof. That (ii) implies (i) is a consequence of Corollary 3.2.5. Assume that $t(P) = 3$. By Proposition 1.8, $3 \geqslant a(p-1) + 1$ and hence $a = 1, p = 3$, or $a = 2$ and $p = 2$, or $a = 1, p = 2$. In the latter case $t(P) = 2$, and so we must have $P \cong \mathbb{Z}_3$ or $|P| = 4$. Since $P \cong \mathbb{Z}_4$ implies $t(P) = 4$, the result follows. ■

4.2. PROPOSITION. The following conditions are equivalent:

(i) $t(P) = 4$

(ii) P is one of the following types:

(a) $p = 2$ and P is cyclic of order 4

(b) $p = 2$ and $P \cong \mathbb{Z}_2 \times \mathbb{Z}_2 \times \mathbb{Z}_2$

Proof. The implication (ii) $\Rightarrow$ (i) follows by virtue of Corollary 3.2.5. Assume that $t(P) = 4$. By Proposition 1.8, $4 \geqslant a(p-1) + 1$ and hence $a = 1, p = 2$, or $a = 1, p = 3$, or $a = 2, p = 2$, or $a = 3, p = 2$. In the first two cases we have $t(P) \leqslant 3$ and thus $|P| = 4$ or $|P| = 8$. The desired conclusion now follows by appealing to Corollary 3.2.5 and Corollary 2.4(i),(ii). ■

4.3. PROPOSITION. The following conditions are equivalent:

(i) $t(P) = 5$

(ii) P is one of the following types:

(a) $p = 2$ and $P \cong \mathbb{Z}_4 \times \mathbb{Z}_2$

(b) $p = 2$ and $P \cong D_3$

(c) $p = 2$ and $P \cong Q_3$

(d) $p = 2$ and $P \cong \mathbb{Z}_2 \times \mathbb{Z}_2 \times \mathbb{Z}_2 \times \mathbb{Z}_2$

(e) $p = 3$ and $P \cong \mathbb{Z}_3 \times \mathbb{Z}_3$

(f) $p = 5$ and $P \cong \mathbb{Z}_5$

Proof. The implication (ii) $\Rightarrow$ (i) is a consequence of Corollary 3.2.5 and Corollary 2.4(i),(ii). Assume that $t(P) = 5$. By Proposition 1.8, $5 \geqslant a(p-1)+1$ and, by Jennings' formula $p-1$ divides 4. Thus $a(p-1) \leqslant 4$ and $p \in \{2,3,5\}$. If $p = 5$ then $a = 1$ and so $P \cong \mathbb{Z}_5$. If $p = 3$, then $a \leqslant 2$. Since $p = 3$, $a = 1$ implies $t(P) = 3$, we have $p = 3, a = 2$ in which case $P \cong \mathbb{Z}_3 \times \mathbb{Z}_3$, by virtue of Corollary 3.2.5. Finally, assume that $p = 2$. Then $a \leqslant 4$ and the result follows by Corollary 2.4, Corollary 3.2.5 and Table 2.6. ∎

4.4. PROPOSITION. The following conditions are equivalent:

(i) $t(P) = 6$

(ii) P is one of the following types:

(a) $p = 2$ and $P \cong \mathbb{Z}_4 \times \mathbb{Z}_2 \times \mathbb{Z}_2$

(b) $p = 2$ and $P \cong D_3 \times \mathbb{Z}_2$

(c) $p = 2$ and $P \cong Q_3 \times \mathbb{Z}_2$

(d) $p = 2$ and $P \cong \langle a,b,c \mid a^2 = b^2 = c^4 = 1, a^{-1}ba = bc^2, a^{-1}ca = c, b^{-1}cb = c \rangle$

(e) $p = 2$ and P is elementary abelian of order 2^5

Proof. The implication (ii) $\Rightarrow$ (i) follows by virtue of Corollary 3.2.5 and Table 2.6. By Proposition 1.8, $6 \geqslant a(p-1)+1$ and, by Jennings' formula $p-1$ divides 5. Hence $p = 2$ and $a \leqslant 5$. If $a \leqslant 4$, then the result is true by virtue of Table 2.6, Corollary 3.2.5 and Corollary 2.4(i),(ii). Finally, assume that $p = 2$ and $a = 5$. If P contains an element of order 4, then there is a normal subgroup Q of P of index 2 such that Q contains an element of order 4. By Theorem 1.9, $t(P) \geqslant t(Q) + t(Q/P) - 1 = t(Q) + 1$ and hence $t(Q) \leqslant 5$. But $|Q| = 2^4$ which contradicts Propositions 4.1 - 4.4. Thus P is elementary abelian and the result follows. ∎

4.5. PROPOSITION. The following conditions are equivalent:

(i) $t(P) = 7$

(ii) P is one of the following types:

(a) $P \cong \mathbb{Z}_7$

(b) P is elementary abelian of order 3^3 or 2^6

(c) $P \cong \mathbb{Z}_4 \times \mathbb{Z}_4$

(d) P is of order 2^5, P is not elementary abelian and P contains a central involution z such that $P/\langle z \rangle$ is elementary abelian

(e) $P \cong \langle a,b \mid a^4 = b^4 = 1, a^{-1}ba = b^3 \rangle$

(f) $P \cong \langle a,b,c \mid a^2 = b^2 = c^4 = 1, a^{-1}ba = b, a^{-1}ca = bc, b^{-1}cb = c \rangle$

Proof. By Proposition 1.8, $6 \geqslant a(p-1)$ and, by Jennings' formula $p-1$ divides 6. Hence the order of P is 7 or 3^a $(a \leqslant 3)$ or 2^a $(a \leqslant 6)$. The desired conclusion is now an easy consequence of Corollary 2.2, Theorem 3.4, Theorem 3.6 and Table 2.6. ∎

5
The Jacobson radical and induced modules

The aim of this chapter is to exhibit some close connections between the Jacobson radical and induced modules. We start by describing $\mathrm{ann}(V^G)$ exclusively in terms of $\mathrm{ann}(V)$. We then examine simple induction and restriction pairs and prove a number of their important properties contained in a work of Knörr (1977). As one of the applications, we provide necessary and sufficient conditions for $J(FG) = FG{\cdot}J(FG)$, where H is a subgroup of G. Further applications are exhibited in the last section, where a detailed analysis of p-radical groups is presented. As a consequence of one of our results, we establish a purely group theoretic fact: the number of p-regular classes of G does not exceed the number of double (P,P)-cosets in G, where $P \in \mathrm{Syl}_p(G)$.

1. ANNIHILATORS OF INDUCED MODULES

Throughout this section, F denotes an arbitrary field and H a subgroup of a finite group G. Unless explicitely stated otherwise, all modules are assumed to be left and finitely generated. Given an FG-module V, we write $\mathrm{ann}(V)$ for the annihilator of V defined by

$$\mathrm{ann}(V) = \{x \in FG \mid xV = 0\}$$

The annihilator of a right FG-module is defined similarly. Note that in both cases $\mathrm{ann}(V)$ is an ideal of FG. Our main result in this section describes $\mathrm{ann}(V^G)$ where V is an FH-module.

Let M be a right FG-module. Then M^* denotes the (left) FG-module with underlying vector space $\mathrm{Hom}_F(M,F)$ and group action

$$(g\phi)(m) = \phi(mg) \qquad (g \in G, m \in M, \phi \in \mathrm{Hom}_F(M,F))$$

If M is a (left) module, then a similar definition gives the right module M^*.

1.1. LEMMA. Let M be a right FG-module and V a right FH-module. Then

(i) $\operatorname{ann}(M) = \operatorname{ann}(M^*)$

(ii) $(V^G)^* \cong (V^*)^G$

Proof. (i) Assume that $x \in \operatorname{ann}(M)$ and let ϕ be an arbitrary element of M^*. Then, for all $m \in M$, $(x\phi)(m) = \phi(mx) = \phi(0) = 0$, so $x \in \operatorname{ann}(M^*)$. Conversely, suppose that $x \in \operatorname{ann}(M^*)$. Then, for all $\phi \in M^*$ and all $m \in M$,

$$\phi(mx) = (x\phi)(m) = 0$$

Hence $mx = 0$ for all $m \in M$ so that $x \in \operatorname{ann}(M)$, as required.

(ii) Let $\{g_1, g_2, \ldots, g_n\}$ be a left transversal for H in G. Then a typical element of $(V^*)^G$ can be uniquely written in the form

$$\sum_{i=1}^{n} g_i \otimes \psi_i \qquad \text{with } \psi_i \in V^*$$

Because $\{g_1^{-1}, g_2^{-1}, \ldots, g_n^{-1}\}$ is a right transversal for H in G, each element of V^G can be uniquely written in the form

$$\sum_{i=1}^{n} v_i \otimes g_i^{-1} \qquad \text{with } v_i \in V$$

It follows that the mapping $\theta : (V^*)^G \longrightarrow (V^G)^*$ defined by

$$[\theta(\sum_{i=1}^{n} g_i \otimes \psi_i)](\sum_{i=1}^{n} v_i \otimes g_i^{-1}) = \sum_{i=1}^{n} \psi_i(v_i)$$

is an injective F-homomorphism. Since the F-dimensions of $(V^*)^G$ and $(V^G)^*$ are the same, we deduce that θ is in fact an F-isomorphism. We are therefore left to verify that for $x = g_i \otimes \psi_i$ and $g \in G, \theta(gx) = g\theta(x)$. To this end, write $gg_i = g_j h$ for some $j \in \{1,2,\ldots,n\}$ and $h \in H$. Then $gx = g_j \otimes h\psi_i$ and so

$$\begin{aligned}(g\theta(x))(v \otimes g_j^{-1}) &= \theta(x)(v \otimes g_j^{-1}g)\\ &= \theta(x)(v \otimes hg_i^{-1})\\ &= \theta(x)(vh \otimes g_i^{-1})\\ &= \psi_i(vh)\\ &= (h\psi_i)(v)\\ &= (\theta(gx)(v \otimes g_j^{-1})\end{aligned}$$

Since for $k \neq j, (g\theta(x))(v \otimes g_k^{-1}) = \theta(gx)(v \otimes g_k^{-1}) = 0$, the result follows. ■

For any additive subgroup X of FG, let $Id(X)$ denote the sum of all two-sided ideals of FG contained in X. Then $Id(X)$ is the largest ideal of FG contained X. We are now ready to prove

1.2. THEOREM. (Knorr (1977)). Let H be a subgroup of a finite group G and let V be an FH-module

(i) If $A = \mathrm{ann}(V)$, then $\mathrm{ann}(V^G) = Id(FG\cdot A)$

(ii) If A is an ideal of FH, then $Id(FG\cdot A) = Id(A\cdot FG)$

Proof. (i) Let $K \subseteq FG\cdot A$ be an ideal of FG. Since $AV = 0$, Proposition 2.4.3 (ii) tells us that $K(V^G) = 0$. Hence $K \subseteq \mathrm{ann}(V^G)$ and therefore

$$Id(FG\cdot A) \subseteq \mathrm{ann}(V^G)$$

To prove the opposite containment, fix an F-basis, say $v_1,\ldots,v_n$ of V. Then the map

$$\begin{cases} FH \longrightarrow V^n = V \times V \times \ldots \times V & (n \text{ times}) \\ x \longmapsto (xv_1, xv_2, \ldots, xv_n) \end{cases}$$

is an FH-homomorphism whose kernel is A. Hence there is an exact sequence

$$0 \longrightarrow FH/A \longrightarrow V^n$$

Tensoring this sequence with FG, using Proposition 2.4.3(i), and taking into account that FG is a flat FH-module, one obtains an exact sequence

$$0 \longrightarrow FG/FG\cdot A \longrightarrow V^G \oplus V^G \oplus \ldots \oplus V^G \qquad (n \text{ times})$$

This shows that $\mathrm{ann}(V^G) \subseteq FG\cdot A$, proving (i).

(ii) Put $V = FH/A$. Then $A = \mathrm{ann}(V)$ and

$$\mathrm{ann}[(V^G)^*] = \mathrm{ann}(V^G) = Id(FG\cdot A)$$

by Lemma 1.1(i) and (i). Again using Lemma 1.1(i) and the right analog of (i), one has

$$\mathrm{ann}[(V^*)^G] = Id[(\mathrm{ann}V^*)FG] = Id(A\cdot FG)$$

The desired assertion now follows by appealing to Lemma 1.1(ii). ∎

1.3. COROLLARY. Let $\{I_\gamma \mid \gamma \in \Gamma\}$ be a set of ideals of FH for some index set Γ, and let $I = \bigcap_{\gamma\in\Gamma} I_\gamma$. Then

$$Id(FG\cdot I) = \bigcap_{\gamma} Id(FG\cdot I_{\gamma})$$

Proof. Put $V_{\gamma} = FH/I_{\gamma}$. Because $\underset{\gamma}{\oplus} V_{\gamma}^{G} \cong (\underset{\gamma}{\oplus} V_{\gamma})^{G}$, the required assertion follows from Theorem 1.2(i). ■

2. SIMPLE INDUCTION AND RESTRICTION PAIRS

Throughout this section, F denotes an arbitrary field of characteristic $p > 0$, H a subgroup of a finite group G and e and f central idempotents of FG and FH, respectively. We recall that by definition 0 is a completely reducible module. Our aim is to present some general results to be used later. These results are extracted from a work of Knorr (1977).

2.1. LEMMA. Let $V \subseteq W$ be FG-modules. Then $eV \subseteq eW$ are FG-modules such that $eW/eV \cong e(W/V)$.

Proof. The fact that $eV \subseteq eW$ are FG-modules is a consequence of the assumption that e is a central idempotent. Let $f : W \longrightarrow W/V$ be the natural homomorphism. Then f restricts to a surjective homomorphism

$$f^* : eW \longrightarrow e(W/V)$$

and $\mathrm{Ker} f^* = eW \cap V$. But $eW \cap V = eV$, hence the result. ■

Following Knorr (1977), we define simple induction and restriction pairs as follows. We say that (e,f) is a *simple induction pair* if eV^G is a completely reducible FG-module for all irreducible FH-modules $V = fV$. The pair (e,f) is called a *simple restriction pair*, if fM_H is a completely reducible FH-module for all irreducible FG-modules $M = eM$. Not surprising, the above definitions may be rephrased in terms of $J(FG)$ and $J(FH)$, as shown below.

2.2. PROPOSITION. (i) The following conditions are equivalent:

(a) (e,f) is a simple induction pair

(b) $eJ(FG)f \subseteq FG\cdot J(FH)$

(c) $fJ(FG)e \subseteq J(FH)FG$

(ii) (e,f) is a simple restriction pair if and only if

$$eJ(FH)f \subseteq J(FG)$$

(iii) Let $E \subseteq F$ be the smallest field containing the coefficients of e and f. Then (e,f) is a simple induction (respectively, restriction) pair for FG and FH if and only if it is for EG and EH.

Proof. (i) Put $V = FHf/J(FH)f$. Because $V = fV$ is completely reducible and contains a copy of every irreducible FH-module $S = fS$, it follows that (e,f) is a simple induction pair if and only if eV^G is completely reducible. Owing to Proposition 2.4.3(i),

$$V^G \cong FGf/FG\cdot J(FH)f$$

and so, by Lemma 2.1,

$$eV^G \cong eFGf/eFG\cdot J(FH)f$$

Thus eV^G is completely reducible if and only if $eJ(FG)f \subseteq eFGJ(FH)f$. Since the latter containment is equivalent to $eJ(FG)f \subseteq FG\cdot J(FH)$, the equivalence of (a) and (b) follows.

Assume that $eJ(FG)f \subseteq FG\cdot J(FH)$. Then

$$\begin{aligned} eJ(FG) &= eJ(FG)f \oplus eJ(FG)(1-f) \\ &\subseteq FG\cdot J(FH) + FG(1-f) \\ &= FG[J(FH) + FH(1-f)] \end{aligned}$$

An application of Theorem 1.2(ii) shows that

$$\begin{aligned} J(FG)e &= eJ(FG) \\ &\subseteq [J(FH) + FH(1-f)]FG \\ &= J(FH)FG + (1-f)FG \end{aligned}$$

Thus $fJ(FG)e \subseteq J(FH)FG$. The converse follows by symmetry.

(ii) Put $M = FGe/J(FG)e$. Since $M = eM$ is completely reducible and contains a copy of every irreducible FG-module $W = eW$, it follows that (e,f) is a simple restriction pair if and only if fM_H is completely reducible. Taking into account that, by Lemma 2.1,

$$fM \cong fFGe/fJ(FG)e \qquad \text{as } FH\text{-modules}$$

we see that fM_H is completely reducible if and only if

$$fJ(FH)FGe \subseteq fJ(FG)e$$

Since the latter containment is equivalent to $eJ(FH)f \subseteq J(FG)$, (ii) follows

(iii) Direct consequence of (i), (ii) and Corollary. 3.1.18. ■

2.3. COROLLARY. The following conditions are equivalent:

(i) For every irreducible FH-module V, V^G is completely reducible

(ii) $J(FG) \subseteq FG \cdot J(FH)$

(iii) $J(FG) \subseteq J(FH)FG$

Proof. Apply Proposition 2.2(i) for the case $e = f = 1$. ■

2.4. COROLLARY. The following conditions are equivalent:

(i) For every irreducible FG-module M, M_H is completely reducible

(ii) $J(FH) \subseteq J(FG)$

Proof. Apply Proposition 2.2(ii) for the case $e = f = 1$. ■

It is obvious that (e,f) is a simple induction (restriction) pair if $ef = 0$. To investigate the case $ef \neq 0$, we need the following preliminary observation.

2.5. LEMMA. The following conditions are equivalent:

(i) $ef \neq 0$

(ii) There exists a projective indecomposable FH-module $V = fV$ such that $eV^G \neq 0$.

(iii) There is an FH-module $V = fV$ such that $eV^G \neq 0$

(iv) There is a projective indecomposable FG-module $P = eP$ such that $fP \neq 0$.

(v) There is an FG-module $M = eM$ such that $fM \neq 0$.

Proof. (i) ⇒ (ii): Owing to Proposition 2.4.3 (i), we have

$$e(FHf)^G \cong eFGf = FGef \neq 0$$

Therefore there exists at least one indecomposable component V of FHf such that $eV^G \neq 0$.

(ii) ⇒ (iii): Trivial

(iii) ⇒ (iv): Let M be an irreducible FG-module in $\mathrm{Soc}(eV^G)$ and let $P = P(M)$ be the projective cover of M. Since $V = fV$, it follows from Theorem 2.4.9(ii) that

$$0 \neq \operatorname{Hom}_{FG}(P, V^G) \cong \operatorname{Hom}_{FH}(P_H, V) \cong \operatorname{Hom}_{FH}(fP_H, V)$$

and thus $fP \neq 0$.

(iv) $\Rightarrow$ (v): Trivial

(v) $\Rightarrow$ (i): By hypothesis, $0 \neq fM = feM = efM$ and so $ef \neq 0$. ■

If e and f are block idempotents, then $\delta(e)$ and $\delta(f)$ denote the defect groups of corresponding blocks; the vertex of a module M is denoted by $vx(M)$. We are now ready to prove the main result of this section.

2.6. THEOREM. Let e and f be block idempotents of FG and FH, respectively, and assume that $ef \neq 0$.

(i) If (e,g) is a simple induction pair, then $fM_H \neq 0$ for all FG-modules $0 \neq M = eM$; moreover, $\delta(f) \geqslant_G \delta(e)$.

(ii) If (e,f) is a simple restriction pair, then $eV^G \neq 0$ for all FH-modules $0 \neq V = fV$; moreover, $\delta(f) \leqslant_G \delta(e)$.

Proof. (i) Let $V = eV$ be an irreducible FG-module such that $fV_H = 0$. We wish to show that $fP_H = 0$ for the projective cover P of V. Assume the contrary. Since $eP = P$, it follows from Theorem 2.4.9(ii) that there exists an irreducible FH-module $W = fW$ such that

$$0 \neq \operatorname{Hom}_{FH}(P_H, W) \cong \operatorname{Hom}_{FG}(P, W^G) \cong \operatorname{Hom}_{.FG}(P, eW^G)$$

But eW^G is completely reducible by hypothesis, so

$$0 \neq \operatorname{Hom}_{FG}(V, W^G) \cong \operatorname{Hom}_{FH}(V_H, W) = \operatorname{Hom}_{FH}(fV_H, W) = 0,$$

a contradiction.

Now assume that $eM = M \neq 0$ and $fM_H = 0$ for some FG-module M. Then $fV_H = 0$ for any composition factor V of M. Since e is a block idempotent, all indecomposable projective modules in the corresponding block are linked (Proposition 1.10.14). Hence by repeated use of the above argument, $fP_H = 0$ for all projective modules $P = eP$, contradicting Lemma 2.5(iv). The first assertion is therefore established.

By Theorem 3.16.14, there exists an irreducible FG-module $V = eV$ such that

$$vx(V) \underset{G}{=} \delta(e)$$

Then, by the foregoing, $fV_H \neq 0$ and so there is an irreducible FH-module W in $\mathrm{Soc}(fV_H)$. Since $eV = V$, it follows from Theorem 2.4.9(i) that

$$0 \neq \mathrm{Hom}_{FH}(W, V_H) \cong \mathrm{Hom}_{FG}(W^G, V) \cong \mathrm{Hom}_{FG}(eW^G, V)$$

But eW^G is completely reducible by hypothesis, and eW^G is a direct summand of W^G so V is a direct summand of W^G. Therefore, by Theorem 3.16.14,

$$\delta(e) \underset{G}{=} vx(V) \leqslant_G vx(W) \leqslant_H \delta(f)$$

(ii) The first statement is an easy adaptation of the ideas in the proof of (i). To prove the assertion on the defect groups, choose an irreducible FH-module $W = fW$ such that

$$vx(W) \underset{H}{=} \delta(f)$$

Let $V = eV$ be an irreducible FG-module in $\mathrm{Soc}(eW^G)$. Then

$$0 \neq \mathrm{Hom}_{FG}(V, W^G) \cong \mathrm{Hom}_{FH}(V_H, W) \cong \mathrm{Hom}_{FH}(fV_H, W)$$

Since fV_H is completely reducible and fV_H is a direct summand of V_H, we deduce that W is a direct summand of V_H. Let A be the vertex of V. Then V is a direct summand of $(V_A)^G$, so V_H is a direct summand of $[(V_A)^G]_H$ and hence W is a direct summand of $[(V_A)^G]_H$. It follows from Mackey's Decomposition (Theorem 2.6.1) that W is a direct summand of

$$FH \otimes_{F(H \cap gAg^{-1})} {}^{g}V \qquad \text{for some } g \in G$$

and thus

$$\delta(f) \underset{H}{=} vx(W) \leqslant_H H \cap gAg^{-1} \leqslant_G A \leqslant_G \delta(e),$$

as required. ∎

2.7. REMARK. The following assertions are direct consequences of the above proof.

(i) Assume the conditions of Theorem 2.6(i). Then

$$\mathrm{Soc}(fV_H) \cong fV_H / J(FH) fV_H$$

for all irreducible FG-modules $V = eV$.

(ii) Assume the conditions of Theorem 2.6(ii). Then

$$\mathrm{Soc}(eW^G) \cong eW^G/J(FG)eW^G$$

for all irreducible FH-modules $W = fW$. ■

2.8. REMARK. Write $e = \Sigma e_i$ and $f = \Sigma f_j$ as a sum of block idempotents of FG and FH, respectively. Then, by the definition, (e,f) is a simple restriction (induction) pair if and only if (e_i, f_j) are simple restriction (induction) pairs for all i and j. ■

We close by providing some applications of Theorem 2.6. The following preliminary observation will be very useful.

2.9. LEMMA. Let $H \subseteq S$ be subgroups of G and let W be an irreducible FS-module.

(i) If (e,f) is a simple induction pair and $fW_H \neq 0$, then eW^G is completely reducible.

(ii) If (e,f) is a simple restriction pair and $eW^G \neq 0$, then fW_H is completely reducible.

Proof. (i) By hypothesis, $fW_H \neq 0$, so we may choose an irreducible FH-module V in $\mathrm{Soc}(fW_H)$. Then $fV = V$ and

$$0 \neq \mathrm{Hom}_{FH}(V, W_H) \cong \mathrm{Hom}_{FS}(V^S, W)$$

Because W is irreducible, there is an exact sequence

$$V^S \longrightarrow W \longrightarrow 0$$

Therefore there exists an exact sequence

$$eV^G \longrightarrow eW^G \longrightarrow 0$$

Since (e,f) is a simple induction pair, eV^G is completely reducible. Then eW^G is also completely reducible.

(ii) The proof is analogous and therefore will be omitted. ■

We are now ready to provide a number of consequences of Theorem 2.6.

2.10. COROLLARY. Let $H \subseteq S$ be subgroups of G and let e and f be central idempotents of FG and FH, respectively. Let $\{\varepsilon_i\}$ denote all block idempotents of FS such that $\varepsilon_i f \neq 0$. Put $\varepsilon = \Sigma\varepsilon_i$ and assume that (ε, f) is a simple induction pair. Then the following conditions are equivalent:

(i) (e,f) is a simple induction pair

(ii) (e,ε) is a simple induction pair.

Proof. (i) $\Rightarrow$ (ii): Let $W = \varepsilon W$ be an irreducible FS-module. By Remark 2.8 and Theorem 2.6(i), $fW_H \neq 0$ since (ε,f) is a simple induction pair. Hence, by Lemma 2.9(i), eW^G is completely reducible which means that (e,ε) is a simple induction pair.

(ii) $\Rightarrow$ (i): Let $V = fV$ be an irreducible FH-module. Then εV^S and $e(\varepsilon V^S)^G$ are completely reducible, since (ε,f) and (e,ε) are simple induction pairs. Invoking Lemma 2.5, it follows from the construction of ε that $V^S = \varepsilon V^S$. Hence eV^G is completely reducible and so (e,f) is a simple induction pair. ∎

2.11. COROLLARY. Let $H \subseteq S$ be subgroups of G and let e and f be central idempotents of FG and FH, respectively. Let $\{\varepsilon_i\}$ denote all block idempotents of FS such that $\varepsilon_i e \neq 0$. Put $\varepsilon = \Sigma\varepsilon_i$ and assume that (e,ε) is a simple restriction pair. Then the following conditions are equivalent:

(i) (e,f) is a simple restriction pair .

(ii) (ε,f) is a simple restriction pair.

Proof. The proof is analogous and therefore will be ommitted. ∎

2.12. PROPOSITION. Let N be a normal subgroup of G, let $b = b(f)$ be a block of FN and let $B_i = B_i(e_i)$, $1 \leqslant i \leqslant m$, be the blocks of FG covering b. Let $V_1,V_2,\ldots,V_t$ be all nonisomorphic irreducible FN-modules in b, and let H_i be the inertia group of V_i, $1 \leqslant i \leqslant t$.

(i) $m \leqslant \min\{(H_1:N),(H_2:N),\ldots,(H_t:N)\}$

(ii) If $m = |G(f):N|$. where $G(f)$ is the inertia group of f, then

$$J(FG)f \subseteq FG\cdot J(FN) \quad \text{and} \quad \delta(e_i) \underset{G}{=} \delta(f)$$

Proof. (i) Let V be an irreducible module in b and H its inertia group. Since $N \lhd G$, it follows from Clifford's theorem that (1,1) is a simple restriction pair. Therefore, by Remark 2.8, (e_i,f) is a simple restriction pair, $1 \leqslant i \leqslant m$. Also, by Lemma 3.10.2(iii), $e_i f \neq 0$ for all $i \in \{1,\ldots,m\}$.

Invoking Theorem 2.6(ii), we conclude that $W_i = e_i V^G \neq 0$, $1 \leqslant i \leqslant m$. Let $\ell_i > 0$ be the composition length of W_i and M_{ij} its composition factors. Because

$$(V^G)_N \cong \bigoplus_{g \in T} {}^g V$$

where T is a transversal for N in G, it follows from Clifford's theorem that

$$\dim_F M_{ij} = r_{ij} |G:H| \dim_F V$$

for some integer $r_{ij} > 0$. Therefore

$$\begin{aligned} |G:N| \dim_F V = \dim_F V^G &= \sum_{i=1}^{m} \dim_F W_i \\ &= \sum_{i=1}^{m} \sum_{j=1}^{\ell_i} r_{ij} |G:H| \dim_F V \end{aligned}$$

and

$$|H:N| = \sum_{i=1}^{m} \sum_{j=1}^{\ell_i} r_{ij} \geqslant m, \tag{*}$$

as required.

(ii) Since $G(f) \supseteq H$, it follows that

$$m = |G(f):N| \geqslant |H:N| \geqslant \min\{(H_1:N), \ldots, (H_t:N)\} \geqslant m$$

and thus $|H:N| = m$. By (*) this implies $\ell_i = 1$ for all i, so $V^G = \oplus\, e_i V^G$ is completely reducible. Hence $(1,f)$ is a simple induction pair and therefore, by Proposition 2.2, $J(FG)f \subseteq FG \cdot J(FN)$. The assertion on the defect groups follows from Theorem 2.6 and Remark 2.8. ■

3. APPLICATIONS

Throughout this section, F denotes an arbitrary field of characteristic $p > 0$, H a subgroup of a finite group G and e and f central idempotents of FG and FH, respectively. Our aim is to tie together the concepts of induction, restriction and the Jacobson radical. All the results presented are contained in a work of Knörr (1977). Some particular cases of these results may be found in Motose and Ninomiya (1975b) and Khatri (1973).

3.1. PROPOSITION. The following conditions are equivalent:

(i) $(eV^G)_H$ is completely reducible for all irreducible FH-modules V

(ii) $eJ(FH)FG \subseteq FG{\cdot}J(FH)$

(iii) $eJ(FH)FG = FG{\cdot}J(FH)e$

Moreover, if these conditions are satisfied, then

$$eJ(FH)^n FG = (eJ(FH){\cdot}FG)^n \qquad \text{for all } n \geqslant 1$$

Proof. (i) $\Rightarrow$ (ii): Put $W = FH/J(FH)$. Then W is completely reducible, so by assumption $(eW^G)_H$ is completely reducible. Therefore $eJ(FH) \subseteq \text{ann}(W^G)$. But

$$\text{ann}(W^G) = Id(FG{\cdot}J(FH))$$

by Theorem 1.2(i), so $eJ(FH)FG \subseteq FG{\cdot}J(FH)$.

(ii) $\Rightarrow$ (iii): It is clear from (ii) that

$$eFGJ(FH)FG \subseteq FG{\cdot}J(FH)$$

and therefore, by Theorem 1.2(ii),

$$\begin{aligned} eFG{\cdot}J(FH) &\subseteq eFGJ(FH)FG \\ &\subseteq Id(FG{\cdot}J(FH)) \\ &= Id(J(FH){\cdot}FG) \end{aligned}$$

It follows that $eFG{\cdot}J(FH) \subseteq eJ(FH)FG$. The opposite containment is obvious from (ii).

(iii) $\Rightarrow$ (i): it is obvious from (iii) that $FG{\cdot}J(FH)e$ is an ideal of FG. Let V be an irreducible FH-module and let $A = \text{ann}(V)$. Then $J(FH) \subseteq A$ and $FG{\cdot}J(FH)e \subseteq FG{\cdot}A$. Invoking Theorem 1.2(i), we deduce that

$$J(FH)e \subseteq FG{\cdot}J(FH)e \subseteq Id(FG{\cdot}A) = \text{ann}(V^G)$$

Thus $J(FH)(eV^G) = 0$ and therefore $(eV^G)_H$ is completely reducible. The final assertion follows by induction on n, using (iii). ∎

3.2. PROPOSITION. The following conditions are equivalent:

(i) $(fM_H)^G$ is completely reducible for all irreducible FG-modules M

(ii) V^G is completely reducible for all irreducible FH-modules $V = fV$, and fM_H is completely reducible for all irreducible FG-modules M.

(iii) $J(FG)f = FG\cdot J(FH)f$

(iv) $fJ(FG) = fJ(FH)FG$

Proof. Condition (ii) may be reformulated as: $(1,f)$ is a simple induction pair and a simple restriction pair. The equivalence of (ii), (iii), and (iv) is therefore a consequence of Proposition 2.2. Since (ii) obviously implies (i), we are left to verify that (i) implies (ii).

Let M be an irreducible FG-module. Then, by hypothesis, $(fM_H)^G$ is completely reducible and so, by Corollary 2.4.5, fM_H is completely reducible.

Let $V = fV$ be an irreducible FH-module and let M be an irreducible FG-module in $\mathrm{Soc}(V^G)$. Then

$$0 \neq \operatorname{Hom}_{FG}(M,V^G) \cong \operatorname{Hom}_{FH}(M_H,V) \cong \operatorname{Hom}_{FH}(fM_H,V)$$

Hence there is an exact sequence

$$fM_H \longrightarrow V \longrightarrow 0$$

which gives rise to an exact sequence

$$(fM_H)^G \longrightarrow V^G \longrightarrow 0$$

Because $(fM_H)^G$ is completely reducible, V^G is also completely reducible. So the proposition is verified. ■

To prove our next result, we need two preliminary observations.

3.3. LEMMA. The following conditions are equivalent:

(i) For any FG-module $M = eM$, if M_H is completely reducible, then so is M.

(ii) $FGJ(FH)FG \supseteq J(FG)e$

Proof. (i) ⇒ (ii): Put $M = FGe/FGJ(FH)FGe$. Then $M = eM$ and M_H is completely reducible. Hence, by assumption, M is completely reducible. Thus $J(FG)e \subseteq FGJ(FH)FGe \subseteq FGJ(FH)FG$.

(ii) ⇒ (i): Let $M = eM$ be an FG-module such that M_H is completely reducible. Then $J(FH)M = 0$ and if $FGJ(FH)FG \supseteq J(FG)e$, then

$$\begin{aligned} J(FG)M = J(FG)eM &\subseteq FGJ(FH)FGM \\ &= FGJ(FH)M \\ &= 0 \end{aligned}$$

Thus M is completely reducible. ∎

3.4. LEMMA. Let M be an FG-module such that $M/J(FG)M$ is H-projective. Then

$$FGJ(FH)M \supseteq J(FG)M$$

Proof. Put $X = FGJ(FH)M$. Then $Y = M/(X+J(FG)M)$ is a direct summand of the completely reducible and H-projective module $M/J(FG)M$, so Y is completely reducible and H-projective. Consider the exact sequence

$$0 \longrightarrow (X+J(FG)M)/X \longrightarrow M/X \longrightarrow Y \longrightarrow 0 \qquad (*)$$

Since $J(FH)M \subseteq X$, it follows that $(M/X)_H$ is completely reducible. Thus (*) splits as an FH-sequence. But Y is H-projective, so (*) splits as an FG-sequence. Hence

$$M/X \cong [(X+J(FG)M)/X] \oplus Y$$

and multiplying both sides by $J(FG)$ gives

$$(J(FG)M+X)/X \cong J(FG)[(X+J(FG)M)/X]$$

since Y is completely reducible. Thus

$$(X+J(FG)M)/X = 0$$

and therefore $J(FG)M \subseteq X$, as asserted. ∎

We are now ready to prove

3.5. PROPOSITION. (i) If $J(FG)e \subseteq FG\cdot J(FH)$, then all irreducible FG-modules $M = eM$ are H-projective.

(ii) If all irreducible FG-modules $M = eM$ are H-projective, then V_H completely reducible implies V completely reducible for any FG-module $V = eV$.

Proof. (i) By Proposition 2.2(i), $(e,1)$ is a simple induction pair. Write $e = \Sigma e_i$ as a sum of block idempotents of FG. Then, by Remark 2.8, each $(e_i,1)$ is a simple induction pair. Hence, by Theorem 2.6(i), $\delta(e_i) \leqslant_G H$. Let $M = eM$ be an irreducible FG-module. Then $M = e_iM$ for some i and hence, by Theorem 3.16.14, M is $\delta(e_i)$-projective. Therefore, M is H-projective, by virtue of Lemma 2.5.8(ii).

(ii) Assume that all irreducible FG-modules $M = eM$ are H-projective. By Lemma 3.3(ii), we need only verify that $FGJ(FH)FG \supseteq J(FG)e$. By hypothesis and by Lemma 2.5.8(i), $FGe/J(FG)e$ is H-projective. Therefore, by Lemma 3.4,

$$FGJ(FH)FG \supseteq FGJ(FH)FGe \supseteq J(FG)e$$

as required. ∎

3.6. LEMMA. If $(eV^G)_H$ is completely reducible for all irreducible FH-modules V, then $J(FH)e \subseteq J(FG)$.

Proof. Assume that for every irreducible FH-module V, $(eV^G)_H$ is completely reducible. Then, by Proposition 3.1, $eJ(FH)FG$ is a nilpotent ideal of FG. Therefore

$$eJ(FH) \subseteq eJ(FH)FG \subseteq J(FG),$$

as required. ∎

3.7. THEOREM. The following conditions are equivalent:

(i) $J(FG)e = FG\cdot J(FH)e$

(ii) $J(FG)e = J(FH)FGe$

(iii) eV^G is completely reducible for any irreducible FH-module V and M_H is completely reducible for any irreducible FG-module $M = eM$

(iv) eV^G and $(eV^G)_H$ are completely reducible for all irreducible FH-modules V

(v) $(eV^G)_H$ is completely reducible for any irreducible FH-module V and all irreducible FG-modules $M = eM$ are H-projective

(vi) $(eV^G)_H$ is completely reducible for any irreducible FH-module V and M_H completely reducible implies M completely reducible for any FG-module $M = eM$.

Proof. Condition (iii) may be reformulated as follows: $(e,1)$ is a simple induction and a simple restriction pair. Hence, by Proposition 2.2, (iii) is equivalent to either of the following conditions:

$$J(FG)e \subseteq FG\cdot J(FH) \quad \text{and} \quad J(FH)e \subseteq J(FG) \tag{1}$$

$$J(FG)e \subseteq J(FH)FG \quad \text{and} \quad J(FH)e \subseteq J(FG) \tag{2}$$

But (1) is equivalent to (i) and (2) is equivalent to (ii), hence (i), (ii), and (iii) are equivalent.

The implications (iv) $\Rightarrow$ (v) $\Rightarrow$ (vi) follow by Propositions 3.5 and 2.2, while the implication (iv) $\Rightarrow$ (iii) is a consequence of Lemma 3.6 and Proposition 2.2. We are therefore left to verify that (vi) implies (iv) and (ii) implies (iv).

(vi) $\Rightarrow$ (iv): By Proposition 3.1, $FGJ(FH)e = J(FH)FGe$ and, by Lemma 3.3,

$$FGJ(FH)FG \supseteq J(FG)e$$

It follows that

$$J(FG)e \subseteq FGJ(FH)FGe = FGJ(FH)e \subseteq FG\cdot J(FH)$$

and hence, by Proposition 2.2, eV^G is completely reducible for any irreducible FH-module V.

(ii) $\Rightarrow$ (iv): Direct consequence of implications (ii) $\Rightarrow$ (iii), (ii) $\Rightarrow$ (i) and Proposition 3.1. ■

For any subgroup H of G, the *normal closure* of H in G is defined as the intersection of all normal subgroups of G containing H.

3.8. LEMMA. Let P be a p-subgroup of G, let N be its normal closure in G and let Q be a Sylow p-subgroup of N. Then

$$FG\cdot I(P)\cdot FG = FG\cdot I(Q)\cdot FG$$

Proof. By Sylow's theorem, we may assume that $Q \supseteq P$ since

$$FG\cdot I(Q)\cdot FG = FG\cdot I(Q^g)FG \qquad \text{for all } g \in G$$

Then $I(Q) \supseteq I(P)$ and one inclusion follows.

On the other hand, put

$$V = FG/FG\cdot I(P)\cdot FG$$

Given $x \in P, g \in G$ we have $(x-1)g \in I(P)FG \subseteq FG\cdot I(P)\cdot FG$, so

$$xg \equiv g(\text{mod } FG\cdot I(P)\cdot FG)$$

and hence $P \subseteq \text{Ker}V$. Thus $N \subseteq \text{Ker}V$ and $Q \subseteq \text{Ker}V$. It follows that $FG\cdot I(Q)V = 0$ and therefore $FG\cdot I(Q)\cdot FG \subseteq FG\cdot I(P)FG$, as required. ■

3.9. LEMMA. Let X be an indecomposable N-projective FG-module with $N \lhd G$. Then X_N is a direct sum of G-conjugate indecomposable FN-modules.

Proof. Because X is N-projective, there exists an indecomposable FN-module U such that X is a direct summand of U^G. As $N \lhd G$, we have

$$(U^G)_N = \bigoplus_{t \in T} (t \otimes U)$$

where T is a transversal for N in G. The $t \otimes U$ are indecomposable FN-modules which are conjugate under the action of G. Since X_N is a direct summand of $(U^G)_N$, the result follows. ■

3.10. THEOREM. Let P be a p-subgroup of G and let N be its normal closure in G. Then the following conditions are equivalent:

(i) For any FG-module $V = eV$, if V_P is completely reducible, then so is V.

(ii) If $V = eV$ is an N-projective FG-module such that $V/J(FN)V$ is indecomposable and $FG \cdot I(P)V \subset V$, then $V/J(FG)V$ is N-projective

(iii) If $M = eM$ is an irreducible FG-module such that $P \subseteq \text{Ker}M$, then M is N-projective.

Proof. (i) $\Rightarrow$ (ii): Assume that V is as in (ii). By Lemma 3.16.15, it suffices to show that

$$J(FG)V = J(FN)V$$

The inclusion $J(FN)V \subseteq J(FG)V$ being obvious, put $W = V/J(FN)V$. Then W_N is completely reducible, say

$$W_N = S_1 \oplus S_2 \oplus \dots \oplus S_r$$

where the S_i are irreducible. By Lemma 3.16.13, W is N-projective. Since W is indecomposable by assumption, it follows from Lemma 3.9 that

$$S_i \cong g_i \otimes S \qquad (1 \leqslant i \leqslant r)$$

for some $g_i \in G$ and some irreducible FN-module S.

Now let Q be a Sylow p-subgroup of N and choose $n_i \in N$ such that $Q^{g_i} = Q^{n_i}$. Then

$$\begin{aligned} I(Q)S_i &\cong I(Q)g_i \otimes S \\ &= g_i \otimes I(Q)^{g_i} S \end{aligned}$$

$$= g_i \otimes I(Q)^{n_i} S$$
$$= g_i^{-1} n_i^{-1} \otimes I(Q)S$$

It will next be shown that $I(Q)S = 0$. Assume the contrary. Then $I(Q)S_i \neq 0$ for all i and therefore $FNI(Q)S_i = S_i$, $1 \leqslant i \leqslant r$. It follows that

$$\begin{aligned} W &\supseteq FG\cdot I(Q)\cdot FGW \\ &= FG\cdot I(Q)W \\ &\supseteq FN\cdot I(Q)\,(\bigoplus_{i=1}^{r} S_i) \\ &= \bigoplus_{i=1}^{r} S_i = W \end{aligned}$$

and so $W = FG\cdot I(Q)FGW$. By Lemma 3.8 then, $FG\cdot I(P)FGW = W$ and therefore

$$FG\cdot I(P)\cdot FGV + J(FN)V = V$$

by definition of W. Since $J(FN)V \subseteq J(FG)V$, it follows from Proposition 1.6.9 that

$$FG\cdot I(P)V = FG\cdot I(P)\cdot FGV = V$$

contradicting the hypothesis.

Thus $I(Q)S = 0$ and so $I(Q)S_i = 0$ for all $i \in \{1,\dots,r\}$. This implies $I(Q)W = 0$ and hence $I(P)W = 0$. Therefore W_P is completely reducible, so W is completely reducible by hypothesis. Hence $J(FG)W = 0$ and $J(FG)V \subseteq J(FN)V$, as required.

(ii) ⇒ (iii): Let M be as in (iii) and L a projective cover of M. Then $eL = L$ is N-projective and $L/J(FN)L$ is indecomposable. Furthermore, $I(P)M = 0$ since $P \subseteq \operatorname{Ker}M$. Hence $I(P)L \subseteq J(FG)L$ and therefore

$$FG\cdot I(P)L \subseteq J(FG)L \subset L$$

By assumption then, $M \cong L/J(FG)L$ is N-projective.

(iii) ⇒ (i): Assume $V = eV$ is an FG-module such that V_P is completely reducible. If $W = V/J(FG)V$, then

$$W = M_1 \oplus M_2 \oplus \dots \oplus M_n$$

for some $n \geqslant 1$ and irreducible FG-modules M_i. Since $P \subseteq \operatorname{Ker}V$, we have

$P \subseteq \mathrm{Ker}M_i$ for all $i \in \{1,\dots,n\}$. Hence, by hypothesis, each M_i is N-projective and therefore Q-projective, if again Q denotes a Sylow p-subgroup of N. Thus W is Q-projective and

$$FG \cdot I(Q)V \supseteq J(FG)V$$

by Lemma 3.4. But $Q \subseteq N \subseteq \mathrm{Ker}V$, so $I(Q)V = 0$ and, by the above, $J(FG)V = 0$. Thus V is completely reducible, as required. ∎

3.11. COROLLARY. Let P be a p-subgroup of G and let N be its normal closure in G.

(i) If all irreducible FG-modules $M = eM$ are N-projective, then

$$FG \cdot I(P)FG \supseteq J(FG)e$$

(ii) If $FG \cdot I(P) \cdot FG \supseteq J(FG)e$ and $M = eM$ is an irreducible FG-module such that $P \subseteq \mathrm{Ker}M$, then $vx(M) = Q \in \mathrm{Syl}_p(N)$ and M has trivial source.

(iii) If $I(P)e \subseteq J(FG)$ and $FG \cdot I(P) \cdot FG \supseteq J(FG)e$, then all irreducible FG-modules $M = eM$ have the same vertices and trivial source.

Proof. (i) By Lemma 3.3, $FG \cdot I(P) \cdot FG \supseteq I(FG)e$ is equivalent to the condition (i) of Theorem 3.10. Since our assumption implies condition (iii) of Theorem 3.10, the assertion follows.

(ii) By Theorem 3.10(iii), M is N-projective, hence Q-projective, and clearly M_Q is completely reducible. Setting $X = vx(M)$, we may therefore assume that $X \subseteq Q$. Now M is a direct summand of $(M_X)^G$ and M_X is completely reducible, hence M is a direct summand of $(1_X)^G$, where 1_X is the trivial FX-module. Therefore M_Q is a direct summand of $[(1_X)^G]_Q$. By Mackey's decomposition,

$$[(1_X)^G]_Q \cong \oplus (1_{X^g \cap Q})^Q$$

and since 1_Q is a direct summand of M_Q, we conclude that

$$1_Q \cong (1_{X^g \cap Q})^Q \qquad \text{for some } g \in G$$

since each $(1_{X^g \cap Q})^Q$ is indecomposable by Theorem 3.4.24. Thus $X^g \cap Q = Q$ and $X = Q$, proving both assertions.

(iii) If $I(P)e \subseteq J(FG)$, then by Proposition 2.2 M_P is completely reducible for all irreducible FG-modules $M = eM$ and therefore $P \subseteq \mathrm{Ker} M$. Now apply (ii). ■

3.12. LEMMA. Assume that V^G is completely reducible for any irreducible FH-module V. Then H is of p'-index in G.

Proof. Let $B = B(e)$ be the principal block of FG. Then a Sylow p-subgroup P of G is a defect group of B. By Theorem 3.16.14, we may choose an irreducible FG-module V in B such that P is a vertex of V. Let W be an irreducible submodule of V_H. Since $eV = V$, we have

$$0 \neq \mathrm{Hom}_{FH}(W, V_H) \cong \mathrm{Hom}_{FG}(W^G, V) \cong \mathrm{Hom}_{FG}(eW^G, V)$$

But eW^G is a submodule of W^G and W^G is completely reducible by assumption. Hence eW^G is completely reducible and therefore V is a direct summand of W^G. It follows that V is H-projective and so H contains a conjugate of P. Thus H is of p'-index in G. ■

3.13. LEMMA. The following conditions are equivalent:

(i) All irreducible FG-modules are H-projective

(ii) H is of p'-index in G

Proof. The implication (ii) $\Rightarrow$ (i) is a consequence of Corollary 2.5.5. Conversely, assume that (i) holds. Let $B = B(e)$ be the principal block of FG. By Theorem 3.16.14, we may choose an irreducible FG-module V in B such that $P \in \mathrm{Syl}_p(G)$ is a vertex of V. By hypothesis, V is H-projective and so H contains a conjugate of P. Thus H is of p'-index in G, as required. ■

We close this section by providing the following application of Theorem 3.7.

3.14. THEOREM. Let H be a subgroup of a finite group G and let F be a field of characteristic $p > 0$. Then the following conditions are equivalent:

(i) $J(FG) = FG \cdot J(FH)$

(ii) $J(FG) = J(FH) \cdot FG$

(iii) There exists a normal subgroup N of G such that $N \subseteq H$ and N has p'-index in G

(iv) V^G is completely reducible for any irreducible FH-module V and M_H is completely reducible for any irreducible FG-module M.

(v) V^G and $(V^G)_H$ are completely reducible for all irreducible FH-modules V

(vi) $(V^G)_H$ is completely reducible for all irreducible FH-modules V and all irreducible FG-modules are H-projective

(vii) $(V^G)_H$ is completely reducible for any irreducible FH-module V and M_H completely reducible implies M completely reducible for any FG-module M.

(viii) $(V^G)_H$ is completely reducible for all irreducible FH-modules V and H is of p'-index in G.

(ix) $(M_H)^G$ is completely reducible for any irreducible FG-module M.

Proof. Applying Proposition 3.2 for $f = 1$ and Theorem 3.7 for $e = 1$, we see that the conditions (i),(ii),(iv),(v),(vi),(vii), and (ix) are equivalent. Furthermore, by Lemma 3.13, (vi) and (viii) are equivalent. We are therefore left to verify that (iii) is equivalent to one of the conditions (i),(iv),(viii). For the sake of clarity, we divide the rest of the proof into two steps.

Step 1. Assume that for any irreducible FH-module V, $(V^G)_H$ is completely reducible. Our aim is to prove that for all $g \in G$,

$$J(FH) = FH \cdot J(FL) \quad \text{where} \quad L = H \cap gHg^{-1}$$

Put $V = FH/J(FH)$. Then V is a completely reducible FH-module, hence so is $(V^G)_H$, by hypothesis. Setting W to be the restriction of ${}^gV$ to FL, it follows from Mackey's decomposition that W^H is completely reducible. Put $A = \text{ann}(W)$ and $S = FL/A$. By Theorem 1.2(i), $J(FH) \subseteq FH \cdot A$ and so by Proposition 2.4.3(i), S^H is completely reducible. Hence by Corollary 2.4.5, S is completely reducible and thus $J(FL) \subseteq A$. But

$$\begin{aligned} A = FL \cap g(\text{ann}V)g^{-1} &= FH \cap F(gHg^{-1}) \cap gJ(FH)g^{-1} \\ &= FH \cap gJ(FH)g^{-1} \\ &\subseteq J(FL) \end{aligned}$$

since $FH \cap gJ(FH)g^{-1}$ is a nilpotent ideal of FL. Hence

$$A = FH \cap gJ(FH)g^{-1} = J(FL)$$

and therefore

$$J(FH) \subseteq FH \cdot J(FL)$$

Replacing g by g^{-1}, it follows by the same argument that

$$FH \cap g^{-1}J(FH)g = J(F(H \cap g^{-1}Hg))$$

Since

$$gFHg^{-1} = F(gHg^{-1}) \quad \text{and} \quad gJ(FH)g^{-1} = J(F(gHg^{-1}))$$

the latter implies that

$$\begin{aligned} F(gHg^{-1}) \cap J(FH) &= gJ(F(H \cap g^{-1}Hg))g^{-1} \\ &= J(F(H \cap gHg^{-1})) \\ &= J(FL) \end{aligned}$$

Thus $J(FH) \supseteq J(FL)$ and therefore $J(FH) = FH \cdot J(FL)$, as asserted.

Step 2. Completion of the proof. Assume that (viii) holds. To prove (iii), we argue by induction on $|H|$. If $H \lhd G$ then we are done with $N = H$. Suppose that $L = H \cap gHg^{-1}$ is a proper subgroup of H for some $g \in G$. Since (i) and (viii) are equivalent, it follows from Step 1 that

$$J(FG) = FG \cdot J(FH) = FG \cdot FH \cdot J(FL) = FG \cdot J(FL)$$

Hence, by induction, there is a normal subgroup N of G such that $N \subseteq L$ and N has p'-index in G. Since $L \subseteq H$, (iii) follows.

Finally, assume that (iii) holds. Denote by V an irreducible FH-module. Since $(V^G)_N$ is a direct sum of FN-modules of the form ${}^gV, g \in G$, $(V^G)_N$ is completely reducible. Because N has p'-index in both H and G, it follows from the equivalence of (i) and (viii) that

$$J(FG) = FG \cdot J(FN) \quad \text{and} \quad J(FH) = FH \cdot J(FN)$$

Thus

$$FG \cdot J(FH) = FG \cdot J(FN) = J(FG),$$

proving (i) and hence the result. ∎

4. p-RADICAL GROUPS

Throughout this section, G denotes a finite group and F a field of characteristic $p > 0$. If X is a subset of FG, we write $r(X) = r_G(X)$ and

$\ell(X) = \ell_G(X)$ for the right and left annihilators of X in FG, respectively. As usual, if X is a subset of G, then X^+ denotes the sum of elements in X.

4.1. LEMMA. Let H be a subgroup of G and let I be a nilpotent left ideal of FH. Let $I^g = g^{-1}Ig$ and let $\bar{I} = \bigcap_{g\in G} FG\cdot I^g$.

(i) $\bar{I}$ is a nilpotent ideal of FG

(ii) $\bar{I} = \ell_G\{\sum_{g\in G} r_H(I)^g FG\}$

(iii) $r_g(\bar{I}) = \sum_{g\in G} r_H(I)^g FG$

(iv) If I is an ideal of FH, then

$$\bar{I} = \bigcap_{g\in G} \bar{I}^g FG$$

Proof. Fix x in G. Then

$$\begin{aligned}\bar{I}x &\subseteq \bigcap_{g\in G} [FG\cdot I^g x] \\ &= \bigcap_{g\in G} [FGxI^{gx}] \\ &\subseteq \bigcap_{g\in G} FG\cdot I^{gx} \\ &= \bar{I}\end{aligned}$$

and thus $\bar{I}$ is an ideal of FG. It follows that

$$\bar{I}^{m+1} \subseteq \bar{I}^m (FG\cdot I) = \bar{I}^m\cdot I$$

for any integer $m \geqslant 1$. Hence, by induction,

$$\bar{I}^{m+2} \subseteq \bar{I}^2\cdot I^m$$

for any integer $m \geqslant 0$. Since I is nilpotent, we deduce that $\bar{I}$ is a nilpotent ideal of FG.

(ii) By Theorem 3.3.6(iii), $I = \ell_H(r_H(I))$ which implies that

$$FG\cdot I = \ell_G(r_H(I)FG)$$

since FG is a free FH-module. Hence, if $g \in G$, then

$$FG\cdot I^g = \ell_G(r_{H^g}(I^g)FG) = \ell_G(\{r_H(I)\}^g FG)$$

It follows that

$$\bar{I} = \bigcap_{g\in G} \ell_G(\{r_H(I)\}^g FG)$$

$$= \ell_G(\sum_{g\in G}\{r_H(I)\}^g FG),$$

as asserted.

(iii) Applying (ii) and Theorem 3.3.6 (iii), we have

$$\begin{aligned} r_G(\bar{I}) &= r_G\{\ell_G(\sum_{g\in G} r_H(I)^g FG)\} \\ &= \sum_{g\in G} r_H(I)^g FG, \end{aligned}$$

as desired.

(iv) Owing to Theorem 3.3.6(v), the left and right annihilators of an ideal in FG (or FH) coincide. Invoking (ii), we therefore derive

$$\begin{aligned} \bar{I} &= \ell_G(\sum_{g\in G} r_H(I)^g FG) \\ &= r_G(\sum_{g\in G} \ell_H(I)^g FG) \\ &= r_G(\sum_{g\in G} FG\ell_H(I)^g) \\ &= \bigcap_{g\in G} r_G(FG\ell_H(I)^g) \\ &= \bigcap_{g\in G} I^g FG \end{aligned}$$

as required. ∎

4.2. LEMMA. Let H be a subgroup of G. Then

(i) $FG\cdot I(H) = \ell_G(H^+) = \{\sum_{g\in G} x_g g \in FG \mid \sum_{h\in H} x_{gh} = 0 \text{ for all } g\in G\}$

(ii) $I(H)\cdot FG = r_G(H^+) = \{\sum_{g\in G} x_g g \in FG \mid \sum_{h\in H} x_{hg} = 0 \text{ for all } g\in G\}$

Proof. (i) The second equality is a consequence of the fact that $(\Sigma x_g g)H^+ = 0$ if and only if $\sum_{h\in H} x_{gh} = 0$ for all $g\in G$. Since $h-1\in \ell_H(H^+)$ it follows that $I(H)\subseteq \ell_H(H^+)$. But $I(H)$ is of codimension 1, hence $I(H) = \ell_H(H^+)$. Since FG is a free FH-module, we conclude that

$$\ell_G(H^+) = FG\ell_H(H^+) = FG\cdot I(H),$$

as required.

(ii) The proof is similar and therefore will be omitted. ∎

4.3. LEMMA. Let P be a Sylow p-subgroup of G.

(i) $\bigcap_{g\in G} FG\cdot I(P^g) = \bigcap_{g\in G} I(P^g)FG$ is a nilpotent ideal of FG

(ii) $\bigcap_{g\in G} FG\cdot I(P^g) = \{ \sum_{x\in G} a_x x \mid \sum_{y\in S} a_{xy} = 0$ for all $x \in G$ and all $S \in \mathrm{Syl}_p(G)\}$

$= \{ \sum_{x\in G} a_x x \mid \sum_{y\in S} a_{yx} = 0$ for all $x \in G$ and all $S \in \mathrm{Syl}_p(G)\}$

Proof. (i) Since $I(P)$ is a nilpotent ideal of FP, the required assertion is a consequence of Lemma 4.1.

(ii) Put $X = \bigcap_{g\in G} FG\cdot I(P^g)$ and $Y = \bigcap_{g\in G} I(P^g)FG$. Then, by (i),

$$X = Y = \bigcap_{S\in \mathrm{Syl}_p(G)} FG\cdot I(S) = \bigcap_{S\in \mathrm{Syl}_p(G)} I(S)FG$$

and so the desired conclusion follows by virtue of Lemma 4.2. ■

Following Motose and Ninomiya (1975b), we say that G is *p-radical* if for any subgroup H of p'-index in G, $J(FG) \subseteq J(FH)FG$. Note that, by Corollary 2.3, $J(FG) \subseteq J(FH)FG$ is equivalent to $J(FG) \subseteq FG\cdot J(FH)$, and that, by Lemma 3.12, and Corollary 2.3, if $J(FG) \subseteq J(FH)FG$, then H is of p'-index.

4.4. THEOREM. (Motose and Ninomiya (1975b)). The following conditions are equivalent:

(i) G is p-radical

(ii) $J(FG) \subseteq FG\cdot I(P)$ for some (and hence all) $P \in \mathrm{Syl}_p(G)$

(iii) $J(FG) = \bigcap_{S\in \mathrm{Syl}_p(G)} FG\cdot I(S)$

(iv) $J(FG) = \bigcap_{S\in \mathrm{Syl}_p(G)} I(S)\cdot FG$

(v) $\mathrm{Soc}(FG) = \sum_{S\in \mathrm{Syl}_p(G)} FG\cdot (S^+)$

(vi) $\mathrm{Soc}(FG) = \sum_{S\in \mathrm{Syl}_p(G)} (S^+)\cdot FG$

(vii) $J(FG) = \{ \sum_{g\in G} x_g g \mid \sum_{s\in S} x_{gs} = 0$ for all $g \in G$ and all $S \in \mathrm{Syl}_p(G)\}$

(viii) $J(FG) = \{ \sum_{g\in G} x_g g \mid \sum_{s\in S} x_{sg} = 0$ for all $g \in G$ and all $S \in \mathrm{Syl}_p(G)\}$

Proof. The equivalence of (iii), (iv), (vii) and (viii) follows from Lemma 4.3.

(i) ⇒ (ii): If $P \in \mathrm{Syl}_p(G)$ then P is of p'-index, so by hypothesis $J(FG) \subseteq FG\cdot I(P)$.

(ii) ⇒ (iii): If $J(FG) \subseteq FG\cdot I(P)$ for some $P \in \mathrm{Syl}_p(G)$, then

$$J(FG) = J(FG)^g \subseteq FG\cdot I(P)^g = FG\cdot I(P^g) \qquad \text{for all } g \in G$$

and hence

$$J(FG) \subseteq \bigcap_{S\in \mathrm{Syl}_p(G)} FG\cdot I(S)$$

Thus, $J(FG) = \bigcap_{S\in \mathrm{Syl}_p(G)} FG\cdot I(S)$, by virtue of Lemma 4.3(i).

(iii) $\Rightarrow$ (i): Let H be a subgroup of G of p'-index. Then there exists $S \in \mathrm{Syl}_p(G)$ such that $S \subseteq H$. By hypothesis, $J(FG) \subseteq FG\cdot I(S)$, so V^G is completely reducible for every irreducible FS-module V, by Corollary 2.3. Since $V^G \cong (V^H)^G$, it follows from Corollary 2.4.5 that V^H is completely reducible, for every irreducible FS-module V. Hence, by Corollary 2.3, $J(FH) \subseteq FH\cdot I(S)$ and therefore, by the implication (ii) $\Rightarrow$ (iii),

$$J(FH) = \bigcap_{S\in \mathrm{Syl}_p(H)} FH\cdot I(S)$$

Thus

$$\begin{aligned} FG\cdot J(FH) &\supseteq \bigcap_{S\in \mathrm{Syl}_p(H)} FG\cdot FH\cdot I(S) \\ &\supseteq \bigcap_{S\in \mathrm{Syl}_p(G)} FG\cdot I(S) \\ &= J(FG), \end{aligned}$$

as required.

(iv) $\Leftrightarrow$ (v) and (iii) $\Leftrightarrow$ (vi): Each of the statements follows from the other by taking annihilators. ∎

As an application of Theorem 4.4, we derive the following two results.

4.5. COROLLARY. Let H be a subgroup of G of p'-index. If G is p-radical, then so is H.

Proof. Let S be a Sylow p-subgroup of G with $S \subseteq H$. Then, by the proof of the implication (iii) $\Rightarrow$ (i) of Theorem 4.4,

$$J(FG) \subseteq FG\cdot I(S) \Rightarrow J(FH) \subseteq FH\cdot I(S)$$

Hence, by Theorem 4.4, if G is p-radical, then so is H. ∎

4.6. COROLLARY. Let P be a Sylow p-subgroup of G and let 1_P be the trivial FP-module. Then G is p-radical if and only if $(1_P)^G$ is completely reducible.

Proof. If $V = FP/I(P)$, then $V \cong 1_P$ and, by Proposition 2.4.3(i),

$$V^G \cong FG/FG\cdot I(P)$$

Hence V^G is completely reducible if and only if $J(FG) \subseteq FG\cdot I(P)$. Since the latter, by Theorem 4.4, is equivalent to G being p-radical, the result follows. ∎

The next result, with the exception of the first property of (i), is due to Khatri (1973).

4.7. THEOREM. Let N be a normal subgroup of G.

(i) If G is p-radical, then so are N and G/N.

(ii) If N is a p-group and G/N is p-radical, then G is p-radical

(iii) If G/N is a p'-group, then G is p-radical if and only if N is p-radical.

Proof. (i) Assume that G is p-radical and let $P \in \mathrm{Syl}_p(G)$. We may choose a Sylow p-subgroup Q of N with $Q \subseteq P$. By Proposition 3.1.8(i), $J(FN) \subseteq J(FG)$ and so, by Theorem 4.4,

$$J(FN) \subseteq FN \cap FG\cdot I(P)$$

If $x \in FN \cap FG\cdot I(P)$, then by Lemma 4.2(i),

$$0 = xP^+ = x(Q^+ + (P-Q)^+) = xQ^+ + x(P-Q)^+$$

But xQ^+ and $x(P-Q)^+$ have disjoint supports, so $xQ^+ = 0$ and therefore, by Lemma 4.2(i), $x \in FN\cdot I(Q)$. Thus $J(FN) \subseteq FN\cdot I(Q)$ and so, by Theorem 4.4, N is p-radical.

If $H = PN$, then H is of p'-index in G, so by hypothesis

$$J(FG) \subseteq FG\cdot J(FH)$$

Thus, by Corollary 2.3, V^G is completely reducible for all irreducible FH-modules V. In particular, $(1_H)^G$ is completely reducible. Hence $(1_{H/N})^{G/N}$ is completely reducible. But H/N is a Sylow p-subgroup of G/N, so G/N is p-radical by Corollary 4.6.

(ii) Assume that N is a normal p-subgroup of G such that G/N is p-radical.

Since $N \subseteq P$, we have $H/N = P/N$. By Corollary 4.6, $(1_{P/N})^{G/N}$ is completely reducible. Hence $(1_P)^G$ is completely reducible and therefore, by Corollary 4.6, G is p-radical.

(iii) Assume that G/N is a p'-group. If G is p-radical, then so is N, by Corollary 4.5. Conversely, suppose that N is p-radical. Since G/N is a p'-group, there is a $P \in \mathrm{Syl}_p(G)$ with $P \subseteq N$. Owing to Theorem 4.4, $J(FN) \subseteq FN\cdot I(P)$. On the other hand, by Proposition 3.1.8, $J(FG) = FG\cdot J(FN)$. Thus

$$J(FG) \subseteq FG\cdot FN\cdot I(P) = FG\cdot I(P)$$

and so G is p-radical, by virtue of Theorem 4.4. ∎

Turning our attention to Frobenius groups, we now prove the following result.

4.8. THEOREM. (Khatri (1973), Motose and Ninomiya (1975b)). Let G be a Frobenius group with kernel N and complement H.

(i) If p divides the order of N, then G is p-radical

(ii) If p divides the order of H, then G is p-radical if and only if so is H.

Proof. (i) By Lemma 3.7.1(iv), N is nilpotent and so N is p-radical, by Theorem 4.7(i). But, by Lemma 3.7.1(iii), $G/N \cong H$ is a p'-group. Hence G is p-radical, by Theorem 4.7(ii).

(ii) By Theorem 4.7(i), we need only verify that if H is p-radical, then so is G. If P is a Sylow p-subgroup of H, then $J(FH) \subseteq FH\cdot I(P)$ by Theorem 4.4. By Theorem 3.7.7,

$$J(FG) = J(FH)(N^+)$$

and therefore

$$J(FG) \subseteq FH\cdot I(P)(N^+) \subseteq FG\cdot I(P)$$

Since $P \in \mathrm{Syl}_p(G)$ we conclude, from Theorem 4.4, that G is p-radical. ∎

4.9. THEOREM. (Khatri (1973)). If $G = G_1 \times G_2$ then G is p-radical if and only if so are G_1 and G_2

Proof. Assume that G_1 and G_2 are p-radical. Then, by Theorem 4.4,

$$J(FG_i) \subseteq FG_i \cdot I(P_i)$$

where $P_i \in \mathrm{Syl}_p(G_i)$, $i = 1,2$. Since, by Theorem 3.2.3,

$$J(FG) = J(FG_1)FG_2 + J(FG_2)FG_1,$$

we have

$$J(FG) \subseteq FG \cdot I(P_1) + FG \cdot I(P_2)$$

If we put $P = P_1 \times P_2$, then $I(P_i) \subseteq I(P)$ and so by the above

$$J(FG) \subseteq FG \cdot I(P)$$

Thus G is p-radical, by Theorem 4.4. The converse being a consequence of Theorem 4.7(ii), the result follows. ■

4.10. LEMMA. Let E be a field extension of F. Then G is p-radical with respect to E if and only if it is p-radical with respect to F.

Proof. If $P \in \mathrm{Syl}_p(N)$, then by Theorem 4.4, it suffices to verify that $J(FG) \subseteq FG \cdot I(P)$ if and only if $J(EG) \subseteq EG \cdot I(P)$. If $J(FG) \subseteq FG \cdot I(P)$, then $J(EG) \subseteq EG \cdot I(P)$ by Corollary 3.1.18. Conversely, assume that $J(EG) \subseteq EG \cdot I(P)$. Then $J(FG) \subseteq EG \cdot I(P) \cap FG = FG \cdot I(P)$ by Lemma 4.2(i). So the Lemma is true. ■

The next result provides another class of p-radical groups.

4.11. Theorem. (Tshushima (1986)). Let G be a p-nilpotent group with abelian p-complement N. Then G is p-radical.

Proof. Owing to Lemma 4.10, we may harmlessly assume that F is algebraically closed. Let $P \in \mathrm{Syl}_p(G)$ and let

$$1 = e_1 + e_2 + \ldots + e_n$$

be a decomposition of 1 as a sum of orthogonal primitive idempotents of FN. Since F is a splitting field for FN, each e_iFN is one-dimensional. Hence $e_iFG \cong (e_iFN)^G$ is of dimension $|P|$. Now $G = NP$, so $FG = (FN)(FP)$ and therefore

$$e_iFG = (e_iFN)(FP) = e_iFP$$

It follows that the map

$$\begin{cases} FP \longrightarrow e_i FG \\ x \longmapsto e_i x \end{cases}$$

is a surjective homomorphism of right FP-modules of the same F-dimension. Thus

$$FP \cong e_i FG \qquad \text{as } FP\text{-modules} \quad (1 \leqslant i \leqslant n)$$

and so $e_i FG \cdot I(P)$ is a unique maximal FP-submodule of $e_i FG$. But $e_i J(FG)$ is a proper FP-submodule of $e_i FG$, hence

$$e_i FG \cdot I(P) \supseteq e_i J(FG) \qquad (1 \leqslant i \leqslant n)$$

and therefore $FG \cdot I(P) \supseteq J(FG)$. The desired conclusion now follows by virtue of Theorem 4.4. ■

The result above need not be true for an arbitrary p-nilpotent group G. The following example is due to Saksonov (1971) (see also Motose and Ninomiya (1975b)). Let $p = 3$ and let F be an arbitrary field of characteristic 3. Let G be the group of order 24 defined by

$$G = \langle x,y,z \mid x^4 = 1,\ x^2 = y^2, yxy^{-1} = x^{-1}, z^3 = 1,\ zxz^{-1} = y, zyz^{-1} = xy \rangle$$

Let C_g be the class sum of the conjugacy class containing $g \in G$. Then we have

$$c_x = x + x^3 + y + xy + x^2y + x^3y$$

$$c_z = z + zx^2y + zx^3 + zx^3y$$

$$c_z^3 = 1 - c_x$$

$$c_x^2 = c_x$$

and therefore

$$(c_x + c_z - 1)^3 = c_x^3 + c_z^3 - 1 = 0$$

Hence

$$c_x + c_z - 1 = x+x^3+y+xy+x^2y+x^3y+z+zx^2y+zx^3+zx^3y - 1$$

is a central nilpotent element of FG and so $c_x + c_z - 1 \in J(FG)$. However, $c_x + c_z - 1$ does not satisfy condition (viii) of Theorem 4.4, since the sum of the coefficients of $c_x + c_z - 1$ over the coset $\{x, zx, z^2x\}$ of $\langle z \rangle$ is $1 \neq 0$.

Thus G is not p-radical.

Remark. According to Tshushima (1986), a p-nilpotent group G is p-radical if and only if

$$[O_{p'}(G), D] \cap C_{O_{p'}(G)}(D) = 1$$

for any p-subgroup D of G.

Let U and V be any FG-modules. Recall the *intertwinning number* for U and V, written $i(U,V)$, is defined to be

$$i(U,V) = \dim_F \operatorname{Hom}_{FG}(U,V)$$

Since

$$\operatorname{Hom}_{FG}(U \oplus V, W) \cong \operatorname{Hom}_{FG}(U,W) \oplus \operatorname{Hom}_{FG}(V,W)$$

we have

$$i(U \oplus V, W) = i(U,W) + i(V,W)$$

for any FG-modules U,V and W. Similarly

$$i(U, V \oplus W) = i(U,V) + i(U,W)$$

and therefore

$$i\left(\bigoplus_{n=1}^{r} U_n, \bigoplus_{m=1}^{k} V_m\right) = \sum_{n=1}^{r} \sum_{m=1}^{k} i(U_n, V_m) \tag{1}$$

For the rest of this section, E denotes the algebraic closure of F and $\{V_1, \ldots, V_r\}$ a full set of nonisomorphic irreducible EG-modules. Given an EG-module M, we write

$$M \approx \sum_{k=1}^{r} a_k V_k \qquad (a_k \geqslant 0)$$

if V_k appears a_k times as a composition factor of M. Note that if V is an irreducible EG-module, then $i(V,V) = 1$. Hence if M is completely reducible, then by (1), $i(M,M) = \sum_{k=1}^{r} a_k^2$. With this information at our disposal, we now prove

4.12. LEMMA. Let $M \approx \sum_{k=1}^{r} a_k V_k$ and $N \approx \sum_{k=1}^{r} b_k V_k$ be two EG-modules.

(i) $i(M,N) \leqslant \sum_{k=1}^{r} a_k b_k$

(ii) M is completely reducible if and only if $i(M,M) = \sum_{k=1}^{r} a_k^2$

Proof. (i) We argue by induction on the value $\sum_{k=1}^{r} a_k$. If $\sum_{k=1}^{r} a_k = 1$, then M is irreducible and so $M \cong V_j$ for some $j \in \{1,\ldots,r\}$. Hence

$$\mathrm{Hom}_{EG}(M,N) = \mathrm{Hom}_{EG}(M,\mathrm{Soc}N)$$

and therefore, by (1),

$$i(M,N) = i(M,\mathrm{Soc}N) \leqslant b_j = \sum_{k=1}^{r} a_k b_k$$

Now assume that M_0 is a submodule of M. By looking at the exact sequence

$$0 \longrightarrow \mathrm{Hom}_{EG}(M/M_0,N) \longrightarrow \mathrm{Hom}_{EG}(M,N) \longrightarrow \mathrm{Hom}_{EG}(M_0,N)$$

we deduce that

$$i(M/M_0,N) + i(M_0,N) \geqslant i(M,N) \tag{2}$$

If M_0 is a nontrivial submodule of M with $M/M_0 \approx \sum_{k=1}^{r} a_k' V_k$ and $M_0 \approx \sum_{k=1}^{r} a_k'' V$, then by the induction hypothesis

$$i(M/M_0,N) \leqslant \sum_{k=1}^{r} a_k' b_k \tag{3}$$

and

$$i(M_0,N) \leqslant \sum_{k=1}^{r} a_k'' b_k \tag{4}$$

But $M \approx \sum_{k=1}^{r} (a_k' + a_k'') V_k$, so it follows from (2), (3) and (4) that

$$\begin{aligned}\sum_{k=1}^{r} a_k b_k &= \sum_{k=1}^{r} a_k' b_k + \sum_{k=1}^{r} a_k'' b_k \\ &\geqslant i(M/M_0,N) + i(M_0,N) \\ &\geqslant i(M,N)\end{aligned}$$

as required.

(iii) If M is completely reducible, then as has been observed earlier $i(M,M) = \sum_{k=1}^{r} a_k^2$. Conversely, assume that $i(M,M) = \sum_{k=1}^{r} a_k^2$ and write

$$M/\mathrm{Soc}M \approx \sum_{k=1}^{r} a_k' V_k, \quad \mathrm{Soc}M \approx \sum_{k=1}^{r} a_k'' V_k$$

Then, by (i), we have

$$i(M/\mathrm{Soc}M,M) \leqslant \sum_{k=1}^{r} a_k' a_k$$

$$i(\mathrm{Soc}M,M) \leqslant \sum_{k=1}^{r} a_k'' a_k$$

and so applying (2), we derive

$$i(M,M) = \sum_{k=1}^{r} a_k^2 = \sum_{k=1}^{r} a_k'a_k + \sum_{k=1}^{r} a_k''a_k \geq i(M/\mathrm{Soc}M,M) + i(\mathrm{Soc}M,M) \geq i(M,M)$$

Thus

$$i(M/\mathrm{Soc}M,M) = \sum_{k=1}^{r} a_k'a_k \tag{5}$$

and

$$i(\mathrm{Soc}M,M) = \sum_{k=1}^{r} a_k''a_k \tag{6}$$

Since

$$i(\mathrm{Soc}M,M) = i(\mathrm{Soc}M,\mathrm{Soc}M) = \sum_{k=1}^{r} (a_k'')^2$$

it follows from (6) that $a_k'' = a_k$ if $a_k'' \neq 0$. Now assume that $a_k'' = 0$ and $a_k \neq 0$. Then $a_k = a_k' \neq 0$ and so by (5), $i(M/\mathrm{Soc}M,M) \neq 0$. Hence there exists a submodule L with $\mathrm{Soc}M \subseteq L \subset M$ such that M/L is isomorphic to a submodule of M. The latter implies that there exists $j \in \{1,\ldots,r\}$ such that $a_j' \neq 0$ and $a_j'' \neq 0$. But if $a_j'' \neq 0$, then $a_j = a_j''$ and hence $a_j' = 0$, a contradiction. Thus $a_k = a_k''$ for all $k \in \{1,\ldots,r\}$ and therefore $M = \mathrm{Soc}M$, as required. ∎

4.13. LEMMA. Let $U_k = P(V_k)$ be a projective cover of V_k, $1 \leq k \leq r$, let $P \in \mathrm{Syl}_p(G)$ and put $u_k = \dim_E U_k$. Then

$$(1_P)^G \approx \sum_{k=1}^{r} (u_k/|P|)V_k$$

Proof. If $(1_P)^G \approx \sum_{k=1}^{r} a_kV_k$, then by Lemma 1.10.12,

$$a_k = i(U_k,(1_P)^G) \qquad (1 \leq k \leq r)$$

On the other hand, by Theorem 2.4.9,

$$\mathrm{Hom}_{EG}(U_k,(1_P)^G) \cong \mathrm{Hom}_{EP}((U_k)_P, 1_P) \qquad (1 \leq k \leq r)$$

and so $a_k = i((U_k)_P,1_P)$. Now $(U_k)_P$ is EP-free of rank $u_k/|P|$, so

$$a_k = (u_k/|P|)i(EP,1_P) = u_k/|P| \qquad (1 \leq k \leq r)$$

as required. ∎

We are now ready to provide another characterization of p-radical groups.

4.14. THEOREM. (Saksonov (1971)). Let E be the algebraic closure of F, let

$\{V_1,\dots,V_r\}$ be a full set of nonisomorphic irreducible EG-modules and, for each $k\in\{1,\dots,r\}$, let $u_k=\dim_E P(V_k)$. Let $P\in \mathrm{Syl}_p(G)$ and let d_P be the number of double (P,P)-cosets in G. Then

$$\sum_{k=1}^{r}(u_k/|P|)^2\geqslant d_P$$

with equality if and only if G is p-radical.

Proof. (Tsushima (1986)). By Lemma 4.10, we may harmlessly assume that $E=F$. Then, by Lemmas 4.12(ii) and 4.13, $(1_P)^G$ is completely reducible if and only if

$$\sum_{k=1}^{r}(u_k/|P|)^2 = i(1_P^G,\ 1_P^G)$$

Let T be a set of double coset representatives for (P,P) in G. In view of the above equality, Lemma 4.12(i) and Corollary 4.6, we are left to verify that

$$|T| = i(1_P^G,\ 1_P^G) \tag{7}$$

By Theorem 2.4.9 and Mackey decomposition (Theorem 2.6.1), we have

$$\begin{aligned}\mathrm{Hom}_{FG}(1_P^G,\ 1_P^G) &\cong \mathrm{Hom}_{FP}((1_P^G)_P,\ 1_P)\\ &\cong \mathrm{Hom}_{FP}(\bigoplus_{t\in T}(1_{tPt^{-1}\cap P})^P,\ 1_P)\\ &\cong \bigoplus_{t\in T}\mathrm{Hom}_{FP}((1_{tPt^{-1}\cap P})^P,\ 1_P)\\ &\cong \bigoplus_{t\in T}\mathrm{Hom}_{F(tPt^{-1}\cap P)}(1_{tPt^{-1}\cap P},1_{tPt^{-1}\cap P})\end{aligned}$$

and thus $i(1_P^G,1_P^G)=|T|$, proving (7). This completes the proof of the theorem.∎

4.15. LEMMA. Let A be a finite dimensional algebra over a field and let V be a (finitely generated) A-module. Then

$$I=\{\phi\in \mathrm{End}_A(V)\,|\,\phi(V)\subseteq J(A)V\}$$

is a nilpotent ideal of $\mathrm{End}_A(V)$.

Proof. It is clear that I is a subspace of the finite-dimensional algebra $\mathrm{End}_A(V)$. If $\phi\in I$ and $\psi\in\mathrm{End}_A(V)$, then

$$(\phi\psi)(V)=\phi(\psi(V))\subseteq\phi(V)\subseteq J(A)V$$

and

$$(\psi\phi)(V) = \psi(\phi(V)) \subseteq \psi(J(A)V) = J(A)\psi(V) \subseteq J(A)V$$

so that $\phi\psi, \psi\phi \in I$. Thus I is an ideal of $\mathrm{End}_A(V)$. Let $n \geqslant 1$ be such that $J(A)^n = 0$. Then, for all $\phi \in I$

$$\phi^n(V) \subseteq J(A)^n V = 0$$

and so $\phi^n = 0$. Thus I is nilpotent, by virtue of Proposition 1.10.5. ∎

The next result is extracted from Tsushima (1986).

14.16. THEOREM. Let $P \in \mathrm{Syl}_p(G)$ and let d_P be the number of double (P,P)-cosets in G. Then the algebra $\mathrm{End}_{FG}[(1_P)^G]$ is of F-dimension d_P and $\mathrm{End}_{FG}[(1_P)^G]$ is semisimple if and only if G is p-radical.

Proof. Put $V = (1_P)^G$ and $A = \mathrm{End}_{FG}(V)$. Since in the proof of (7) we did not use the assumption that F is algebraically closed, we have

$$\dim_F A = i(1_P^G, 1_P^G) = d_P$$

Owing to Corollary 4.6, we are left to verify that V is completely reducible if and only if A is semisimple.

If V is completely reducible, then A is semisimple by Corollary 1.3.8 and Proposition 1.5.2. Conversely, assume that A is semisimple. Let $\{V_1,\ldots,V_r\}$ be a full set of nonisomorphic irreducible FG-modules. Since for all $k \in \{1,\ldots,r\}$, 1_P is an irreducible constituent of $\mathrm{Soc}((V_k)_P)$, it follows from Nakayama reciprocity (Theorem 2.4.10) that each V_k is an irreducible constituent of $V/J(FG)V$. Hence

$$V/J(FG)V \cong \bigoplus_{k=1}^{r} a_k V_k \tag{8}$$

for some positive integers a_k.

Assume by way of contradiction that $J(FG)V \neq 0$. Then $J(FG)V$ has an irreducible submodule W which is a copy of V_j for some $j \in \{1,\ldots,r\}$. Hence, by (8), there is a surjective FG-homomorphism

$$\pi : V/J(FG)V \longrightarrow W$$

But then the composite map

$$\phi : V \longrightarrow V/J(FG)V \xrightarrow{\pi} W \subseteq J(FG)V$$

is a nonzero element of A such that $\phi(V) \subseteq J(FG)V$. Invoking Lemma 4.15, we conclude that A is not semisimple, a contradiction. ■

As an application of the result above, we now prove the following group-theoretic fact.

14.17. COROLLARY. (Saksonov (1971)). Let r be the number of p-regular classes of G, let $P \in \mathrm{Syl}_p(G)$ and let d_P be the number of double (P,P)-cosets in G. Then

$$d_P \geqslant r$$

Proof. Let F be an algebraically closed field of characteristic p. Then, by Theorem 2.3.2, there exists exactly r, say $V_1,\dots,V_r$, of nonisomorphic irreducible FG-modules. Keeping the notation of Theorem 14.16, write

$$\mathrm{Soc}V \cong \bigoplus_{k=1}^{r} b_k V_k \tag{9}$$

Then we have

$$b_k = i(V_k, \mathrm{Soc}V) = i(V_k, V) = i((N_k)_P, 1_P) \neq 0$$

for all $k \in \{1,\dots,r\}$. Since

$$\mathrm{Hom}_{FG}(V/J(FG)V,\ \mathrm{Soc}V) \subseteq A \tag{10}$$

and, by (8) and (9),

$$i(V/J(FG)V,\ \mathrm{Soc}V) = \sum_{k=1}^{r} a_k b_k \tag{11}$$

we conclude that

$$d_P = \dim_F A \geqslant \sum_{k=1}^{r} a_k b_k \geqslant r\ , \tag{12}$$

as required.

14.18. COROLLARY. (Saksonov (1971)). Let r be the number of p-regular classes of G, let $P \in \mathrm{Syl}_p(G)$ and let d_P be the number of double (P,P)-cosets in G. Let E be the algebraic closure of F and let $\{V_1,\dots,V_r\}$ be a full set of nonisomorphic irreducible EG-modules. Then the following conditions are equivalent:

(i) $d_P = r$

(ii) $(1_P)^G \cong V_1 \oplus V_2 \oplus \ldots \oplus V_r$

(iii) G is p-radical and $\dim_E P(V_k) = |P|$ for all $k \in \{1,2,\ldots,r\}$.

Proof. (i) $\Rightarrow$ (ii): Put $V = (1_P)^G$ and $A = \mathrm{End}_{FG}(V)$. It follows from (10), (11) and (12) that $A = \mathrm{Hom}_{FG}(V/J(FG)V, \mathrm{Soc}V)$ and

$$a_k = b_k = 1 \qquad \text{for all } k \in \{1,2,\ldots,r\}$$

It follows from (8) and (9) that $V/J(FG)V \cong \mathrm{Soc}V$ and hence $A \cong \mathrm{End}_{FG}(\mathrm{Soc}V)$. But $\mathrm{Soc}V$ is completely reducible, so A is semisimple and therefore, by Theorem 14.16, G is p-radical. Hence, by Corollary 4.6, $V = \mathrm{Soc}V$ and the required assertion follows by virtue of (9).

(ii) $\Rightarrow$ (iii): By Corollary 4.6, G is p-radical, while by Lemma 4.13, $\dim_E P(V_k) = |P|$ for all $k \in \{1,2,\ldots,r\}$.

(iii) $\Rightarrow$ (i): If $u_k = \dim_E P(V_k)$, then $(u_k/|P|) = 1$ for all $k \in \{1,\ldots,r\}$. Since G is p-radical,

$$d_P = \sum_{k=1}^{r} (u_k/|P|)^2 = r$$

by virtue of Theorem 4.14. ■

Remark. Owing to Saksonov (1971), all groups satisfying the equivalent conditions of the above corollary are solvable.

14.19. LEMMA. Let $A = M_n(F)$ and let $\phi,\psi \in A^* = \mathrm{Hom}_F(A,F)$ be such that (A,ϕ) and (A,ψ) are symmetric algebras. Then there exists a nonzero λ in F such that

$$\phi(a) = \lambda\psi(a) \qquad \text{for all } a \in A$$

Proof. Owing to Lemma 3.3.2(i), $f : A \longrightarrow A^*$ and $g : A \longrightarrow A^*$ where

$$f(a)(x) = \phi(xa) \quad \text{and} \quad g(a)(x) = \psi(xa) \qquad (x,a \in A)$$

are isomorphisms of (A,A)-bimodules. Hence $g^{-1}f : A \longrightarrow A$ is an (A,A)-bimodule automorphism. Let $A°$ be the opposite ring to A so that A is an $A \otimes_F A°$ module via

$$(a_1 \otimes a_2)a = a_1 a a_2 \qquad (a_1, a \in A, a_2 \in A°)$$

Then $g^{-1}f$ is in fact an automorphism of $A \underset{F}{\otimes} A°$-modules. Since $A \underset{F}{\otimes} A°$ is a simple algebra with F as a splitting field and A is an irreducible $A \underset{F}{\otimes} A°$-module, we conclude that there exists $0 \neq \lambda \in F$ such that $(g^{-1}f)(a) = \lambda a$ for all $a \in A$. Hence $f(a) = \lambda g(a)$ for all $a \in A$. In particular, for all $a \in A$ we have

$$\phi(a) = f(a)(1) = \lambda g(a)(1) = \lambda\psi(a)$$

as required. ■

14.20. LEMMA. Let $e = \sum_{g\in G} e_g g$ be an idempotent of FG. Then, for any p-singular $C \in C\ell(G)$, $\sum_{g\in C} e_g = tr(eC^+) = 0$. Furthermore, if $eFGe$ is simple, then $eC^+ = 0$ for any p-singular $C \in C\ell(G)$.

Proof. Choose a positive integer n such that for all $g \in G$, g^{p^n} is a p'-element. Owing to Lemma 2.2.1, we have

$$e = e^{p^n} \equiv \sum_{g\in G} e_g^{p^n} g^{p^n} \pmod{[FG,FG]}$$

Our choice of n ensures that for all $g \in C$, e_g is the coefficient of g in

$$e - \sum_{g\in G} e_g^{p^n} g^{p^n} \in [FG,FG]$$

Therefore, by Lemma 2.2.2, $\sum_{g\in C} e_g = 0$. Since C^{-1} is p-singular, we also have

$$tr(eC^+) = \sum_{g\in C^{-1}} e_g = 0$$

Now assume that the algebra $eFGe$ is simple, say $eFGe \cong M_n(D)$ for a suitable division ring D. If E is a maximal subfield of D, then

$$eEGe \cong E \underset{F}{\otimes} M_n(D) \cong M_n(E)$$

Hence we may assume that $eFGe = \sum_{1\leqslant i,j\leqslant n} Fe_{ij}$ with matrix units e_{ij}. The map

$$\begin{cases} eFGe \xrightarrow{\psi} F \\ \sum_{i,j} \lambda_{ij} e_{ij} \longmapsto \sum_i \lambda_{ii} \end{cases}$$

is such that $(eFGe,\psi)$ is a symmetric algebra. On the other hand, by Lemma 3.3.11, $(eFGe,tr)$ is also a symmetric algebra. Thus, by Lemma 14.19, there exists $\lambda \in F$ such that $c = \lambda e$, $tr(ac) = \psi(a)$ for all $a \in eFGe$. Since $eC^+ = eC^+e \in Z(eFGe)$, we have $eC^+ = \mu e$ for some $\mu \in F$. It follows that

$$\mu = \psi(\mu e e_{11}) = \psi(eC^+ e_{11}) = tr(C^+ e_{11} c)$$
$$= tr(\lambda C^+ e_{11}) = \lambda tr(e_{11} C^+)$$

But e_{11} is an idempotent of FG, so $tr(e_{11}C^+) = 0$ and thus $eC^+ = 0$, as required. ■

We are now ready to prove

14.21. PROPOSITION. (Tsushima (1971b)). Let c be the sum of all p-elements of G including 1. Then

$$c^2 = \sum_{i=1}^{n} e_i$$

where the e_i are all block idempotents of FG of defect zero.

Proof. Let e be a block idempotent of FG whose defect is not zero. Then, by Theorem 3.6.4(ii), $c^2 e = 0$. Now assume that e is a block idempotent of FG of defect zero. By Theorem 3.6.4(i), FGe is a simple F-algebra. Hence, by Lemma 14.20, $ce = e$ and so $c^2 e = e$, proving the assertion. ■

14.22. LEMMA. Let G be a permutation group on a set $X = \{1,2,\ldots,n\}$, and let $G_{i_1 i_2 \ldots i_k}$ be the stabilizer of the points $i_1, i_2, \ldots, i_k$ in X. Then the following conditions are equivalent:

(i) For every $i,j \in X$, p divides the order of G_{ij}

(ii) $(\sum_{k=1}^{n} FG\cdot G_k^+)^2 = 0$

Proof. (i) ⇒ (ii): Fix $i,j \in \{1,2,\ldots,n\}$ and let

$$G_j = \bigcup_{k=1}^{s} G_{ij} x_k$$

be a coset decomposition of G_j over G_{ij}. By hypothesis, p divides the order of G_{ij} and so

$$G_i^+ G_j^+ = G_i^+ G_{ij}^+ (x_1 + \ldots + x_s)$$
$$= |G_{ij}| G_i^+ (x_1 + \ldots + x_s)$$
$$= 0$$

Given $g \in G$, we have $gG_j g^{-1} = G_{g(j)}$, so $gG_j^+ g^{-1} = G_{g(j)}^+$ and thus

$$G_i^+ \, g \, G_j^+ = G_i^+ G_{g(j)}^+ g = 0$$

for all $g \in G$ and all $i,j \in \{1,\dots,n\}$. Therefore (ii) holds.

(ii) $\Rightarrow$ (i): If $\left(\sum_{k=1}^{n} FG\cdot G_k^+\right)^2 = 0$, then given $i,j \in \{1,2,\dots,n\}$,

$$G_i^+ G_j^+ = |G_{ij}|G_i^+(x_1 + \dots + x_s) = 0$$

Since $G_i x_m \cap G_i x_n = \emptyset$ for $m \neq n$, p divides the order of G_{ij}, as required. ■

With this information at our disposal, we now establish the following result.

14.23. THEOREM. (Motose and Ninomiya (1975b)). Assume that a group G is both p-solvable and p-radical, and let F be an arbitrary field of characteristic p. Then FG has no blocks of defect zero if and only if each pair of Sylow p-subgroups of G has a nontrivial intersection.

Proof. Let c be the sum of all p-elements of G including 1. Then, by Proposition 14.21, c^2 is equal to the sum of all block idempotents of FG of defect zero. Note also that, since G is p-solvable, $\mathrm{Soc}(FG) = FGc$ by virtue of Corollary 3.5.7. On the other hand, because G is p-radical, we have

$$\mathrm{Soc}(FG) = \sum_{S\in \mathrm{Syl}_p(G)} FG\cdot(S^+)$$

by Theorem 4.4. Thus FG has no blocks of defect zero if and only if

$$\left[\sum_{S\in \mathrm{Syl}_p(G)} FG\cdot(S^+)\right]^2 = 0 \tag{13}$$

Fix $P \in \mathrm{Syl}_p(G)$ and let $G = \bigcup_{i=1}^{n} Pg_i$, $g_1 = 1$, be a coset decomposition of G over P. Regarding G as a permutation group on the set $X = \{P, Pg_2, \dots, Pg_n\}$, we see that the stabilizer of Pg_i is $g_i^{-1}Pg_i$. Invoking Lemma 14.22, we infer that (13) holds if and only if $g_i^{-1}Pg_i \cap g_j^{-1}Pg_j \neq 1$ for all $i,j \in \{1,2,\dots,n\}$. So the theorem is true. ■

Remark. According to Tsushima (1986), Okuyama has proved that any p-radical group is p-solvable.

As has been previously illustrated, there is a p-nilpotent group which is not supersolvable and not p-radical. It is therefore appropriate to ask: Is any supersolvable group p-radical? The negative answer to the above question was provided by Saksonov (1971) who exhibited a supersolvable group of order 108 which is not p-radical. A simpler example, due to Passman, is found in Bedi

(1979) and will be considered below.

Let G be a split extension of $M(3)$ by a group of automorphisms of order 2:

$$G = \langle a,b,c,d \mid a^3=b^3=c^3=d^2=1, b^{-1}ab=ac, a^{-1}ca=c, b^{-1}cb=c, d^{-1}ad=a^2, d^{-1}bd=b^2, d^{-1}cd=c\rangle$$

Then G is of order 54 and is supersolvable since it has an invariant series

$$\langle 1\rangle \subset \langle c\rangle \subset \langle a\ c\rangle \subset \langle a,b,c\rangle \subset G$$

with cyclic factors. Let F be an arbitrary field of characteristic $p = 2$. By the definition of G, we have $G' = \langle a,b,c\rangle$ and thus

$$G = G'P \qquad \text{where} \quad P = \langle d\rangle$$

Since $c \in G - P$ and c is central, we have $N_G(P) \neq P$ and thus G is not a Frobenius group with P as a complement. Invoking Theorem 3.13.4, we conclude that

$$J(FG) \not\subseteq Z(FG)$$

It therefore suffices to show that

$$\bigcap_{g\in G} FG\cdot I(P^g) \qquad \text{is central}$$

To this end, put $I = \bigcap_{g\in G} FG\cdot I(P^g)$ and let X be the set of all elements of G of order 2. Given $x\in X$, there exists $g \in G$ such that $P^g = \langle x\rangle$ and hence

$$I \subseteq FG\cdot I(P^g) = FG(x-1) \tag{14}$$

Therefore $I(x-1) \subseteq FG(x-1)^2 = 0$, so $I(x-1) = 0$ for all $x \in X$. Since $\{d,ad,bd\} \subseteq X$ and since $G = \langle d,ad,bd\rangle$, we see that $G = \langle X\rangle$. Hence, by Proposition 2.1.10,

$$I(G) = \sum_{x\in X} (x-1)FG$$

Applying (14), we conclude that $I\cdot I(G) = 0$ and thus, by Lemma 3.11.1,

$$I \subseteq FG\cdot(G^+) = FG^+ \subseteq Z(FG)$$

as required. ∎

6
The Loewy length of projective modules

The aim of this chapter is to provide some general information pertaining to the Loewy length of projective modules. As a preparation for the proof of the main theorem (Theorem 4.6), we establish a number of important results of independent interest. These results are directly connected with the nilpotency index of $J(FG)$ to be examined in the next chapter.

1. PRELIMINARY RESULTS

Throughout this section, G denotes a finite group and F a field of characteristic $p > 0$. As usual, all modules are assumed to be finitely generated left modules and 1_G denotes the trivial FG-module. For any FG-module $V, L(V)$ is the *Loewy length* of V, i.e. the smallest integer n such that $J(FG)^n V = 0$. Finally omitting reference to p which is fixed in the following, we set $t(G) = L(FG)$, the nilpotency index of $J(FG)$.

1.1. LEMMA. Let R be an artinian ring.

(i) If R is semisimple and I is a left ideal of R, then $I = Re$ for some idempotent e of R. In particular, $I^n = I$ for all $n \geqslant 1$.

(ii) If R is semisimple, then so is R/I for any ideal I of R.

(iii) If $f : R \longrightarrow S$ is a surjective homomorphism of rings, then

$$J(S) = f(J(R))$$

(iv) For any ideal I of R,

$$J(R/I) = (J(R)+I)/I$$

Proof. (i) Assume that R is semisimple. Since R is artinian, ${}_R R$ is completely reducible, by Proposition 1.6.2. Hence ${}_R R = I \oplus J$ for some left ideal J of R. If $1 = e_1 + e_2$ with $e_1 \in I$ and $e_2 \in J$, then e_1 is an idempotent of R such that $I = Re_1$.

(ii) Let $J/I = J(R/I)$. Since R is artinian, so is R/I and therefore J/I is nilpotent, say $J^n \subseteq I$, for some $n \geqslant 1$. By (i), $J^n = J$ and hence $J/I = 0$, as required.

(iii) Since $f(J(R))$ is a nilpotent ideal of S, we certainly have $f(J(R)) \subseteq J(S)$. Hence, by Corollary 1.6.17(iv), it suffices to verify that $J(S/f(J(R))) = 0$. But $S/f(J(R))$ is isomorphic to a factor ring of $R/J(R)$ hence $S/f(J(R))$ is semisimple, by virtue of (ii).

(iv) Direct consequence of (iii). ■

1.2. COROLLARY. Let N be a normal subgroup of a group G. Then

$$J(F(G/N)) \cong (J(FG) + FG\cdot I(N))/FG\cdot I(N)$$

Proof. Let $f : FG \longrightarrow F(G/N)$ be the natural homomorphism. Then $\mathrm{Ker}f = FG\cdot I(N)$ and so the required assertion follows by virtue of Lemma 1.1(iv).■

Let R be an artinian ring, let V be an R-module and let $P(V)$ be a projective cover of V. Then, by definition of $P(V)$, there is an exact sequence

$$0 \longrightarrow \Omega(V) \longrightarrow P(V) \longrightarrow V \longrightarrow 0$$

where $\Omega(V)$ is a superfluous submodule of $P(V)$ and hence

$$\Omega(V) \subseteq J(R)P(V)$$

We shall refer to $\Omega(V)$ as the *Heller module* of V. As we shall see below, V determines the isomorphism class of $\Omega(V)$.

1.3. LEMMA. (Schanuel's Lemma). Let R be an artinian ring, let V, W_1, W_2 be R-modules and let P_1, P_2 be projective R-modules such that the sequences

$$0 \longrightarrow W_1 \longrightarrow P_1 \xrightarrow{f_1} V \longrightarrow 0$$

$$0 \longrightarrow W_2 \longrightarrow P_2 \xrightarrow{f_2} V \longrightarrow 0$$

are exact. Then

$$W_1 \oplus P_2 \cong W_2 \oplus P_1$$

Proof. Let X be the submodule of $P_1 \oplus P_2$ defined by

$$X = \{(x_1, x_2) \mid f_1(x_1) = f_2(x_2)\}$$

If $\pi : P_1 \oplus P_2 \longrightarrow P_1$ is the projection map, then $\pi(P_1) = P_1$ and

$$X \cap \mathrm{Ker}\pi = \{(0,x_2)\,|\,x_2 \in \mathrm{Ker}f_2\} \cong W_2$$

Hence we have an exact sequence

$$0 \longrightarrow W_2 \longrightarrow X \longrightarrow P_1 \longrightarrow 0$$

Since P_1 is projective, we infer that $X \cong W_2 \oplus P_1$. By the same argument, $X \cong W_1 \oplus P_2$ and the result follows. ■

1.4. LEMMA. Let R be an artinian ring, let V be an R-module and let P be a projective R-module.

(i) V determines the isomorphism class of $\Omega(V)$

(ii) Assume that $f : P \longrightarrow V$ is a surjective homomorphism. Then

$$P \cong P(V) \oplus P' \quad \text{and} \quad \mathrm{Ker}f \cong \Omega(V) \oplus P'$$

for some R-module P'.

Proof. (i) Let P_1 and P_2 be two projective covers of V and let

$$0 \longrightarrow \Omega_i(V) \longrightarrow P_i \longrightarrow V \longrightarrow 0 \qquad (i = 1,2)$$

be the corresponding exact sequences. By Lemma 1.3,

$$\Omega_1(V) \oplus P_2 \cong \Omega_2(V) \oplus P_1$$

while by Lemma 1.9.2, $P_1 \cong P_2$. Hence, by the Krull-Schmidt theorem, $\Omega_1(V) \cong \Omega_2(V)$.

(ii) By Lemma 1.3, $\mathrm{Ker}f \oplus P(V) \cong \Omega(V) \oplus P$ and, by Lemma 1.9.2(ii), $P \cong P(V) \oplus P'$ for some R-module P'. Hence

$$\mathrm{Ker}f \oplus P(V) \cong \Omega(V) \oplus P' \oplus P(V)$$

and so, by the Krull-Schmidt theorem, $\mathrm{Ker}f \cong \Omega(V) \oplus P'$. ■

1.5. LEMMA. Let 1_G be the trivial FG-module. Then $P(1_G) = FG$ if and only if G is a p-group.

Proof. Consider the exact sequence $0 \longrightarrow I(G) \longrightarrow FG \longrightarrow FG/I(G) \longrightarrow 0$. Since $FG/I(G) \cong 1_G$, it follows from the definition of $P(1_G)$ that $P(1_G) = FG$ if and only if $I(G) \subseteq J(FG)$. If G is a p-group, then $J(FG) = I(G)$ by Corollary

3.1.2. Conversely, if $I(G) \subseteq J(FG)$ then each $g-1$, $g \in G$, is nilpotent. The desired conclusion now follows by virtue of

$$(g-1)^{p^n} = g^{p^n} - 1 \qquad \text{for all } n \geqslant 1 \quad ■$$

Let R be an artinian ring and let V an R-module. Put

$$S_o(V) = 0, \; S_1(V) = \text{Soc}(V)$$

and define the *i-th socle* $S_i(V)$ of V by

$$S_i(V)/S_{i-1}(V) = \text{Soc}(V/S_{i-1}(V)) \qquad (i \geqslant 2)$$

If $V = {}_RR$ then we put $S_i(R) = S_i(V)$ and refer to $S_i(R)$ as the *i-th socle* of R. For any ideal I of R, we put $I^o = R$.

1.6. LEMMA. Let R be an artinian ring. Then for all $i \geqslant 0$

$$S_i(R) = r(J(R)^i)$$

Proof. By Proposition 1.6.29, $\text{Soc}(R) = r(J(R))$. Assume, by induction, that

$$S_{i-1}(R) = r(J(R)^{i-1})$$

Since $S_i(R)/S_{i-1}(R)$ is a completely reducible R-module, we have

$$J(R)S_i(R) \subseteq S_{i-1}(R)$$

and hence

$$J(R)^i S_i(R) \subseteq J(R)^{i-1} S_{i-1}(R) = 0$$

Thus $S_i(R) \subseteq r(J(R)^i)$. Conversely, assume that $J(R)^i x = 0$ for some $x \in R$. Then

$$J(R)x \subseteq r(J(R)^{i-1}) = S_{i-1}(R)$$

and therefore $(Rx + S_{i-1}(R))/S_{i-1}(R)$ is a completely reducible R-module. Thus

$$Rx + S_{i-1}(R) \subseteq S_i(R)$$

proving that $x \in S_i(R)$, as required. ■

Let V be an A-module, where A is a finite-dimensional algebra over a field F. Recall that the dual module $V^* = \text{Hom}_F(V,F)$ is a right A-module with the action of A on V^* given by

$$(\psi a)(v) = \psi(av) \qquad (a \in A, v \in V, \psi \in V^*)$$

In the case $A = FG$ there is a simple device which allows us to make any right A-module V into a left A-module in the following natural way.

The F-linear extension of the map $g \longmapsto g^{-1}, g \in G$, is an anti-automorphism of the group algebra FG. Therefore, if W is a right FG-module, then W becomes a left FG-module under the action

$$gw = wg^{-1} \qquad (g \in G, w \in W)$$

In particular, for every left FG-module V, V^* is a left FG-module under the action

$$(g\psi)(v) = \psi(g^{-1}v) \qquad (g \in G, v \in V, \psi \in V^*)$$

We shall refer to the left FG-module V^* as the *contragradient* of V. In what follows, for any (left) FG-module V, V^* denotes the FG-module which is the contragradient of V.

1.7. LEMMA. Let V be an FG-module

(i) If W is a submodule of V, then

$$W^{\perp} = \{f \in V^* | f(W) = 0\}$$

is a submodule of V^* such that $W^* \cong V^*/W^{\perp}$.

(ii) If $W_2 \subseteq W_1$ are submodules of V, then

$$(W_1/W_2)^* \cong W_2^{\perp}/W_1^{\perp}$$

(iii) If V and W are FG-modules and $\alpha \in \mathrm{Hom}_F(V,W)$, then the map $\alpha^* : W^* \longrightarrow V^*$ defined by

$$[\alpha^*(f)](v) = f(\alpha(v)) \qquad (v \in V, f \in W^*)$$

is an FG-homomorphism if and only if so is α.

(iv) If $\alpha \in \mathrm{Hom}_{FG}(V,W)$ and $\beta \in \mathrm{Hom}(W,X)$, then $(\beta\alpha)^* = \alpha^*\beta^*$. If

$$0 \longrightarrow V \xrightarrow{\alpha} W \xrightarrow{\beta} X \longrightarrow 0$$

is an exact sequence of FG-modules, then so is

$$0 \longrightarrow X^* \xrightarrow{\beta^*} W^* \xrightarrow{\alpha^*} V^* \longrightarrow 0$$

Proof. (i) Let π be the restriction mapping of V^* into W^*. Since

$$[\pi(gf)](w) = (gf)(w) = f(g^{-1}w) = (\pi f)(g^{-1}w) = [g(\pi f)](w)$$

for all $w \in W, f \in V^*$ and $g \in G$, it follows that π is a surjective FG-homomorphism with $\mathrm{Ker}\pi = W^{\perp}$. Thus $W^{\perp}$ is a submodule of V^* and W^* $V^*/W^{\perp}$.

(ii) Note that $(W_1/W_2)^*$ is isomorphic to the FG-submodule

$$\{f \in W_1^* \mid f(W_2) = 0\}$$

of W_1^*, and in the isomorphism $W_1^* \cong V^*/W_1^{\perp}$, this corresponds to $W_2^{\perp}/W_1^{\perp}$.

(iii) and (iv). The required properties are standard for F-spaces and F-homomorphisms. It therefore suffices to verify that α^* is an FG-homomorphism if and only if so is α. If $\alpha \in \mathrm{Hom}_{FG}(V,W)$, then

$$\begin{aligned}[\alpha^*(gf)](v) &= (gf)(\alpha(v)) = f(g^{-1}(\alpha(v))) = f(\alpha(g^{-1}v)) \\ &= (\alpha^* f)(g^{-1}v) = g(\alpha^* f)(v)\end{aligned}$$

for all $v \in V, f \in W^*$ and $g \in G$. The converse is true by a similar argument. ■

1.8. COROLLARY. Let V and W be FG-modules. Then

$$\mathrm{Hom}_{FG}(V,W) \cong \mathrm{Hom}_{FG}(W^*,V^*) \qquad \text{(as } F\text{-spaces)}$$

Proof. The map $\alpha \mapsto \alpha^*$ provides an isomorphism of $\mathrm{Hom}_F(V,W)$ onto $\mathrm{Hom}_F(W^*,V^*)$. Since, by Lemma 1.7, $\alpha \in \mathrm{Hom}_{FG}(V,W)$ if and only if $\alpha^* \in \mathrm{Hom}_{FG}(W^*,V^*)$, the assertion follows. ■

1.9. LEMMA. Let V be an FG-module

(i) The canonical mapping $\pi : V \longrightarrow (V^*)^*$ defined by

$$\pi(v)(f) = f(v) \qquad (v \in V, f \in V^*)$$

is an FG-isomorphism

(ii) If $V = V_1 \oplus V_2$ is a direct decomposition of V into FG-submodules V_1 and V_2, then $V^* = V_1^{\perp} \oplus V_2^{\perp} \cong V_1^* \oplus V_2^*$

(iii) V is irreducible (indecomposable) if and only if V^* is irreducible (indecomposable).

Proof. (i) It is a standard fact that π is an F-isomorphism. Since

$$[g(\pi(v))](f) = \pi(v)(g^{-1}f) = (g^{-1}f)(v) = f(gv) = [\pi(gv)](f)$$

for all $v \in V, f \in V^*$ and $g \in G$, π is an FG-isomorphism.

(ii) The direct decomposition $V^* = V_1^{\perp} \oplus V_2^{\perp}$ of the F-space V^* is well known. The desired conclusion now follows from Lemma 1.7(i).

(iii) If V^* is irreducible, then V is irreducible by Lemma 1.7(i). If V is irreducible, then $(V^*)^*$ is irreducible by (i). Hence V^* is irreducible by Lemma 1.7(i). Properties (i) and (ii) immediately imply that V^* is indecomposable if and only if so is V. ■

1.10. LEMMA. For any $g \in G$, let $\phi_g \in (FG)^*$ be defined by

$$\phi_g(\sum_{t\in G} x_t t) = x_g$$

Then the map

$$\begin{cases} FG \longrightarrow (FG)^* \\ \Sigma x_g g \longmapsto \Sigma x_g \phi_g \end{cases}$$

is an FG-isomorphism.

Proof. Since $\{\phi_g | g \in G\}$ is obviously an F-basis for V^*, the given map is at least an F-isomorphism. Taking into account that for all $g,h \in G$

$$h\phi_g = \phi_{hg}$$

the result follows. ■

1.11. LEMMA. Let V be an FG-module.

(i) V is projective if and only if so is V^*

(ii) If V is irreducible, then $V^* \cong \text{Soc}(P(V)^*)$

(iii) If W is a completely reducible FG-module and V is irreducible, then

$$\text{Hom}_{FG}(W, P(V)^*) \cong \text{Hom}_{FG}(W, V^*)$$

Proof. (i) Assume that V is projective, say

$$FG \oplus \ldots \oplus FG \cong V \oplus W$$

for some FG-module W. Then, applying Lemma 1.9(ii) and Lemma 1.10, we have

$$FG \oplus \ldots \oplus FG \cong (FG)^* \oplus \ldots \oplus (FG)^*$$
$$\cong V^* \oplus W^*$$

proving that V^* is projective. The converse is a consequence of Lemma 1.9(i).

(ii) By definition of $P(V)$, there is a surjective homomorphism $f : P(V) \longrightarrow V$. Hence, by Lemma 1.7(iv), $f^* : V^* \longrightarrow P(V)^*$ is an injective homomorphism. By (i) and Lemma 1.9(iii), $P(V)^*$ is projective indecomposable and V^* is irreducible. Since $\mathrm{Soc}(P(V)^*)$ is irreducible (Theorem 3.3.8), the assertion follows.

(iii) Since W is completely reducible, $\mathrm{Hom}_{FG}(W,P(V)^*) \cong \mathrm{Hom}_{FG}(W, \mathrm{Soc}(P(V)^*))$. Now apply (ii). ■

Given $x = \Sigma x_g g \in FG$, let

$$x^* = \Sigma x_g g^{-1}$$

Then, as has been previously observed, the map $x \longmapsto x^*$ is an anti-automorphism of FG. Thus, if I is a left ideal of FG, then

$$\hat{I} = \{x^* | x \in I\}$$

is a right ideal of FG.

1.12. LEMMA. Let $I_1 \subseteq I$ be left ideals of FG. Then

$$I^{\perp} \cong \ell(\hat{I})$$

and

$$I_1^{\perp}/I_2^{\perp} \cong \ell(\hat{I}_1)/\ell(\hat{I})$$

as FG-modules.

Proof. Fix $x = \Sigma x_g g \in FG$. By Lemma 1.10, it suffices to verify that

$$x \in \ell(\hat{I}) \text{ if and only if } \Sigma x_g \phi_g \in I^{\perp}$$

To this end, we first observe that given $y = \Sigma y_g g \in FG$,

$$(\sum_{g\in G} x_g \phi_g)(y) = \sum_{g\in G} x_g y_g = tr(xy^*)$$

Hence $\Sigma x_g \phi_g \in I^{\perp}$ if and only if

$$tr(xy^*) = 0 \qquad \text{for all } y \in I$$

The desired conclusion is therefore a consequence of Theorem 3.3.6(ii). ■

1.13. LEMMA. For all $i \geqslant 0$

(i) $S_i(FG)^* \cong FG/J(FG)^i$ as FG-modules

(ii) $[S_{i+1}(FG)/S_i(FG)]^* \cong J(FG)^i/J(FG)^{i+1}$ as FG-modules

Proof. Put $J = J(FG)$. Then $\hat{J} = J$ and so $\widehat{J^i} = J^i$ for all $i \geqslant 1$. Thus, by Lemma 1.12 and Theorem 3.3.6(v),

$$(J^i)^{\perp} \cong \ell(J^i) = r(J^i) \tag{1}$$

and

$$(J^{i+1})^{\perp}/(J^i)^{\perp} \cong \ell(J^{i+1})/\ell(J^i) = r(J^{i+1})/r(J^i) \tag{2}$$

(i) We have

$$\begin{aligned}(FG/J^i)^* &\cong (J^i)^{\perp}/(FG)^{\perp} && \text{(by Lemma 1.7(ii))}\\ &= (J^i)^{\perp} && \text{(since } (FG)^{\perp} = 0)\\ &\cong r(J^i) && \text{(by (1))}\\ &= S_i(FG) && \text{(by Lemma 1.6)}\end{aligned}$$

Hence the required assertion is a consequence of Lemma 1.9(i)

(ii) We have

$$\begin{aligned}(J^i/J^{i+1})^* &\cong (J^{i+1})^{\perp}/(J^i)^{\perp} && \text{(by Lemma 1.7(ii))}\\ &\cong r(J^{i+1})/r(J^i) && \text{(by (2))}\\ &= S_{i+1}(FG)/S_i(FG) && \text{(by Lemma 1.6)}\end{aligned}$$

Now apply Lemma 1.9(i). ■

1.14. LEMMA. Let V be an FG-module. Then

$$\dim_F V = \dim_F \mathrm{Hom}_{FG}(V,FG)$$

Proof. It is obvious that $\dim_F V = \dim_F \mathrm{Hom}_F(V,F)$. If $H = 1$, then $FG \cong (1_H)^G$. Now apply Theorem 2.4.9 for the case $H = 1$. ■

1.15. LEMMA. Let P be a projective indecomposable FG-module. Then for all $i \geqslant 0$,

$$\mathrm{Hom}_{FG}(P/J(FG)^i P,FG) \cong \mathrm{Hom}_{FG}(P,S_i(FG))$$

Proof. We may assume that $P = FGe$ for some primitive idempotent e of FG.

Assume that $f : P \longrightarrow FG$ is an FG-homomorphism. It suffices to show that

$$\mathrm{Ker} f \supseteq J(FG)^i P \text{ if and only if } f(P) \subseteq S_i(FG)$$

Assume that $\mathrm{Ker} f \supseteq J(FG)^i P$. Then

$$f(J(FG)^i e) = J(FG)^i f(e) = 0$$

so $f(e) \in S_i(FG)$ by Lemma 1.6. Hence

$$f(P) = FGf(e) \subseteq S_i(FG)$$

Conversely, assume that $f(P) \subseteq S_i(FG)$. Then, by Lemma 1.6, $J(FG)^i f(P) = 0$ and so $f(J(FG)^i P) = 0$, as required. ■

We have now accumulated all the information necessary to prove the following useful result.

1.16. PROPOSITION. Let $P = P(V)$ be a projective cover of an irreducible FG-module V. Then for all $i \geqslant 0$

$$\dim_F (J(FG)^i P/J(FG)^{i+1}P) = \dim_F \mathrm{Hom}_{FG}(J(FG)^i/J(FG)^{i+1}, V^*)$$

Proof. Put $S_i = S_i(FG)$ and $J = J(FG)$. Then

$$\begin{aligned}
\dim_F(J^iP/J^{i+1}P) &= \dim_F(P/J^{i+1}P) - \dim_F(P/J^iP)\\
&= \dim_F \mathrm{Hom}_{FG}(P/J^{i+1}P, FG) - \dim_F \mathrm{Hom}_{FG}(P/J^iP, FG) \quad \text{(by Lemma 1.14)}\\
&= \dim_F \mathrm{Hom}_{FG}(P, S_{i+1}) - \dim_F \mathrm{Hom}_{FG}(P, S_i) \quad \text{(by Lemma 1.15)}\\
&= \dim_F \mathrm{Hom}_{FG}(P, S_{i+1}/S_i)\\
&= \dim_F \mathrm{Hom}_{FG}((S_{i+1}/S_i)^*, P^*) \quad \text{(by Corollary 1.8)}\\
&= \dim_F \mathrm{Hom}_{FG}((S_{i+1}/S_i)^*, V^*) \quad \text{(by Lemma 1.11(iii))}\\
&= \dim_F \mathrm{Hom}_{FG}(J^i/J^{i+1}, V^*) \quad \text{(by Lemma 1.13(ii))}
\end{aligned}$$

as asserted. ■

For later use, we next record.

1.17. LEMMA. Let P be a projective FG-module and let $V_1, V_2, \ldots, V_n$ be composition factors of an FG-module V. Then

$$P \underset{F}{\otimes} V \cong \bigoplus_{i=1}^{n} P \otimes V_i \qquad \text{as } FG\text{-modules}$$

Proof. It suffices to show that for any submodule W of V,

$$P \underset{F}{\otimes} V \cong P \underset{F}{\otimes} W \oplus P \underset{F}{\otimes} (V/W) \qquad \text{as } FG\text{-modules}$$

To this end, consider the natural exact sequence

$$0 \longrightarrow W \longrightarrow V \longrightarrow V/W \longrightarrow 0$$

Tensoring with P we obtain an exact sequence

$$0 \longrightarrow P \underset{F}{\otimes} W \longrightarrow P \underset{F}{\otimes} V \longrightarrow P \underset{F}{\otimes} (V/W) \longrightarrow 0$$

But P is projective, hence so is $P \underset{F}{\otimes} (V/W)$ (see Proposition 2.5.7(ii)). Thus the above sequence splits and the result follows. ∎

1.18. LEMMA. Let H be a subgroup of G, let U be an FH-module, and let V be an FG-module. Then

$$(U \underset{F}{\otimes} V_H)^G \cong U^G \underset{F}{\otimes} V \qquad \text{as } FG\text{-modules}$$

Proof. Let $u_1, u_2, \ldots, u_m$ and $v_1, v_2, \ldots, v_n$ be F-bases of U and V, respectively. Choose a left transversal $g_1, g_2, \ldots, g_k$ for H in G. Then $g_t \otimes (u_i \otimes v_j)$, $1 \leqslant t \leqslant k$, $1 \leqslant i \leqslant m$, $1 \leqslant j \leqslant n$, is an F-basis of $(U \underset{F}{\otimes} V_H)^G$. As an F-basis of $U^G \underset{F}{\otimes} V$, choose the elements $(g_t \otimes u_i) \otimes g_t v_j$, $1 \leqslant t \leqslant k$, $1 \leqslant i \leqslant m$, $1 \leqslant j \leqslant n$. Then the mapping

$$f : g_t \otimes (u_i \otimes v_j) \longmapsto (g_t \otimes u_i) \otimes g_t v_j$$

defines an F-isomorphism of $(U \underset{F}{\otimes} V_H)^G$ onto $U^G \underset{F}{\otimes} V$. We are therefore left to prove that, given $x = g_t \otimes (u_i \otimes v_j)$ and $g \in G$, $f(gx) = gf(x)$. Using the fact that $\{u_i\}$ and $\{v_j\}$ are bases of U and V, respectively, it is straightforward to verify that for all $u \in U, v \in V$

$$f(g_t \otimes (u \otimes v)) = (g_t \otimes u) \otimes g_t v$$

Write $gg_t = g_s h, h \in H$, so that

$$gx = gg_t \otimes (u_i \otimes v_j) = g_s \otimes (hu_i \otimes hv_j)$$

Then we have

$$f(gx) = (g_x \otimes hu_i) \otimes (g_s hv_j) = (g_s h \otimes u_j) \otimes (g_s hv_j)$$

and, on the other hand,

$$gf(x) = (gg_t \otimes u_i) \otimes (gg_t v_j) = (g_s h \otimes u_i) \otimes (g_s hv_j)$$

So the lemma is true. ■

2. THE LOEWY LENGTH OF PROJECTIVE COVERS

Throughout this section, G denotes a finite group, N a normal subgroup of G, F a field of characteristic $p > 0$, and $P_G(V)$ (or simply $P(V)$) a projective cover of the FG-module V. As usual, we write 1_G for the trivial FG-module. Finally, $L(V)$ and $\Omega(V)$ denote the Loewy length and the Heller module of V, respectively.

Our aim is to examine the situation where V is an FG-module and N acts trivially on V. Thus V can be also viewed as an $F(G/N)$-module via

$$(gN)v = gv \qquad (g \in G, v \in V)$$

and we compute the Loewy lengths of the corresponding projective covers $P_G(V)$ and $P_{G/N}(V)$. Let

$$\pi : FG \to F(G/N)$$

be the natural homomorphism. Then, by the definition of the action of G/N on V, we have

$$\pi(x)v = xv \qquad \text{for all } x \in FG \tag{1}$$

Let V be an FG-module. Since

$$FG\cdot I(N) = I(N)\cdot FG$$

we have

$$FG\cdot I(N)V = I(N)V$$

Thus $I(N)V$ is a submodule of V such that N acts trivially on $V/I(N)V$. Therefore $V/I(N)V$ can be viewed as an $F(G/N)$-module. Moreover, if N acts trivially on V, then $I(N)V = 0$ and so

$$I(N)P(V) \subseteq \Omega(V) \tag{2}$$

2.1. LEMMA. Let N act trivially on FG-modules V and W.

(i) If $V = FG/FG\cdot I(N)$, then $V \cong F(G/N)$ as $F(G/N)$-modules

(ii) A map $f : V \longrightarrow W$ is an FG-homomorphism if and only if f is an $F(G/N)$-homomorphism. Furthermore, the kernels of both homomorphisms are the same.

(iii) If $f : M \longrightarrow V \oplus W$ is an FG-isomorphism, then f is also an $F(G/N)$-isomorphism.

(iv) $J(FG)V = J(F(G/N))V$

Proof. (i) Assume that $V = FG/FG\cdot I(N)$. Since $F(G/N) \cong FG/FG\cdot I(N)$ as F-algebras, we certainly have $\dim_F V = \dim_F F(G/N)$. Hence the map

$$\begin{cases} F(G/N) \longrightarrow V \\ gN \longmapsto g + FG\cdot I(N) \end{cases} \qquad (g \in G)$$

is at least an F-isomorphism. Since this map obviously preserves the action of G/N, it is in fact an $F(G/N)$-isomorphism.

(ii) Direct consequence of the definition of the action of G/N on V.

(iii) Direct consequence of (ii)

(iv) Let $\pi : FG \longrightarrow F(G/N)$ be the natural homomorphism. By Lemma 1.1(iii),

$$\pi(J(FG)) = J(F(G/N))$$

and so by (1) we have

$$J(FG)V = \pi(J(FG))V = J(F(G/N))V$$

as required. ∎

2.2. LEMMA. Let V be a projective FG-module. Then $V/I(N)V$ is a projective $F(G/N)$-module.

Proof. By hypothesis,

$$FG \oplus \ldots \oplus FG \cong V \oplus X$$

for some FG-module X. Hence

$$I(N)FG \oplus \ldots \oplus I(N)FG \cong I(N)V \oplus I(N)X$$

and therefore

$$FG/FG\cdot I(N) \oplus \dots \oplus FG/FG\cdot I(N) \cong V/I(N)V \oplus X/I(N)X$$

Now apply Lemma 2.1(i),(iii). ∎

Let V be an FG-module and let $x \in FG$. Then we put

$$\operatorname{ann}_V(x) = \{v \in V \mid xv = 0\}$$

It is clear that if $x \in Z(FG)$, then $\operatorname{ann}_V(x)$ is a submodule of V.

2.3. LEMMA. Let V be an FG-module such that V_N is a projective FN-module. Then

$$I(N)V = \operatorname{ann}_V(N^+)$$

Proof. We may harmlessly assume that

$$FN \oplus \dots \oplus FN = V \oplus X$$

for some FN-module X, in which case

$$I(N) \oplus \dots \oplus I(N) = I(N)V \oplus I(N)X$$

Thus $I(N)V = V \cap (I(N) \oplus \dots \oplus I(N))$. Fix $v \in V$ and write $v = v_1 + \dots + v_n$, $v_i \in FN$. Then $N^+v = 0$ if and only if $N^+v_i = 0$ for all i. But $N^+v_i = 0$ if and only if $v_i \in I(N)$ by Lemma 3.3.12(ii). Hence $v \in \operatorname{ann}_V(N^+)$ if and only if $v \in I(N)V$, as required. ∎

2.4. LEMMA. Let $V = P(1_G)$ and let n be the Loewy length of V. Then

(i) $\operatorname{Soc}V = G^+V$

(ii) $G^+ \in J(FG)^{n-1}$

(iii) $J(FG)V = I(G)V$

Proof. (i) Let e be a primitive idempotent of FG such that $FGe/J(FG)e \cong 1_G$. Then V is identifiable with FGe and, by Theorem 3.3.8(iii), $\operatorname{Soc}(FGe) \cong 1_G$. If $0 \neq x \in \operatorname{Soc}(FGe)$, then $gx = x$ for all $g \in G$, so $x = \lambda G^+$ for some $0 \neq \lambda \in F$. Thus

$$\operatorname{Soc}(FGe) = FG^+ = G^+(FGe)$$

as asserted.

(ii) By hypothesis, $J(FG)^{n-1}e$ is a nonzero submodule of FGe, so $G^+ \in J(FG)^{n-1}e \subseteq J(FG)^{n-1}$.

(iii) If $I(G)e = FGe$, then $\text{Soc}(FGe) = G^+(I(G)e) = 0$, a contradiction. Thus $I(G)e$ is a proper submodule of FGe. But $I(G)e \supseteq J(FG)e$ and $J(FG)e$ is a maximal submodule of FGe. Hence $I(G)e = J(FG)e$ as required. ∎

2.5. LEMMA. Let V be an FG-module on which N acts trivially. Then

(i) $P(V)_N \cong eP_N(1_N)$ for some $e \geq 1$

(ii) $I(N)\cdot P(V) = J(FN)\cdot P(V)$

Proof. (i) Write $P(V)/J(FG)P(V) = V_1 \oplus \ldots \oplus V_n$ for some irreducible FG-modules V_i and some $n \geq 1$. Then, by Theorem 1.9.3, $P(V) = P(V_1) \oplus \ldots \oplus P(V_n)$ and thus we may assume that V is irreducible. Write

$$P(V)_N = X_1 \oplus \ldots \oplus X_m$$

where the X_i are indecomposable FN-modules. Since $P(V)_N$ is a projective FN-module, each X_i is projective. By Theorem 3.3.8(iii), $V \cong \text{Soc}P(V)$, and hence N acts trivially on $\text{Soc}(P(V))$. Since $(\text{Soc}(P(V)))_N \subseteq \text{Soc}(P(V)_N)$, it follows that $\text{Soc}(X_i) \cong 1_N$ for some $i \in \{1,2,\ldots,m\}$. But, by Lemma 5.3.9, all the X_i are G-conjugate. Hence $\text{Soc}(X_i) \cong 1_N \cong \text{Soc}(P_N(1_N))$ for all i, so $X_i \cong P_N(1_N)$ for all i, as required.

(ii) Direct consequence of (i) and Lemma 2.4(iii). ∎

We are now ready to prove the following result.

2.6. THEOREM. (Lorenz (1985)). Let V be an FG-module on which N acts trivially.

(i) $P_{G/N}(V) \cong P_G(V)/I(N)P_G(V)$

(ii) $L(P_G(V)) \geq L(P_{G/N}(V)) + L(P_N(1_N)) - 1$

(iii) If $J(FG)\cdot J(FN) = J(FN)\cdot J(FG)$, then

$$L(P_G(V)) \leq L(P_{G/N}(V)L(P_N(1_N))$$

(iv) If N is a p'-group, then

$$L(P_G(V)) = L(P_{G/N}(V))$$

(v) If G/N is a p'-group, then

$$L(P_G(V)) = L(P_N(1_N))$$

Proof. (i) Set $P = P_G(V)$ and $H = G/N$. Since N acts trivially on V, it follows from (2) that

$$J(FG)\cdot P \supseteq \Omega(V) \supseteq I(N)\cdot P$$

Thus, by Lemma 2.1(ii), we obtain an exact sequence of FH-modules

$$0 \longrightarrow \Omega(V)/I(N)P \longrightarrow P/I(N)P \longrightarrow V \rightarrow 0 \qquad (4)$$

Moreover, by (3) and Lemma 2.1(iv), we have

$$\begin{aligned}\Omega(V)/I(N)P &\subseteq J(FG)(P/I(N)P)\\ &= J(FH)(P/I(N)P)\end{aligned} \qquad (5)$$

But P is projective as an FG-module, so by Lemma 2.2 $P/I(N)P$ is projective as an FH-module. Thus, by (4) and (5),

$$P_H(V) \cong P/I(N)P$$

as required.

(ii) Put $n = L(P_H(V))$ and $m = L(P_N(1_N))$. Since $P_H(V) \cong P/I(N)P$, it follows from (5) that

$$J(FG)^{n-1}\cdot P \not\subseteq I(N)\cdot P$$

Now P is projective, hence so is P_N and so, by Lemma 2.3, $I(N)P = \operatorname{ann}_P(N^+)$. Thus $N^+J(FG)^{n-1}\cdot P \neq 0$. But, by Lemma 2.4, $N^+ \in J(FN)^{m-1}$ while by Proposition 3.1.9 $J(FN)^{m-1} \subseteq J(FG)^{m-1}$. Hence

$$J(FG)^{m+n-2}P = J(FG)^{m-1}J(FG)^{n-1}P \neq 0$$

and therefore $L(P) \geqslant m+n-1$.

(iii) Assume that $J(FG)\cdot J(FN) = J(FN)\cdot J(FG)$. Keeping the notation of (ii), we have

$$J(FG)^n\cdot P \subseteq I(N)P$$

by virtue of (5). Applying Lemma 2.5, we deduce that

$$J(FG)^n\cdot P \subseteq J(FN)\cdot P \qquad (6)$$

and that

$$J(FN)^m P = 0 \tag{7}$$

We now claim that for any $k \geqslant 1$,

$$J(FG)^{nk} P \subseteq J(FN)^k P \tag{8}$$

If sustained, the required assertion will follow from (7) by taking $k = m$. In view of (6), (8) is true for $k = 1$. Hence

$$\begin{aligned} J(FG)^{n(k+1)} P &= J(FG)^n \, J(FG)^{nk} P \\ &\subseteq J(FG)^n \, J(FN)^k P && \text{(by induction hypothesis)} \\ &= J(FN)^k \, J(FG)^n P && \text{(since } J(FG)\cdot J(FN) = J(FN)\cdot J(FG)\text{)} \\ &\subseteq J(FN)^k \, J(FN) P && \text{(by (6))} \\ &= J(FN)^{k+1} P \end{aligned}$$

as required.

(iv) Assume that N is a p'-group. Then $L(P_N(1_N)) = 1$ and hence by (ii)

$$L(P_G(V)) \geqslant L(P_{G/N}(V))$$

Since $J(FG)\cdot J(FN) = J(FN)\cdot J(FG) = 0$, it follows from (iii) that

$$L(P_G(V)) \leqslant L(P_{G/N}(V))$$

as required.

(v) Assume that G/N is a p'-group. Then $L(P_{G/N}(V)) = 1$ and hence by (ii)

$$L(P_G(V)) \geqslant L(P_N(1_N))$$

By Proposition 3.1.8(ii), $J(FG) = FG\cdot J(FN) = J(FN)\, FG$ and hence

$$J(FG)\cdot J(FN) = J(FN)\cdot J(FG).$$

Applying (iii), we conclude that

$$L(P_G(V) \leqslant L(P_N(1_N))$$

as required. ■

As an application of Theorem 2.6, we now prove

2.7. THEOREM. (Willems (1980)). Let V be an irreducible FG-module such that $\mathrm{Ker}V$ is p-solvable of order $p^a m, a \geqslant 0, (p,m) = 1$. Then

$$L(P(V)) \geqslant a(p-1) + 1.$$

Proof. We argue by induction on the order of G. Set $K = \mathrm{Ker}V$. If $0_{p',p}(K) = 1$ then $a = 0$ and there is nothing to prove. Assume that $N = 0_{p'}(K) \neq 1$. Then, by Theorem 2.6(iv), $L(P(V)) = L(P_{G/N}(V))$. Thus the assertion follows by the induction hypothesis. Finally, assume that $N = 0_p(K) \neq 1$. Then, by Theorem 2.6(ii), we have

$$L(P_{G/N}(V)) + L(P_N(1_N)) - 1 \leqslant L(P(V))$$

Using the induction hypothesis for $L(P_{G/N}(V))$ and the estimation

$$L(P(1_N)) \geqslant b(p-1) + 1$$

where $|N| = p^b$ (see Lemma 1.5 and Proposition 4.1.8), we get

$$\begin{aligned} a(p-1) + 1 &= (a-b)(p-1) + 1 + b(p-1) \\ &\leqslant L(P_{G/N}(V)) + L(P_N(1_N)) - 1 \\ &\leqslant L(P(V)) \end{aligned}$$

This completes the proof of the theorem. ∎

2.8. COROLLARY. (Wallace (1968)). Let G be a p-solvable group of order $p^a m, (p,m) = 1$. Then $t(G) \geqslant a(p-1) + 1$.

Proof. Applying Theorem 2.7 for $V = 1_G$, we get

$$t(G) \geqslant L(P(1_G)) \geqslant a(p-1) + 1,$$

as required. ∎

2.9. LEMMA. Let V be an FG-module. Then

$$\mathrm{Hom}_{FG}(V,1_G) \cong \mathrm{Hom}_F(V/I(G)V,1_G)$$

Proof. An F-linear map $f : V \longrightarrow 1_G$ is an FG-homomorphism if and only if $I(G)V \subseteq \mathrm{Ker}f$. Thus $\mathrm{Hom}_{FG}(V,1_G)$ is a subspace of $\mathrm{Hom}_F(V,1_G)$ consisting of those $f \in \mathrm{Hom}_F(V,1_G)$ for which $\mathrm{Ker}f \supseteq I(G)V$. Since the latter subspace is isomorphic to $\mathrm{Hom}_F(V/I(G)V,1_G)$, the result follows. ∎

Next we establish the following important result.

2.10. THEOREM. (Alperin, Collins and Sibley (1984)). Let V be an FG-module on which N acts trivially. View FN as an FG-module via conjugation of G on N.

Then for all $i \geqslant 0$ we have FG-isomorphisms

$$P_{G/N}(V) \otimes_F J(FN)^i/I(N)J(FN)^i \cong J(FN)^i P_G(V)/J(FN)^{i+1}P_G(V)$$

where $P_{G/N}(V)$ is viewed as an FG-module by letting N act trivially. (By convention, $J(FN)^\circ = FN$).

Proof. By Theorem 2.6(i) and Lemma 2.5(ii) and 2.1(ii)

$$P_{G/N}(V) \cong P_G(V)/J(FN)P_G(V) \qquad \text{as } FG\text{-modules} \tag{9}$$

Put $P = P_G(V), J = J(FN)$ and consider the map

$$\begin{cases} P/JP \otimes J^i/I(N)J^i \longrightarrow J^iP/J^{i+1}P \\ (x+JP) \otimes (y+I(N)J^i) \longmapsto yx + J^{i+1}P \end{cases} \tag{10}$$

This is obviously well defined, F-linear and surjective. Moreover, for all $g \in G$,

$$g((x+JP) \otimes (y+I(N)J^i) = (gx+JP) \otimes (gyg^{-1}+I(N)J^i)$$

is mapped to

$$(gyg^{-1})(gx) + J^{i+1}P = g(yx+J^{i+1}P)$$

Thus the map (10) is a surjective FG-homomorphism. In view of (9), we are left to verify that the modules in (10) are of the same F-dimension. Put $Q = P_N(1_N)$ and observe that, by Lemma 2.5(i), we need only show that

$$\dim_F(Q/JQ \otimes J^i/I(N)J^i) = \dim_F(J^iQ/J^{i+1}Q) \tag{11}$$

But $Q/JQ \cong 1_N$, so $\dim_F Q/JQ = 1$ and therefore

$$\begin{aligned} \dim_F(Q/JQ \otimes J^i/I(N)J^i) &= \dim_F(J^i/I(N)J^i) \\ &= \dim_F \mathrm{Hom}_F(J^i/I(N)J^i, 1_N) \\ &= \dim_F \mathrm{Hom}_{FN}(J^i/J^{i+1}, 1_N) \quad \text{(by Lemma 2.9)} \end{aligned}$$

which equals the right-hand side of (11), by virtue of Proposition 1.16. ■

Returning to the discussion of Loewy lengths, we now prove

2.11. THEOREM. (Lorenz (1985)). Let N be a normal p-subgroup of G and let V be an FG-module on which N acts trivially. View FN as an FG-module via conjugation of G on N. Then

$$L(P_G(V)) \geqslant t(N) - 1 + \max_X L(P_{G/N}(V) \underset{F}{\otimes} X)$$
$$\geqslant t(N) - 1 + \max_X L(P_{G/N}(V \underset{F}{\otimes} X)),$$

where X runs over the FG-composition factors of FN.

Proof. Set $P = P_G(V)$, $H = G/N$ and $\ell_i = L(I(N)^i P/I(N)^{i+1})$, $i \geqslant 0$. By Theorem 2.10,

$$P_H(V) \underset{F}{\otimes} I(N)^i/I(N)^{i+1} \cong I(N)^i P/I(N)^{i+1}P \qquad (i \geqslant 0)$$

and hence

$$L(P_H(V) \underset{F}{\otimes} I(N)^i/I(N)^{i+1}) = \ell_i \qquad (i \geqslant 0)$$

If $m < \ell_i$, then

$$J(FG)^m I(N)^i P \not\subseteq I(N)^{i+1}P$$
$$= \mathrm{ann}_P I(N)^{t(N)-i-1},$$

where the latter equality follows by virtue of Corollary 4.1.7, since P is free over FN. It therefore follows that

$$I(N)^{t(N)-i-1} \cdot J(FG)^m \cdot I(N)^i P \neq 0$$

and thus

$$L(P) > t(N) + m - 1.$$

Therefore, $L(P) \geqslant t(N) + \ell - 1$, where $\ell = \max_i \ell_i$. Note that $P_H(V)$ is projective over FH and conjugation action of N on each factor $I(N)^i/I(N)^{i+1}$ is trivial. Hence, by Lemma 1.17,

$$\bigoplus_i (P_H(V) \underset{F}{\otimes} I(N)^i/I(N)^{i+1}) \cong \bigoplus_X (P_H(V) \otimes X)$$

where X runs over the FH (and hence FG)-composition factors of FN. Thus

$$\ell = \max_X L(P_H(V) \otimes X)$$

Since $P_H(V \underset{F}{\otimes} X)$ is a summand of $P_H(V) \underset{F}{\otimes} X$, we also have

$$\ell \geqslant L(P_H(V \underset{F}{\otimes} X))$$

as required. ■

2.12. THEOREM. (Lorenz (1985)). Let N be a normal p-subgroup of G, let V

be an irreducible FG-module and assume that $G = NH$ for some subgroup H of G with $N \cap H = 1$. For each $i \geqslant 0$, let V_i denote the FG-module

$$(I(N)^i/I(N)^{i+1}) \otimes_F V$$

where G acts by conjugation on $I(N)^i/I(N)^{i+1}$.

(i) $P_G(V) \cong (1_H)^G \otimes_F P_H(V)$, where N acts trivially on $P_H(V)$.

(ii) $L(P_G(V)) \geqslant L(P_H(V)) + L((V_H)^G) - 1$

(iii) $L((V_H)^G) \geqslant t(N) - 1 + \max_i L(V_i)$

In particular,

$$L((V_H)^G) = t(N)$$

if and only if all V_i are completely reducible.

Proof. (i) Put $X = (1_H)^G \otimes_F P_H(V)$, $J = J(FG)$ and note that, by Lemma 1.18, $X \cong [(P_H(V))_H]^G$. Then we have

$$\begin{aligned} X/J\cdot X &\cong X/I(N)\cdot X/J\cdot X/I(N)\cdot X \\ &\cong P_H(V)/J(FH)\cdot P_H(V) \\ &\cong V \end{aligned}$$

Because X is projective, it follows that $X \cong P_G(V)$ as required.

(ii) If $n = L(P_H(V))$, then by (i),

$$\begin{aligned} J^{n-1}\cdot P_G(V) &= J^{n-1}(P_H(V) \otimes (1_H)^G) \\ &\supseteq FN\cdot J(FH)^{n-1}(P_H(V) \otimes (1_H)) \\ &= FN\cdot (J(FH)^{n-1}P_H(V) \otimes 1_H) \\ &= FN\cdot (V \otimes 1_H) \\ &\cong (V_H)^G \end{aligned}$$

proving (ii).

(iii) We first observe that

$$V_i \cong I(N)^i\cdot (V_H)^G/I(N)^{i+1}\cdot (V_H)^G$$

Thus if $m < n_i = L(V_i)$, then

$$\begin{aligned} J^m\cdot I(N)^i\cdot (V_H)^G &\not\subseteq I(N)^{i+1}\cdot (V_H)^G \\ &= \mathrm{ann}_{(V_H)^G} I(N)^{t(N)-i-1} \end{aligned}$$

Here the latter equality follows from Corollary 4.1.7, since $(V_H)^G$ is free over FN. Therefore, for all $i \geqslant 0$

$$L((V_H)^G) \geqslant t(N) + n_i - 1$$

Since the last assertion is obvious, the result is established. ■

3. THE LOEWY LENGTH OF INDUCED MODULES

Throughout this section, F denotes a field of characteristic $p > 0$ and G a finite group. As usual, all modules over a ring R are assumed to be left and finitely generated. If W is an R-module and $E = \mathrm{End}_R(W)$, then W will also be regarded as an E-module via

$$\phi w = \phi(w) \qquad \text{for all } \phi \in E, w \in W$$

3.1. LEMMA. Let $W = Re$ for some idempotent e of R and let $E = \mathrm{End}_R(W)$. Then

$$J(E)^i W = Re(eJ(R)e)^i \qquad \text{for all } i \geqslant 1$$

In particular, the Loewy length of W as an E-module is equal to the nilpotency index of $J(E)$.

Proof. For each $x \in eRe$, let $f_x \in E$ be defined by $f_x(w) = wx$ for all $w \in W$. By Proposition 1.5.6, the map

$$\begin{cases} eRe \longrightarrow E \\ x \longmapsto f_x \end{cases}$$

is an anti-isomorphism of rings. Furthermore, by Proposition 1.6.35, $J(eRe) = eJ(R)e$ and so $f_x \in J(E)^i$ if and only if $x \in (eJ(R)e)^i$. Thus

$$J(E)^i W = Re(eJ(R)e)^i$$

Finally, $J(E)^i W = 0$ if and only if $(eJ(R)e)^i = 0$. Hence the Loewy length of the E-module W is equal to the nilpotency index of $J(eRe) = eJ(R)e$. Since $eRe \cong E^\circ$, the result follows. ■

3.2. LEMMA. Let N be a normal subgroup of G, let V be an irreducible FN-module and let H be the inertia group of V. Then each FH-homomorphism

$\theta : V^H \longrightarrow V^H$ extends to a unique FG-homomorphism $\theta' : V^G \longrightarrow V^G$ and the map $\theta \mapsto \theta'$ is an F-algebra isomorphism of $\mathrm{End}_{FH}(V^H)$ onto $\mathrm{End}_{FG}(V^G)$.

Proof. Let $g_1, g_2, \ldots, g_s$ be a left transversal for N in H and $g_1, \ldots, g_k$ a left transversal for N in G. Then

$$V^H = \bigoplus_{i=1}^{s} g_i \otimes V \quad \text{and} \quad V^G = \bigoplus_{i=1}^{k} g_i \otimes V$$

Given $\theta \in \mathrm{End}_{FH}(V^H)$, define $\theta' \in \mathrm{End}_F(V^G)$ by

$$\theta'(g_i \otimes v) = g_i\, \theta\,(1 \otimes v) \qquad (v \in V)$$

Then it is immediate that θ' is a unique element of $\mathrm{End}_{FG}(V^G)$ extending θ. Hence the map $\theta \mapsto \theta'$ is an injective homomorphism $\mathrm{End}_{FH}(V^H) \to \mathrm{End}_{FG}(V^G)$ of F-algebras. Note that $(V^H)_N$ is the sum of all submodules of $(V^G)_N$ isomorphic to V. Hence for any $\psi \in \mathrm{End}_{FG}(V^G)$, $\psi(V^H) \subseteq V^H$. This proves that the given map is surjective and the result follows. ∎

We have at our disposal all the information necessary to prove the following result in which $L(V^G)$ denotes the Loewy length of the FG-module V^G.

3.3. THEOREM. (Clarke (1972)). Let N be a normal p'-subgroup of G, let H be the inertia group of an irreducible FN-module V and let $E = \mathrm{End}_{FG}(V^G)$. Then

(i) $J(FG)^n V^G = FG\cdot J(FH)^n V^H = J(E)^n V^G$ for all $n \geqslant 1$

(ii) $L(V^G)$ is equal to the nilpotency index of $J(\mathrm{End}_{FH}(V^H))$.

Proof. We may take $V = FNe$ for some primitive idempotent e of FN in which case $V^G = FGe$. Hence, by Lemma 3.1, the Loewy length of V^G as an E-module is equal to the nilpotency index of $J(E)$. Thus (ii) is a consequence of (i) and Lemma 3.2.

Write

$$1 = e_1 + e_2 + \ldots + e_m$$

as a sum of primitive idempotents of FN with $e = e_1$. Then we have

$$\begin{aligned} J(FG)V^G &= J(FG)e = FG(J(FG)e) \\ &= FGe(J(FG)e) + FGe_2(J(FG)e) + \ldots + FGe_m(J(FG)e) \qquad (1) \end{aligned}$$

as left FG-modules, where the sum is not necessarily direct. For each $a \in e_i FGe$, let $f_a \in \text{Hom}_{FG}(FGe_i, FGe)$ be defined by $f_a(x) = xa$ for all $x \in FGe_i$. Then, as can be seen from Lemma 1.5.5, the map

$$\begin{cases} e_iFGe \longrightarrow \text{Hom}_{FG}(FGe_i, FGe) \\ \quad a \longmapsto f_a \end{cases}$$

is an F-isomorphism. In particular, if $FGe_i \cong FGe$, then there is an $a \in e_iFGe$ such that $f_a : FGe_i \longrightarrow FGe$ is an FG-isomorphism.

For the sake of clarity, we divide the rest of the proof into three steps.

Step 1. Here we prove that $J(FG)e = FGeJ(FG)e$.

Since $FGeJ(FG)e \subseteq J(FG)e$, it follows from (1) that we need only verify that

$$e_iJ(FG)e \subseteq e_iFGeJ(FG)e \qquad (1 \leqslant i \leqslant m) \tag{2}$$

Let f_i be the block idempotent of FN with $e_if_i = e_i$, $1 \leqslant i \leqslant m$, and let f_i^* be the sum of G-conjugates of f_i. Then f_i^* is a central idempotent of FG. Now if f_1 and f_i are not G-conjugate, then $f_i^*f_1^* = 0$ and hence

$$e_iFGe = e_if_if_i^* \, FGf_1^*f_1e_1 = 0$$

We may therefore assume that f_1 and f_i are G-conjugate, say $f_1 = g^{-1}f_ig$. Then

$$(g^{-1}e_ig)f_1 = g^{-1}(e_if_i)g = g^{-1}e_ig$$

and so the irreducible FN-modules $FN(g^{-1}e_ig)$ and FNe are in the same block FNf_1. But N is a p'-group, so $FNe \cong FN(g^{-1}e_ig)$ and hence

$$FGe \cong FG(g^{-1}e_ig) \cong FGe_i$$

By the foregoing, there is an $a \in e_iFGe$ such that $f_a : FGe_i \longrightarrow FGe$ is an FG-isomorphism. Hence there is a $b \in eFGe_i$ such that $f_a^{-1}(y) = yb$ for all $y \in FGe$. Therefore $xab = x$ for all x in FGe_i. Thus

$$e_i = e_iab = (e_ia)b = ab$$

and so for any $c \in e_iJ(FG)e$,

$$c = e_ic = (ab)c = a(bc) \in e_iFGeJ(FG)e$$

This proves (2) and hence the required assertion.

Step 2. Here we prove that $J(FG)^n V^G = J(E)^n V^G$ *for all* $n \geqslant 1$.

By Lemma 3.1, it suffices to verify that

$$J(FG)^n e = (FGe)(eJ(FG)e)^n \qquad \text{for all } n \geqslant 1$$

The case $n = 1$ being proved in Step 1, we argue by induction on n. So assume that

$$J(FG)^k e = (FGe)(eJ(FG)e)^k \qquad \text{for all } k \leqslant n \tag{3}$$

Multiplying (3) on the left by $J(FG)$ gives

$$J(FG)^{k+1} e = (J(FG)e)^{k+1} \qquad \text{for all } k \leqslant n \tag{4}$$

whereas multiplying (3) on the right by $J(FG)e$ gives

$$(J(FG)^k e)(J(FG)e) = (FGe)(eJ(FG)e)^{k+1} \qquad \text{for all } k \leqslant n \tag{5}$$

Thus we have

$$\begin{aligned} J(FG)^{n+1} e &= (J(FG)e)^{n+1} && \text{(using (4) with } k = n) \\ &= (J(FG)e)^n (J(FG)e) \\ &= (J(FG)^n e)(J(FG)e) && \text{(using (4) with } k = n-1) \\ &= (FGe)(eJ(FG)e)^{n+1} && \text{(using (5) with } k = n) \end{aligned}$$

proving that (3) holds for $k = n + 1$

Step 3. We now complete the proof by showing that

$$J(FG)^n V^G = FG\cdot J(FH)^n V^H \qquad \text{for all } n \geqslant 1$$

We keep the notation of Lemma 3.2 and put $E_1 = \operatorname{End}_{FH}(V^H)$. Then

$$\begin{aligned} J(FG)^n V^G &= J(E)^n \Big(\sum_{i=1}^{k} g_i \otimes V\Big) && \text{(by Step 2)} \\ &= \sum_{i=1}^{k} g_i J(E)^n (1 \otimes V) \\ &= \sum_{i=1}^{k} g_i J(E_1)^n (1 \otimes V) && \text{(by Lemma 3.2)} \\ &= \sum_{i=1}^{s} g_i J(E_1)^n (1 \otimes V) + \sum_{i=s+1}^{k} g_i J(E_1)^n (1 \otimes V) \\ &= J(E_1)^n V^H + \sum_{i=s+1}^{k} g_i J(E_1)^n (1 \otimes V) \\ &= J(FH)^n V^H + \sum_{i=s+1}^{k} g_i J(E_1)^n (1 \otimes V) && \text{(by Step 2)} \\ &= FG\cdot J(FH)^n V^H \end{aligned}$$

as required. ■

3.4. COROLLARY. Let N be a p-nilpotent normal subgroup of G, let F be an algebraically closed field of characteristic p and let V be a principal indecomposable FN-module. Then $L(V^G)$ is equal to the nilpotency index of $J(\underset{FG}{\mathrm{End}}(V^G))$.

Proof. Our proof of the equality $J(FG)^n V^G = J(E)^n V^G$ in Theorem 3.3 relied only on the fact that if N is a p'-group then any two principal indecomposable FN-modules in the same block are isomorphic. Since the latter property also holds under present hypothesis (see Corollary 3.10.10), we deduce that $J(FG)^n V^G = J(E)^n V^G$ for all $n \geqslant 1$. The desired conclusion now follows by virtue of Lemma 3.1. ∎

3.5. COROLLARY. Further to the assumptions and notation of Theorem 3.3, assume that F is an algebraically closed field of characteristic p. Then there exists $\alpha \in Z^2(H/N,F^*)$ such that $L(V^G)$ is equal to the nilpotency index of $F^\alpha(H/N)$. Furthermore, if for all $q \neq p$ the Sylow q-subgroups of H/N are cyclic, then $L(V^G)$ is equal to the nilpotency index of $F(H/N)$.

Proof. By Theorem 3.4.21, $\underset{FH}{\mathrm{End}}(V^H) \cong F^\alpha(H/N)$ for some $\alpha \in Z^2(H/N,F^*)$. Hence the first assertion follows from Theorem 3.3(ii). The second assertion is a consequence of the first and Lemma 3.4.9. ∎

We close by providing a sufficient condition under which $L(V^G)$ is equal to the nilpotency index of FP, where P is a Sylow p-subgroup of the group H in Theorem 3.3. To achieve this, we need to establish some preliminary results concerning twisted group algebras.

3.6. LEMMA. Let K be an arbitrary field whose characteristic does not divide the order of G. Then, for any $\alpha \in Z^2(G,K^*)$, $K^\alpha G$ is a semisimple K-algebra.

Proof. It suffices to verify that every $K^\alpha G$-module V is completely reducible. Assume that W is a submodule of V. Since V is a vector space over K, its subspace W has a complement in V, say

$$V = W \oplus W'$$

Let $\theta : V \longrightarrow W$ be the projection map, and let $\psi : V \to V$ be defined by

$$\psi(v) = |G|^{-1} \sum_{x\in G} \bar{x}\theta\bar{x}^{-1} v$$

Bearing in mind that for all $v \in V$ and $y \in G$,

$$\begin{aligned}\psi(\bar{y}v) &= |G|^{-1} \sum_{x\in G} \bar{x}\theta\bar{x}^{-1}\bar{y}v = |G|^{-1} \sum_{z\in G} \overline{yz}\ \theta\ \overline{yz}^{-1}\bar{y}v \\ &= |G|^{-1} \sum_{z\in G} \overline{yz}\ \alpha(y,z)^{-1}\ \theta\ \alpha(y,z)\bar{z}^{-1}\bar{y}^{-1}\bar{y}v \\ &= |G|^{-1} \sum_{z\in G} \overline{yz}\ \theta\ \bar{z}^{-1} v \\ &= \bar{y}\psi(v),\end{aligned}$$

we deduce that ψ is a $K^{\alpha}G$-homomorphism.

Assume that $v \in W$. Then, for any $x \in G$, $\bar{x}^{-1}v \in W$ and so $\theta(\bar{x}^{-1}v) = \bar{x}^{-1}v$. Accordingly, $\bar{x}\theta\bar{x}^{-1}v = v$ and $\psi(v) = v$. Setting $W'' = \mathrm{Ker}\psi$, it follows that W'' is an $K^{\alpha}G$-submodule of V such that $W'' \cap W = 0$. Finally, suppose that $v \in V$. Then, by the above, $v - \psi(v) \in W''$ and so $v = \psi(v) + (v-\psi(v)) \in W + W''$. Thus $V = W \oplus W''$ and the result follows. ∎

Let N be a normal subgroup of G and let K be any field. For any $\beta \in Z^2(G/N,K^*)$, let $\inf\beta$ be the element of $Z^2(G,K^*)$ defined by

$$(\inf\beta)(x,y) = \beta(xN,yN)$$

Note that $\inf\beta(x,y) = 1$ for all $x,y \in N$ and hence $K^{\inf\beta}N$ and KN are identifiable. In what follows we write $I(N)$ for the augmentation ideal of $K^{\inf\beta}N$ so that $I(N)$ is the K-linear span of $\{\bar{n}-\bar{1} \mid 1 \neq n \in N\}$.

3.7. LEMMA. Let N be a normal subgroup of G, let $\beta \in Z^2(G/N,K^*)$ and let $\alpha = \inf\beta$.

(i) $K^{\alpha}G\cdot I(N)$ is an ideal of $K^{\alpha}G$ such that $K^{\alpha}G/K^{\alpha}G\cdot I(N) \cong K^{\beta}(G/N)$

(ii) If N is a normal p-subgroup of G and $\mathrm{char}K = p$, then

$$K^{\alpha}G\cdot I(N) \subseteq J(K^{\alpha}G)$$

and the nilpotency index of $K^{\alpha}G\cdot I(N)$ is equal to that of $J(KN)$

(iii) If N is a normal Sylow p-subgroup of G and $\mathrm{char}K = p$, then

$$J(K^{\alpha}G) = K^{\alpha}G\cdot I(N)$$

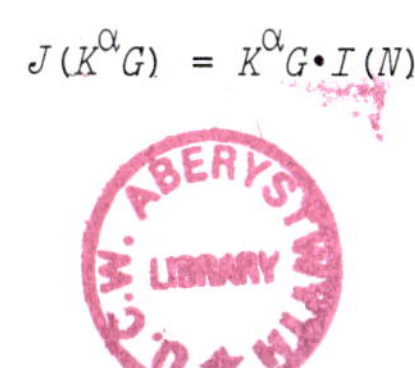

In particular, by (ii), the nilpotency indices of $J(K^{\alpha}G)$ and $J(KN)$ coincide.

Proof. (i) Define a surjective K-homomorphism $f : K^{\alpha}G \rightarrow K^{\beta}(G/N)$ by

$$f(\bar{g}) = \overline{gN} \qquad (g \in G)$$

Then, for all $x,y \in G$, we have

$$\begin{aligned} f(\bar{x}\bar{y}) &= \alpha(x,y)f(\overline{xy}) = \alpha(x,y)\overline{xyN} \\ &= \beta(xN,yN)\overline{xyN} = \overline{xN}\ \overline{yN} \\ &= f(\bar{x})f(\bar{y}) \end{aligned}$$

Thus f is a surjective homomorphism of F-algebras. We are therefore left to verify that $\text{Ker}f = K^{\alpha}G\cdot I(N)$.

If $n \in N$, then $\bar{n} - \bar{1} \in \text{Ker}f$, so $I(N) \subseteq \text{Ker}f$ and thus

$$K^{\alpha}G\cdot I(N) \subseteq \text{Ker}f$$

Let T be a transversal for N in G and let S be the F-linear span of the elements $\bar{t}, t \in T$. It will next be shown that

$$K^{\alpha}G = S + K^{\alpha}G\cdot I(N) \tag{6}$$

and for this it suffices to verify that each $\bar{g} \in S + K^{\alpha}G\cdot I(N), g \in G$. Write $g = tn$ for some $t \in T$ and $n \in N$. Then

$$\begin{aligned} \bar{g} &= \overline{tn} = \alpha^{-1}(t,n)\bar{t}\bar{n} \\ &= \alpha^{-1}(t,n)\bar{t} + \alpha^{-1}(t,n)\bar{t}(\bar{n} - \bar{1}), \end{aligned}$$

proving that $\bar{g} \in S + K^{\alpha}G\cdot I(N)$ and hence (6) is established.

Fix $x \in \text{Ker}f$. Then, by (6), x can be written in the form

$$x = \lambda_1\bar{t}_1 + \ldots + \lambda_n\bar{t}_n + y$$

where $\lambda_i \in F, t_i \in T, 1 \leqslant i \leqslant n$, and $y \in K^{\alpha}G\cdot I(N)$. Because $y \in \text{Ker}f$, we have

$$f(x) = \lambda_1\overline{t_1N} + \ldots + \lambda_n\overline{t_nN} = 0,$$

which implies that $\lambda_1 = \ldots = \lambda_n = 0$ and so $x \in K^{\alpha}G\cdot I(N)$. This shows that $\text{Ker}f \subseteq K^{\alpha}G\cdot I(N)$ and hence that $\text{Ker}f = K^{\alpha}G\cdot I(N)$, as required.

(ii) Fix $n \in N$ and write $|N| = p^d$ for some $d \geqslant 1$. Since $\alpha(x,y) = 1$ for all $x,y \in N$, we have

$$(\bar{n}-\bar{1})^{p^d} = \bar{n}^{p^d} - \bar{1} = \overline{n^{p^d}} - \bar{1} = 0$$

Bearing in mind that the elements $\bar{n}-\bar{1}$, $n \neq 1$, constitute a K-basis for $I(N)$, we conclude that $I(N)^m = 0$ for some $m \geqslant 1$. By (i), $K^{\alpha}G\cdot I(N)$ is an ideal of $K^{\alpha}G$, so $K^{\alpha}G\cdot I(N) = I(N)K^{\alpha}G$ and therefore

$$[K^{\alpha}G\cdot I(N)]^m = K^{\alpha}G\cdot I(N)^m = 0$$

Thus $K^{\alpha}G\cdot I(N) \subseteq J(K^{\alpha}G)$. Furthermore, the above equality shows that the nilpotency index of $J(KN) = I(N)$ coincides with that of $K^{\alpha}G\cdot I(N)$.

(iii) Direct consequence of (i), (ii) and Lemma 3.6. ∎

3.8. PROPOSITION. Let N be a normal p-subgroup of G, let F be an algebraically closed field of characteristic p and let $\alpha \in Z^2(G,F^*)$.

(i) There exists $\beta \in Z^2(G/N,F^*)$ such that α is cohomologous to $\inf\beta$.

(ii) If N is a normal Sylow p-subgroup of G, then

(a) $F^{\alpha}G/J(F^{\alpha}G) \cong F^{\beta}(G/N)$

(b) The nilpotency index of $F^{\alpha}G$ is equal to that of $J(FN)$.

Proof. (i) By Lemma 3.4.9, we may harmlessly assume that

$$\bar{x}\bar{y} = \overline{xy} \qquad \text{for all } x,y \in N \tag{7}$$

Fix $g \in G, n \in N$ and write $p^d = |N|$. Then

$$\bar{g}\bar{n}\bar{g}^{-1} = \lambda\,\overline{gng^{-1}}$$

for some $\lambda \in F^*$ and so by (7),

$$\bar{1} = \overline{n^{p^d}} = \bar{n}^{p^d} = \bar{g}\bar{n}^{p^d}\bar{g}^{-1} = \lambda^{p^d}\,\overline{gng^{-1}}^{p^d} = \lambda^{p^d}\cdot\bar{1}$$

Hence $\lambda^{p^d} = 1$, so $\lambda = 1$ and thus

$$\bar{g}\bar{n}\bar{g}^{-1} = \overline{gng^{-1}} \qquad \text{for all } g \in G, n \in N \tag{8}$$

Clearly $F^{\alpha}G\cdot I(N)$ is a left ideal of $F^{\alpha}G$. Because $I(N)$ is the F-linear span of the set $\{\bar{n}-\bar{1} \mid n \in N\}$, $F^{\alpha}G\cdot I(N)$ is the F-linear span of the set

$$\{\bar{g}(\bar{n}-\bar{1}) \mid g \in G\ n \in N\}$$

We claim that $F^{\alpha}G\cdot I(N)$ is an ideal. To substantiate our claim, we need only show that $\bar{g}(\bar{n}-\bar{1})\bar{x} \in F^{\alpha}G\cdot I(N)$ for all $g,x \in G, n \in N$. Invoking (8), we have

$$\begin{aligned}\bar{g}(\bar{n}-\bar{1})\bar{x} &= \bar{g}\,\bar{x}\,(\bar{x}^{-1}\bar{n}\,\bar{x}-\bar{1})\\ &= \bar{g}\,\bar{x}\,(\overline{x^{-1}nx}-\bar{1}) \in F^{\alpha}G\cdot I(N),\end{aligned}$$

as claimed.

Let T be a transversal for N in G containing 1, and let S be the F-linear span of $\{\bar{t}\,|\,t \in T\}$. We now prove that

$$F^{\alpha}G = S \oplus F^{\alpha}G\cdot I(N) \qquad \text{as } F\text{-spaces} \tag{9}$$

Given $t \in T$ and $x,y \in N$, we have

$$\begin{aligned}\overline{tx}(\bar{y}-\bar{1}) &= \alpha^{-1}(t,x)\bar{t}\,\bar{x}\,(\bar{y}-\bar{1})\\ &= \alpha^{-1}(t,x)\bar{t}(\overline{xy}-\bar{x})\\ &= \alpha^{-1}(t,x)\bar{t}(\overline{xy}-\bar{1}) - (\bar{x}-\bar{1})\end{aligned}$$

and so $F^{\alpha}G\cdot I(N)$ is in fact the F-linear span of

$$\{\bar{t}(\bar{n}-\bar{1})\,|\,t \in T,\ 1 \neq n \in N\}$$

Observe that the latter set combined with T consists of $|G|$ elements. The equality

$$\overline{tn} = \alpha^{-1}(t,n)\bar{t}(\bar{n}-\bar{1}+\bar{1}) = \alpha^{-1}(t,n)\bar{t}(\bar{n}-\bar{1}) + \alpha^{-1}(t,n)\bar{t}$$

now proves (9).

Setting $\overline{tN} = \bar{t} + F^{\alpha}G\cdot I(N)$, it follows from (9) that $F^{\alpha}G/F^{\alpha}G\cdot I(N)$ is an F-algebra with the elements $\{\overline{tN}\,|\,t \in T\}$ as a basis. Given $t_1,t_2 \in T$, there exist unique $t \in T, n \in N$ with $t_1t_2 = tn$. Setting

$$\beta(t_1N,t_2N) = \alpha(t_1,t_2)\alpha^{-1}(t,n)$$

we have

$$\bar{t}_1\bar{t}_2 \equiv \beta(t_1N,t_2N)\bar{t} \pmod{F^{\alpha}G\cdot I(N)}$$

and so

$$\overline{t_1N}\cdot\overline{t_2N} = \beta(t_1N,t_2N)\overline{t_1t_2N}$$

proving that $F^{\alpha}G/F^{\alpha}G\cdot I(N) \cong F^{\beta}(G/N)$.

Given $g \in G$, there exist unique $t(g) \in T$ and $n(g) \in N$ with $g = t(g)n(g)$. Define $\lambda : G \longrightarrow F^*$ by $\lambda(g) = \alpha(t(g),n(g))$, set $\gamma = \alpha(\delta\lambda)$ and $\tilde{g} = \lambda(g)\bar{g}, g \in G$. Then $\{\tilde{g} | g \in G\}$ is an F-basis of $F^{\gamma}G$ with $\tilde{x}\tilde{y} = \gamma(x,y)\widetilde{xy}$ and

$$\tilde{g} \equiv \overline{t(g)} \pmod{F^{\alpha}G \cdot I(N)}$$

Hence the map $f : F^{\alpha}G \longrightarrow F^{\beta}(G/N)$ defined by $f(\tilde{g}) = \overline{gN}$ determines an F-algebra homomorphism. Thus, given $x,y \in G$, we have

$$\begin{aligned} \beta(xN,yN)\overline{xyN} &= \overline{xN}\ \overline{yN} = f(\tilde{x})f(\tilde{y}) \\ &= f(\tilde{x}\tilde{y}) = \gamma(x,y)f(\widetilde{xy}) \\ &= \gamma(x,y)\overline{xyN} \end{aligned}$$

proving that $\gamma(x,y) = \beta(xN,yN)$. Hence $\gamma = \inf\beta$ as required.

(ii) By (i), replacing α by a cohomologous cocycle, if necessary, we may assume that $\alpha = \inf\beta$. Now apply Lemma 3.7. ∎

3.9. COROLLARY. Let F be an algebraically closed field of characteristic $p > 0$, let N be a normal p'-subgroup of G and let H be the inertia group of an irreducible FN-module V. If $P \in \mathrm{Syl}_p(H)$ assume that $PN \triangleleft H$. Then the Loewy length of V^G is equal to the nilpotency index of $J(FP)$.

Proof. Apply Corollary 3.5 and Proposition 3.8(ii)(b). ∎

4. GROUPS OF p-LENGTH 2

Throughout this section, G denotes a finite group and F a field of characteristic p. All conventions and notations adopted in the previous section remain in force.

Our aim is to provide circumstances under which the inequality of Theorem 2.12(ii) becomes an equality. One of our results will prove that this is always the case if H is p-nilpotent with elementary abelian Sylow p-subgroups.

4.1. LEMMA. Let V and W be FG-modules. Set $n = L(V), m = L(W)$ and

$$X_{ij} = (J^{i}V/J^{i+1}V) \otimes_F (J^{j}W/J^{j+1}W)$$

$(0 \leqslant i \leqslant n-1, 0 \leqslant j \leqslant m-1)$, where $J = J(FG)$. Then

$$L(V \underset{F}{\otimes} W) \leqslant \sum_{\ell=0}^{n+m-2} \max\{L(X_{ij}) \mid i+j = \ell\}$$

Proof. Put $Y_{ij} = J^i V \underset{F}{\otimes} J^j W$. Then

$$Y_{ij} \supseteq Y_{i+1,j} + Y_{i,j+1}$$

and

$$Y_{ij}/(Y_{i+1,j} + Y_{i,j+1}) \cong X_{ij}$$

Now let $Y_\ell = \sum_{i+j=\ell} Y_{ij}$ for $0 \leqslant \ell \leqslant n+m-1$. Then

$$0 = Y_{n+m-1} \subseteq Y_{n+m-2} \subseteq \dots \subseteq Y_0 = V \underset{F}{\otimes} W$$

and the canonical map

$$\bigoplus_{i+j=\ell} Y_{ij} \longrightarrow Y_\ell$$

yields a surjective homomorphism

$$\bigoplus_{i+j=\ell} X_{ij} \cong \bigoplus_{i+j=\ell} Y_{ij}/(Y_{i+1,j} + Y_{i,j+1}) \longrightarrow X_\ell/X_{\ell+1}$$

where $X_\ell = \sum_{i+j=\ell} X_{ij}$. It follows that

$$L(X_\ell/X_{\ell+1}) \leqslant \max\{L(X_{ij}) \mid i+j = \ell\}$$

thus proving the assertion. ■

4.2. LEMMA. Let N be a normal subgroup of G such that G/N is a p-group. Let W be an FG-module and set $V = (W_N)^G \cong W \underset{F}{\otimes} (1_N)^G$. Then

$$L(V) \leqslant t(G/N) + L(W) - 1$$

Proof. Put $M = G/N$ and view FM as an FG-module via $FM \cong (1_N)^G$. Then $I(M)^i = J^i \cdot FM$, where $J = J(FG)$, and

$$I(M)^i/I(M)^{i+1} \cong n_i 1_G$$

for suitable integers n_i. In the notation of Lemma 4.1, we therefore have

$$\begin{aligned} X_{ij} &= (I(M)^i/I(M)^{i+1}) \underset{F}{\otimes} (J^j W/J^{j+1} W) \\ &\cong n_i (J^j \cdot W/J^{j+1} \cdot W) \end{aligned}$$

and so $L(X_{ij}) = 1$ for all i,j. Hence, by Lemma 3.1,

$$L(V) \leqslant L(FM) + L(W) - 1$$

as required. ■

4.3. LEMMA. Let G be a p-nilpotent group, let F be an algebraically closed field of characteristic p and let W be an irreducible FN-module with inertia group H where $N = O_{p'}(G)$

(i) If V is an irreducible FG-module such that W is a submodule of V_N, then

$$P(V) \cong W^G \quad \text{and} \quad L(P(V)) = t(H/N)$$

(ii) There exists a unique irreducible FG-module X such that $X_N \cong W$.

Proof. That $P(V) \cong W^G$ is a particular case of Lemma 3.4.26(i). Let X and Y be irreducible FG-modules with $X_N \cong Y_N \cong W$. Then $P(X) \cong P(Y) \cong W^G$ and hence $X \cong Y$. The existence of an irreducible FG-module X with $X_N \cong W$ follows from Theorem 3.4.5 and Lemma 3.4.9. Finally, the equality $L(P(V)) = t(H/N)$ is a particular case of Corollary 3.5. ■

4.4. THEOREM. (Lorenz (1985)). Let F be an algebraically closed field of characteristic $p > 0$ and let V be an irreducible FG-module. Assume that G has a normal p-subgroup N with $G = NH$ and $N \cap H = 1$ for some p-nilpotent subgroup H of G. Let W be an irreducible submodule of V_Q, where $Q = O_{p'}(H)$, and let T be the inertia group of W in H.

(i) $P_G(V) \cong W^G$

(ii) $L(P_G(V)) \geqslant t(T/Q) + L((V_H)^G) - 1$ with equality if $T \cap S$ has a normal complement in $S \in \mathrm{Syl}_p(H)$

(iii) If S is elementary abelian of order p^n and p^d is the p-part of $\dim_F V$, then

$$L(P_G(V)) = (n-d)(p-1) + L((V_H)^G)$$

(iv) If V_Q is irreducible, then

$$L(P_G(V)) \geqslant t(S) + t(N) - 1$$

and equality holds if and only if the FG-module

$$V \otimes_F \left(\bigoplus_{i\geqslant 0} [I(N)^i/I(N)^{i+1}]\right)$$

is completely reducible. Here we view $I(N)^i/I(N)^{i+1}$ as an FG-module by letting

G act by conjugation.

Proof. (i) and (ii). By Lemma 4.3(i), we have $P_H(V) \cong W^H$ and $L(W^H) = t(T/Q)$. Invoking Theorem 2.12, we infer that

$$L(P_G(V)) \geqslant t(T/Q) + L((V_H)^G) - 1$$

and

$$P_G(V) \cong (1_H)^G \underset{F}{\otimes} W^H \cong W^G$$

By Lemma 4.3(ii), there exists a unique FT-module U such that $U_Q \cong W$. The induced module U^H is irreducible and $U^H \cong V_H$, since both have a common FQ-component. Assume that S_1 is a normal complement for $T \cap S$ in S and put $X = \langle Q, S_1 \rangle \subseteq H$. Then X is normal in H and

$$V_X \cong (U^H)_X \cong W^X$$

Thus we have

$$\begin{aligned} P_G(V) &\cong (1_H)^G \underset{F}{\otimes} W^H \\ &\cong (1_H)^G \underset{F}{\otimes} (V_X)^H \\ &\cong (1_H)^G \otimes (V_{\langle N,X\rangle})^G \\ &\cong ((V_H)^G|_{\langle N,X\rangle})^G \end{aligned}$$

It follows from Lemma 4.2 that

$$L(P_G(V)) \leqslant t(G/\langle N,X\rangle) + L((V_H)^G) - 1$$

Since $G/\langle N,X\rangle \cong T/Q$, the required assertion follows.

(iii) Assume that S is elementary abelian of order p^n and let p^d be the p-part of $\dim_F V$. By assumption on S, $T \cap S$ has a normal complement in S, and by Corollary 3.2.5,

$$t(T/Q) = k(p-1) + 1$$

where $p^k = |T \cap S|$. Let U be as in the proof of (ii) so that $U^H \cong V_H$. Then $\dim_F U$ is not divisible by p and hence the p-part of $\dim_F V$ equals $p^d = |H/T|$. Thus $k = n - d$ which proves the assertion by applying (ii).

(iv) By (ii), we have

$$L(P_G(V)) = t(S) + L((V_H)^G) - 1$$

and, by Theorem 2.12(iii), $L((V_H)^G) \geqslant t(N)$ with equality if and only if

$$V \underset{F}{\otimes} (\bigoplus_{i \geqslant 0} [I(N)^i/I(N)^{i+1}]$$

is completely reducible. This completes the proof of the theorem. ∎

4.5. LEMMA. Let N be a normal p-subgroup of G with $G = NH$ and $N \cap H = 1$ for some subgroup H of G. Then

$$J(FG) = FG \cdot I(N) + J(FH)$$

Proof. Let $\pi : FG \longrightarrow FH$ be the natural projection. Then $\text{Ker}\pi = FG \cdot I(N)$, π is the identity map on FH and, by Lemma 1.1(iii), $J(FH) = \pi(J(FG))$. Furthermore, by Proposition 3.1.1(i), we have $FG \cdot I(N) \subseteq J(FG)$. Now, if $x \in J(FG)$, then $\pi(x) = y = \pi(y)$ for some $y \in J(FH)$. Thus $x - y \in FG \cdot I(N)$ and so $x \in FG \cdot I(N) + J(FH)$. Conversely, let $x \in FG \cdot I(N) + J(FH)$. Then $\pi(x) = y$ for some $y \in J(FH)$ and hence $\pi(x) = \pi(z)$ for some $z \in J(FG)$. Hence $x - z \in FG \cdot I(N) \subseteq J(FG)$ and therefore $x \in J(FG)$. ∎

We are now ready to establish the final result of this section.

4.6. THEOREM. (Lorenz (1985)). Let N be a normal p-subgroup of G with $G = NH$ and $N \cap H = 1$ for some Frobenius group H with kernel Q and complement S, where Q and S are p' and p-groups, respectively. Then the following conditions are equivalent:

(i) $t(G) = t(S) + t(N) - 1$

(ii) $\bigoplus_{i \geqslant 0} (I(N)^i/I(N)^{i+1})$ is a completely reducible FG-module, where G acts by conjugation.

(iii) $\sum_{q \in Q} x^{qs} - \sum_{q \in Q} x^q \in I(N)^{i+1}$ for all $i \geqslant 0$, all $x \in I(N)^i$ and all $s \in S$.

Proof. (i) ⇒ (ii): By Corollary 3.1.18, we may harmlessly assume that F is algebraically closed. By taking $V = 1_G$ in Theorem 4.4, we have

$$t(G) \geqslant L(P_G(V)) \geqslant t(S) + t(N) - 1$$

and so $L(P_G(V)) = t(S) + t(N) - 1$. Now apply Theorem 4.4(iv).

(ii) ⇒ (iii): Owing to Theorem 3.7.7, our assumption on H implies that

$$J(FH) = I(S)e \tag{1}$$

where

$$e = |Q|^{-1} \sum_{q\in Q} q$$

Set $X = FG\cdot I(N)$ and $Y = I(S)e$. Then, by (1) and Lemma 4.5.

$$J(FG) = X + Y \tag{2}$$

By hypothesis, $J(FG)$ annihilates each $I(N)^i/I(N)^{i+1}$. But X also annihilates $I(N)^i/I(N)^{i+1}$, so Y annihilates each $I(N)^i$. Therefore $(s-1)\sum_{q\in Q} q$ annihilates $J(N)^i/I(N)^{i+1}$ for all $i \geq 0$ and all $s \in S$. Since for any $x \in I(N)$

$$(s-1)(\sum_{q\in Q} q)x = \sum_{q\in Q} x^{qs} - \sum_{q\in Q} x^q$$

the required assertion follows.

(iii) ⇒ (i): An easy calculation shows that for all $i \geq 0$

$$e\cdot I(S)\cdot I(N)^i e + eI(N)^{i+1}e = e\cdot I(N)^i I(S)e + eI(N)^{i+1}e \tag{3}$$

Set $\ell = t(N) + t(S) - 1$. In view of (2), we have to show that if $\alpha \in FG$ can be written as a product of ℓ factors each of which belongs to either X or Y then $\alpha = 0$.

We argue by descending induction on the number $\ell_x = \ell_x(\alpha)$ of factors from X involved in α. If $\ell_x \geq t(N)$, then $\alpha \in X^{t(N)} = 0$. We may therefore assume that $\ell_x < t(N)$. Then the number of factors from Y involved in α is at least $t(S)$. Let $n_y = n_y(\alpha)$ denote the length of the longest consecutive subproduct of α consisting entirely from factors in Y. Clearly, if $n_y \geq t(S)$ then $\alpha = 0$. So assume that $n_y < t(S)$. Then α contains a subproduct which either belongs to $YX^iY^{n_y}$ or to $Y^{n_y}X^iY (i > 0)$. We examine the first case, the second being entirely analogous. Now

$$\begin{aligned} YX^iY^{n_y} &= e\cdot I(S)\cdot I(N)^i\cdot eI(S)^{n_y} \\ &\subseteq e\, I(N)^i I(S)eI(S)^{n_y} + eI(N)^{i+1}eI(S)^{n_y} \quad \text{(by (3))} \\ &\subseteq X^iY^{n_y+1} + X^{i+1}Y^{n_y} \end{aligned}$$

Thus we have $\alpha = \alpha_1 + \alpha_2$ with $\alpha_1,\alpha_2 \in J(FG)^\ell$, but $\ell_x(\alpha_2) > \ell_x(\alpha)$ and

$\ell_x(\alpha_1) = \ell_x(\alpha), n_y(\alpha_1) > n_y(\alpha)$. By induction, we deduce that $\alpha_1 = \alpha_2 = 0$ and hence $\alpha = 0$. This proves the theorem. ∎

7
The nilpotency index

Let G be a finite group, let F be a field of characteristic p and let $t(G)$ be the nilpotency index of $J(FG)$. Our aim in this chapter is to provide some important information on the number $t(G)$. After examining an upper and lower bounds for $t(G)$, we investigate circumstances under which the equalities hold. Almost all of the results presented pertain to p-solvable groups since virtually nothing is known in the general case. A number of interesting examples is also provided.

1. SOME RESULTS ON p-SOLVABLE GROUPS

Throughout this section, G denotes a finite group and p a fixed prime. We write $GL(n,p)$ and $SL(n,p)$ for the general linear group and the special linear group of degree n over the field $\mathbb{F}_p$ of p-elements, respectively. Our aim is to record some group-theoretic results required for subsequent investigations.

First we record the following conventions. If $M \subseteq N$ are normal subgroups of G, we set

$$C_G(N/M) = \{g \in G \mid gM \in C_{G/M}(N/M)\}$$

so that

$$C_{G/M}(N/M) = C_G(N/M)/M$$

Expressed otherwise, $C_G(N/M)$ is the kernel of the homomorphism $G \longrightarrow \mathrm{Aut}(N/M)$, where the image of $g \in G$ is the automorphism of N/M given by

$$nM \longmapsto g^{-1}ngM$$

1.1. LEMMA. Let Φ be the Frattini subgroup of a p-group G, let $(G:\Phi) = p^d$ and consider $V = G/\Phi$ to be a d-dimensional vector space over the field $\mathbb{F}_p$ of p elements (see Lemma 4.3.1). Then

(i) $G = \langle x_1, x_2, \dots, x_n\rangle$ if and only if V is spanned by $\Phi x_1, \Phi x_2, \dots, \Phi x_n$. In particular, if $G = \langle x_1, x_2, \dots, x_n\rangle$, then $n \geqslant d$.

(ii) The group G can be generated by exactly d elements. The subset $\{x_1, x_2, \dots, x_d\}$ generates G if and only if $\{\Phi x_1, \dots, \Phi x_d\}$ is a basis of V.

Proof. (i) If $G = \langle x_1, x_2, \dots, x_n\rangle$, then V is generated by $\Phi x_1, \dots, \Phi x_n$. Since a d-dimensional vector space cannot be spanned by less than d elements, we have $n \geqslant d$. Conversely, assume that $V = \langle \Phi x_1, \dots, \Phi x_n\rangle$. Setting $X = \{x_1, \dots, x_n\}$, we have $G = \langle X, \Phi\rangle$, so $G = \langle X\rangle$ since Φ can be excluded from any generating set of G.

(ii) If $\{\Phi x_1, \dots, \Phi x_d\}$ is a basis of V, then by (i), we have $G = \langle x_1, \dots, x_d\rangle$. Conversely, if $G = \langle x_1, \dots, x_d\rangle$, then by (i) we get $V = \langle \Phi x_1, \dots, \Phi x_d\rangle$. In this case $\Phi x_1, \dots, \Phi x_d$ are linearly independent and so form a basis of V. ■

As an application, we prove the following result of P. Hall (1933).

1.2. THEOREM. Let Φ be the Frattini subgroup of a p-group G, and let $(G:\Phi) = p^d$ and $|G| = p^n$. Denote by I the set of all automorphisms of G which leave every element of G/Φ invariant.

(i) The set I is a normal subgroup of $\mathrm{Aut}(G)$ and the factor group $\mathrm{Aut}(G)/I$ is isomorphic to a subgroup of $GL(d,p)$.

(ii) The subgroup I is a p-group of order less than $p^{(n-d)d}$. Thus

$$|\mathrm{Aut}(G)| \text{ divides } p^m \prod_{i=1}^{d} (p^i - 1)$$

where $m = nd - d(d+1)/2$.

Proof. (i) Put $V = G/\Phi$. Since Φ is a characteristic subgroup of G, every automorphism σ of G induces an automorphism $f(\sigma)$ of V. The map $\sigma \longrightarrow f(\sigma)$ is obviously a homomorphism for $\mathrm{Aut}(G)$ into $\mathrm{Aut}(V) \cong GL(d,p)$. By definition, we have $I = \mathrm{Ker}$ and (i) follows.

(ii) Fix a basis $\{\Phi x_1, \dots, \Phi x_d\}$ of V over $\mathbb{F}_p$. Let X be the totality of all $(y_1, y_2, \dots, y_d)$ such that $y_i \in G$ and $\Phi y_i = \Phi x_i$ for all $i \in \{1,2,\dots,d\}$. Clearly we get $|X| = p^{(n-d)d}$. Because any element ψ of I satisfies $\psi(\Phi x_i) = \Phi x_i$ for all i, the group I acts on X by the rule

$$(y_1, y_2, \ldots, y_d) \longmapsto (\psi(y_1), \ldots, \psi(y_d))$$

Let S be an orbit under this action. Then the length of S is the index of the stabilizer $I(Y)$ of an element $Y = (y_1, y_2, \ldots, y_d)$. If $\psi \in I(Y)$, then $\psi(y_i) = y_i$ for all $i \in \{1,2,\ldots,d\}$. Because $G = \langle y_1, y_2, \ldots, y_d \rangle$ by Lemma 1.1, we get $\psi = 1$. Hence $I(Y) = \{1\}$ and $|S| = |I|$. Since S is an arbitrary orbit, each orbit is of length $|I|$; hence $|I|$ divides $|X|$. This shows that I is a p-group whose order is a divisor of $p^{(n-d)d}$. The last statement is a consequence of the fact that $|GL(d,p)| = p^{d(d-1)/2} \prod_{i=1}^{d} (p^i - 1)$. ■

For future use, we need an important property of the upper p-series of a p-solvable group, known as the Hall-Higman's lemma (Hall and Higman (1956)). This is recorded in

1.3. LEMMA. Let G be a p-solvable group and let $\Phi/O_{p'}(G)$ be the Frattini subgroup of $O_{p',p}(G)/O_{p'}(G)$. Then

$$C_G(O_{p',p}(G)/\Phi) = O_{p',p}(G)$$

In particular,

(a) $C_G(O_{p',p}(G)/O_{p'}(G)) \subseteq O_{p',p}(G)$

(b) $C_G(O_p(G)) \subseteq O_p(G)$ if $O_{p'}(G) = 1$

(c) The group $G/O_{p',p}(G)$ is isomorphic to a subgroup of $\mathrm{Aut}(O_{p',p}(G)/\Phi)$ and

$$\mathrm{Aut}(O_{p',p}(G)/\Phi) \cong GL(d,p)$$

where $p^d = |O_{p',p}(G)/\Phi|$

Proof. Put $K = C_G(O_{p',p}(G)/\Phi)$ and observe that, by Lemma 4.3.1, $O_{p',p}(G) \subseteq K$. By the definition of $O_{p',p}(G)$, $K/O_{p',p}(G)$ cannot be a nontrivial p-group, so that, if $K \supset O_{p',p}(G)$, K must contain an element g not in $O_{p',p}(G)$ of order prime to p. Then g induces an automorphism of $O_{p',p}(G)/O_{p'}(G)$ of order prime to p. But an automorphism of a p-group which is the identity modulo the Frattini subgroup has order a power of p (Theorem 1.2(ii)). Hence g induces in $O_{p',p}(G)/\Phi$ an automorphism which is not the identity, contrary to the definition of K. Thus $K = O_{p',p}(G)$, as required. ■

We close this section by recording some standard group-theoretic facts required for subsequent investigations. The proofs of these facts are rather involved and therefore will be omitted.

It will be convenient to introduce the following terminology. We say that a group K is *involved* in G provided K is isomorphic to a homomorphic image of a subgroup H of G.

1.4. THEOREM. Let G be a p-solvable group and assume that either $p \geq 5$ or $p = 3$ and $SL(2,3)$ is not involved in G.

(i) If P is a Sylow p-subgroup of G, then every normal abelian subgroup of P is contained in $O_{p',p}(G)$.

(ii) If a Sylow p-subgroup of G is of class at most 2, then G has p-length 1.

Proof. For the proof of (i) refer to Gorenstein (1968,p.234). Property (ii) is an immediate consequence of (i). ∎

For a given prime p, we define $Qd(p)$ to be the semidirect product of $\mathbb{Z}_p \times \mathbb{Z}_p$ and $SL(2,p)$ with respect to the natural action of $SL(2,p)$ on $\mathbb{Z}_p \times \mathbb{Z}_p$.

1.5. THEOREM. (i) If p is odd, then the Sylow p-subgroups of $Qd(p)$ are non-abelian of order p^3 and exponent p.

(ii) If $G = Qd(3)$, then G is of 3-length 2.

Proof. For the proof of (i), refer to Glauberman (1971, Example 11.4). Property (ii) is an easy exercise based on Glauberman (1971,pp.32-33) and Gorenstein (1968,p.203) (see also Koshitani (1982a,p.33)). ∎

1.6. THEOREM. Let G be a p-solvable group with Sylow p-subgroup P of class $c(P)$ and let $\ell_p(G)$ be the p-length of G.

(i) $C_G(P \cap O_{p',p}(G)) \subseteq O_{p',p}(G)$ and, in particular, $Z(P) \subseteq O_{p',p}(G)$

(ii) $\ell_p(G) \leq c(P)$

Proof. See Gorenstein (1968,p.228) and Huppert (1967,p.691) ∎

1.7. THEOREM. Let p be an odd prime, and let G be a finite group with a Sylow p-subgroup P such that $O_p(G) \neq 1$, G is p-solvable and $Qd(p)$ is not involved in G. If A is an abelian normal subgroup P, then $A \subseteq O_{p',p}(G)$.

Proof. See Glauberman (1986, Lemma 6.3) and Gorenstein (1986, Theorem 8.1.3) (for a short proof refer also to Koshitani (1982a, Lemma 2). ■

1.8. THEOREM. (Hall-Higman). Let G be a p-solvable group of linear transformations in which $O_p(G) = 1$ acting on a vector space over a field F of characteristic p. Let g be an element of G of order p^n. Then the minimal polynomial of g in V is $(X-1)^r$, where either

(i) $r = p^n$, or

(ii) There exists an integer $n_o \leqslant n$ such that $p^{n_o} - 1$ is a power of a prime q and the Sylow q-subgroups of G are nonabelian. In this case, if n_o is the least such integer, then

$$p^{n-n_o}(p^{n_o}-1) \leqslant r \leqslant p^n$$

Proof. See Gorenstein (1968, p. 359). ■

For future use, we also record the following general group-theoretic facts.

1.9. THEOREM. Let A be an abelian normal subgroup of G with $|G/A| = \prod_{i=1}^{s} p_i^{\alpha_i}$, where the p_i are distinct primes, and let P_i/A be a Sylow p_i-subgroup of G/A. Then A has a complement in G if and only if A has a complement in all P_i.

Proof. See Huppert (1967, p. 124). ■

Recall that the group $M_n(p)$ is defined for $n \geqslant 4$ if $p = 2$ and $n \geqslant 3$ if $p \neq 2$ as follows:

$$M_n(p) = \langle a,b \mid a^{p^{n-1}} = 1, b^p = 1, b^{-1}ab = a^{1+p^{n-2}} \rangle$$

Let D_n and Q_n be the dihedral and generalized quaternion groups of order 2^n. Here $n \geqslant 2$ for D_n and $n \geqslant 3$ for Q_n. Finally, for $n > 3$, let $S(n)$ be the semidihedral group of order 2^n defined by

$$S(n) = \langle a,b \mid a^{2^{n-1}} = 1,\ b^2 = 1,\ b^{-1}ab = a^{-1+2^{n-2}} \rangle$$

1.10. THEOREM. Let P be a nonabelian p-group of order p^n which contains a cyclic subgroup of order p^{n-1}. Then

(i) If p is odd, P is isomorphic to $M_n(P)$

(ii) If $p = 2$ and $n = 3$, then P is isomorphic to D_3 or Q_3

(iii) If $p = 2$ and $n > 3$, then P is isomorphic to $M_n(2)$, D_n, Q_n or $S(n)$

Proof. See Gorenstein (1968, p.193). ∎

A p-group P is said to be *regular*, if for any $x,y \in P$ there exist $x_1, x_2, \ldots, x_n$ in $\langle x,y\rangle'$ such that

$$x^p y^p = (xy)^p x_1^p x_2^p \ldots x_n^p$$

1.11. THEOREM. Let P be a p-group.

(i) If the nilpotency class of P is less than p, then P is regular

(ii) If P is of exponent p, then P is regular

(iii) If $p = 2$ and P is regular, then P is abelian.

Proof. See Huppert (1967, p.322). ∎

2. UPPER AND LOWER BOUNDS FOR $t(G)$.

Throughout this section, G denotes a finite group and F an arbitrary field of characteristic $p > 0$. As usual, $t(G)$ is the nilpotency index of $J(FG)$. If B is a block of FG, then we write $t(B)$ for the nilpotency index of $J(B)$.

We start by recording some properties which are immediate consequences of previously established results.

2.1. PROPOSITION. (Passman (1970a), Tshushima (1967), Wallace (1968)). Let G be a p-solvable group of order $p^a m$, where $p \nmid m$ and $a \geqslant 1$. Then

$$a(p-1) + 1 \leqslant t(G) \leqslant p^a$$

Proof. Apply Proposition 3.1.9 and Corollary 6.2.8. ∎

2.2. LEMMA. Let B be the principal block of FG, where G is a p-solvable group. Then

$$t(B) = t(G/O_{p'}(G))$$

Proof. Direct consequence of Proposition 3.1.20. ∎

The first equality in the result below is contained in a work of Clarke (1972).

2.3. PROPOSITION. Let G be a p-solvable group of p-length 1, let $P \in Syl_p(G)$ and let B be the principal block of FG. Then

$$t(G) = t(P) = t(B)$$

Proof. Note that $P \cong O_{p'}(G)P/O_{p'}(G) \in Syl_p(G/O_{p'}(G))$ and $G/O_{p'}(G)$ is of p-length 1. Hence, by Lemma 2.2, it suffices to prove that $t(G) = t(P)$. By Corollary 3.1.18, we may harmlessly assume that F is algebraically closed. Furthermore, by Proposition 3.1.8(ii), we may also assume that G is p-nilpotent. The desired assertion is now a consequence of Morita's result (Corollary 3.10.10). An alternative proof, which avoids the use of Proposition 3.1.8(ii) and Corollary 3.10.10, can be obtained by applying Corollary 6.3.9. ∎

2.4. PROPOSITION. (Koshitani (1977b), Tsushima (1978b)). Let G be a p-solvable group of order $p^a m$, where $p \nmid m$, and let $P \in Syl_p(G)$. Then $t(G) = p^a$ if and only if P is cyclic.

Proof. Assume that P is cyclic. Then, by Proposition 3.16.20, G is of p-length 1. Hence, by Proposition 2.3, $t(G) = t(P)$ and therefore, by Theorem 4.3.2, $t(G) = p^a$.

Conversely, suppose that $t(G) = p^a$. Put $M = O_{p'}(G), H = O_{p',p}(G), p^r = |H/M|$ and Φ/M the Frattini subgroup of H/M. Then, by Proposition 3.1.9,

$$J(FG)^{p^{a-r}} \subseteq FG \cdot J(FH)$$

and so

$$0 \neq J(FG)^{p^a - p^{a-r}} \subseteq FG \cdot J(FH)^{p^r - 1}$$

Hence, by Propositions 2.1 and 2.3, $t(H) = t(H/M) = p^r$. Since H/M is of order p^r, it follows from Theorem 4.3.2 that H/M is cyclic. Hence, by Lemma 4.3.1, H/Φ is cyclic of order p. Therefore $|Aut(H/\Phi)| = p-1$ and, by Lemma 1.3, G/H is a subgroup of $Aut(H/\Phi)$. Thus G/H is a p'-group, proving that G is of p-length 1. Hence $P \cong H/M$ is cyclic as asserted. ∎

2.5. PROPOSITION. (Motose and Ninomiya (1975a)). Let G be a p-solvable group of p-length 1 and of order $p^a m$, $a \geqslant 1$, $p \nmid m$. Then $t(G) = a(p-1) + 1$ if and only if the Sylow p-subgroups of G are elementary abelian.

Proof. If $P \in Syl_p(G)$, then $t(G) = t(P)$ by Proposition 2.3. Now apply Theorem 4.3.2. ∎

2.6. PROPOSITION. (Motose and Ninomiya (1975a), Wallace (1968)). Let G be a

p-solvable group of order $p^a m$, $a \geq 1$, $p \nmid m$. Then the following conditions are equivalent:

(i) $t(G) = 3$

(ii) $p^a = 3$ or $p^a = 4$ and the Sylow 2-subgroups of G are elementary abelian.

Proof. (i) ⇒ (ii): Let P be a Sylow p-subgroup of G. By Proposition 1.1,

$$3 = t(G) \geq a(p-1) + 1$$

This implies that $a = 1$ and $p = 3$ or that $a = 2$ and $p = 2$. In both cases G is of p-length 1 since P is abelian (see Proposition 3.16.20). Hence $t(G) = t(P)$ by Proposition 2.3. Thus, in case $p^a = 4$, P is elementary abelian by Proposition 2.4.

(iii) ⇒ (i): By the foregoing, G is of p-length 1. Now apply Propositions 2.5 and 2.4. ∎

Let G be a finite group of order $p^a m$, $a \geq 1$, $p \nmid m$, not necessarily p-solvable. It will next be shown that the inequalities

$$t(G) \geq a(p-1) + 1 \quad \text{and} \quad t(G) \leq p^a$$

given in Proposition 2.1, no longer hold. In order to accomplish this, we first prove two auxiliary assertions.

2.7. LEMMA. Let $1 = e_1 + \ldots + e_n$, where the e_i are primitive orthogonal idempotents of FG. For each $i \in \{1,2,\ldots,n\}$, let ℓ_i and r_i be the Loewy and composition lengths of FGe_i, respectively. Then

$$t(G) = \max\{\ell_1,\ldots,\ell_n\} \leq \max\{r_1 \ldots r_n\}$$

Proof. Since $FG = FGe_1 \oplus \ldots \oplus FGe_n$ and $J(FGe_i) = J(FG)e_i$, we have

$$J(FG)^k = J(FG)^k e_1 \oplus \ldots \oplus J(FG)^k e_n \qquad \text{for all } k \geq 1$$

Hence $t(G) = \max\{\ell_1,\ldots,\ell_n\}$. Since $\ell_i \leq r_i$ for all i, the assertion follows. ∎

2.8. LEMMA. Let $C = (c_{ij})$, $1 \leq i,j \leq k$, be the Cartan matrix of FG. Then

$$t(G) \leq \max_{1\leq i\leq k} \{\sum_{j=1}^{k} c_{ij}\}$$

Proof. Let $FGe_1, FGe_2, \ldots, FGe_k$ be all nonisomorphic principal indecomposable FG-modules. Then, by the definition of C, c_{ij} is the multiplicity of the irreducible FG-module $FGe_j/J(FG)e_j$ as a composition factor of FGe_i, $1 \leqslant i \leqslant k$. Hence, the composition length of FGe_i is equal to

$$c_{i1} + c_{i2} + \ldots + c_{ik}$$

Now apply Lemma 2.7. ■

2.9. EXAMPLE. (Wallace (1968)). There exists a group G of order $p^a m$, $a \geqslant 1$, $p \nmid m$, such that

$$t(G) < a(p-1) + 1$$

Proof. Let $G = A_5$, the alternating group of degree 5, and let $p = 5$. Since $|G| = 60$, we have

$$a(p-1) + 1 = 1(5-1) + 1 = 5$$

It will be shown that $t(G) \leqslant 4$. We may harmlessly assume that F is algebraically closed. Then it is well known (see Brauer and Nesbitt (1941)) that the Cartan matrix of FG is

$$\begin{bmatrix} 2 & 1 & 0 \\ 1 & 3 & 0 \\ 0 & 0 & 1 \end{bmatrix}$$

Invoking Lemma 2.8, we conclude that $t(G) \leqslant 4$ as required. ■

2.10. EXAMPLE. Let q be a power of a prime with $q \equiv 1 (\text{mod } 4)$, let $G = PSL(2,q)$ and let F be a field of characteristic 2. Then

(i) The Sylow 2-subgroups of G are dihedral of order 2^a for some $a \geqslant 2$

(ii) $t(G) > 2^a$

In particular (by Proposition 2.1), $t(G) > t(P)$ where P is a Sylow 2-subgroup of G. Indeed, property (i) is a consequence of a general group-theoretic fact (see Gorenstein (1968), p.418). To prove (ii), we may assume that F is algebraically closed. Owing to Erdmann (1977, Theorem 2(a), p.667), there is a projective indecomposable module V in the principal block of FG such that the Loewy length $L(V)$ of V is given by

$$L(V) = 2^a + 1$$

Hence, by Lemma 2.7

$$t(G) \geqslant 2^a + 1$$

as required. ■

Our next aim is to prove that

$$t(G) \leqslant \dim_F J(FG) + 1$$

and to provide necessary and sufficient conditions under which the equality holds.

In what follows, we choose a complete set

$$FGe_1, FGe_2, \ldots, FGe_k$$

of nonisomorphic principal indecomposable FG-modules, where $FGe_1/J(FG)e_1$ is a trivial FG-module. We denote by $C = (c_{ij})$, $1 \leqslant i,j \leqslant k$, the Cartan matrix of FG.

2.11. LEMMA. With the notation above,

$$t(G) \leqslant \max_{1\leqslant i\leqslant k} \{\sum_{j=1}^{k} c_{ij}\} \leqslant \max_{1\leqslant i\leqslant k} \{\dim_F J(FG)e_i + 1\} \leqslant \dim_F J(FG) + 1$$

Proof. The inequality on the left was established in Lemma 2.8. Since $\sum_{j=1}^{k} c_{ij}$ is equal to the composition length of FGe_i, we have

$$\sum_{j=1}^{k} c_{ij} \leqslant \dim_F J(FG)e_i + 1 \qquad (1 \leqslant i \leqslant k)$$

proving that

$$\max_{1\leqslant i\leqslant k} \{\sum_{j=1}^{k} c_{ij}\} \leqslant \max\{\dim_F J(FG)e_i + 1\}$$

The inequality on the right being obvious, the result follows. ■

2.12. THEOREM. (Motose and Ninomiya (1975a)). Let P be a Sylow p-subgroup of G. Then the following conditions are equivalent:

(i) $t(G) = \dim_F J(FG) + 1$

(ii) $G = P$ is cyclic or G is a Frobenius group with cyclic complement P.

Proof. (i) ⇒ (ii): Assume that $\dim_F J(FG) + 1 = t(G)$. Then, by Lemma 2.11,

$$1 + \dim_F J(FG) = \max_{1\leqslant i\leqslant k} \{\dim_F J(FG)e_i + 1\}$$

Since $\dim_F J(FG)e_1 \geqslant |P| - 1$ (by equality (1) in the proof of Theorem 3.8.4), it follows that

$$J(FG) = J(FG)e_1 \quad \text{and} \quad J(FG)e_i = 0 \qquad \text{for all } i \neq 1$$

This implies that the principal block B of FG contains only one irreducible module. Hence, by Proposition 3.8.10, G is p-nilpotent. Thus, by Proposition 3.1.20, $B \cong FP$. Since 1 is the only nonzero idempotent of FP, it follows that $B = FGe_1$ and hence that $J(B) = J(FG)e_1 = J(FG)$. Thus

$$\dim_F J(FG) = \dim_F J(FP) = |P| - 1$$

and so, by Theorem 3.8.11, $G = P$ or G is a Frobenius group with complement P. But $t(G) = t(P)$ by Proposition 2.3, so $t(P) = |P|$ and hence, by Theorem 4.3.2, P is cyclic.

(ii) $\Rightarrow$ (i): By Theorem 3.8.11, $\dim_F J(FG) = |P| - 1$, while by Theorem 4.3.2, $t(P) = |P|$. Hence, by Proposition 2.3,

$$t(G) = t(P) = |P| = \dim_F J(FG) + 1$$

as asserted. ∎

We next quote the following result, whose proof is almost identical to that of Proposition 3.1.8. For details refer to Passman (1970a) or to Karpilovsky (1985).

2.13. PROPOSITION. (Passman (1970a)). Let N be a normal subgroup of G of index n and let F be an arbitrary field. Then, for all $\alpha \in Z^2(G,F^*)$,

(i) $J(F^\alpha G)^n \subseteq F^\alpha G \cdot J(F^\alpha N) \subseteq J(F^\alpha G)$

(ii) If $n \neq 0$ in F, then $J(F^\alpha G) = F^\alpha G \cdot J(F^\alpha N) = J(F^\alpha N) \cdot F^\alpha G$. In particular $J(F^\alpha G)$ and $J(F^\alpha N)$ have the same nilpotency index. ∎

Repeating the argument of Proposition 3.1.9, we immediately derive

2.14. COROLLARY. (Passman (1970a)). Let N be a normal subgroup of G such that G/N is p-solvable and let $\alpha \in Z^2(G,F^*)$. If $|G/N| = np^a$ where $(n,p)=1$, then

$$J(F^\alpha G)^{p^a} \subseteq F^\alpha G \cdot I(N) \subseteq J(F^\alpha G)$$

In particular, if G is p-solvable of order np^a where $(n,p) = 1$, then

$$J(F^{\alpha}G)^{p^a} = 0 \quad \blacksquare$$

2.15. PROPOSITION. Let F be an algebraically closed field of characteristic $p > 0$, let $N = O_{p'}(G)$ and let $\{e_1,\ldots,e_s\}$ be a full set of representatives of G-conjugacy classes of block idempotents of FN. For each $i \in \{1,2,\ldots,s\}$, let G_i be the inertia group of e_i and let $H_i = G_i/N$. Then there exists $\alpha_i \in Z^2(H_i,F^*)$ such that

$$t(G) = \max_{1\leqslant i\leqslant s} \{t(F^{\alpha_i}H_i)\}$$

where $t(F^{\alpha_i}H_i)$ is the nilpotency index of $J(F^{\alpha_i}H_i)$.

Proof. Direct consequence of Morita's Theorem (Theorem 3.10.9). $\blacksquare$

2.16. COROLLARY. Further to the assumptions and notation of Proposition 2.15, assume that G is p-solvable and let p^{n_i} be the order of Sylow p-subgroups of H_i, $1 \leqslant i \leqslant s$. Then

$$t(F^{\alpha_i}H_i) \leqslant p^{n_i}$$

and

$$t(G) \leqslant \max_{1\leqslant i\leqslant s} \{p^{n_i}\} .$$

Proof. Apply Corollary 2.14 and Proposition 2.15. $\blacksquare$

3. GROUPS G WITH $t(G) = a(p-1) + 1$.

Throughout this section, F is an arbitrary field of characteristic p, G a finite group and $t(G)$ is the nilpotency index of $J(FG)$.

Assume that G is a p-solvable group of order $p^a m$, $p \nmid m$, $a \geqslant 1$. Then, by Proposition 2.1, we have

$$t(G) \geqslant a(p-1) + 1$$

Furthermore, by Proposition 2.5, if G is of p-length 1 then $t(G) = a(p-1) + 1$ if and only if the Sylow p-subgroups of G are elementary abelian. Our aim in this section is to provide further information on the structure of G with $t(G) = a(p-1) + 1$. We first illustrate that Proposition 2.5 is no longer true without the assumption that G is of p-length 1.

Let $G = S_4$ be the symmetric group of degree 4 and let $p = 2$. Obviously G is a solvable group whose 2-length is greater than 1 and whose Sylow 2-subgroups are not elementary abelian. It will be shown that

$$t(G) = 4 (= a(p-1) + 1)$$

Actually this fact can be immediately deduced from Lorentz's theorem (Theorem 6.4.6). However, rather than appeal to the general result, we give a simple ad hoc argument which is applicable to the present very special situation. As usual, for any subset X of G, we write X^+ for the sum of elements in X.

3.1. PROPOSITION. (Motose and Ninomiya (1975a)). Let $G = S_4$, let G_i be the stabilizer of $i \in \{1,2,3,4\}$ and let F be an arbitrary field of characteristic 2. Then

(i) $J(FG) = FG_1^+ + FG \cdot I(N)$

where N is the Klein's four group contained in G.

(ii) $t(G) = 4$

Proof. (i) Since N is a normal 2-subgroup of G with $G = NG_1$ and $G_1 \cap N = 1$, it follows from Lemma 6.4.5 that

$$J(FG) = J(FG_1) + FG \cdot I(N)$$

But $G_1 \cong S_3$ and so $J(FG_1) = FG_1^+$ by virtue of Example 3.7.9. This proves (i).

(ii) We first show that

$$G_i^+ g G_j^+ = 0 \qquad \text{for all } i,j \in \{1,2,3,4\} \text{ and all } g \in G \tag{1}$$

Since $G_i^+ g G_j^+ g^{-1} = G_i^+ G_{g(j)}^+$, it suffices to verify that $G_i^+ G_j^+ = 0$. In case $i \neq j, (G_i^+)^2 = 6G_i^+ = 0$. Assume that $i \neq j$ and write $\{1,2,3,4\} = \{i,j\} \cup \{k,\ell\}$. Then

$$G_i^+ + G_i^+(k,\ell),\ G_i^+(k,i) = G_i^+(k,\ell,i) \text{ and } G_i^+(\ell,i) = G_i^+(k,i,\ell)$$

and so

$$\begin{aligned} G_i^+ G_j^+ &= G_i^+(1+(k,\ell) + (k,i) + (k,\ell,i) + (\ell,i) + (k,i,\ell)) \\ &= 2G_i^+ + 2G_i^+(k,i) + 2G_i^+(\ell,i) \\ &= 0 \end{aligned}$$

proving (1).

Since $I(N)^2 = FN^+$, it follows from (i) that

$$J(FG)^2 = (FG_1^+ + FG\cdot I(N))^2$$
$$= (FG_1^+)^2 + FG\cdot I(N)^2 + I(N)G_1^+ + G_1^+\cdot I(N)$$
$$= FG\cdot N^+ + I(N)G_1^+ + G_1^+\cdot I(N)$$

But, by (1), $(G_1^+)^2 = 0$ and $G_1^+\cdot I(N)G_1^+ = 0$. Since $I(N)\cdot N^+ = 0$ and $N^+ \in Z(FG)$, we therefore conclude that

$$J(FG)^4 = (FG\cdot N^+ + I(N)G_1^+ + G_1^+\cdot I(N))^2$$
$$= (FG\cdot N^+)^2 + (I(N)G_1^+)^2 + (G_1^+ I(N))^2 + N^+G_1^+\cdot I(N) + FG(G_1^+ I(N)N^+)$$
$$+ G_1^+ I(N)^2 G_1^+ + I(N)(G_1^+)^2\cdot I(N) + FG\cdot N^+(I(N)\cdot G_1^+) + FG(I(N)G_1^+ N^+)$$
$$= 0$$

Hence $t(G) \leqslant 4 = 3(2-1) + 1$ and so, by Proposition 1.1, $t(G) = 4$. ∎

We next provide further circumstances under which $t(G) = a(p-1) + 1$ implies that the Sylow p-subgroups of G are elementary abelian. The following preliminary observation will clear our path.

3.2. LEMMA. Let p be an odd prime, let G be a p-solvable group and let $P \in \mathrm{Syl}_p(G)$. If P is of order p^3 and exponent p^2, then G is of p-length 1.

Proof. If $G/0_{p'}(G)$ is of p-length 1, then so is G. Since $P \in \mathrm{Syl}_p(G/0_{p'}(G))$, we may assume that $0_{p'}(G) = 1$ in which case $0_p(G) \neq 1$ since G is p-solvable. We first show that $Qd(p)$ is not involved in G. Assume that $X/Y \cong Qd(p)$ for some subgroup X of G and some $Y \lhd X$. By Theorem 1.5(i), the Sylow p-subgroups of $Qd(p)$ are nonabelian of order p^3 and exponent p. Hence $p \nmid (G:X)$ and $p \nmid |Y|$. Let Q be a Sylow p-subgroup of X. Then Q is a Sylow p-subgroup of G and so $Q \cong P$. On the other hand, $QY/Y \cong Q/(Q \cap Y) \cong Q$ is a Sylow p-subgroup of X/Y, a contradiction. Thus $Qd(p)$ is not involved in G and so, by Theorem 1.7, $A \subseteq 0_{p',p}(G)$ for any abelian normal subgroup A of P.

Since $|P| = p^3$, $P/Z(P)$ is abelian. Thus, if $x \in P$, then $x \in \langle Z(p),x\rangle$ which is a normal abelian subgroup of P. Hence, by the foregoing, $x \in 0_{p',p}(G)$ and therefore $P \subseteq 0_{p',p}(G)$, proving that G has p-length 1. ∎

We are now ready to prove the following result established by Motose (1980) for $p = 2$ and by Koshitani (1982a) for $p \neq 2$.

3.3. THEOREM. Let G be a p-solvable group with a metacyclic Sylow p-subgroup P of order p^a, $a \geqslant 1$. If $p = 2$, assume that $G/O_{p'}(G) \not\cong S_4$. Then $t(G) = a(p-1) + 1$ if and only if P is elementary abelian.

Proof. If P is elementary abelian, then G is of p-length 1 (Proposition 3.16.20), so $t(G) = a(p-1) + 1$ by Proposition 2.5. Conversely assume that $t(G) = a(p-1) + 1$. By Proposition 2.5, it suffices to verify that G is of p-length 1. Let G be a counter-example of minimal order. If $O_{p'}(G) \neq 1$, then by Proposition 2.1,

$$a(p-1) + 1 = t(G) \geqslant t(G/O_{p'}(G)) \geqslant a(p-1) + 1$$

and so $t(G/O_{p'}(G)) = a(p-1) + 1$. But the group $G/O_{p'}(G)$ satisfies the same hypotheses as G, a contradiction. Thus we must have $O_{p'}(G) = 1$, in which case $O_p(G) \neq 1$, say $|O_p(G)| = p^b, b \geqslant 1$. If $t(O_p(G)) > b(p-1) + 1$, then by Theorem 4.1.9 and Proposition 2.1,

$$\begin{aligned} a(p-1) + 1 = t(G) &\geqslant t(O_p(G)) + t(G/O_p(G)) - 1 \\ &> b(p-1) + 1 + (a-b)(p-1) + 1 - 1 \\ &= a(p-1) + 1, \end{aligned}$$

a contradiction. Thus $t(O_p(G)) = b(p-1) + 1$, in which case $O_p(G)$ is elementary abelian by Proposition 2.5. But P is metacyclic, hence so is $O_p(G)$ and therefore $O_p(G)$ is elementary abelian of order $\leqslant p^2$. Furthermore, by Lemma 1.3, $C_G(O_p(G)) = O_p(G)$. Thus $G/O_p(G)$ is isomorphic to a subgroup of $\mathrm{Aut}(O_p(G))$. Because G is not of p-length 1, the group P is nonabelian. We now distinguish two cases.

Case 1. Assume that $p = 2$. By the foregoing, $|P| \geqslant 8$, $|O_p(G)| \leqslant 4$ and $|G/O_p(G)| \leqslant 6$. Since G is not of p-length 1, we have $|G| = 24$, in which case $O_p(G) = C_G(O_p(G))$ is elementary abelian of order 4 and $G/O_p(G)$ acts faithfully on $O_p(G)$ by conjugation.

Let $S = \langle a \rangle$ be a Sylow 3-subgroup of G and $O_p(G) = \{1, x, y, xy\}$. Since

$C_G(O_p(G)) = O_p(G)$, we may assume that $a^{-1}xa = y$ and $a^{-1}ya = xy$. Consequently, $N_G(S) \cap O_p(G) = 1$ and $|N_G(S)| = 6$ by Sylow's theorem. We conclude therefore that G is a semidirect product of $O_p(G)$ and $N_G(S)$ and so $N_G(S)$ is the automorphism group of $O_p(G)$. Since the latter obviously implies that $G \cong S_4$, we derive a desired contradiction.

Case 2. We now assume that p is odd. If $O_p(G)$ is cyclic of order p, then $G/O_p(G)$ is a p'-group, contrary to the assumption that G is not of p-length 1. Thus $O_p(G)$ is elementary abelian of order p^2. Since

$$|\mathrm{Aut}(O_p(G))| = |GL(2,p)| = p(p-1)^2(p+1),$$

we have $|P/O_p(G)| = 1$ or p. But P is nonabelian and metacyclic, so P is of order p^3 and exponent p^2. It follows from Lemma 3.2 that G is of p-length 1, a contradiction. ∎

In view of Theorem 3.3, the case where $p = 2$ and $G/O_{p'}(G) \cong S_4$ deserves a separate treatment. *For the rest of this section F is assumed to be algebraically closed.*

3.4. THEOREM. (Motose (1980)). Let $p = 2$, let $G/O_{p'}(G) \cong S_4$ and let $e_1, e_2, \ldots, e_t$ be all block idempotents of $FO_{p'}(G)$. For each $i \in \{1,\ldots,t\}$, let G_i be the inertia group of e_i and put $H_i = G_i/O_{p'}(G)$. Then

$$t(G) = \begin{cases} 5 & \text{if } |H_i| = 8 \text{ for some } i \in \{1,2,\ldots,t\} \\ 4 & \text{otherwise} \end{cases}$$

Proof. Since the order of Sylow 2-subgroups of G is 8, it follows from Proposition 2.1, that $t(G) \geqslant 4$. On the other hand, if 2^{n_i} is the order of Sylow 2-subgroups of H_i, then by Corollary 2.16

$$t(G) \leqslant \max_{1 \leqslant i \leqslant t} \{2^{n_i}\}$$

Assume that $|H_i| \neq 8$ and $|H_i| \neq 24$ for all $i \in \{1,\ldots,t\}$. Then each $n_i \leqslant 2$ and hence $t(G) = 4$. Suppose that $|H_i| \neq 8$ for all $i \in \{1,\ldots,t\}$, but $|H_k| = 24$ for some k. Then $H_k \cong S_4$ and so, by Lemma 3.4.9, $F^{\alpha_k}H_k \cong FS_4$ for all $\alpha_k \in Z^2(H_k, F^*)$. Hence, by Proposition 2.15 and Corollary 2.16, $t(G) = t(S_4)$. Thus $t(G) = 4$ by Proposition 3.1.

Finally, assume that $|H_i| = 8$ for some $i \in \{1,2,\dots,t\}$. Then H_i is a dihedral group of order 8 and by Lemma 3.4.9, $F^{\alpha_i}H_i \cong FH_i$ for all $\alpha_i \in Z^2(H_i,F^*)$. Hence $t(H_i) = 5$ by Theorem 4.3.4. By the foregoing, this implies that $t(G) = 5$ as required. ∎

We now present two examples which illustrate the following facts:

(i) There exists a group G and a central p'-subgroup N such that $t(G) \neq t(G/N)$

(ii) If G is as in (i), then $t(B) \neq t(G)$ where B is the principal block of FG.

3.5. EXAMPLE. (Motose (1980)). Assume that $p = 2$ and let $\langle d \rangle$ be a cyclic group of order 3. Denote by G the semidirect product of $\langle d \rangle$ by S_4 such that $(12)^{-1}d(12) = d^{-1}$ and A_4 acts trivially on $\langle d \rangle$. Then $t(G) = 4$ by Theorem 3.4. ∎

3.6. EXAMPLE. (Motose (1980)). Assume that $p = 2$ and let $\langle c \rangle, \langle d \rangle$ be cyclic groups of order 3. Let a homomorphism $f : S_3 \longrightarrow GL(2,3)$ be defined by

$$a = (123) \longmapsto \begin{bmatrix} 1 & 0 \\ 1 & 1 \end{bmatrix} \quad \text{and} \quad b = (12) \longmapsto \begin{bmatrix} 1 & 0 \\ 0 & -1 \end{bmatrix}$$

Then f can be regarded as a homomorphism of S_4 into $GL(2,3)$, since S_4 is a semidirect product of $\mathbb{Z}_2 \times \mathbb{Z}_2$ and S_3. Let G be the semidirect product of $M = \langle c \rangle \times \langle d \rangle$ and S_4 with respect to f. Then the following relations hold:

$$a^{-1}ca = c\ ,\ a^{-1}da = cd\ ,\ b^{-1}cb = c \quad \text{and} \quad b^{-1}db = d^{-1}$$

Let ε be a primitive 3rd-root of 1 in F and put

$$e = (1+\varepsilon c+\varepsilon^2 c^2)(1+d+d^2)$$

Then e is a block idempotent of FM such that

$$a^{-1}ea \neq a \quad \text{and} \quad b^{-1}eb = e$$

Hence, if $G(e)$ is the inertia group of e, then $|G(e)/M| = 8$. Thus, by Theorem 3.4 and Example 3.5, we derive

$$t(G) = 5 \quad \text{and} \quad t(G/\langle c \rangle) = 4$$

Furthermore, if B is the principal block of FG, then by Lemma 2.2, $t(B) = t(S_4)$. Hence, by Proposition 3.1, $t(B) = 4$. ∎

3.7. *Remark.* Similar examples can be constructed for an arbitrary prime p (see Motose (1982)).

4. COMPUTATION OF $t(G)$ WITH $M(p) \in \mathrm{Syl}_p(G)$

Throughout this section, $M(p)$ denotes an extra-special p-group of order p^3, p odd, defined in Sec. 2 of Chapter 4. Note that, by Corollary 4.2.4, $t(M(p)) = 4p-3$. Our aim is to extend this result to p-solvable groups having $M(p)$ as a Sylow p-subgroup. The information obtained will be important for subsequent investigations. Note that in case $p \geqslant 5$ any p-solvable group G with $M(p) \in \mathrm{Syl}_p(G)$ is of p-length 1 (Theorem 1.4). Hence, by Proposition 2.3, $t(G) = 4p-3$. In case $p = 3$, there exist 3-solvable groups G with $M(3) \in \mathrm{Syl}_3(G)$ which are not of 3-length 1. The most important example of such a group is the group $Qd(3)$ (see Theorem 1.5). As we shall see below, the case where $G = Qd(3)$ is crucial for the computation of $t(G)$ with $M(3) \in \mathrm{Syl}_3(G)$. For this reason, from now on we assume that F is a field of characteristic 3 and begin by attacking the case where $G = Qd(3)$.

Recall that the group $M(3)$ is defined by

$$M(3) = \langle c,x,y \mid c^3 = x^3 = y^3 = 1,\ x^c = x,\ y^c = xy, x^y = x\rangle$$

The group $M = SL(2,3)$ acts naturally on the elementary abelian group

$$U = \langle x,y \mid x^3 = y^3 = 1,\ xy = yx\rangle$$

where U is regarded as a vector space over the field $\mathbb{F}_3$ of 3 elements. The group $Qd(3)$ is defined to be a semidirect product of U and M with respect to this action. It will be convenient to fix the following notation:

$$H = Qd(3)$$
$$a = \begin{bmatrix} 0 & 2 \\ 1 & 0 \end{bmatrix},\ b = \begin{bmatrix} 1 & 2 \\ 2 & 2 \end{bmatrix} \text{ and } c = \begin{bmatrix} 1 & 0 \\ 1 & 1 \end{bmatrix} \text{ in } M = SL(2,3)$$

Note that

$$a^4 = 1,\ b^2 = a^2,\ b^a (= a^{-1}ba) = b^{-1}$$

$$c^3 = 1,\ a^c = a^3 b,\ b^c = a$$

$$x^a = y^2,\ y^a = x,\ x^b = xy^2,\ y^b = x^2y^2$$

$x^c = x$, $y^c = xy$ and $Q = \langle a,b \rangle$ is a normal 2-subgroup of M

Moreover, we put

$$S = \langle c \rangle,\ a \circ b = a + b - ab$$

$$f = a^2 - 1,\ \tau = c(1 + a \circ b)f$$

$$T = \{f, \tau, \tau^2\},\ \tilde{u} = 1 + u + u^2 \quad \text{for} \quad u \in U$$

$$X^+ = \sum_{h \in X} h, \text{ where } X \subseteq FH$$

$$A = J(FS)Q^+FH,\ B = J(FT)FH \quad \text{and} \quad C = J(FU)FH$$

where FT is the group algebra of the group T over F (see Lemma 4.1(iv) below).

4.1. LEMMA. With the notation above, the following properties hold:

(i) f is a central idempotent of FM

(ii) $a^2 f = -f$ and $fuf = -f(u+u^2) = fu^2f$ for all $u \in U$

(iii) τ is central in FM. In particular, c commutes with $(a \circ b)f$.

(iv) $((a \circ b)f)^2 = 0$. Hence $\tau^2 = c^2(1 - a \circ b)f$ and $\tau^3 = f$ which implies that T is a cyclic group of order 3 consisting of F-linearly independent elements.

(v) $(f-\tau)u(f-\tau) = -(f-\tau)\tilde{u}(f-\tau) + T^+$ for all $u \in U$

(vi) $Q^+uQ^+ = (U^+-1)Q^+$ for all $u \in U-1$

(vii) $f(a \circ b)xQ^+ = \tilde{x}(y^2-y)Q^+$ and $Q^+x(a \circ b)f = Q^+(y-y^2)\tilde{x}$

(viii) $(f-\tau)uM^+ \in C^3M^+$ and $M^+u(f-\tau) \in M^+C^3$ for all $u \in U$

(ix) $T^+uQ^+(1-c) \in C^3M^+$ and $(1-c)Q^+uT^+ \in M^+C^3$ for all $u \in U$

(x) $T^+uM^+ = M^+uT^+ = 0$ for all $u \in U$

(xi) $T^+u(f-\tau)vQ^+(1-c) = (1-c)Q^+v(f-\tau)uT^+ = 0$ for all $u,v \in U$

Proof. The validity of (i), (ii), (iv) and (v) follows by straightforward calculation. Note also that (x) is an easy consequence of (ix).

(iii) Invoking (i) and (ii), we derive

$$\begin{aligned}\tau^{a} &= c^{a}(1+a+b^{a}-ab^{a})f\\ &= cb(1+a+b^{3}-ab^{3})f\\ &= \tau\end{aligned}$$

A similar argument shows that $\tau^{b} = \tau$ and $\tau^{c} = \tau$.

(vi) Because Q is both transitive and regular on $U-1$, we have

$$Q^{+}uQ^{+} = \sum_{q\in Q} u^{q}Q^{+} = (U^{+}-1)Q^{+}$$

(vii) Applying (ii), we have

$$\begin{aligned}f(a\circ b)xQ^{+} &= -f(x^{a}+x^{b}-x^{ab})Q^{+}\\ &= (1-a^{2})(y^{2}+xy^{2}-xy)Q^{+}\\ &= (y^{2}+xy^{2}-xy-y-x^{2}y+x^{2}y^{2})Q^{+}\\ &\tilde{x}(y^{2}-y)Q^{+}\end{aligned}$$

A similar argument proves that $Q^{+}x(a\circ b)f = Q^{+}(y-y^{2})\tilde{x}$

(viii) Invoking (i), (iii), and (vii), we derive

$$\begin{aligned}(f-\tau)xM^{+} &= (f(1-c) - (a\circ b)fc)xM^{+}\\ &= (fx(1-c) - (a\circ b)fxc)M^{+}\\ &= -f(a\circ b)xM^{+}\\ &= -f(a\circ b)xQ^{+}\tau\\ &= \tilde{x}(y-y^{2})M^{+}\end{aligned}$$

and a similar argument shows that

$$M^{+}x(f-\tau) = M^{+}(y^{2}-y)\tilde{x}$$

Because Q is transitive on $U-1$, the assertion follows.

(ix) Owing to (i), (iii), and (vii) it follows that

$$\begin{aligned}T^{+}xQ^{+}(1-c) &= f(\tau + (a\circ b)(c-c^{2}))(1-c)xQ^{+}\\ &= f(a\circ b)xQ^{+}c^{+}\\ &= \tilde{x}(y^{2}-y)M^{+}\end{aligned}$$

Similarly we have

$$(1-c)Q^{+}xT^{+} = M^{+}(y-y^{2})\tilde{x}$$

Since Q is transitive on U-1, the assertion follows.

(xi) The proof of (ix) together with (vii) and (x) ensures that

$$\begin{aligned}
&T^+y(f-\tau)xQ^+(1-c)\\
&\quad = T^+y(f(1-c) - (a\circ b)fc)xQ^+(1-c)\\
&\quad = T^+yfxQ^+\tau - T^+y\cdot f(a\circ b)xQ^+\cdot(c-c^2)\\
&\quad = T^+y(x^2-x)M^+ - T^+y\tilde{x}(y^2-y)Q^+(c-c^2)\\
&\quad = T^+(y+xy+x^2y)Q^+(c^2-c)\\
&\quad = \{(T^+xQ^+(1-c))^{a^3}+(T^+xQ^+(1-c)^{a^3c}+(T^+xQ^+(1-c))^{a^3c^2}\}(-c)\\
&\quad = (\tilde{y}+\widetilde{xy}+\widetilde{x^2y})(x^2-x)M^+\\
&\quad = 0
\end{aligned}$$

A similar argument shows that

$$\begin{aligned}
&T^+x(f-\tau)xQ^+(1-c)\\
&\quad = (1-c)Q^+x(f-\tau)yT^+\\
&\quad = (1-c)Q^+x(f-\tau)xT^+\\
&\quad = 0
\end{aligned}$$

Because Q is transitive on U-1 and $M/\langle a^2\rangle = PSL(2,3)$ is transitive on the projective space $\{\{x,x^2\}, \{y,y^2\}, \{xy,x^2y^2\}, \{x^2y,xy^2\}\}$, the desired assertion follows. ∎

4.2. LEMMA. With the notation above,

$$J(FM) = J(FT)FM + J(FS)Q^+$$

and

$$J(FH) = A + B + C$$

Proof. Note that $J(FS)Q^+$ is a nilpotent ideal of FM and that, by Lemma 4.1(i), (iii), FT is central in FM. Thus $J(FM)$ contains both $J(FT)FM$ and $J(FS)Q^+$. Since $\{1,2,3\}$ is the set of degrees of all irreducible $\bar{F}$-representations of M, where $\bar{F}$ is a splitting field for M, the radical $J(FM)$ is of F-dimension 10. On the other hand, one easily verifies that the set

$$\{f-\tau^k, (f-\tau^k)a, (f-\tau^k)b, (f-\tau^k)ab, (1-c^k)Q^+ \mid k=1,2\}$$

consists of some F-linearly independent elements in $J(FT)FM + J(FS)Q^+$. This establishes the first assertion. Taking into account that the kernel of the natrual map $FH \longrightarrow F(H/U) = FM$ is contained in $J(FH)$, we obtain $J(FH) = A+B+C$. ∎

The next lemma is crucial.

4.3. LEMMA. The following properties hold:

(i) $C^4 = U^+FM$ and $C^5 = 0$

(ii) $A^2 = M^+FU$ and $A^3 = 0$

(iii) $B^2 \subseteq T^+FH + (f-\tau)C^2(f-\tau)FH$

(iv) $B^3 \subseteq T^+C^2(f-\tau)FH + (f-\tau)C^2T^+FH + T^+C^4$

(v) $B^5 = 0$

(vi) $BA^2B = 0$

(vii) $B^3A = AB^3 = 0$

(viii) $B^2AB^2 = 0$

Proof. (i) Straightforward

(ii) Applying Lemma 4.1(vi), we derive

$$\begin{aligned} A^2 &= J(FS)Q^+ \cdot FH \cdot J(FS) \cdot Q^+ \cdot FH \\ &= J(FS)Q^+ \cdot FH \cdot Q^+ J(FS)FH \\ &= J(FS)(FU^+ + F)Q^+ J(FS)FH \\ &\subseteq J(FS)^2 Q^+ \cdot FH \\ &= M^+ \cdot FU \\ &\subseteq A^2 \end{aligned}$$

and

$$\begin{aligned} A^3 &= S^+Q^+ \cdot FU \cdot Q^+ J(FS)FH \\ &= S^+(FU^+ + F)Q^+ J(FS)FH \\ &\subseteq S^+J(FS)FH \\ &= 0 \end{aligned}$$

as asserted.

(iii) By Lemma 4.1(i),(iii), $f-\tau$ is central in FM. Applying Lemma 4.1(v), we derive

$$\begin{aligned} B^2 &= J(FT)FH\,J(FT)FH \\ &\subseteq (f-\tau)FU(f-\tau)FH \\ &\subseteq T^+\cdot FH + (f-\tau)C^2(f-\tau)FH \end{aligned}$$

as required.

(iv) Invoking Lemma 4.1(v) together with (i) and (iii), we obtain

$$\begin{aligned} B^3 &\subseteq T^+\cdot FU(f-\tau)FH + (f-\tau)C^2(f-\tau)FU(f-\tau)FH \\ &\subseteq T^+C^2(f-\tau)FH + (f-\tau)C^2T^+\cdot FH + T^+C^4 \end{aligned}$$

as desired.

(v) Applying (i), (iii), (iv) and Lemma 4.1(v), it follows that

$$\begin{aligned} B^5 &= B^2B^3 \\ &\subseteq T^+\cdot FU\cdot T^+C^2(f-\tau)FH + T^+\cdot FU\cdot(f-\tau)C^2T^+\cdot FH \\ &+ T^+\cdot FU\cdot T^+C^4 + (f-\tau)C^2(f-\tau)FU\cdot T^+C^2 \\ &+ (f-\tau)C^4T^+\cdot FH + C^6 \\ &\subseteq T^+C^4(f-\tau)FH + T^+C^4T^+\cdot FH + C^6 + C^6 + 0 + 0 \\ &= 0 \end{aligned}$$

as required.

(vi) By Lemma 4.1(viii) and (ii), we have

$$\begin{aligned} BA^2B &\subseteq (f-\tau)FU\cdot M^+\cdot FU(f-\tau)FH \\ &\subset \sum_{u,v\in U}(f-\tau)uM^+v(f-\tau)FH \\ &\subseteq \sum_{v\in U} C^3M^+v(f-\tau)FH \\ &\subseteq C^6 = 0 \end{aligned}$$

as asserted.

(vii) Invoking (iv) together with Lemma 4.1(ix), (xi), we obtain

$$\begin{aligned} AB^3 &\subseteq J(FS)Q^+\cdot FU\cdot T^+C^2 + J(FS)Q^+\cdot FU\cdot(f-\tau)FU\cdot T^+\cdot FH \\ &+ J(FS)Q^+\cdot FU\cdot T^+C^4 \\ &\subseteq \sum_{u\in U} FH(1-c)Q^+uT^+C^2 + \sum_{u,v\in U} FH(1-c)Q^+u(f-\tau)v\cdot T^+\cdot FH \\ &+ \sum_{u\in U} FH\cdot Q^+(1-u)T^+C^4 \end{aligned}$$

$$\subseteq C^3C^2 + 0 + C^5$$
$$= 0$$

Similarly we have $B^3A = 0$.

(viii) Applying (iii) and Lemma 4.1(ix),(x), we infer that

$$B^2AB^2 \subseteq (C^2 + T^+\cdot FH)J(FS)Q^+\cdot FH((f-\tau)C^2 + T^+\cdot FH)$$
$$\subseteq \sum_{u\in U} C^2J(FS)Q^+(1-u)fC^2 + \sum_{u\in U} T^+uQ^+(1-c)FH\cdot C^2$$
$$+ \sum_{u\in U} C^2(1-c)Q^+uT^+\cdot FH + \sum_{u\in U} T^+\cdot FU(1-c)Q^+uT^+\cdot FH$$
$$\subseteq C^5 + C^5 + C^5 + T^+\cdot FU\cdot M^+\cdot C^3$$
$$= 0$$

as required. ■

4.4. LEMMA. If $H = Qd(3)$ and $p = 3$, then $t(H) = 9$.

Proof. Applying Lemma 4.3, we see that

$$(A+B)^5 \subseteq FH\cdot A^3 + FH\cdot AB^3 + FH\cdot B^3A$$
$$+ B^2AB^2 + B^5 + FH\cdot BA^2B$$
$$= 0$$

Since $C^5 = 0$, it follows from Lemma 4.2 that

$$J(FH)^9 = ((A+B) + C)^9 = 0,$$

proving that $t(H) \leqslant 9$. ■

Put $\chi = 1 + x + x^2$ and $\nu = 1 + y + y^2$. Applying Lemma 4.1(i) - (iv) together with Lemma 4.2, we obtain

$$\chi f\tau y\tau^2 f = \chi c(1+ a\circ b)fyc^2(1-a\circ b)f$$
$$= \chi(1+a\circ b)fc^{-2}yc^2(1-a\circ b)f$$
$$= \chi f(1-b^{-1}\circ a^{-1})x^2yf(1-a\circ b)$$
$$= \chi f(x^2y+(x^2y)^a a + (x^2y)^b b - (x^2y)^{ab}ab)f(1-a\circ b)$$
$$= f\chi(x^2y+xya+xb - yab)f(1-a\circ b)$$
$$= \chi f(y+ya+b - yab)(1-a-b+ab)f$$
$$= \chi f(1+a+b+ab-ya-yb-yab)f$$

and

$$\begin{aligned}
\chi f\tau^2 y\tau f &= \chi c^2(1-a\circ b)fyc(1+a\circ b)f \\
&= \chi(1-a\circ b)fy^c(1+a\circ b)f \\
&= \chi f(1+b^{-1}\circ a^{-1})xyf(1+a\circ b) \\
&= \chi f(xy-(xy)^a a - (xy)^b b + (xy)^{ab}ab)f(1+a\circ b) \\
&= f\chi(xy-xy^2a-yb+x^2ab)f(1+a\circ b) \\
&= \chi f(y-y^2a-yb+ab)f(1+a\circ b) \\
&= \chi f(y-ya-yb+ab)(1+a+b-ab)f \\
&= \chi f(1-a+b+ab+ya-yb-yab)f
\end{aligned}$$

whence

$$\begin{aligned}
\chi f(y+\tau y\tau^2+\tau^2 y\tau)f &= \chi f(y-1)(1+b+ab)f \\
&= -\chi f\nu f(1+b+ab) \\
&= -U^+f(1+b+ab) \qquad (1)
\end{aligned}$$

By Lemma 4.2, $J(FH)^8$ contains $\chi T^+\nu T^+$. Applying (1), we have

$$\begin{aligned}
\chi T^+\nu T^+ &= \chi T^+(fyf + fy^2f)T^+ \\
&= -\chi T^+fyfT^+ \\
&= -\chi fT^+yT^+f \\
&= -\chi f(y+\tau y\tau^2+\tau^2y\tau)T^+f \\
&= U^+f(1+b+ab)T^+ \\
&= U^+T^+(1+b+ab) \\
&= U^+(1+c+c^2)(1+b+ab)f \\
&\neq 0
\end{aligned}$$

thus completing the proof. ∎

We have now come to the demonstration for which this section has been developed. The following result is essentially due to Motose (1984a) who established the crucial case where $p = 3$.

4.5. THEOREM. Let G be a p-solvable group, p odd, with a Sylow p-subgroup P isomorphic to $M(p)$. Then

$$t(G) = 4p - 3$$

Proof. Assume that G is of p-length 1. Then, by Corollary 4.2.4 and Proposition 2.3, $t(G) = 4p-3$. If $p \geqslant 5$ then G is of p-length 1 (Theorem 1.4), so we may assume that $p = 3$ and that G is not of p-length 1. By Corollary 3.1.18, we may also assume that the field F is algebraically closed. Assume by way of contradiction that $t(G) \neq 9$ and let G be a counter-example of minimal order.

We set

$$W = O_{3'}(G),\ G_1 = G/W,\ U_1 = O_3(G_1)$$

and

$$P_1 = PW/W \cong P,\ G_2 = G_1/U_1$$

Let g_1 and g_2 be the images of an element $g \in G$ in G_1 and G_2, respectively.

We first observe that G_2 is contained in $\mathrm{Aut}(U_1)$ by Lemma 1.3. If $|U_1| = 3$, then $|G_2| \leqslant |\mathrm{Aut}(U_1)| = 2$ and so $U_1 = P_1$, contrary to the fact that $|P_1| = 3^3$. If $|U_1| = 3^3$, then $U_1 = P_1$ and so G is of 3-length 1, again a contradiction. Thus we may assume that U_1 is elementary abelian of order 3^2 and G_2 is a subgroup of $\mathrm{Aut}(U_1) = GL(2,3)$.

By Theorem 1.6(i), U_1 contains x and so $U_1 = \langle x_1, c_1^k y_1^\ell \rangle$, $k \geqslant 2, \ell \geqslant 0$. We may replace generators of P_1 and U_1 preserving the same relations as those between x_1, y_1 and c_1. Hence we may assume that $U_1 = \langle x_1, y_1 \rangle$.

Let X denote the inverse image of $SL(2,3) \cap G_2$ under the natural homomorphism $G \longrightarrow G_2$. Then $(G:X) \leqslant 2$ and so $t(G) = t(X)$ by Proposition 3.1.8 (ii). The latter implies that $G = X$ and hence $G_2 \subseteq SL(2,3)$.

The group $Q = \langle a,b \rangle$ is normal in $SL(2,3)$. Hence, if $V \in \mathrm{Syl}_2(G_2)$, then $V \subseteq Q$. If $|V| \leqslant 2$, then G_1 has a normal Sylow 3-subgroup P_1, contrary to the fact that $|U_1| = 3^2$. If $|V| \geqslant 4$, then $\langle V, c_2 \rangle$ contains the Sylow 2-subgroup Q of $SL(2,3)$ since $c_2 = \begin{bmatrix} 1 & 0 \\ 1 & 1 \end{bmatrix}$ and $V \subseteq Q = \langle a,b \rangle$. Thus we must have $G = SL(2,3)$.

Let Q_1 be a Sylow 2-subgroup of G_1 such that $U_1 Q_1 = O_{3,3'}(G_1)$. Because

$$<c_1> \cap U_1 = 1 \text{ and } Q_1 \cap U_1 = 1,$$

it follows from Theorem 1.9 that G_1 is a semidirect product of U_1 and $SL(2,3)$. Taking into account that $SL(2,3) \subseteq \text{Aut}(U_1)$ acts naturally on U_1, we deduce that $G_1 = Qd(3)$.

Since every 2-subgroup of G_1 is a quaternion group or a cyclic group and since $|G_1| = 2^3 \cdot 3^3$, it follows from Lemma 3.4.9 that for any subgroup H of G_1 and any $\alpha \in Z^2(H, F^*)$,

$$F^{\alpha}H \cong FH \tag{2}$$

Let $\{e_s\}$ be the set of all block idempotents of FW and let I_s be the inertia group of e_s in G. Assume that $W \neq 1$. Then $|G/W| < |G|$ and the Sylow 3-subgroups of G and G/W are isomorphic. Hence, by assumption $9 = t(G/W) = t(I_s/W) < t(G)$ for all s such that the Sylow 3-subgroups of I_s/W are isomorphic to $M(3)$. If the Sylow 3-subgroups of I_s/W are of order $\leqslant 3^2$, then I_s is of 3-length 1 (Proposition 3.16.20), so by Proposition 2.3 and Corollary 3.2.5, $t(I_s/W) < 9$. Invoking (2) and Proposition 2.15, we therefore conclude that

$$9 < t(G) = \max_s \{t(I_s/W)\} \leqslant 9,$$

a contradiction. Thus $W = 1$ and $G = Qd(3)$, contrary to Lemma 4.4. This completes the proof of the theorem. ■

5. CHARACTERIZATIONS OF GROUPS G WITH $t(G) = p^{a-1} + p - 1$

Throughout this section, G denotes a finite group and F an arbitrary field of characteristic p. Recall that, by Theorem 4.3.4, if P is a noncyclic group of order p^a, then the following conditions are equivalent:

(i) $t(P) = p^{a-1} + p - 1$

(ii) $t(P) > p^{a-1}$

(iii) P contains an element of order p^{a-1}

Our aim is to generalize the above fact to p-solvable groups. In future we shall restrict ourselves to the case where $p \geqslant 3$. The reason for this is as follows.

Let $p = 2$ and let $G = S_4$. Then the Sylow 2-subgroups of G are dihedral of order 8 and hence satisfy condition (iii). On the other hand, by Proposition 3.1.

$$t(G) = 4 \neq 2^2 + 2 - 1$$

Thus for $p = 2$, Theorem 4.3.4 cannot be extended to arbitrary p-solvable groups. The following result is of independent interest and is the key to the solution of our problem.

5.1. THEOREM. (Koshitani (1983),(1985a), Motose (1985)). *Let G be a p-solvable group with a Sylow p-subgroup P of order p^a and let $\Phi/O_{p'}(G)$ be the Frattini subgroup of $O_{p',p}(G)/O_{p'}(G)$. Assume that $t(G) > p^{a-1}$ and that G is not of p-length 1. Then*

$$p = 2 \text{ , } G/\Phi \cong S_4 \text{ and } \Phi/O_{p'}(G) \text{ is cyclic}$$

Proof. Set $M = O_{p'}(G), N = O_{p',p}(G)$, and $p^r = |N/M|$. Owing to Corollary 3.1.18, we may harmlessly assume that F is algebraically closed. For the sake of clarity, we divide the proof into four steps.

Step 1. Here we prove that G/N is isomorphic to a subgroup of $GL(2,p)$ and that Φ/M is cyclic. Since $|GL(2,p)| = p(p-1)(p^2-1)$, the first assertion will imply that G is of p-length 2. By Proposition 3.1.9, we have

$$J(FG)^{p^{a-r}} \subseteq FG \cdot J(FN)$$

and hence

$$0 \neq J(FG)^{p^{a-1}} \subseteq FG \cdot J(FN)^{p^{r-1}}$$

Thus $t(N) > p^{r-1}$ and hence, by Proposition 2.3,

$$t(N/M) = t(N) > p^{r-1} \tag{1}$$

If N/M is cyclic, then N/Φ is of order p and hence G is of p-length 1 since G/N is a subgroup of $\text{Aut}(N/\Phi)$ (Lemma 3.1). Thus N/M is noncyclic, in which case $t(N/M) < p^r$ by Proposition 2.4. It follows from (1) that exactly one of the following holds:

$$p^{r-1} < t(N/M) < p^r \text{ and } r \geqslant 3 \tag{2}$$

or

$$N/M \text{ is elementary abelian of order } p^2 \tag{3}$$

If (2) holds, then by Theorems 4.3.4 and 1.10, N/Φ is elementary abelian of order p^2 and Φ/M is cyclic. Thus in any case G/N is a subgroup of $\mathrm{Aut}(N/\Phi) \cong GL(2,p)$ and Φ/M is cyclic.

Step 2. We now prove that $p = 2$ or $p = 3$.

In view of Step 1, every p-element of G/N is of order p and is conjugate to

$$\begin{bmatrix} 1 & 0 \\ \lambda & 1 \end{bmatrix} \quad \text{for some} \quad 0 \neq \lambda \in \mathbb{F}_p$$

Thus $(X-1)^2$ is the minimal polynomial of every p-element of G/N. Invoking Theorem 1.8, we therefore derive that $p \in \{2,3\}$.

Step 3. The aim of this step is to prove that $p = 2$.

Assume by way of contradiction that $p \neq 2$ so that, by Step 2, $p = 3$. It follows from Theorem 1.9 that G/Φ is a semidirect product of N/Φ and G/N. Hence G/Φ is a subgroup of T where T is the semidirect product of $\mathbb{Z}_3 \times \mathbb{Z}_3$ and $GL(2,3)$ with respect to the natural action of $GL(2,3)$ on $\mathbb{Z}_3 \times \mathbb{Z}_3$. Let $e_1, e_2, \ldots, e_s$ be all block idempotents of FM and let G_i be the inertia group of e_i in G, $1 \leqslant i \leqslant s$. Then, by Proposition 2.15, we have

$$t(G) = \max_{1 \leqslant i \leqslant s} \{t(F^{\alpha_i} H_i)\}$$

where $H_i = G_i/M$, $\alpha_i \in Z^2(H_i, F^*)$ and $t(F^{\alpha_i} H_i)$ is the nilpotency index of $J(F^{\alpha_i} H_i)$.

Let S be the semidirect product of $\mathbb{Z}_3 \times \mathbb{Z}_3$ and $SL(2,3)$ with respect to the natural action of $SL(2,3)$ on $\mathbb{Z}_3 \times \mathbb{Z}_3$. Let $f : G/M \longrightarrow G/\Phi$ be the natural homomorphism and let L_i be the inverse image of $f(H_i) \cap S$. Since $[H_i : (L_i \cap H_i)] \leqslant 2$, it follows from Lemma 3.4.9 and Proposition 2.13(ii) that

$$t(G) = \max_{1 \leqslant i \leqslant s} \{t(F^{\alpha_i} H_i)\} = \max_{1 \leqslant i \leqslant s} \{t(L_i \cap H_i)\}$$

Now $U_i = (L_i \cap H_i)/((\Phi/M) \cap H_i)$ is isomorphic to a subgroup of S and, by Theorem 4.5 and Proposition 3.1.9, $t(U_i) \leqslant 9$. Hence, applying Theorem 4.1.9 and Proposition 4.1.8, we derive

$$t(L_i \cap H_i) \leqslant |H_i \cap (\Phi/M)|\, t(U_i)$$
$$\leqslant |\Phi/M| \cdot 9$$
$$= 3^r$$

contrary to the fact that $3^r = 3^{a-1} < t(G)$.

Step 4. Completion of the proof.

By Step 1, the group Φ/M is cyclic and G is of 2-length 2. Since G/N is a subgroup of $GL(2,2)$, we have $G/N = GL(2,2)$. Applying Theorem 1.9, we deduce that G/Φ is a semidirect product of $\mathbb{Z}_2 \times \mathbb{Z}_2$ and $GL(2,2)$ with respect to the natural action of $GL(2,2)$ on $\mathbb{Z}_2 \times \mathbb{Z}_2$. Thus $G/\Phi \cong S_4$ and the result follows. ∎

Let p be an odd prime and let $n \geqslant 3$. Recall that the group $M_n(p)$ is defined by

$$M_n(p) = \langle a,b \mid a^{p^{n-1}} = 1,\ b^p = 1,\ b^{-1}ab = a^{1+p^{n-2}} \rangle$$

5.2. THEOREM. (Koshitani (1985a)). Let p be an odd prime, let G be a p-solvable group and let P be a Sylow p-subgroup of G of order p^a with $a \geqslant 2$. Then the following conditions are equivalent:

(i) $t(G) = p^{a-1} + p - 1$

(ii) $t(B) = p^{a-1} + p - 1$, where B is the principal block of FG

(iii) $p^{a-1} < t(G) < p^a$

(iv) G has p-length 1 and P is isomorphic to either $\mathbb{Z}_{p^{a-1}} \times \mathbb{Z}_p$ or $M_a(p)$ for $a \geqslant 3$

(v) $t(P) = p^{a-1} + p - 1$

(vi) $p^{a-1} < t(P) < p^a$

(vii) P is noncyclic but has a cyclic subgroup of index p

(viii) P is isomorphic to either $\mathbb{Z}_{p^{a-1}} \times \mathbb{Z}_p$ or $M_a(p)$ for $a \geqslant 3$

Proof. Conditions (v) - (viii) are equivalent by virtue of Theorems 4.3.4 and 1.10. Hence, if G is of p-length 1, then conditions (i) - (viii) are equivalent by Proposition 2.3. We are therefore left to verify that each of the conditions (i), (ii), (iii) and (viii) imply that G is of p-length 1.

Now (i) obviously implies (iii) and (iii) implies that G is of p-length 1, by virtue of Theorem 5.1. Assume that (ii) holds. Then $t(B) > p^{a-1}$ and by Lemma 2.2, $t(B) = t(G/O_{p'}(G))$. Hence, by Theorem 5.1, $G/O_{p'}(G)$ is of p-length 1 and therefore G is of p-length 1.

Finally assume that (viii) holds. If $P \cong \mathbb{Z}_{p^{a-1}} \times \mathbb{Z}_p$, then G is of p-length 1 by Proposition 3.16.20. Assume that $P \cong M_a(p)$ for some $a \geqslant 3$. To prove that G is of p-length 1, we may harmlessly assume that $O_{p'}(G) = 1$ in which case $O_p(G) \neq 1$ since G is p-solvable.

Suppose that G has subgroups X and Y such that $Y \lhd X$ and $X/Y \cong Qd(p)$. Let Q be a Sylow p-subgroup of X/Y. Then Q is metacyclic since P is metacyclic. On the other hand, Q is of order p^3 and exponent p, by virtue of Theorem 1.5(i). This is a contradiction. Therefore, by Theorem 1.7, if A is an abelian normal subgroup of P, then $A \subseteq O_{p',p}(G)$. Since $a^p \in Z(P)$, $\langle a^p, b\rangle \cong \mathbb{Z}_{p^{a-2}} \times \mathbb{Z}_p$ and hence $\langle a^p, b\rangle \lhd P$. Clearly $\langle a\rangle \lhd P$ and thus $P \subseteq O_{p',p}(G)$. Hence G is of p-length 1 as desired. ∎

We close this section by illustrating the fact that the converse of Theorem 5.1 does not hold. Namely, we give examples of groups G such that

(i) $G/\Phi \cong S_4$, $\Phi/O_{2'}(G) \cong \mathbb{Z}_2$ and $t(G) = 9 > 2^{4-1}$

(ii) $G/\Phi \cong S_4$, $\Phi/O_{2'}(G) \cong \mathbb{Z}_2$ and $t(G) = 7 < 2^{4-1}$

The above examples will indicate the difficulty in characterizing all 2-solvable groups G with $t(G) > 2^{a-1}$ where 2^a is the order of Sylow 2-subgroups of G and the underlying field F is of characteristic 2.

Let $Q = \langle a,b \mid a^4 = 1,\ b^2 = a^2,\ bab^{-1} = a^{-1}\rangle$ be the quaternion group and let σ and τ be automorphisms defined by

$$\sigma(a) = b,\ \sigma(b) = ab,\ \tau(a) = a^{-1},\ \tau(b) = a^{-1}b$$

Then $\sigma^3 = 1$, $\tau^2 = 1$, $\tau\sigma\tau^{-1} = \sigma^2$ and so $R = \langle \sigma,\tau\rangle$ is isomorphic to S_3. Let W be the semidirect product of Q and R with respect to this action of R on Q. Then we have

$$O_{2'}(W) = 1,\ O_2(W) = Q$$

$\langle a^2\rangle$ is the Frattini subgroup of $O_2(W)$

$$W/\langle a^2\rangle \cong S_4$$

For the rest of this section, we assume that $p = 2$ *and* F *is algebraically closed.* The latter condition does not impose any restriction, by virtue of Corollary 3.1.18. We need the following preliminary result.

5.3. LEMMA. (Motose (1985)). With the notation above, $t(W) = 7$.

Proof. Put $X = R^{+}\cdot FW$ and $Y = J(FQ)FW$. Since $J(FQ)FW \subseteq J(FW)$, it follows from Lemma 4.2 that

$$\begin{aligned} J(FW) &= J(FR)FW + J(FQ)FW \\ &= X + Y \end{aligned}$$

We shall show that $X^3 = 0$. Since $Y^5 = 0$ this will imply that

$$J(FW)^7 = (X+Y)^7 = 0$$

It suffices to verify that $R^{+}xR^{+}yR^{+} = 0$ for all $x,y \in Q$, and the latter will follow if we show that $R^{+}xR^{+}yR^{+} = 0$ for all $x,y \in C = Q - \{1,a^2\}$. Fix x,y in C. Since $\{1\}$, $\{a^2\}$ and C are all orbits of Q under the action of R, we have

$$\begin{aligned} R^{+}xR^{+}yR^{+} &= C^{+}R^{+}yR^{+} = (C^{+})^2R^{+} \\ &= (a^2+1)^2(a+b+ab)^2R^{+} \\ &= 0 \end{aligned}$$

proving that $t(W) \leqslant 7$.

Since $R^{+}xR^{+} = C^{+}R^{+}$ for $x \in C$ and a^2+1 and Q^{+} are central in FW, we also have

$$\begin{aligned} (R^{+}(a+1)(b+1))^2 &= R^{+}(ab+a+b+1)R^{+}(a+1)(b+1) \\ &= C^{+}R^{+}(ab+a+b+1) \\ &= (a+b+ab)(a^2+1)R^{+}(ab+a+b+1) \\ &= (a+b+ab)R^{+}Q^{+} \\ &= Q^{+}R^{+} \\ &\neq 0 \end{aligned}$$

This implies that $t(W) > 6$ and hence that $t(W) = 7$. ∎

5.4. EXAMPLE. (Motose (1985)). Keeping the notation of Theorem 5.1, we shall present an example of a group G with $G/\Phi \cong S_4$, $\Phi/O_{2'}(G) \cong \mathbb{Z}_2$ and $t(G) = 9 > 8$. Consider the homomorphism $R \longrightarrow GL(2,3)$ defined by

$$\sigma \longmapsto \begin{bmatrix} 1 & 0 \\ 1 & 1 \end{bmatrix} \text{ and } \tau \longmapsto \begin{bmatrix} 1 & 0 \\ 0 & 1 \end{bmatrix}$$

In what follows, we shall regard this as a homomorphsim from W to $GL(2,3)$. Let G denote the semidirect product of the elementary abelian group $M = \langle c\rangle \times \langle d\rangle$ of order 3^2 and W with respect to the above homomorphism. Then clearly $M = O_{2'}(G)$, $G/\Phi \cong S_4$ and $\Phi/M \cong \mathbb{Z}_2$. Let ε be a primitive 3rd root of 1 and let e be a block idempotent of FM defined by

$$e = (1+\varepsilon c+\varepsilon^2 c^2)(1+d+d^2)$$

Then $\sigma^{-1}e\sigma \neq e$ and $\tau^{-1}e\tau = e$ and so the inertia group $G(e)$ of e in G is given by

$$G(e) = \langle \tau, Q\rangle \cdot M$$

Let $\{e_1 = M^+, e_2 = e, e_3, \ldots, e_9\}$ be the set of all block idempotents of FM and let $H_i = G(e_i)/M$, $1 \leqslant i \leqslant 9$. Owing to Proposition 2.15, there exists $\alpha_i \in Z^2(H_i, F^*)$ such that

$$t(G) = \max_{1\leqslant i\leqslant 9} \{t(F^{\alpha_i}H_i)\}$$

where $t(F^{\alpha_i}H_i)$ is the nilpotency index of $J(F^{\alpha_i}H_i)$. Since $t(F^{\alpha_i}H_i) = t(H_i)$ by Lemma 3.4.9, we have

$$t(G) = \max_{1\leqslant i\leqslant 9} \{t(H_i)\}$$

If H_i contains no Sylow 2-subgroups of G, then $t(H_i) \leqslant 8$ by Proposition 3.1.9. On the other hand, $t(H_1) = t(W) = 7$ by Lemma 5.3. Since $\langle \tau, Q\rangle$ contains an element τb of order 8, we also have

$$t(H_2) = t(\langle \tau, Q\rangle) = 9$$

by Corollary 4.2.2(ii). Consequently, $t(G) = 9 > 8$. ∎

5.5. EXAMPLE. (Motose (1985)). Again, we preserve the previous notation. Our aim is to present an example of a group G with $G/\Phi \cong S_4$, $\Phi/O_{2'}(G) \cong \mathbb{Z}_2$ and $t(G) = 7 < 8$.

Consider the homomorphism $<\tau> \to GL(2,3)$ defined by

$$\tau \longmapsto \begin{bmatrix} 1 & 0 \\ 0 & -1 \end{bmatrix}$$

and regard it as a homomorphism from W to $GL(2,3)$. Let G be the semidirect product of the elementary abelian group $M = <c> \times <d>$ and W with respect to this homomorphism. Then clearly $M = O_{2'}(G)$, $G/\Phi \cong S_4$ and $\Phi/M \cong \mathbb{Z}_2$. Let $\{e_1 = M^+, e_2, \ldots, e_9\}$ be the set of all block idempotents of FM. Then $G(e_i) = G$ or $<\sigma, Q>\cdot M$ and so

$$t(G) = \{\max t(W), t(<\sigma,Q>)\}$$

by virtue of Proposition 2.15 and Lemma 3.4.9. Now $t(W) = 7$ by Lemma 5.3 and

$$t(<\sigma,Q>) = t(Q) = 5$$

by Proposition 3.1.8(ii) and Corollary 4.2.4(ii). Thus $t(G) = t(W) = 7$, as required. ■

6. GROUPS G WITH $t(G) < (a+2)(p-1) + 1$

Throughout this section, G is a finite group and F is an arbitrary field of characteristic p. We know, from Theorem 4.3.6, that if P is a group of order p^a which is not elementary abelian, then the following conditions are equivalent:

(i) $t(P) = (a+1)(p-1) + 1$

(ii) $t(P) < (a+2)(p-1) + 1$

(iii) There exists a central element z of order p in P such that $P/<z>$ is elementary abelian.

Our aim is to extend this result to p-solvable groups under certain conditions. Recall that the prime p is said to be a *Fermat prime* if $p = 2^n + 1$ for some $n \geq 1$ (the latter, of course, implies that n is a power of 2).

6.1. LEMMA. (Motose (1984d)). Let G be a p-solvable group with a Sylow p-subgroup P of order p^a, $a \geq 1$. Assume that G is of p-length ≥ 2 and P is regular. If $t(G) < (a+2)(p-1) + 1$, then P is a Fermat prime and the Sylow 2-subgroups of $G/O_{p'}(G)$ are nonabelian.

Proof. Owing to Proposition 3.16.20, P is nonabelian. Hence, by Theorem

1.11(iii), p is odd. Since $t(G) \geqslant t(G/O_{p'}(G))$ (Lemma 2.2), we may harmlessly assume that $O_{p'}(G) = 1$. Put $U = O_p(G)$, Φ the Frattini subgroup of U and write $|U| = p^s$, $s \geqslant 1$. By Theorem 4.1.9, we have

$$t(G) \geqslant t(U) + t(G/U) - 1$$

while by Proposition 2.1,

$$t(G/U) \geqslant (a-s)(p-1) + 1$$

Hence we deduce that

$$\begin{aligned} t(U) &\leqslant t(G) - t(G/U) + 1 \\ &< (a+2)(p-1) + 1 - (a-s)(p-1) - 1 + 1 \\ &= (s+2)(p-1) + 1 \end{aligned}$$

It follows, from Theorem 4.3.6, that U is of exponent p. Because P is regular and $\langle x,U\rangle' \subseteq U$ for all $x \in P$, it follows that

$$(xu)^p = x^p \qquad \text{for all } x \in P, u \in U$$

Hence we have

$$u^{x^{p-1}+\dots+x+1} = u^{x^{p-1}} \cdots u^x u = x^{-p}(xu)^p = 1 \tag{1}$$

for all $x \in P$ and $u \in U$, where $u^{x^s} = x^{-s}ux^s$ and $x^{p-1} + \dots + x + 1$ is the sum of endomorphisms $x^{p-1},\dots,x,1$ of U. By Lemma 1.3, G/U is a subgroup of $GL(U/\Phi)$. Invoking (1) together with Theorem 1.8, the result follows. ■

6.2. THEOREM. (Motose (1984d)). Let G be a p-solvable group with a regular Sylow p-subgroup P of order $p^a, a \geqslant 1$. If p is not a Fermat prime or the Sylow 2-subgroups of $G/O_{p'}(G)$ are abelian, then the following are equivalent:

(i) $t(G) = a(p-1) + 1$ or $t(G) = (a+1)(p-1) + 1$

(ii) $t(G) < (a+2)(p-1) + 1$

(iii) G is of p-length 1 and P has a central element z of order p such that $P/\langle z\rangle$ is elementary abelian

In particular, by Proposition 2.5, $t(G) = a(p-1) + 1$ if and only if P is elementary abelian.

Proof. By Lemma 6.1, our assumption on G forces G to be of p-length 1.

Hence, by Proposition 2.3, $t(G) = t(P)$. If P is not elementary abelian, then $t(G) \neq a(p-1) + 1$ by Proposition 2.5. Hence in this case (i), (ii) and (iii) are equivalent by Theorem 4.3.6. In case P is elementary abelian, then (i), (ii) and (iii) are always true, again by Proposition 2.5. This completes the proof of the theorem. ■

7. CHARACTERIZATIONS OF GROUPS G WITH $t(G) = p^{a-1}$, p ODD.

Let G be a finite group and F an arbitrary field of characteristic p. Owing to Theorem 4.3.5, if P is a group of order p^a, $a \geqslant 1$, then the following conditions are equivalent:

(i) $t(P) = p^{a-1}$

(ii) $P \cong M(3)$ or $P \cong \mathbb{Z}_2 \times \mathbb{Z}_2 \times \mathbb{Z}_2$

Our aim in this section is to extend this result to p-solvable groups where p is odd. The case $p = 2$ seems formidable for the following reason. Assume that $p = 2$ and consider the following three groups:

$$G_1 = \mathbb{Z}_2 \times \mathbb{Z}_2 \times \mathbb{Z}_2 \times \mathbb{Z}_3 \times \mathbb{Z}_3 \times \mathbb{Z}_3$$

$$G_2 = S_4 \times \mathbb{Z}_3 \times \mathbb{Z}_3$$

G_3 is the group constructed in Example 3.6.

These three groups have the same order $2^3 3^3$. Furthermore, by Corollary 3.2.5, Proposition 3.1 and Example 3.6, we have

$$t(G_1) = 4 \qquad t(G_2) = 4 \qquad t(G_3) = 5$$

The above shows that there exist two groups, namely G_1 and G_2, with the same nilpotency index whose Sylow 2-subgroups have different structures. On the other hand, there are two groups, namely G_2 and G_3, with different nilpotency indices whose Sylow 2-subgroups have identical structure. This explains why we shall restrict ourselves to the case $p \geqslant 3$.

7.1. LEMMA. Let G be a p-solvable group of p-length 1 and of order $p^a, a \geqslant 1$. Then $t(G) = p^{a-1}$ if and only if $P \cong M(3)$ or $P \cong \mathbb{Z}_2 \times \mathbb{Z}_2 \times \mathbb{Z}_2$.

Proof. Direct consequence of Theorem 4.3.5 and Proposition 2.3. ■

7.2. THEOREM. (Motose (1986a)). Let G be a p-solvable group with a Sylow p-subgroup P of order $p^a, a \geqslant 1, p$ odd. Then $t(G) = p^{a-1}$ if and only if $p = 3$ and $P \cong M(3)$.

Proof. If $p = 3$ and $P \cong M(3)$, then $t(G) = 9 = 3^{3-1}$ by Theorem 4.5. Conversely, assume that $t(G) = p^{a-1}$. To prove that $p = 3$ and $P \cong M(3)$, we argue by induction on $|G|$. Owing to Lemma 7.1, we may harmlessly assume that G has p-length at least 2. Furthermore, by Corollary 3.1.18, we may also assume that F is algebraically closed. Put $M = O_{p'}(G)$, $N = O_{p',p}(G)$, $p^r = |N/M|$ and let Φ/M by the Frattini subgroup of N/M. Then G/N is isomorphic to a subgroup of $GL(N/\Phi)$, where N/Φ is regarded as a vector space over $\mathbb{F}_p$ (Lemma 1.3). Since G has p-length at least two, the group N/M is not cyclic. Indeed, if N/M is cyclic, then N/Φ is cyclic of order p and hence G/N is a p'-group, contrary to the assumption that G is of p-length 2.

Every subgroup of $GL(2,3)$ acts naturally on the elementary abelian group E of order 9. Let T (respectively, S) be the semidirect product of E and $GL(2,3)$ (respectively, $SL(2,3)$) with respect to this action. We divide the rest of the proof into a number of steps.

Step 1. Here we prove that if $J(FN)^{p^{r-1}} \neq 0$, *then* $p = 3$, G/N *is a subgroup of* $GL(2,3)$ *(hence* $a = r+1$*) and* $|\Phi/M| = 3^{r-2}$.

Assume that $J(FN)^{p^{r-1}} \neq 0$. Then, by Proposition 2.3,

$$t(N/M) = t(N) > p^{r-1}$$

It follows, from Theorem 5.2, that

$$t(N) = p^{r-1} + p - 1 \tag{1}$$

and that $N/M \cong \mathbb{Z}_{p^{r-1}} \times \mathbb{Z}_p$ or $N/M \cong M_r(p)$ for $1 \geqslant 3$. Since in both cases the minimal number of generators of N/M is two, we conclude, from Lemma 4.3.1, that N/Φ is an elementary abelian group of order p^2. In particular, $|\Phi/M| = p^{r-2}$ Now G/N is a subgroup of $GL(N/\Phi) = GL(2,p)$, and every element of order p in $GL(2,p)$ is conjugate to

$$\begin{bmatrix} 1 & 0 \\ \lambda & 1 \end{bmatrix} \qquad \text{for some } 0 \neq \lambda \in \mathbb{F}_p$$

Thus $(X-1)^2$ is the minimal polynomial of every element of order p in G/N. Since p is odd, Theorem 1.8 implies that $p = 3$.

Step 2. *Here we prove that $p = 3$, G/N is a subgroup of $GL(2,3)$ (hence $a = r+1$), $|\Phi/M| = 3^{r-2}$ and $t(N) \leqslant 3^{r-1} + 2$.*

Assume that $J(FN)^{p^{r-1}} = 0$. Then $t(N) \leqslant p^{r-1}$ and, by Proposition 3.1.9,

$$J(FG)^{p^{a-r}} \subseteq FG \cdot J(FN)$$

But, by hypothesis, $t(G) = p^{a-1}$ and so, raising both sides to the power of $p^{r-1} - 1$ yields

$$0 \neq J(FG)^{p^{a-1} - p^{a-r}} \subseteq FG \cdot J(FN)^{p^{r-1} - 1}$$

Thus we must have

$$t(N/M) = t(N) = p^{r-1} \tag{2}$$

Since p is odd, Theorem 4.3.5 implies that $N/M \cong M(3)$. Hence $r = p = 3$, $|\Phi/M| = 3^{r-2}$, N/Φ is elementary abelian of order 3^2 and G/N is a subgroup of $GL(2,3)$.

Invoking Step 1, we deduce that $p = 3$, G/N is a subgroup of $GL(2,3)$ and $|\Phi/M| = 3^{r-2}$. Furthermore, by (1) and (2), we also have $t(N) \leqslant 3^{r-1} + 2$.

Step 3. *Reduction to the case where $G/\Phi = S$.*

By Step 2 and Theorem 1.9, G/Φ is a semidirect product of N/Φ and G/N and hence G/Φ is a subgroup of T. By Proposition 2.15, there exists a block idempotent e of FM such that for $H = G(e)/M$,

$$t(G) = t(F^{\alpha}H))$$

for some $\alpha \in Z^2(H,F^*)$. Let $f : G/M \longrightarrow G/\Phi$ be the natural homomorphism, let L be the inverse image of $f(H) \cap S$ and let $V = L \cap H$. Then $2 \geqslant (H:V)$. Invoking Lemma 3.4.9 together with Proposition 3.1.8(ii), we derive

$$3^r = 3^{a-1} = t(G) = t(F^{\alpha}H) = t(V)$$

Set $U = V/((F/M) \cap H)$ and observe that, by Theorem 4.5 and Proposition 2.1, $t(U) \leqslant 9$. Since U is isomorphic to a subgroup of S, it follows from Theorem 4.1.9 and Proposition 4.1.8 that

$$\begin{aligned} 3^r &= t(V) \leqslant t(U) \cdot |H \cap (\Phi/M)| \\ &\leqslant 9|\Phi/M| \\ &= 3^r \end{aligned}$$

where the latter equality follows from Step 2. Thus $H \supseteq \Phi/M$ and $t(U) = 9$.

Since $t(U) = 9$ and $M(3)$ is of exponent 3, it follows from Propositions 2.1 and 2.4 that U contains a Sylow 3-subgroup of S. Hence a 3-Sylow subgroup D of V is of order $27|\Phi/M| = 3^{a+1} = 3^a$ and so D is isomorphic to P.

Assume that either $M \neq 1$ or $V \neq G/M$. Then, by induction, $t(V) = 3^r$ implies that D is isomorphic to $M(3)$. Thus we may assume that N is a 3-group, $V = G$ and $U = G/\Phi$ is a subgroup of S which contains a Sylow 3-subgroup of S. But G has 3-length at least 2, so G/N must contain an element of order 4. Thus $G/N = SL(2,3)$ since G/N contains an element of order 3. Hence $G/\Phi = S$, as required.

Step 4. Completion of the proof.

Let $f : FG \longrightarrow F(G/\Phi) = FS$ be the natural homomorphism. Then, by Lemma 4.2 and the proof of Lemma 4.4,

$$f(J(FG)) = J(FS) = Z + f(J(FN)FG)$$

where Z is a right ideal of FS such that $Z^5 = 0$. Since

$$J(FG) \supseteq J(FN)FG \supseteq J(F\Phi)FG = \mathrm{Ker} f$$

it follows that

$$J(FG) = f^{-1}(Z) + J(FN)FG$$

On the other hand, it is easy to see that

$$(f^{-1}(Z))^5 \subseteq \mathrm{Ker} f = J(F\Phi)FG$$

and so

$$(f^{-1}(Z))^{5|\Phi|} = 0$$

Taking into account that $t(N) \leqslant 3^{r-1} + 2$ (Step 2), we derive

$$\begin{aligned} 3^r &= t(G) \leqslant 5|\Phi| + t(N) - 1 \\ &\leqslant 5 \cdot 3^{r-2} + 3^{r-1} + 1 \end{aligned}$$

Thus $r = 2$ and $\Phi = 1$, which by Step 3 implies that $G = S$. Hence, by Theorem 1.5(i), $P \cong M(3)$ and the result follows. ■

8
Radicals of blocks

Let B be a block of a group algebra FG with defect d. The problem that motivates this chapter is to discover relations between d and the nilpotency index $t(B)$ of $J(B)$. The first two sections provide some general information on lower bound for $t(B)$ and upper bound for $t(Z(B))$. The rest is based on the Fong correspondence and the Külshammer's structure theorem for blocks of p-solvable groups. The last section provides a number of important applications. Among other results, we prove that if G is p-solvable, then $d(p-1)+1 \leqslant t(B) \leqslant p^d$ and $t(B) = p^d$ if and only if $\delta(B)$ is cyclic.

1. A LOWER BOUND FOR $t(B)$ IN TERMS OF THE EXPONENT OF $\delta(B)$.

Throughout this section, G denotes a finite group and F a field of characteristic $p > 0$. Given a block B of FG, we write $t(B)$ for the nilpotency index of the Jacobson radical $J(B)$ of B and $\delta(B)$ for the defect group of B. As usual, we write $C\ell(G)$ and $\exp(G)$ for the set of conjugacy classes of G and the exponent of G, respectively. Let $\mathbb{N}_0$ denote the set of all nonnegative integers. Given a subset X of G, $n \in \mathbb{N}_0$ and a finite-dimensional F-algebra A, we write

$$T(A) = \{a \in A \mid a^{p^m} \in [A,A]\} \text{ for some } m \geqslant 1$$

$$T_n(A) = \{a \in A \mid a^{p^n} \in [A,A]\}$$

$$X^+ = \sum_{x \in X} x \qquad \text{(if } X = \emptyset, \text{ then by definition } X^+ = 0)$$

$$X^{p^n} = \{x^{p^n} \mid x \in X\}$$

$$X^{1/p^n} = \{g \in G \mid g^{p^n} \in X\}$$

1.1. LEMMA. Let $n \in \mathbb{N}_0$ and let A be a finite-dimensional F-algebra.

(i) $[A,A] \subseteq T_n(A)$, $T_n(A) \subseteq T_{n+1}(A)$ and $T(A) = \bigcup_{i=0}^{\infty} T_i(A)$

(ii) There exists $m \in \mathbb{N}_0$ such that $T_m(A) = T(A)$.

Proof. (i) Direct consequence of the property

$$x \in [A,A] \quad \text{implies} \quad x^p \in [A,A]$$

established in the proof of Lemma 2.2.1.

(ii) Choose a basis $a_1, a_2, \ldots, a_s$ for the F-space $T(A)$. For each $i \in \{1,2,\ldots,s\}$, there exists $m_i \geq 1$ such that $a_i^{p^{m_i}} \in [A,A]$. Hence, given $x = \sum_{i=1}^{s} \lambda_i a_i$ with $\lambda_i \in F$, we have for $m = \max\{m_1,\ldots,m_s\}$,

$$x^{p^m} \equiv \sum_{i=1}^{s} \lambda_i^{p^m} a_i^{p^m} \equiv 0 \pmod{[A,A]}$$

This proves that $T(A) \subseteq T_m(A)$ and hence that $T_m(A) = A$. ∎

For any p-subgroup D of G, let $I_D(G)$ be the F-linear span of all C^+, $C \in C\ell(G)$ with $\delta(C) \underset{G}{\subseteq} D$. Observe that, by Lemma 2.2.5, $I_D(G)$ is an ideal of $Z(FG)$.

1.2. LEMMA. Let D be a p-subgroup of G.

(i) $I_D(G)^{\perp}$ is a $Z(FG)$-submodule of FG with $[FG,FG] \subseteq I_D(G)^{\perp}$

(ii) Given $x = \Sigma x_g g \in T(FG)$ and $n \in \mathbb{N}_0$ with $q = p^n \geq \exp(D)$, we have

$$x^q \in I_D(G)^{\perp}$$

Proof. (i) Assume that $y \in I_D(G)^{\perp}$ and $z \in Z(FG)$. Since $I_D(G)$ is an ideal of $Z(FG)$, we have

$$zt \in I_D(G) \qquad \text{for all } t \in I_D(G)$$

It follows that

$$tr((yz)t) = tr(y(zt)) = 0 \qquad \text{for all } t \in I_D(G)$$

proving that $yz \in I_D(G)^{\perp}$. Hence $I_D(G)^{\perp}$ is a $Z(FG)$-submodule of FG. Owing to Lemma 3.3.9(ii), $Z(FG) = [FG,FG]^{\perp}$ and so, by Theorem 3.3.6(i),

$$[FG,FG] = Z(FG)^{\perp}$$

Bearing in mind that $I_D(G) \subseteq Z(FG)$, we derive

$$I_D(G)^{\perp} \supseteq Z(FG)^{\perp} = [FG,FG]$$

as required.

(ii) Assume by way of contradiction that $tr(x^q C^+) \neq 0$ for some $C \in C\ell(G)$

with $\delta(C) \subseteq_G D$. Then, $q \geqslant \exp(\delta(C))$, so by Lemma 3.9.2(ii), C is p-regular and $tr(xS^+) \neq 0$ for some p-regular section S of G. Because

$$tr(xS^+) = \sum_{g \in S^{-1}} x_g$$

we have $\sum_{g \in S^{-1}} x_g \neq 0$ for the p-regular section S^{-1} of G. Hence, by Lemma 3.3.13, $x \notin T(FG)$, a contradiction. So the lemma is true. ∎

Let A be a finite-dimensional F-algebra. Owing to Lemma 1.1, there exists $m \in \mathbb{N}_0$ such that $T_m(A) = T(A)$. We now define the *index* of $T(A)$ as the least nonnegative integer n such that $T_n(A) = T(A)$.

1.3. THEOREM. (Külshammer (1982)). Let $B = B(e)$ be a block of FG and let n be the index of $T(B)$. Then p^n is the exponent of $\delta(B)$

Proof. Put $D = \delta(B)$ and let $m \in \mathbb{N}_0$. It suffices to show that

(i) If $p^m \geqslant \exp(D)$, then $T(B) = T_m(B)$

(ii) If $p^m < \exp(D)$, then $T(B) \neq T_m(B)$

To prove (i), assume that $p^m \geqslant \exp(D)$ and let $q = p^m$. By Proposition 2.2.6, $e \in I_D(G)$ and so $Z(B) = Z(B)e \subseteq I_D(G)$. Hence, by Lemma 1.2(ii), for all $x \in T(B)$ we have

$$tr(x^q Z(B)) \subseteq tr(x^q I_D(G)) = 0,$$

proving that $x^q \in Z(B)^\perp = [B,B]$. Consequently $x \in T_m(B)$ and therefore $T(B) = T_m(B)$.

To prove (ii), assume that $q = p^m < \exp(D)$ and write

$$e = \sum_{g \in G} e_g g \qquad (e_g \in F)$$

Applying the Brauer homomorphism $Z(FG) \longrightarrow Z(FC_G(D))$ together with Theorem 2.2.4, we infer that Supp$e$ contains a p-regular element $g \in C_G(D)$. Choose an element $d \in D$ of maximal order and a p-regular element $h \in C_G(D)$ with $g^{-1} = h$. Then $d-1$ is nilpotent and $dh = hd$. Thus $(d-1)he$ is nilpotent and so $(d-1)he \in T(B)$. We are therefore left to verify that $(d-1)he \notin T_m(B)$.

To this end, we first note that

$$\begin{aligned}((d-1)he)^q &= (dhe-he)^q \\ &= d^q h^q e - h^q e \\ &= d^q g^{-1} e - g^{-1} e \\ &= \sum_{y\in G} e_y d^q g^{-1} y - \sum_{y\in G} e_y g^{-1} y\end{aligned}$$

and that, by our choice of q and d, $d^q \neq 1$. Since $dg = gd$, the element $d^q g^{-1}$ must be p-singular. If $y \in \mathrm{Supp}e$, then by Theorem 2.2.4 y is p-regular and thus $tr(d^q g^{-1} y) = 0$. The conclusion is that

$$\begin{aligned}tr[((d-1)he)^q] &= -\sum_{y\in G} e_y tr(g^{-1}y) \\ &= -e_g \neq 0\end{aligned}$$

But then

$$((d-1)he)^q \notin [B,B]$$

and hence

$$(d-1)he \notin T_m(B),$$

as required. ■

As an application of Theorem 1.3, we now prove the following result.

1.4. THEOREM. (Külshammer (1982)). Let $B = B(e)$ be a block of FG with defect group D and let $q = p^n$ for some $n \in \mathbb{N}_0$.

(i) If $q < \exp(D)$, then there exists $x \in J(B)$ with $x^q \notin [B,B]$. In particular,

$$J(B)^q \neq 0$$

(ii) If $q = \exp(D)$, then $z^q = 0$ for all $z \in J(Z(B))$.

Proof. (i) By Theorem 1.3, $T(B) \neq T_n(B)$ and so there exists $y \in T(B)$ with $y^q \notin [B,B]$. Applying Theorem 3.3.10(i), we also have $T(B) = [B,B] + J(B)$. Therefore $y = k + x$ for some $k \in [B,B]$ and some $x \in J(B)$. Because $k^q \in [B,B]$, we deduce that

$$0 \not\equiv y^q \equiv k^q + x^q \equiv x^q (\mathrm{mod}\ [B,B])$$

as we wished to show

(iii) Owing to Theorem 1.3, we have

$$gz \in FG \cdot J(Z(B)) \subseteq J(B) \subseteq T(B) = T_n(B)$$

for all $g \in G$ and all $z \in J(Z(B))$. Applying Lemma 3.9.2(i) together with the fact that $[B,B]$ is a $Z(FG)$-module, we derive

$$0 = tr((gz)^q C^+) = [tr(gz(C^{1/q})^+)]^q$$

for all $C \in C\ell(G)$. Thus $tr(FGz(C^{1/q})^+) = 0$ and so

$$z(C^{1/q})^+ = 0 \qquad \text{for all } C \in C\ell(G)$$

By Theorem 3.9.3, we infer that

$$z^q C^+ = 0 \quad \text{for all } p\text{-regular } C \in C\ell(G) \text{ with } \delta(C) \underset{G}{\subseteq} D$$

Since each $C \in C\ell(G)$ with $C \subseteq \text{Supp} e$ is p-regular and $\delta(C) \underset{G}{\subseteq} D$ (Theorem 2.2.4 and Proposition 2.2.6) it follows that

$$z^q = z^q e = 0$$

This completes the proof of the theorem. ∎

1.5. COROLLARY. Let B be a block of FG and let $p^n, n > 0$, be the exponent of defect groups of B. Then $t(B) \geqslant p^{n-1} + 1$.

Proof. Direct consequence of Theorem 1.4(i). ∎

1.6. COROLLARY. Suppose that G is abelian and let p^n be the exponent of the Sylow p-subgroup of G. Then $x^{p^n} = 0$ for all $x \in J(FG)$.

Proof. Apply Theorem 1.4(ii). ∎

1.7. COROLLARY. If t is the exponent of Sylow p-subgroups of G, then

$$z^t = 0 \quad \text{for all } z \in J(Z(FG))$$

Proof. Apply Theorem 1.4(ii). ∎

We next illustrate that Theorem 1.4(ii) need not be true for $z \in J(B)$. The following two observations will clear our path.

1.8. LEMMA. Let $R, R_1, R_2, \ldots, R_n$ be rings and let G, G_1, G_2 be groups.

(i) $(R_1 \times R_2 \times \ldots \times R_n)G \cong R_1G \times R_2G \times \ldots \times R_nG$ as rings and $R_1 \times \ldots \times R_n$-modules.

(ii) $R(G_1 \times G_2) \cong (RG_1)G_2$ as rings and R-modules

(iii) For any $k \geqslant 1$, $M_k(R)G \cong M_k(RG)$ as rings and R-modules

Proof. (i) Put $S = R_1 \times R_2 \times \ldots \times R_n$ and let $f_i : S \longrightarrow R_i$ be the projection map. Denote by $\pi_i : SG \longrightarrow R_iG$ the ring homomorphism induced by f_i. Then $\bigcap_{i=1}^{n} \mathrm{Ker} f_i = 0$ and $\mathrm{Ker}\pi_i = (\mathrm{Ker} f_i)G$, $1 \leqslant i \leqslant n$. Consider the map

$$\begin{cases} SG \longrightarrow R_1G \times R_2G \times \ldots \times R_nG \\ x \longmapsto (\pi_1(x), \pi_2(x), \ldots, \pi_n(x)) \end{cases}$$

It is clear that π is a homomorphism of rings and S-modules. Since

$$\mathrm{Ker}\pi = \bigcap_{i=1}^{n} \mathrm{Ker}\pi_i = \bigcap_{i=1}^{n} (\mathrm{Ker} f_i)G = (\bigcap_{i=1}^{n} \mathrm{Ker} f_i)G = 0,$$

π is injective. To prove that π is surjective, fix

$$(y_1, y_2, \ldots, y_n) \in R_1G \times R_2G \times \ldots \times R_nG$$

and write $y_i = \Sigma r_{gi}g$ with $r_{gi} \in R_i$, $1 \leqslant i \leqslant n$. For any $g \in G$, let $r_g \in S$ be such that $f_i(r_g) = r_{gi}$, $1 \leqslant i \leqslant n$. Then $\pi_i(\Sigma r_g g) = y_i$, $1 \leqslant i \leqslant n$, and thus

$$\pi(\Sigma r_g g) = (y_1, y_2, \ldots, y_n),$$

as required.

(ii) By looking at the chain of rings $R \subseteq RG_1 \subseteq (RG_1)G_2$, we infer that $(RG_1)G_2$ is a free R-module with $\{g_1g_2 | g_1 \in G_1, g_2 \in G_2\}$ as a basis. Thus the map $(g_1, g_2) \longmapsto g_1g_2$ induces an isomorphism

$$R(G_1 \times G_2) \longrightarrow (RG_1)G_2$$

of R-modules. Since this induced map obviously preserves multiplication, (ii) is established.

(iii) Let e_{ij} be the $n \times n$ matrix with (i,j)-th entry 1 and zero entry elsewhere, and let $e_{i,j,g}$ be the $n \times n$ matrix with (i,j)-th entry $g \in G$ and zero entry elsewhere. Then the map $e_{ij}g \longmapsto e_{i,j,g}$ induces an isomorphism $M_k(R)G \longrightarrow M_k(RG)$ of R-modules. A routine calculation shows that this map also preserves multiplication. So the lemma is true. ■

1.9. LEMMA. Let $G = H \times P$, where P is a p-group and H is a p'-group. Write $FH \cong \prod_{i=1}^{k} M_{n_i}(D_i)$ for suitable k, n_i and division rings D_i. Then

(i) $FG \cong \prod_{i=1}^{k} M_{n_i}(D_i P)$

(ii) Each $M_{n_i}(D_i P)$ is indecomposable

(iii) P is the defect group of any block of FG.

Proof. (i) Direct consequence of Lemma 1.8.

(ii) It suffices to show that each $D_i P$ is indecomposable. Let e be a non-zero central idempotent of $D_i P$. Then $e \in Z(D_i)P$ and $Z(D_i)$ is a field of characteristic p. Hence, by Theorem 2.2.4, $e = 1$ as required.

(ii) Let e be any block idempotent of FG and let $C \in C\ell(G)$ be such that $C \subseteq \text{Supp}e$. By Theorem 2.2.4, $C \subseteq H$ and hence P is the defect group of C. Thus P is the defect group of e, as required. ■

It is now an easy matter to show that Theorem 1.4(ii), need not be true for $z \in J(B)$.

1.10. EXAMPLE. (Külshammer (1982)). Let F be an algebraically closed field of characteristic 2, let H be a nonabelian group of order 27 and let $P = \langle g \rangle \times \langle h \rangle$ be an elementary abelian group of order 4. It is easy to verify that FH has a block $B_0 \cong M_3(F)$. Hence, by Lemma 1.9, the group algebra $F(H \times P)$ has a block $B \cong M_3(FP)$. Let $I(P)$ be the augmentation ideal of FP. Then $J(FP) = I(P)$ and so $J(B) \cong M_3(I(P))$. Identifying $J(B)$ and $M_3(I(P))$, we see that

$$z = \begin{bmatrix} 0 & g-1 & 0 \\ h-1 & 0 & 0 \\ 0 & 0 & 0 \end{bmatrix} \in J(B)$$

but $z^2 \neq 0$. ■

2. AN UPPER BOUND FOR $t(Z(B))$

Throughout this section, G denotes a finite group and F a field of characteristic p. Our aim is to provide an upper bound for the nilpotency index $t(Z(B))$ of $J(Z(B))$, where B is a block of FG.

2.1. LEMMA. Let P be a p-subgroup of G and let H be a subgroup of G such that $C_G(P) \subseteq H \subseteq N_G(P)$. Let e be a block idempotent of FG with defect group D, let $\rho : Z(FG) \longrightarrow Z(FH)$ be the Brauer homomorphism, and let $\rho(e) = e_1 + \ldots + e_n$, where each e_i is a block idempotent of FH. If D_i is a defect group of e_i, then $D_i \subseteq_G D$.

Proof. Let $C_1, C_2, \ldots, C_t$ be all elements in $C\ell(G)$ with $\delta(C_i) \subseteq_G D$ and let $I_D(G)$ be the F-linear span of $C_1^+, C_2^+, \ldots, C_t^+$. We know, from Lemma 2.2.5 and Proposition 2.2.6, that $I_D(G)$ is an ideal of $Z(FG)$ containing e. If $L \in C\ell(H)$ and $L \subseteq C_i$ for some $i \in \{1,2,\ldots,t\}$, then

$$\delta(L) \subseteq \delta(C_i)^x \cap H \subseteq D^x \cap H$$

for some $x \in G$. Because $e \in I_D(G)$, we conclude that

$$\rho(e) \in \sum_{x \in G} I_{D^x \cap H}(H)$$

Taking into account that $I_{D^x \cap H}(H)$ is an ideal of $Z(FH)$ and that $e_i = e_i \rho(e)$, we see that

$$e_i \in \sum_{x \in G} I_{D^x \cap H}(H) \qquad (1 \leqslant i \leqslant n)$$

Hence, by Lemma 3.16.5, $e_i \in I_{D^x \cap H}(H)$ for some $x \in G$ and therefore $D_i \subseteq_H D^x \cap H \subseteq_G D$, as required. ■

2.2. LEMMA. Let P be a normal p-subgroup of G, let $H = PC_G(P)$, and let $C_1, \ldots, C_r$ be all distinct p-regular classes of H. For each $i \in \{1,2,\ldots,r\}$, let $\bar{C}_i$ be the image of C_i under the natural homomorphism $G \longrightarrow G/P$. Then

(i) $\bar{C}_1, \bar{C}_2, \ldots, \bar{C}_r$ are all distinct p-regular classes of H/P and $|C_i| = |\bar{C}_i|$ for all $i \in \{1,2,\ldots,r\}$.

(ii) For all $C \in \{C_1, \ldots, C_r\}$, $C \subseteq C_G(P)$ and $P \subseteq \delta(C)$

(iii) For all $C \in \{C_1, \ldots, C_r\}$, $\delta(C)/P$ is a defect group of $\bar{C}$.

Proof. Let $|G| = p^n q$, where $(p,q) = 1$. Fix $C \in \{C_1, \ldots, C_r\}$ and let $x \in C$. As x is a p'-element and as

$$x = cd = dc \in C_G(P)P \qquad (c \in C_G(P), d \in P)$$

we have $x \in C_G(P)$. This shows that $C \subseteq C_G(P)$ and hence that $P \subseteq \delta(C)$, proving (ii).

For any $g \in G$, let $\bar{g}$ be the image of g in G/P. We next show that

$$C_{\bar{H}}(\bar{x}) = C_H(x)/P \tag{1}$$

where $\bar{H} = H/P$. By (ii), we have $P \subseteq C_H(x)$ and therefore $C_H(x)/P \subseteq C_{\bar{H}}(\bar{x})$. Let $\bar{h} \in C_{\bar{H}}(\bar{x})$. Then $h^{-1}xh = xd = dx$ for some $d \in P$. Since x is a p'-element, say of order k, and d is a p-element, it follows that

$$1 = (h^{-1}xh)^k = x^k d^k = d^k$$

and hence that $d = 1$. Thus $h \in C_H(x)$ and (1) is established. Since (1) implies (iii), we are left to verify (i).

To this end, we first observe that $|C| = |\bar{C}|$ since by (1)

$$(\bar{H}:C_{\bar{H}}(\bar{x})) = (H:C_H(x))$$

Assume that $X_1, X_2 \in \{C_1,\dots,C_r\}$ are such that $\bar{X}_1 = \bar{X}_2$, and let $x_i \in X_i, i = 1,2$. Then $x_1 = yx_2y^{-1}d$ for some $y \in H$ and $d \in P$. By (ii), $P \subseteq \delta(X_2)$ and so yx_2y^{-1} and d commute. Since x_1 and yx_2y^{-1} are p'-elements, we have $x_1 = yx_2y^{-1}$. Thus $X_1 = X_2$, proving that the $\bar{C}_1,\dots,\bar{C}_r$ are distinct.

Finally, let Y denote a conjugacy class of p'-elements of H/P and let $y \in H$ be such that $\bar{y} \in Y$. Then we may assume that $y \in C_G(P)$. If m is the order of $\bar{y}$, then $y^m \in P$. Because $(p,m) = 1$, there are integers i and j such that $ip^n + jm = 1$. Hence

$$y = y^{ip^n} y^{jm} \in y^{ip^n} P \tag{2}$$

As p^n is the highest power of p dividing the order of G, y^{ip^n} is a p'-element. Let $C \in \{C_1,\dots,C_r\}$ be such that $y^{ip^n} \in C$. Then, by (2), $Y = \bar{C}$ and the result follows. ■

2.3. LEMMA. Let P be a normal p-subgroup of G, let $H = PC_G(P)$ and let $\pi : FG \longrightarrow F(G/P)$ be the natural homomorphism. Let $b_1, b_2, \dots, b_n$ be all blocks of FH and let D_i be a defect group of b_i, $1 \leqslant i \leqslant n$.

(i) $\pi(b_1), \pi(b_2), \dots, \pi(b_n)$ are all distinct blocks of $F(H/P)$

(ii) For each $i \in \{1,2,\dots,n\}, D_i/P$ is a defect group of $\pi(b_i)$.

Proof. (i) Let $T = \pi(Z(FH))$ and let e_i be the block idempotent of FH

contained in b_i, $1 \leqslant i \leqslant n$. By Proposition 3.1.1, $\mathrm{Ker}\pi$ is a nilpotent ideal and therefore, by Lemma 3.4.16, $\pi(e_1),\ldots,\pi(e_n)$ are all distinct primitive idempotents of T. It therefore suffices to prove that T contains all central idempotents of $F(H/P)$.

Let $C_1, C_2,\ldots,C_r$ be all p-regular classes of H. If C_i' denotes the image of C_i under the natural homomorphism $G \longrightarrow G/P$, then by Lemma 2.2(i), $C_1', C_2',\ldots,C_r'$ are all distinct p-regular classes of H/P. Hence, by Theorem 2.2.4, T contains all central idempotents of $F(H/P)$.

(ii) Let e be a block idempotent of FH and let $C_1, C_2,\ldots,C_t$ be all p-regular classes of H such that $\mathrm{Supp}e = C_1 \cup \ldots \cup C_t$. Then

$$\mathrm{Supp}\pi(e) = C_1' \cup \ldots \cup C_t'$$

where C_i' is the image of C_i under the natural homomorphism $G \longrightarrow G/P$. The desired conclusion now follows from Lemma 2.2.(iii). ■

2.4. THEOREM. (Okuyama (1981)). Let $B = B(e)$ be a block of FG with defect d. Then

$$t(Z(B)) \leqslant p^d$$

Proof. We have $Z(B) = Z(FG)e$, so $J(Z(B)) = J(Z(FG))e$ and therefore for all $k \geqslant 1$, $J(Z(B))^k = J(Z(FG))^k e$. Thus we need only verify that

$$J(Z(FG))^{p^d} e = 0 \tag{3}$$

If $d = 0$ then $J(B) = 0$ (Theorem 3.6.4) and therefore $J(Z(FG))e = 0$. Thus (3) is true for $d = 0$ and we argue by induction on $d > 0$. In what follows, we write D for a defect group of B. For the sake of clarity, the rest of the proof is divided into three steps.

Step 1. Let $g \in D$ be an element of order p^n, $n > 0$, let $P = \langle g \rangle$ and let $H = C_G(P)$. Let $\rho : Z(FG) \longrightarrow Z(FH)$ be the Brauer homomorphism. It will be shown that

$$\rho(J(Z(FG))^{p^d-1} e) \subseteq FH \cdot \left(\sum_{i=0}^{p^n-1} g^i\right) \tag{4}$$

By hypothesis, there exists $C \in C\ell(G)$ with $C \subseteq \mathrm{Supp}e$ and $\delta(C) = D$. Hence

$H \cap C \neq \emptyset$ and so, by Theorem 2.2.3, $\rho(e) \neq 0$. Therefore we may write

$$\rho(e) = e_1 + e_2 + \dots + e_m$$

as a sum of block idempotents of FH. Fix $f \in \{e_1, \dots, e_m\}$ and observe that, by Lemma 2.1, the defect of f is at most d. Let $\pi : FH \longrightarrow F(H/P)$ be the natural homomorphism. Then, by Lemma 2.3, $\pi(f)$ is a block idempotent of $F(H/P)$ with defect at most $d-n$, since the defect of f is at most d. Since $\mathrm{Ker}\pi = FH(g-1)$, it follows from (3) by induction hypothesis that

$$J(Z(FH))^{p^{d-n}} f \subseteq FH(g-1) \tag{5}$$

Since $p^d - 1 \geqslant p^{d-n}(p^n - 1)$, raising both sides of (5) to the power of $p^n - 1$ yields

$$\begin{aligned} J(Z(FH))^{p^d-1} f &\subseteq J(Z(FH))^{p^{d-n}(p^n-1)} f \\ &\subseteq [FH(g-1)]^{p^n-1} \\ &= FH(g-1)^{p^n-1} \\ &= FH[FP(g-1)]^{p^n-1} \\ &= FH \cdot J(FP)^{p^n-1} \\ &= FH\left(\sum_{i=o}^{p^n-1} g^i\right) \end{aligned}$$

where the last equality follows by virtue of Lemma 3.11.2. Thus we have

$$J(Z(FH))^{p^d-1}\rho(e) \subseteq FH\left(\sum_{i=o}^{p^n-1} g^i\right)$$

proving (4) by virtue of $\rho(J(Z(FG)) \subseteq J(Z(FH))$.

Step 2. Let $S_1, S_2, \dots, S_r$ be all p-regular sections of G and let $Z_{p'} = \sum_{i=1}^{r} FS_i^+$. We next prove that

$$J(Z(FG))^{p^d-1} e \subseteq Z_{p'} \tag{6}$$

Let x be a p-regular element of G and g a p-element of G of order p^n such that $xg = gx$. It suffices to show that, for any z in $J(Z(FG))^{p^d-1}e$, the coefficients of x and xg in z are equal. This is so since the coefficients of z are constant on conjugacy classes of G.

We may harmlessly assume that either $x \in \mathrm{Supp}z$ or $xg \in \mathrm{Supp}z$. Let D be a defect group of e. Then, by Lemma 2.2.5 and Proposition 2.2.6, $I_D(G)$ is an ideal of $Z(FG)$ containing e. In particular, we must have $z \in I_D(G)$. Since $x \in \text{Supp}z$ or $xg \in \text{Supp}z$, it follows from the definition of $I_D(G)$ that g belongs to a conjugate of D. Thus we may assume that g belongs to a defect group of e. Let $P = \langle g \rangle$, $H = C_G(P)$ and $\rho : Z(FG) \longrightarrow Z(FH)$ the Brauer homomorphism. Since $x, xg \in H$, the coefficients of x and xg in z are the same as the coefficients of x and xg in $\rho(z)$. But, by (4),

$$\rho(z) \in FH\left(\sum_{i=o}^{p^n-1} g^i\right)$$

and so the coefficients of x and xg in $\rho(z)$ are equal, proving (6)

Step 3. Completion of the proof. Owing to Theorem 3.3.14, we have

$$J(Z(FG))Z_{p'} = 0$$

Hence, multiplying both sides of (6) by $J(Z(FG))$, we obtain

$$J(Z(FG))^{p^d} e = 0$$

proving (3) and hence the result. ∎

2.5. COROLLARY. Let G be a group of order $p^a m$, $(p,m) = 1$. Then

$$J(Z(FG))^{p^a} = 0$$

Proof. Let $e_1, e_2, \ldots, e_n$ be all block idempotents of FG. Then, by Theorem 2.4,

$$J(Z(FG))^{p^a} e_i = 0 \qquad \text{for all } i \in \{1,2,\ldots,n\}$$

Since $J(Z(FG))^{p^a} = J(Z(FG))^{p^a} e_1 + \ldots + J(Z(FG))^{p^a} e_n$, the result follows. ∎

A weaker version of Corollary 2.5, namely the equality

$$J(Z(FG))^{(p^{a+1}-1)/(p-1)} = 0$$

was established by Passman (1980).

3. DEFECT GROUPS OF COVERING BLOCKS.

Throughout this section, G denotes a finite group, N a normal subgroup of G and F an arbitrary field of characteristic $p > 0$. Our aim is to prove that if B is a block of FG covering a block b of FN, then $\delta(B) \cap N$ is a defect group of b.

3.1. LEMMA. Let $B = B(e)$ and $b = b(f)$ be blocks of FG and FN, respectively, such that B covers b. Then the following conditions are equivalent:
(i) V is an indecomposable FN-module such that V is a summand of U_N for some projective indecomposable FG-module U lying in B.
(ii) V is a projective indecomposable FN-module such that V lies in $b^g = g^{-1}bg$ for some $g \in G$.

Proof. Let X and Y denote the classes of FN-modules satisfying (i) and (ii), respectively. Let f^* be the sum of all distinct G-conjugates of f. Then f^* is a central idempotent of FG and $ef^* = e$ (see Lemma 3.10.1). Hence $f^*U = U$ for $U \in B$ and $f^*V = V$ for $V \in X$. Because V is indecomposable, there is a $g \in G$ such that $f^gV = V$, so $V \in b^g$. Clearly, V is projective, so $V \in Y$ and thus $X \subseteq Y$.

To prove that $Y \subseteq X$, first assume that $V \in Y$ is such that $L = V/J(FN)V$ is a composition factor of some $V' \in X$. If V' is a summand of U_N for the projective module $U \in B$, then for a suitable $r \in \mathbb{N}_0$ and

$$W = J(FN)^r U/J(FN)^{r+1} U$$

L is a summand of W_N. Let $\pi : Q \longrightarrow J(FN)^r U$ be a projective cover of $J(FN)^r U$. Then π induces a surjective homomorphism

$$\bar{\pi} : Q/J(FN)Q \longrightarrow W,$$

so L is a summand of $(Q/J(FN)Q)_N$. But $Q \in B$, hence $Q_N = \oplus V_i$, all $V_i \in X$, and therefore $L \cong V_i/J(FN)V_i$ for some i and $V \cong V_i \in X$.

Now assume that $V \in Y$ and $V' \in X$ are in the same block. Then V and V' are linked (see Proposition 1.10.14), so we may assume they have a composition factor $Q/J(FN)Q$ in common (for some $Q \in Y$). Then $V/J(FN)V$ is a composition factor of Q as well. Applying the above argument twice, we derive $Q \in X$ and $V \in X$. Since obviously ${}^gV \in X$ if $V \in X$ and X is nonempty, X contains at least one V in each block b^g, so X contains all V in all b^g, that is $X \supseteq Y$. So the lemma is proved. ■

We are now ready to prove the following result.

3.2. THEOREM. (Knörr (1976)). Let F be an arbitrary field of characteristic $p > 0$, let N be a normal subgroup of G and let B and b be blocks of FG and FN, respectively, such that B covers b. If D is a defect group of B, then $D \cap N$ is a defect group of b.

Proof. We denote by $\delta(B)$ and $\delta(b)$ defect groups of blocks B and b, respectively. Let X and Y be the classes of FN-modules satisfying (i) and (ii), respectively, in Lemma 3.1. We know, from Lemma 3.1, that $X = Y$. By Theorem 3.16.14, there exists an irreducible module $M \in b$ such that $vx(M) =_G \delta(b)$, where $vx(M)$ denotes a vertex of M. Now $M \cong V/J(FN)V$ for some $V \in X = Y$. Therefore M is a summand of $(U/J(FN)U)_N$ for some projective indecomposable FG-module $U \in B$. Owing to Lemma 3.16.13, the indecomposable FG-module $U/J(FN)U$ is FN-projective, so by Lemma 3.16.1,

$$\delta(b) =_N vx(M) \subseteq_G vx(U/J(FN)U) \subseteq_G N \cap \delta(B)$$

On the other hand, by Proposition 3.10.6, there is a block B_1 of FG such that $\delta(B) \subseteq_G \delta(B_1)$ and $\delta(b) =_G N \cap \delta(B_1)$. Hence

$$\delta(b) \subseteq_G N \cap \delta(B) \subseteq_G N \cap \delta(B_1) =_G \delta(b)$$

and so $\delta(b) =_G \delta(B) \cap N$, as asserted. ∎

4. REGULAR BLOCKS.

Throughout, G denotes a finite group, H a subgroup of G and F an arbitrary field of characteristic $p > 0$. As usual, $C\ell(G)$ denotes the set of conjugacy classes of G and, for each $C \in C\ell(G)$, $\delta(C)$ is a p-defect group of C. If b is a block of FH, we write $b = b(e,\lambda)$ to indicate that e is a block idempotent of FH lying in b and λ the irreducible representation of $Z(FH)$ associated with b. We denote by

$$\pi : Z(FG) \longrightarrow Z(FH)$$

the natural projection so that

$$\pi(C^+) = (C \cap H)^+ \qquad \text{for all } C \in C\ell(G)$$

Note that π is a surjective F-homomorphism, but need not preserve multiplication.

Let $b = b(e,\lambda)$ be a block of FH and let $\lambda^G = \lambda\circ\pi$. If the map λ^G is an irreducible representation of $Z(FG)$ (which is equivalent to λ^G being a ring homomorphism), we denote by b^G the unique block of FG corresponding to λ^G and say that b^G is *defined*. In what follows $\delta(b)$ always denotes a defect group of b.

4.1. LEMMA. Let $H \subseteq S \subseteq G$ be a chain of subgroups of G and let $b = b(e,\lambda)$ be a block of FH.

(i) If b^S is defined and either one of b^G or $(b^S)^G$ is defined, then so is the other and $(b^S)^G = b^G$.

(ii) If b^G is defined, then $\delta(b) \underset{G}{\subseteq} \delta(b^G)$.

Proof. (i) Consider the sequence of natural projections

$$Z(FG) \xrightarrow{\pi_1} Z(FS) \xrightarrow{\pi_2} Z(FH)$$

Then, for all $C \in C\ell(G)$, we have

$$\begin{aligned}(\pi_2\pi_1)(C^+) &= \pi_2(C \cap S)^+\\ &= (C \cap S \cap H)^+\\ &= (C \cap H)^+\\ &= \pi(C^+)\end{aligned}$$

proving that $\pi = \pi_2\pi_1$. Hence

$$(\lambda^S)^G = (\lambda\circ\pi_2)\circ\pi_1 = \lambda\circ\pi = \lambda^G$$

and therefore b^G is defined if and only if $(b^S)^G$ is defined. Since $(\lambda^S)^G$ and λ^G are the irreducible representations of $Z(FG)$ associated with blocks $(b^S)^G$ and b^G, respectively, the assertion follows.

(ii) By hypothesis, λ^G is an irreducible representation of $Z(FG)$. By Proposition 2.2.6, there exists $C \in C\ell(G)$ with $\delta(C) = \delta(b^G)$ such that $\lambda^G(C^+) \neq 0$. Hence $\lambda(L^+) \neq 0$ for some $L \in C\ell(H)$ with $L \subseteq C$. Then $\delta(L) \underset{G}{\subseteq} \delta(C)$ and, by Proposition 2.2.6, $\delta(b) \underset{H}{\subseteq} \delta(L)$. Thus

$$\delta(b) \underset{G}{\subseteq} \delta(L) \underset{G}{\subseteq} \delta(C) = \delta(b^G)$$

as asserted. ∎

For future use, we next record

4.2. LEMMA. Let N be a normal subgroup of G, and let B and b be blocks of FG and FN, respectively. Then the following conditions are equivalent:

(i) B covers b

(ii) There exist irreducible FG and FN-modules V and W, respectively, such that $V \in B, W \in b$ and W is a submodule of V_N.

Furthermore, if B covers b, then for any irreducible FG-module V in B, there exists an irreducible submodule of V_N belonging to b.

Proof. (i) $\Rightarrow$ (ii): Write $B = B(E)$ and $b = b(e)$ so that, by Lemma 3.10.2, (i) is equivalent to $Ee \neq 0$. If e annihilates all irreducible submodules of V_N, for any irreducible FG-module V in B, then e annihilates all irreducible FG-modules in B. Hence the same is true for $Ee \in B$ and so Ee is an idempotent in $J(B)$. Since $J(B)$ is nilpotent, we conclude that $Ee = 0$. Thus (i) implies (ii).

(ii) $\Rightarrow$ (i): Assume that (ii) holds. Then, by Clifford's theorem,

$$V_N = g_1 W \oplus \ldots \oplus g_n W$$

for some $1 = g_1, g_2, \ldots, g_n$ in G. By hypothesis, $EV = V$ and $eW = W$ and so

$$EeV = eV = W \oplus eg_2 W \oplus \ldots \oplus eg_n W \quad \text{(direct sum of } FN\text{-modules)}$$

This proves that $Ee \neq 0$ and hence that (ii) implies (i).

Assume that B covers b and let V be any irreducible FG-module in B. Denote by W any irreducible submodule of V_N and by $b_1 = b_1(e_1)$ the block of FN containing W. Then, by the above B covers b and hence by Lemma 3.10.2, $e = g^{-1}e_1 g$ for some $g \in G$. It follows that

$$e(g^{-1}W) = g^{-1}e_1 W = g^{-1}W \neq 0$$

so $g^{-1}W$ is an irreducible submodule of V_N belonging to b. ■

4.3. LEMMA. Let $b = b(e,\lambda)$ be a block of FH and let P be a p-subgroup of H. If $C_G(P) \subseteq H \subseteq N_G(P)$, then $\lambda^G = \lambda \circ \rho$, where $\rho : Z(FG) \longrightarrow Z(FH)$ is the Brauer homomorphism with respect to P. In particular, b^G is defined.

Proof. It suffices to show that for each $C \in C\ell(G)$,

$$\lambda^G(C^+) = (\lambda\circ\rho)(C^+)$$

To this end, write $C \cap H = \bigcup_{i=1}^{n} C_i$ with each C_i in $Cl(H)$. Then

$$\pi(C^+) = \sum_{i=1}^{n} C_i^+$$

and

$$\rho(C^+) = \sum_{i=1}^{n} (C_i \cap C_G(P))^+$$

and so we need only verify that

$$\lambda(C_i^+) = \lambda(C_i \cap C_G(P))^+ \qquad \text{for all } i \in \{1,\ldots,n\}$$

Because either $C_i \subseteq C_G(P)$ or $C_i \cap C_G(P) = \emptyset$, we may assume the latter. Let $L \in Cl(H) \cap \text{Supp}e$ be such that $\delta(L) = \delta(b)$. Then $P \subseteq \delta(L)$ since $P \lhd H$. Hence $\delta(C_i) \subset_H \delta(L)$ and so, by Proposition 2.2.6, $\lambda(C_i^+) = 0$, as required. ∎

Let N be a normal subgroup of G. A block $B = B(E,\lambda)$ of FG is said to be *regular with respect to* N if

$$\lambda(C^+) = 0 \quad \text{for all} \quad C \in Cl(G) \quad \text{with} \quad C \not\subseteq N$$

4.4. LEMMA. Let N be a normal subgroup of G and let $B = B(E,\lambda)$ be a block of FG. Then the following conditions are equivalent:

(i) B is regular with respect to N

(ii) $B = b^G$ for every block b of FN which is covered by B

(iii) $B = b^G$ for some block b of FN

Proof. (i) ⇒ (ii): Let $b = b(e,\mu)$ be a block of FN which is covered by B. Owing to Lemma 3.10.4, there exists an F-algebra isomorphism

$$\psi : \mu(Z^*(FN)) \longrightarrow \lambda(Z^*(FN))$$

such that $\psi(\mu(x)) = \lambda(x)$ for all $x \in Z^*(FN)$. From the definition of μ^G, we have $\mu^G(y) = \mu(y)$ for all $y \in FN$, hence

$$\psi(\mu^G(x)) = \lambda(x) \qquad \text{for all } x \in Z^*(FN)$$

If $C \in Cl(G)$ is such that $C \not\subseteq N$, then $\mu^G(C^+) = 0 = \lambda(C^+)$. Hence

$$\psi(\mu^G(z)) = \lambda(z) \qquad \text{for all } z \in Z(FG)$$

and so μ^G is an irreducible representation of $Z(FG)$. Because $\mu^G(E) = \lambda(E) = 1$, E is a block idempotent of b^G and thus $B = b^G$.

(ii) $\Rightarrow$ (iii): Obvious

(iii) $\Rightarrow$ (i): Let $b = b(e,\mu)$ be a block of FN such that $B = b^G$. Then μ^G and λ are equivalent and so it suffices to show that $\mu^G(C^+) = 0$ for all $C \in C\ell(G)$ with $C \not\subseteq N$. The latter being a consequence of the definition of μ^G, the result follows. ∎

We have now come to the demonstration for which this section has been developed.

4.5. THEOREM. Let N be a normal subgroup of G and let B and b be blocks of FG and FN, respectively

(i) If B is regular with respect to N, then B covers b if and only if $B = b^G$.

(ii) If D is a defect group of B and $C_G(D) \subseteq N$, then B is regular with respect to N, and $B = \beta^G$ for any block β of FN which is covered by B.

(iii) The block b^G is defined if and only if there exists a block of FG which is regular with respect to N and covers b.

(iv) If b^G is defined, then b^G is the unique block of FG which is regular with respect to N and covers b.

(v) If $C_G(P) \subseteq N$ for some normal p-subgroup P of G, then B is regular with respect to N. In particular, b^G is defined and is the only block of FG covering b. Furthermore, by Lemma 3.10.2, if b_1 and b_2 are blocks of FN, then $b_1^G = b_2^G$ if and only if b_1 and b_2 are G-conjugate.

Proof. (i) If B covers b, then $B = b^G$ by Lemma 4.4(ii). Conversely, assume that $B = b^G$. If $b = b(e,\mu)$, then $B = B(E,\mu^G)$ for some block idempotent E of FG. Because $\mu^G(x) = \mu(x)$ for all $x \in Z^*(FN)$, we infer, from Lemma 3.10. 4, that B covers b.

(ii) Write $B = B(E,\lambda)$ and assume that $C \in C\ell(G)$ is such that $C \not\subseteq N$. If $\lambda(C^+) \neq 0$, then by Proposition 2.2.6, $D \underset{G}{\subseteq} \delta(C)$ which means that $g \in C_G(D)$ for some $g \in C$. But then $g \in N$ and so $C \subseteq N$, a contradiction.

(iii) If b^G is defined, then b^G is regular (Lemma 4.4) and covers b, by virtue of (i). Conversely, assume that B is a block of FG which is regular and covers b. Then, by (i), $B = b^G$ and so b^G is defined.

(iv) Assume that b^G is defined and let B be any block of FG which is regular and covers b. Then, as we have seen in (iii), b^G is regular, b^G covers b and $B = b^G$.

(v) Write $B = B(E,\lambda)$ and let $C \in C\ell(G)$ be such that $C \not\subseteq N$. If $\lambda(C^+) \neq 0$, then by Proposition 2.2.6, $\delta(B) \underset{G}{\subseteq} \delta(C)$. Because P is a normal p-subgroup of G, $P \subseteq \delta(B)$ and hence $P \underset{G}{\subseteq} \delta(C)$. By the argument of (ii), we deduce that $C \subseteq N$, a contradiction. ■

5. THE FONG CORRESPONDENCE.

Throughout this section, G denotes a finite group, N a normal subgroup of G and F an algebraically closed field of characteristic $p > 0$.

If $\beta = \beta(\varepsilon)$ is a block of FN, then $G(\varepsilon)$ denotes the inertia group of ε. If H is a subgroup of G, then

$$T_{G,H} : C_{FG}(H) \longrightarrow Z(FG)$$

is the relative trace map.

5.1. LEMMA. Let H be a subgroup of G and let V be an irreducible FH-module such that V^G is an irreducible FG-module. If V lies in the block b of FH, then b^G is defined and V^G lies in b^G.

Proof. Let $\pi : FG \longrightarrow FH$ be the natural projection and write

$$V^G = V \oplus g_2 V \oplus \ldots \oplus g_n V \quad \text{(direct sum of } F\text{-spaces)}. \tag{1}$$

where $\{1, g_2, \ldots, g_n\}$ is a left transversal for H in G. Let ϕ and ψ be the central characters of FH and FG afforded by V and V^G, respectively. Owing to Proposition 1.10.15, it suffices to verify that

$$\phi\pi(C^+) = \psi(C^+) \qquad \text{for all } C \in C\ell(G) \tag{2}$$

To this end, write $C \cap H = C_1 \cup C_2 \cup \ldots \cup C_r$ for suitable $C_i \in C\ell(H)$, and put $X = C - (C \cap H)$. Then

$$\psi(C^+)v = C^+v = \sum_{i=1}^{r}(C_i^+)v + X^+v$$
$$= \phi\pi(C^+)v + X^+v \qquad (v \in V)$$

and so $X^+v = 0$, by virtue of (1). This proves (2) and hence the result. ∎

5.2. LEMMA. Let H be a subgroup of G with $N \subseteq H \subseteq G$, and let W be an irreducible FN-module such that H contains the inertia group of W. Denote by $W(H)$ and $W(G)$ the sets of all irreducible FH and FG-modules, respectively, whose restriction to N have W as a summand.

(i) If $V \in W(H)$, then $V^G \in W(G)$

(ii) The map $V \mapsto V^G$ induces a bijective correspondence between the isomorphism classes of FH and FG-modules in $W(H)$ and $W(G)$, respectively.

Proof. Let S denote the inertia group of W and assume that the result is true for S instead of H. Then, for $X \in W(S)$, the maps $X \mapsto X^H$ and $X \mapsto X^G$ satisfy (i) and (ii), with respect to $W(S), W(H)$ and $W(S), W(G)$, respectively. Because $(X^H)^G \cong X^G$, it follows that (i) and (ii) are satisfied with respect to $W(H), W(G)$. Hence we may assume that $S = H$.

(i) Suppose that $V \in W(H)$. Because V is an irreducible submodule of $(V^G)_H$, there exists a composition factor X of V^G such that V is a composition factor of X_H. Since X_N is completely reducible, V_N is a submodule of X_N; in particular, W is a summand of X_N. Let U denote the sum of all submodules of X_N isomorphic to W. Then $V \subseteq U$ and, by Clifford's theorem, U is an irreducible FH-module such that $X \cong U^G$. Since V is irreducible, we conclude that $X \cong V^G$ is also irreducible.

(ii) Assume that $V_1, V_2 \in W(H)$ are such that $V_1^G \cong V_2^G$. Then $(V_1^G)_H \cong (V_2^G)_H$, $(V_1^G)_H = V_1 \oplus V_1'$ and $(V_2^G)_H = V_2 \oplus V_2'$, where the FH-modules V_1' and V_2' do not have W as a summand. Thus $V_1 \cong V_2$ and we are left to verify that each $U \in W(G)$ is isomorphic to V^G for some $V \in W(H)$. The latter being a consequence of Clifford's theorem, the result follows. ∎

5.3. LEMMA. Let $\beta = \beta(\varepsilon)$ be a block of FN, let $H = G(\varepsilon)$ and let $B = B(E)$ be a block of FG covering β. Then there exists exactly one block $b = b(e)$ of FH with $Ee\varepsilon \neq 0$. Moreover,

(i) The restriction of $T_{G,H}$ to $Z(b)$ is an isomorphism $Z(b) \cong Z(B)$ of F-algebras; in particular, $E = T_{G,H}(e)$

(ii) By right multiplication, FGe is a free b-module of rank $n = (G:H)$

(iii) Left multiplication gives an isomorphism

$$B \cong \operatorname{End}_b(FGe) \cong M_n(b) \qquad \text{of } F\text{-algebras}$$

Proof. By definition, ε is a central idempotent of FH, so we have a decomposition

$$\varepsilon = e_1 + e_2 + \ldots + e_k$$

where each e_i is a block idempotent of FH. Since

$$0 \neq E\varepsilon = E\varepsilon^2 = Ee_1\varepsilon + Ee_2\varepsilon + \ldots + Ee_k\varepsilon,$$

there is a block $b = b(e)$ of FH with $Ee\varepsilon \neq 0$ and $e = e_i$ for some $i \in \{1,2,\ldots,k\}$. For a given $g \in G$, put ${}^g e = geg^{-1}$, and let T be a left transversal for H in G. Then, for all distinct u,v in T, we have

$${}^u e\,{}^v e = {}^u e({}^u\varepsilon\,{}^v\varepsilon)\,{}^v e = 0$$

since $\varepsilon e = e\varepsilon = e$ and ${}^u\varepsilon\,{}^v\varepsilon = 0$. Hence $E' = T_{G,H}(e)$ is a central idempotent of FG. Observe also that

$$FGe = \bigoplus_{t \in T} tFHe = \bigoplus_{t\in T} tb \quad \text{(direct sum of right } b\text{-modules)}$$

proving (ii).

Let x be an element of FGE' with $xFGe = 0$. Then, for all $t \in T, xte = 0$ and

$$0 = \sum_{t\in T} xtet^{-1} = xE' = x$$

Thus left multiplication gives an injective homomorphism $FGE' \to \operatorname{End}_b(FGe)$ of F-algebras. Since

$$FGE' = \bigoplus_{t\in T} FG\,{}^t e = \bigoplus_{t\in T} {}^t(FGe),$$

we have

$$\dim_F FGE' = n^2 \dim_F b = \dim_F M_n(b)$$

and thus

$$FGE' \cong \operatorname{End}_b(FGe) \cong M_n(b)$$

In particular, E' is a block idempotent of FG. Now $EE'e = Ee \neq 0$, so $EE' \neq 0$ and therefore $E = E'$, proving (iii).

Given $x,y \in Z(b)$, we have

$$\begin{aligned} T_{G,H}(x)T_{G,H}(y) &= \sum_{u,v\in T} {}^{u}x\,{}^{v}y \\ &= \sum_{u,v\in T} {}^{u}x\,{}^{u}e\,{}^{v}e\,{}^{v}y \\ &= \sum_{t\in T} {}^{t}x\,{}^{t}y \\ &= T_{G,H}(xy) \end{aligned}$$

since ${}^{u}e\,{}^{v}e = 0$ for $u \neq v$ and ${}^{u}x\,{}^{u}e = {}^{u}x$ for all $u \in T$. Since for all $x \in Z(b)$,

$$T_{G,H}(x)e = \sum_{t\in T} {}^{t}xe = \sum_{t\in T} {}^{t}x\,{}^{t}ee = x$$

the map $T_{G,H} : Z(b) \longrightarrow Z(B)$ is an injective homomorphism of F-algebras. Moreover, this map is bijective since $\dim_F Z(b) = \dim_F Z(B)$, by virtue of (iii). This proves (i).

Finally, suppose that $b' = b'(e')$ is another block of FH with $Ee'\varepsilon \neq 0$. Then

$$0 \neq Ee'\varepsilon = \sum_{t\in T} {}^{t}e\,{}^{t}\varepsilon\varepsilon e' = ee'$$

and $e = e'$, thus completing the proof. ■

We are at last in a position to attain our main objective, which is to prove the following important result known as the Fong correspondence.

5.4. THEOREM. (Fong (1961), Reynolds (1963), Külshammer (1981(b)). Let F be an algebraically closed field of characteristic $p > 0$, let N be a normal subgroup of G; let $\beta = \beta(\varepsilon)$ be a block of FN and let $H = G(\varepsilon)$.

(i) If b is a block of FH covering β, then b^G is defined and covers β

(ii) If B is a block of FG covering β, then there exists a unique block b of FH, called the *Fong correspondent* of B in FH (with respect to β) such that b covers β and $B = b^G$. In particular, if $b_1,\ldots,b_s$ are all distinct blocks of FH covering β, then $b_1^G,\dots,b_s^G$ are all distinct blocks of FG covering β.

(iii) If $b = b(e)$ is the Fong correspondent of $B = B(E)$ in FH, then

(a) $E = T_{G,H}(e)$ and e is the unique block idempotent of FH with $Ee\varepsilon \neq 0$

(b) The restriction of $T_{G,H}$ to $Z(b)$ is an isomorphism $Z(b) \cong Z(B)$ of F-algebras.

(c) $B \cong M_n(b)$, where $n = (G : H)$

(d) The map $V \mapsto V^G$ induces a bijective correspondence between the isomorphism classes of irreducible modules in b and B, respectively.

(e) The blocks b and B have a defect group in common

(f) β is the unique block of FN covered by b.

(g) If G is p-solvable and $N = O_{p'}(G)$, then $e = \varepsilon$ and a Sylow p-subgroup of H is a defect group of b. In particular, by Step 1 in the proof of Theorem 3.10.9,

$$b \cong M_k(F^{\alpha}(H/N)) \quad \text{for some } k \geqslant 1 \text{ not divisible by } p \text{ and some } \alpha \in Z^2(H/N,F^*).$$

Proof. (i) Write $b = b(e)$ and fix an irreducible FH-module V in b. Because b covers β, it follows from Lemma 4.2 that there exists an irreducible submodule W of V_N belonging to β. Now, for any $g \in G$, ${}^gW \in \beta(g\,\varepsilon\, g^{-1})$ and therefore the inertia group of W is contained in H. Invoking Lemma 5.2, we infer that V^G is an irreducible FG-module, whose restriction to N has W as a summand. Hence, by Lemma 5.1, b^G is defined and V^G lies in b^G. Therefore, by Lemma 4.2, b^G covers β.

(ii) Let B be a block of FG covering β and let V be an irreducible FG-module in B. Owing to Lemma 4.2, there exists an irreducible FN-module W in β such that W is a summand of V_N. Since the inertia group of W is contained in H, $V \cong U^G$ for some irreducible FH-module U whose restriction to N has W as a summand (Lemma 5.2). If b is the block of FH containing U, then b^G is defined and V lies in b^G, by Lemma 5.1, and hence $B = b^G$. Moreover, by Lemma 4.2, b covers β. Let E be the block idempotent of B. If $Ee\varepsilon\, U^G = 0$, then $e\varepsilon\, U^G = 0$ and so $e\varepsilon U = \varepsilon e U = \varepsilon U = 0$. But then $\varepsilon W = 0$, a contradiction, and hence $Ee\varepsilon \neq 0$. This proves (ii), by appealing to Lemma 5.3.

(iii) Properties (a), (b) and (c) are direct consequence of Lemma 5.3. To prove (d), note that by (c) B and b have the same number of nonisomorphic irreducible modules. Thus, it suffices to show that $V_1^G \cong V_2^G$ implies $V_1 \cong V_2$, where V_i is an irreducible FH-module in b, $i = 1,2$.

Owing to Lemma 4.2, there exists an irreducible FN-module W_i in β such that W_i is a submodule of $(V_i)_N$, $i = 1,2$. If $V_1^G \cong V_2^G$, then $(V_1^G)_N \cong (V_2^G)_N$ and, since $(V_i)_N$ is a submodule of $(V_i^G)_N$, we infer that

$$W_2 \cong {}^gW_1 \in \beta(g\,\varepsilon\, g^{-1}) = \beta(\varepsilon)$$

for some $g \in G$, and hence for $g \in H$. Consequently, $(V_1)_N$ and $(V_2)_N$ have W_2 as a common summand. But H contains the inertia group of W_2, so by Lemma 5.2 we have $V_1 \cong V_2$, proving (d).

By Lemma 4.1, $\delta(b) \underset{G}{\subseteq} \delta(B)$ and, by (d) and Theorem 2.2.3, $\delta(b)$ and $\delta(B)$ have the same defect. Hence $\delta(b) \underset{G}{=} \delta(B)$, proving (e).

If $\beta_1 = \beta_1(\varepsilon_1)$ is another block of FN covered by b, then by Lemma 3.10.2, $\varepsilon_1 = h^{-1}\varepsilon h$ for some $h \in H$. Because ε is a central idempotent of FH, we conclude that $\varepsilon_1 = \varepsilon$, proving (f).

Property (g) is a consequence of a general result to be proved in the next section (see Proposition 6.4(iii),(iv)). ■

5.5. COROLLARY. Let F be an algebraically closed field of characteristic $p > 0$, let G be a p-solvable group and let B be a block of FG. Then there exists a subgroup H of G containing $O_{p'}(G)$ and a block b of FH with defect group D such that

(i) $J(B) \cong M_n(J(b))$, where $n = (G : H)$

(ii) D is a defect group of B and $D \in Syl_p(H)$

(iii) $t(B) = t(b)$, where $t(B)$ and $t(b)$ are the nilpotency indices of $J(B)$ and $J(b)$, respectively.

(iv) $b \cong M_k(F^{\alpha}(H/O_{p'}(G))$ for some $k \geqslant 1$ not divisible by p and some $\alpha \in Z^2(H/O_{p'}(G), F^*)$.

Proof. Property (iii) is a consequence of (i), while properties (i), (ii), and (iv) follow from Theorem 5.4(iii), (c), (e), (g). ∎

Another useful observation is provided by

5.6. PROPOSITION. Let F be an algebraically closed field of characteristic $p > 0$, let G be a finite p-solvable group and let B be a block of FG with defect group $D \neq 1$. Assume that B covers a block $\beta = \beta(\varepsilon)$ of $FO_{p'}(G)$ such that $\varepsilon \in Z(FG)$. Then there exists a p-solvable group $\bar{G}$ with $O_p(\bar{G}) \neq 1$ and a block $\bar{B}$ of $F\bar{G}$ with defect group $\bar{D}$ such that

$$t(B) = t(\bar{B}), \ |D| = |\bar{D}| \quad \text{and} \quad \bar{D} \in Syl_p(\bar{G})$$

Proof. By Theorem 3.10.9(ii) and Step 1 in its proof, we have

$$B \cong M_k(F^{\alpha}(G/O_{p'}(G))) \tag{1}$$

and

$$D \in Syl_p(G) \tag{2}$$

for some $k \geqslant 1$ not divisible by p and some $\alpha \in Z^2(G/O_{p'}(G),F^*)$. It is a standard fact (see Karpilovsky (1985, pp.96,97)) that there is an exact sequence

$$1 \longrightarrow Z \longrightarrow \bar{G} \longrightarrow G/O_{p'}(G) \longrightarrow 1 \tag{3}$$

where Z is a central p'-subgroup of $\bar{G}$ such that $F^{\alpha}(G/O_{p'}(G))$ is a direct factor of $F\bar{G}$. Hence, by (1), B is a direct factor of $M_k(F\bar{G})$. Since

$$M_k(F\bar{G}) \cong M_k(B_1) \times \dots \times M_k(B_s)$$

where $B_1,\dots,B_s$ are all blocks of $F\bar{G}$, we infer that

$$B \cong M_k(\bar{B}) \quad \text{for some block } \bar{B} \text{ of } F\bar{G} \tag{4}$$

Since $p \nmid k$, it follows from (4) and Proposition 1.5.3 that the p-parts of the dimensions of irreducible B and $\bar{B}$-modules coincide. If $\bar{D}$ is a Sylow p-subgroup of $\bar{G}$, then by (2) and (3), $|D| = |\bar{D}|$. Hence, by Theorem 2.2.7, $\bar{D}$ is a defect group of $\bar{B}$. Furthermore, by (4), we also have $t(B) = t(\bar{B})$. Since G is p-solvable, and $D \neq 1$ we have $O_p(G/O_{p'}(G)) \neq 1$. Hence, the inverse image X of $O_p(G/O_{p'}(G))$ in $\bar{G}$ satisfies:

$$X \lhd \bar{G} \quad \text{and} \quad X/Z \text{ is a nontrivial } p\text{-group}$$

Since Z is a central p'-subgroup of $\bar{G}$, we infer that $X = Z \times P$ for some nontrivial normal p-subgroup P of $\bar{G}$. Thus $O_p(\bar{G}) \neq 1$ and the result follows. ∎

6. THE KÜLSHAMMER'S STRUCTURE THEOREM.

Throughout this section, G denotes a finite group, N a normal subgroup of G and F an algebraically closed field of characteristic $p > 0$. If $\beta = \beta(\varepsilon)$ is a block of FN, then $G(\varepsilon)$ denotes the inertia group of ε. If H is a subgroup of G, then

$$T_{G,H} : C_{FG}(H) \longrightarrow Z(FG)$$

is the relative trace map. Given subsets X,Y of FG, we put

$$N_G(X,Y) = N_G(X) \cap N_G(Y)$$

All the results of this section are extracted from an important work of Külshammer (1981b).

6.1. LEMMA. Let $B = B(E)$ be a block of FG covering a block $\beta = \beta(\varepsilon)$ of FN and let $b = b(e)$ be the Fong correspondent of B with respect to β.

(i) For each defect group δ of β, there is a defect group d of b with $\delta = d \cap N$

(ii) If δ is a defect group of β and d is a defect group of b with $\delta = d \cap N$, then for each normal subgroup P of d with $\delta \subseteq P$ there is a block $\gamma = \gamma(\varepsilon)$ of $F(PN)$ with defect group P.

Proof. (i) This is a particular case of Theorem 3.2.

(ii) Since $(PN:N) = (P:(P \cap N))$ is a power of p, all block idempotents of $F(PN)$ lie in FN by Theorem 2.2.4. Hence ε is a block idempotent in $F(PN)$, too, and we are left to verify that P is a defect group of the corresponding block $\gamma = \gamma(\varepsilon)$ of $F(PN)$. Owing to Proposition 2.2.6, there exists $C \in C\ell(H)$, where $H = G(\varepsilon)$, such that

(a) $C \subseteq N$

(b) $C \subseteq \mathrm{Supp}\varepsilon$

(c) d is contained in a defect group of C.

By (b) and (c), P is contained in a defect group Δ of $\gamma = \gamma(\varepsilon)$, and so $PN = \Delta N$. On the other hand, Theorem 3.2 implies that $\delta = \Delta \cap N$. This gives the equality

$$\begin{aligned}|\Delta| &= |\Delta N:N|\,|\Delta \cap N|\\ &= |PN:N|\,|\delta|\\ &= |PN:N|\,|P \cap N|\\ &= |P|\end{aligned}$$

and thus P is a defect group of $\gamma = \gamma(\varepsilon)$. ■

6.2. PROPOSITION. Let $B = B(E)$ be a block of FG covering a block $\beta = \beta(\varepsilon)$ of FN and let $b = b(e)$ be the Fong correspondent of B with respect to β. Suppose δ is a defect group of β and D is a common defect group of B and b with $\delta = D \cap N$. Let P be a normal subgroup of D with $\delta \subseteq P$, let $\pi : FG \longrightarrow FC_G(P)$ be the natural projection and set $H = G(\varepsilon)$. Then

(i) There is a block $b' = b'(e')$ of $FN_H(P)$ with defect group D and $\pi(e)e' = e'$.

(ii) For each block b' in (i), there are blocks $B' = B'(E')$ of $FN_G(P)$ with defect group D and $\beta' = \beta'(\varepsilon')$, $\varepsilon' = \pi(\varepsilon)$, of $FN_N(P)$ with defect group δ such that

(a) b' is the Fong correspondent of B' with respect to β'

(b) $\pi(E)E' = E'$, $(b')^H = b$, and $(B')^G = B$

Proof. (i) Owing to Lemma 6.1, there is a block $\gamma = \gamma(\varepsilon)$ of $F(PN)$ with defect group P. It follows, from the Brauer correspondence (Theorem 3.9.21), that there exists a block $\gamma' = \gamma'(\varepsilon')$ with $\varepsilon' = \pi(\varepsilon)$ of $F(N_{PN}(P)) = F(PN_N(P))$ with defect group P. Clearly ε' is a central idempotent of $F(C_N(P))$. Let $\beta_1 = \beta_1(\varepsilon_1)$ be a block of $F(N_N(P))$ with $\varepsilon'\varepsilon_1 \neq 0$. By Lemma 6.1, ε_1 is a central idempotent of $F(N_{PN}(P))$. Hence $\varepsilon' = \varepsilon_1$ and we have a block $\beta' = \beta'(\varepsilon')$ of $F(N_N(P))$. Again, by Lemma 6.1, this block has defect group $P \cap N_N(P) = \delta$.

If $g \in N_G(P)$, then $\pi({}^g\varepsilon) = {}^g\pi(\varepsilon)$ and, in particular, ${}^g\pi(\varepsilon) = \pi(\varepsilon)$ for all $g \in N_H(P)$. On the other hand, if g is an element in $N_G(P)$ with ${}^g\pi(\varepsilon) = \pi(\varepsilon)$, then

$$\pi(\varepsilon\,{}^{g}\varepsilon) = \pi(\varepsilon)\pi({}^{g}\varepsilon) = \pi(\varepsilon)\,{}^{g}\pi(\varepsilon) \neq 0$$

and so $\varepsilon = {}^{g}\varepsilon$, proving that

$$C_{N_G(P)}(\varepsilon') = N_H(P)$$

Let $\pi' : FG \longrightarrow FC_G(D)$ be the natural projection. Then

$$\pi'(\pi(e)) = \pi'(e) \neq 0$$

and $\pi(e)$ is a nonzero central idempotent of $F(N_H(P))$. Write

$$\pi(e) = e_1 + e_2 + \dots + e_n$$

as a sum of block idempotents in $F(N_H(P))$. Then

$$\pi'(e) = \pi'(e_1) + \pi'(e_2) + \dots + \pi'(e_n)$$

where each $\pi'(e_i)$ is a central idempotent in $F(C_H(D))$. This implies the existence of a block $b' = b'(e')$ of $F(N_H(P))$ with $\pi(e)e' = e'$ and $\pi'(e') \neq 0$. It follows that $b' = b'(e')$ has defect group D, proving (i).

(ii) Assume that $b' = b'(e')$ is a block of $F(N_H(P))$ with defect group D and $\pi(e)e' = e'$. Then we have

$$e'\varepsilon' = e'\pi(e)\pi(\varepsilon) = e'\pi(e\varepsilon) = e'\pi(e) = e'$$

and we get a block $B' = B'(E')$ of $F(N_G(P))$ with $E'e'\varepsilon' \neq 0$. By Theorem 5.4, $b' = b'(e')$ is the Fong correspondent of $B' = B'(E')$ with respect to $\beta' = \beta'(\varepsilon')$ and B' has defect group D. By Lemma 4.3, the blocks $(B')^G$ and $(b')^H$ are defined and for each conjugacy class C of G or H, we have

$$(C \cap (N_G(P))^+E' + J(Z(B')) = \pi(C^+)E' + J(Z(B'))$$

or

$$(C \cap N_H(P))^+e' + J(Z(b')) = \pi(C^+)e' + J(Z(b')),$$

respectively. Now $\pi(e)e' = e'$ implies that $(b')^H = b$. Hence

$$(B')^G = (b')^G = b^G = \bar{B}$$

so $\pi(E)E' \neq 0$ and thus $\pi(E)E' = E'$, proving (ii). ■

The next result provides an important reduction to $C_G(P)$.

6.3. PROPOSITION. Let $B = B(E)$ be a block of FG covering a block $\beta = \beta(\varepsilon)$ of FN, and let $b = b(e)$ be the Fong correspondent of B with respect to β. Assume that δ is a defect group of β and D a common defect group of B and b with $\delta = D \cap N$. Let P be a normal subgroup of D with $\delta \subseteq P$ and put $H = G(\varepsilon)$.

(i) There exist blocks $b_1 = b_1(e_1)$ of $F(C_H(P))$ with $b_1^H = b$ and $a_1 = a_1(e_1)$ of $FN_H(P,b_1)$ with defect group D.

(ii) For each block $b_1 = b_1(e_1)$ in (i), there are blocks $B_1 = B_1(E_1)$ of $F(C_G(P))$ and $A_1 = A_1(E_1)$ of $FN_G(P,B_1)$ with defect group D such that

(a) $(B_1)^G = B$ and $(b_1)^{C_G(P)} = B_1$

(b) $N_G(P,B_1) = N_H(P,b_1)C_G(P)$

Proof. (i) Owing to Proposition 6.2, we may assume that P is a normal subgroup of G. Let $b_1 = b_1(e_1)$ be a block of $F(C_H(P))$ with $ee_1 \neq 0$. Since every block idempotent of $FN_H(P,b_1)$ lies in $FC_H(P)$ (see Lemma 3.1.11) the Fong correspondent of $b = b(e)$ with respect to $b_1 = b_1(e_1)$ is a block $a_1 = a_1(e_1)$ of $FN_H(P,b_1)$. By Theorem 5.4, we may assume that a_1 has defect group D, while by Theorem 4.5, $b_1^H = b$, proving (i).

(ii) Assume that $b_1 = b_1(e_1)$ and $a_1 = a_1(e_1)$ are arbitrary blocks of $F(C_H(P))$ and $FN_H(P,b_1)$, respectively, as stated in the theorem. By Lemma 6.1, there exists a block $\gamma = \gamma(\varepsilon)$ of $F(PN)$ with defect group P. Since

$$\varepsilon e_1 = \varepsilon e e_1 = e e_1 = e_1,$$

there is a block $\gamma_1 = \gamma_1(\varepsilon_1)$ of $F(C_{PN}(P))$ with $\varepsilon_1 \varepsilon e_1 \neq 0$. Because

$$\varepsilon_1 \varepsilon e_1 E = \varepsilon_1 \varepsilon e_1 e E = \varepsilon_1 \varepsilon e_1 e = \varepsilon_1 \varepsilon e_1 \neq 0$$

we get a block $B_1 = B_1(E_1)$ of $F(C_G(P))$ with $\varepsilon_1 \varepsilon e_1 E_1 \neq 0$ and a corresponding block $A_1 = A_1(E_1)$ of $FN_G(P,B_1)$.

Set $S = C_G(\varepsilon_1) \cap C_G(P)$. Then $S \subseteq H$ since for elements $g \in S$, we have

$${}^g\gamma = {}^g((\gamma_1)^{PN}) = ({}^g\gamma_1)^{PN} = \gamma_1^{PN} = \gamma$$

Let $c_1 = c_1(f_1)$ be the Fong correspondent of $B_1 = B_1(E_1)$ with respect to

$\gamma_1 = \gamma_1(\varepsilon_1)$ and $c_2 = c_2(f_2)$ the Fong correspondent of $b_1 = b_1(e_1)$ with respect to $\gamma_1 = \gamma_1(\varepsilon_1)$. We wish to show that $f_1 = f_2$.

Assume by way of contradiction that $f_1 \neq f_2$. For elements $g \in S$, we have

$$^{g}f_1 f_2 = f_1 f_2 = 0,$$

and for elements $g \in C_G(P) - S$, we have

$$^{g}f_1 f_2 = {}^{g}f_1 {}^{g}\varepsilon_1 \varepsilon_1 f_2 = 0$$

Thus

$$E_1 e_1 = T_{C_G(D),S}(f_1) T_{C_H(D),S}(f_2) = 0,$$

contrary to $\varepsilon_1 \varepsilon e_1 E_1 \neq 0$. Hence $f_1 = f_2$ and

$$b_1^{C_G(P)} = C_1^{C_G(P)} = B_1$$

It follows that

$$a_1^{N_G(P,B_1)} = b_1^{N_G(P,B_1)} = B_1^{N_G(P,B_1)} = A_1$$

and $A_1 = A_1(E_1)$ has defect group D. Now (a) follows from

$$B_1^G = b_1^G = b^G = B$$

It follows from (a) that $N_H(P,b_1)C_G(P) \subseteq N_G(P,B_1)$. To prove the reverse containment, assume that $g \in N_G(P,B_1)$. Then

$$0 \neq {}^{g}(\varepsilon_1 E_1) = {}^{g}\varepsilon_1 E_1$$

By Lemma 3.10.2, there exists an element $c \in C_G(P)$ with ${}^{gc}\varepsilon_1 = \varepsilon_1$, in particular, $gc \in H$. Hence ${}^{gc}S = S$ and ${}^{gc}c_1 = {}^{gc}c_1({}^{gc}f_1)$ is a block of FS with

$$0 \neq {}^{gc}(E_1 f_1 \varepsilon_1) = E_1 {}^{gc}f_1 \varepsilon_1$$

The uniqueness of the Fong correspondent implies that

$$^{gc}f_1 = f_1 \quad \text{and} \quad {}^{gc}c_1 = c_1$$

which in turn implies that

$$^{gc}b_1 = {}^{gc}(c_1^{C_H(P)}) = ({}^{gc}c_1)^{C_H(P)} = b_1$$

Hence $gc \in N_H(P,b_1)$ and $g \in N_H(P,b_1)C_G(P)$, as required. ■

6.4. PROPOSITION. Let G be p-solvable, let $N = O_{p'}(G)$ and let $B = B(E)$ be a block of FG with defect group D such that B covers a block $\beta = \beta(\varepsilon)$ of FN with $\varepsilon \in Z(FG)$. Put $P = D \cap O_{p',p}(G)$ and let $\pi : FG \longrightarrow F(C_G(P))$ be the natural projection. Then

(i) P is a Sylow p-subgroup of $O_{p',p}(G)$ such that

$$O_{p'}(N_G(P)) = N_N(P)$$

and

$$O_{p'}(N_G(D)) = N_N(D)$$

(ii) $\pi(\varepsilon)$ is a block idempotent of $F(C_N(P)), F(C_G(P))$ and $F(N_G(P))$.

(iii) $E = \varepsilon$

(iv) D is a Sylow p-subgroup of G

Proof. (i) Choose a Sylow p-subgroup Q of $O_{p',p}(G)$ with $P \subseteq Q$. By Theorem 2.2.4, ε is a block idempotent of $F(QN)$, and by Theorem 3.10.9, the corresponding block $\gamma = \gamma(\varepsilon)$ of $F(QN)$ has defect group Q. By Lemma 6.1, there is a defect group d of B with $Q = d \cap O_{p',p}(G)$. Let $g \in G$ be such that $d = g^{-1}Dg$. Then

$$P \subseteq Q = g^{-1}Dg \cap O_{p',p}(G) = g^{-1}Pg$$

and

$$P = Q = g^{-1}Pg$$

Hence P is a Sylow p-subgroup of $O_{p',p}(G)$. It follows from Lemma 7.1.3 that $O_{p'}(N_G(P)) \subseteq C_G(P) \subseteq PN$. Taking into account that $N_G(D) \subseteq N_G(P)$, (i) is established.

(ii) As we have seen in (i), the block $\gamma = \gamma(\varepsilon)$ of $F(PN)$ has defect group P. By Proposition 6.2, we get a block $\gamma' = \gamma'(\varepsilon')$, $\varepsilon' = \pi(\varepsilon)$, of $F(N_{PN}(P)) = F(PN_N(P)) = F(PC_N(P))$, a block $\beta' = \beta'(\varepsilon')$ of $F(C_N(P))$ and a block $B' = B'(E')$ of $F(N_G(P))$ with $E'\varepsilon' \neq 0$. By Lemmas 5.3 and 3.1.11, $E' \in F(C_G(P)) \subseteq F(PN)$ and even $E' \in F(PC_N(P))$. So we get $E' = E'\varepsilon' = \varepsilon'$ and

(ii) follows

(iii) Let $B_1 = B_1(E_1)$ be a block of FG with $E_1\varepsilon \neq 0$, and let D_1 be a defect group of $B_1 = B_1(E_1)$. As in (i), $D_1 \cap O_{p',p}(G)$ is a Sylow p-subgroup of $O_{p',p}(G)$ and so we may assume that $P = D_1 \cap O_{p',p}(G)$. It is clear that

$$0 \neq \pi(E_1) = \pi(E_1\varepsilon) = \pi(E_1)\pi(\varepsilon),$$

so (ii) implies that $\pi(E_1) = \pi(\varepsilon)$. Thus

$$0 \neq \pi(E_1)\pi(E) = \pi(E_1E)$$

and $E_1 = E$, proving (iii).

(iv) Direct consequence of (iii) and Theorem 3.10.9(ii). ■

The next observation is the so called Frattini argument.

6.5. LEMMA. Let N be a normal subgroup of G and let P be a Sylow p-subgroup of N. Then $G = N_G(P)\cdot N$

Proof. Let g be an arbitrary element of G. Then $g^{-1}Ng \subseteq N$ and so $g^{-1}Pg$ is a Sylow p-subgroup of N. Hence, by Sylow theorem, $g^{-1}Pg = n^{-1}Pn$ for some $n \in N$. It follows that $gn^{-1} \in N_G(P)$ and hence $g \in N_G(P)\cdot N$, as required. ■

Let $1 \to Z \xrightarrow{\alpha} X \xrightarrow{\beta} Y \to 1$ be an extension of finite groups, where Z is not necessarily abelian. This extension defines a homomorphism

$$\theta : Y \longrightarrow \text{Out}\, Z = \text{Aut}Z/\text{Inn}Z$$

in the following way: For any $y \in Y, \theta(y)$ consists of all maps $Z \longrightarrow Z, z \longmapsto xzx^{-1}$, where x ranges over those elements in X for which $\beta(x) = y$.

We come now to the main result of this section.

6.6. THEOREM. (Külshammer (1981b). Let G be a finite p-solvable group, let F be an algebraically closed field of characteristic p, and let $B = B(E)$ be a block of FG with defect group D. Then there are a normal subgroup P of D, blocks $B_1 = B_1(E_1)$ of $F(C_G(P))$ and $A_1 = A_1(E_1)$ of $FN_G(P,B_1)$ with defect group D and a subgroup H of $N_G(P,B_1)$ such that the following properties hold:

(i) $B_1^G = B$

(ii) D is a Sylow p-subgroup of H

(iii) $PO_{p'}(H) = O_{p',p}(H)$

(iv) $B \cong M_n(F^{\alpha}(H/O_{p'}(H))$ for some $n \geqslant 1$ and some $\alpha \in Z^2(H/O_{p'}(H),F^*)$

(v) $H/O_{p'}(H)$ is isomorphic to an extension of P with the group $N_G(P,B_1)/PC_G(P)$, where the corresponding homomorphism

$$N_G(P,B_1)/PC_G(P) \longrightarrow \text{Out}P$$

is induced by the natural homomorphism $N_G(P,B_1) \longrightarrow \text{Aut}P$.

Proof. Put $N = O_{p'}(G)$ and let $\beta = \beta(\varepsilon)$ be a block of FN covered by B. We argue by induction on the order of G and distinguish two cases.

Case 1. $S = G(\varepsilon) \neq G$.

Let $b = b(e)$ be the Fong correspondent of B with respect to β. By Theorem 5.4, we may assume that D is a common defect group of B and b. By induction, there are a normal subgroup P of D, blocks $b_1 = b_1(e_1)$ of $F(C_S(P))$ and $a_1 = a_1(e_1)$ of $FN_S(P,b_1)$ with defect group D, a subgroup H of $N_S(P,b_1)$, a natural number m and $\alpha \in Z^2(\bar{H},F^*)$ where $\bar{H} = H/O_{p'}(H)$, such that the following conditions hold:

(i') $b_1^S = b$

(ii') D is a Sylow p-subgroup of H

(iii') $PO_{p'}(H) = O_{p',p}(H)$

(iv') $b \cong M_m(F^{\alpha}\bar{H})$

(v') $\bar{H}$ is isomorphic to an extension of P with the group $N_S(P,b_1)/PC_S(P)$, where the automorphisms are induced by $N_S(P,b_1)/C_S(P)$.

Put $n = m(G : S)$ and choose blocks $B_1 = B_1(E_1)$ of $F(C_G(P))$ and $A_1 = A_1(E_1)$ of $FN_G(P,B_1)$ as in Proposition 6.3. Then (i) is a consequence of (i') and Proposition 6.3, (iv) follows from Theorem 5.4, and by Proposition 6.3,

$$N_G(P,B_1)/C_G(P) \cong N_S(P,b_1)/C_S(P)$$

which implies (v).

Case 2. $S = G(\varepsilon) = G$

In this case, ε is a central idempotent of FG, so we may apply Proposition 6.4.

Put $P = D \cap O_{p',p}(G)$, $H = N_G(P)$, and let

$$\pi : FG \longrightarrow F(C_G(P))$$

be the natural projection. Denote by $B_1 = B_1(E_1)$ and $A_1 = A_1(E_1)$ the blocks of $F(C_G(P))$ and $F(N_G(P))$ with the block idempotent $E_1 = \pi(\varepsilon)$, respectively. By Proposition 6.4(ii), we have $N_G(P) = H = N_G(P,B_1)$. By Proposition 6.2, the block $A_1 = A_1(E_1)$ of $F(N_G(P))$ has defect group D. Parts (i), (ii), and (iii) are now direct consequence of Proposition 6.4. By Lemma 6.5,

$$G = N_G(P)PN = N_G(P)N$$

and

$$G/N \cong N_G(P)/N_N(P) = N_G(P)/C_N(P)$$

From Lemma 7.1.3, we have $C_G(P) \subseteq PN$. Thus (iv) and (v) are immediate consequences of the isomorphism established in Step 1 of Theorem 3.10.9. This completes the proof of the theorem. ∎

Let G and H be two groups. If a homomorphism $\phi : G \longrightarrow \mathrm{Aut}(H)$ is given, we say that G *acts* on H via ϕ; the homomorphism ϕ is called an *action* of G on H. The action ϕ of G is called *faithful* if $\mathrm{Ker}\phi = 1$.

Let ϕ be an action of a group G on another group H. The *semidirect product* $H \rtimes_\phi G$ (or simply $H \rtimes G$ if the action ϕ is understood) consists of all pairs (h,g), $h \in H$, $g \in G$, with the product defined by the formula

$$(h,g)(h',g') = (h\phi(g)(h'),gg')$$

Usually, the subgroups $(H,1)$ and $(1,G)$ are identified with H and G, respectively, so that

$$H \rtimes G = HG,\ H \lhd H \rtimes G \text{ and } ghg^{-1} = \phi(g)(h)$$

for all $g \in G$, $h \in H$.

6.7. COROLLARY. Further to the assumptions and notation of Theorem 6.6, assume that $P = D$. Then

$$B \cong M_n(F^{\alpha}(D \rtimes S))$$

for some $n \geqslant 1$ and some $\alpha \in Z^2(D \rtimes S, F^*)$, where S is a p'-group acting faithfully on D.

Proof. Let H be as in Theorem 6.6. By Theorem 6.6(iv), it suffices to verify that $H/O_{p'}(H) \cong D \rtimes S$ for a suitable p'-group S acting faithfully on D. By Theorem 6.6(ii), D is a Sylow p-subgroup of H and hence $D \cong DO_{p'}(H)/O_{p'}(H)$ is a Sylow p-subgroup of $H/O_{p'}(H)$. Since $D = P$, Theorem 6.6(v) implies that the Sylow p-subgroup of $H/O_{p'}(H)$ is normal. Thus $H/O_{p'}(H) \cong D \rtimes S$ for some p'-group S. Because $O_{p'}(H/O_{p'}(H)) = 1$, it follows from Lemma 7.1.3 that $C_{DS}(D) \subseteq D$ and hence S acts faithfully on D, as required. ■

We close this section by exhibiting circumstances under which $P = D$.

6.8. THEOREM. (Külshammer (1981b)). Let G be a finite p-solvable group, let F be an algebraically closed field of characteristic p and let $B = B(E)$ be a block of FG with defect group D. Assume that at least one of the following conditions hold:

(a) G has p-length 1

(b) D is abelian

(c) D is a $T.I.$ subgroup of G (i.e. for each $g \in G$, either $g^{-1}Dg = D$ or $g^{-1}Dg \cap D = 1$)

(d) For all normal subgroups Q of D and all blocks $B_Q = B_Q(E_Q)$ of $F(C_G(Q))$ with $B_Q^G = B$, $N_G(Q, B_Q) = DC_G(Q)$

Then $P = D$, where P is as in Theorem 6.6. Hence, by Corollary 6.7,

$$B \cong M_n(F^{\alpha}(D \rtimes S))$$

for some $n \geqslant 1$ and some $\alpha \in Z^2(D \rtimes S, F^*)$, where S is a p'-group acting faithfully on D.

Proof. We keep the notation of Theorem 6.6.

(a) If G has p-length 1, then H is also of p-length 1. Hence, by Theorem 6.6 (ii), (iii),

$$DO_{p'}(H) = O_{p',p}(H) = PO_{p'}(H)$$

and therefore $P = D$.

(b) If D is abelian, then $D \subseteq C_G(P)$. It follows from Lemma 7.1.3 and Theorem

6.6.(iii) that $D \subseteq C_H(P) \subseteq PO_{p'}(H)$ and $P = D$.

(c) Assume that D is a $T.I.$ subgroup of G. If $P = 1$, then $H = O_{p'}(H)$ and $D = 1 = P$. If $P \neq 1$, then $H \subseteq N_G(P) \subseteq N_G(D)$. Hence

$$DO_{p'}(H) \subseteq O_{p',p}(H) = PO_{p'}(H)$$

and so $D = P$.

(d) In this case we have

$$H = H \cap DC_G(P) = DC_H(P) \subseteq DO_{p'}(H)$$

by Lemma 7.1.3. Thus $DO_{p'}(H) = PO_{p'}(H)$ and $D = P$. Note that in this case even $S = 1$. ∎

7. APPLICATIONS.

Throughout this section, G denotes a finite group, F an arbitrary field of characteristic $p > 0$ and $t(B)$ the nilpotency index of the Jacobson radical $J(B)$ of a block B of FG. As usual, we write $t(G)$ for the nilpotency index of $J(FG)$. Given subsets X,Y of FG, we put

$$N_G(X,Y) = N_G(X) \cap N_G(Y)$$

7.1. LEMMA. Let B be a block of FG with defect group D and let E be the algebraic closure of F. Then there exist blocks $B_1,\ldots,B_k$ of EG with defect group D such that

$$E \underset{F}{\otimes} J(B)^m = J(B_1)^m \oplus \ldots \oplus J(B_k)^m \qquad \text{for all } m \geqslant 1$$

In particular,

$$t(B) = \max\{t(B_1),\ldots,t(B_k)\}$$

Proof. By Lemma 3.16.10, we can write

$$E \underset{F}{\otimes} B = B_1 \oplus \ldots \oplus B_k$$

for an integer $k \geqslant 1$, where each B_i is a block of FG with defect group D. Invoking Corollary 3.1.18, we therefore derive

$$E \underset{F}{\otimes} J(B)^m = J(E \underset{F}{\otimes} B)^m = \bigoplus_{i=1}^{k} J(B_i)^m$$

as required. ■

7.2. THEOREM. Let G be a finite p-solvable group and let B be a block of FG with defect group D. Assume that at least one of the following conditions hold:

(a) G has p-length 1

(b) D is abelian

(c) D is a $T.I.$ subgroup of G

(d) For all normal subgroups Q of D and all blocks B_Q of $F(C_G(Q))$ with $B_Q^G = B$, $N_G(Q,B_Q) = DC_G(Q)$

Then

$$t(B) = t(D)$$

Proof. Owing to Lemma 7.1, we may harmlessly assume that F is algebraically closed. Hence, by Corollary 6.7, $t(B) = t(F^{\alpha}H)$ where H is a group having D as a normal Sylow p-subgroup and $\alpha \in Z^2(H,F^*)$. Thus $t(B) = t(D)$ by virtue of Proposition 6.3.8.

7.3. COROLLARY. Let G be a finite p-solvable group and let B be a block of FG with abelian defect group $D \cong \mathbb{Z}_{p^{n_1}} \times \mathbb{Z}_{p^{n_2}} \times \ldots \times \mathbb{Z}_{p^{n_k}}$, $n_i > 0$, $1 \leqslant i \leqslant k$. Then

$$t(B) = 1 - k + \sum_{i=1}^{k} p^{n_i}$$

Proof. Direct consequence of Theorem 7.2 and Corollary 3.2.5. ■

To prove our next result, we require the following preliminary observation.

7.4. LEMMA. Let $B = B(e)$ be a block of FG, let P be a normal p-subgroup of G and let $f : FG \longrightarrow F(G/P)$ be the natural homomorphism. Then

(i) $f(B) = B_1 \oplus \ldots \oplus B_n$ for some $n \geqslant 1$ and some blocks $B_1,\ldots,B_n$ of $F(G/P)$

(ii) $t(B) \geqslant t(P) + t(B_i) - 1$ for all $i \in \{1,\ldots,n\}$.

Proof. Part (i) is a consequence of Lemma 2.3. To prove (ii), fix $\bar{B} \in \{B_1,\ldots,B_n\}$ and put $t = t(\bar{B})$. Let $\bar{f}$ be the restriction of f to B. Then

$$\mathrm{Ker}\bar{f} = \mathrm{Ker}f \cap B = (\mathrm{Ker}f)e = I(P)B$$

and, by Lemma 6.1.1(iii), $\bar{f}(J(B)) = J(\bar{f}(B))$. Since $J(\bar{B})^{t-1} \neq 0$, we derive

$$\bar{f}[J(B)^{t-1}] = [\bar{f}(J(B))]^{t-1} = [J(\bar{f}(B))]^{t-1} = \bigoplus_{i=1}^{n} J(B_i)^{t-1} \neq 0$$

Then

$$J(B)^{t-1} \not\subseteq \mathrm{Ker}\bar{f} = I(P)B$$

and so there exists $w \in J(B)^{t-1} - I(P)B$. Let $G = \bigcup_{j=1}^{q} g_j P$ be a coset decomposition of G. We can write $w = \sum_{j=1}^{q} g_j s_j$ with $s_j \in FP$. Clearly, $w \notin I(P)FG = FG \cdot I(P)$. Thus we may assume that $s_1 \notin I(P)$, in which case we may write

$$s_1 = \sum_{x \in P} c_x x \ , \ c_x \in F \ , \quad \text{and} \quad \sum_{x \in P} c_x \neq 0$$

It will next be shown that $wP^+ \neq 0$. Assume the contrary. Since

$$wP^+ = (\Sigma g_j s_j)P^+ = \Sigma g_j (s_j P^+)$$

and since $s_j P^+ \in FP$ for all j, we have $s_j P^+ = 0$ for all j. Hence

$$0 = s_1 P^+ = (\sum_{x \in P} c_x x)P^+ = (\sum_{x \in P} c_x)P^+$$

and so $\sum_{x \in P} c_x = 0$, a contradiction. Thus $wP^+ \neq 0$.

Since

$$I(P)^{t(P)-1} = FP^+$$

by Lemma 3.11.2, and since

$$e\, I(P)^h = I(P)^h e \subseteq J(B)^h$$

for any integer $h \geqslant 0$, we derive

$$wP^+ \in J(B)^{t+t(P)-2}$$

Thus $t(B) \geqslant t + t(P) - 1$, as asserted. ■

We are now ready to prove

7.5. THEOREM. (Koshitani (1983), (1984)). Let G be a finite p-solvable group and let B be a block of FG with defect group D of order p^d.

(i) $d(p-1) + 1 \leqslant t(B) \leqslant p^d$

(ii) $t(B) = p^d$ if and only if D is cyclic.

Proof. By Lemma 7.1, we may assume that F is algebraically closed. If $d = 0$ then $J(B) = 0$ by Theorem 3.6.4(i), in which case both assertions are trivial. We may therefore assume that $d > 0$.

We first show that $t(B) \leqslant p^d$ with $t(B) = p^d$ if and only if D is cyclic. By Corollary 5.5, we may assume that D is a Sylow p-subgroup of G. Hence, by Proposition 7.2.1, $t(B) \leqslant t(G) \leqslant p^d$. Furthermore, if $t(B) = p^d$ then $t(G) = p^d$ and hence, by Proposition 7.2.4, D is cyclic. Conversely, if D is cyclic then $t(B) = p^d$ by Corollary 7.3.

We are left to verify that $t(B) \geqslant d(p-1) + 1$. We argue by double induction on d and $|G|$. Since $d \geqslant 1$, we have $p \mid |G|$. If $G = \delta(B)$, then $B = FG$ and the required assertion is true by Proposition 7.2.1.

Let $N = O_{p'}(G)$ and let $\beta = \beta(\varepsilon)$ be a block of FN covered by B. By Theorem 5.4, FH, $H = G(\varepsilon)$, has a block b with the same defect d and $t(B) = t(b)$. If $G \neq G(\varepsilon)$, then since $|G(\varepsilon)| < |G|$ we get the result by induction. Hence we may assume that $G = H$ in which case $\varepsilon \in Z(FG)$. Then, by Proposition 5.6, there exists a finite p-solvable group X such that $O_p(X) \neq 1$ and FX has a block B^* with the same defect d and $t(B^*) = t(B)$. Let $P = O_p(X)$, $|P| = p^r$ and $\bar{X} = X/P$. By Lemmas 2.3 and 7.4, $F\bar{X}$ has a block $\bar{B}$ with defect $d-r$ and $t(B^*) \geqslant t(P) + t(\bar{B}) - 1$. By Proposition 4.1.8,

$$t(P) \geqslant r(p-1) + 1$$

Since $d-r < d$, we get by induction that

$$t(\bar{B}) \geqslant (d-r)(p-1) + 1$$

Thus

$$t(B) = t(B^*) \geqslant t(P) + t(\bar{B}) - 1 \geqslant d(p-1) + 1$$

as asserted. ■

7.6. THEOREM. (Koshitani (1985a). Let G be a finite p-solvable group and let B be a block of FG with defect group D of order p^d. If $p \geqslant 3$ and $d \geqslant 2$, then the following are equivalent.

(i) $t(B) = p^{d-1} + p - 1$

(ii) $p^{d-1} < t(B) < p^d$

(iii) $t(D) = p^{d-1} + p - 1$

(iv) $p^{d-1} < t(D) < p^d$

(v) D is noncyclic but has a cyclic subgroup of index p

(vi) D is isomorphic to either $\mathbb{Z}_{p^{d-1}} \times \mathbb{Z}_p$, or $M_d(p)$ for $d \geqslant 3$

Proof. By Lemma 7.1, we may assume that F is algebraically closed. The equivalence of (iii) - (vi) follows from Theorem 7.5.2. The implication (i) $\Rightarrow$ (ii) being trivial, we are left to verify that (ii) implies (iv) and (iii) implies (i). By Corollary 5.5, we assume that D is a Sylow p-subgroup of G.

Assume that (ii) holds. Since $t(B) \leqslant t(G)$, we have $t(G) > p^{d-1}$. Hence, by Theorem 7.5.1, G is of p-length 1. Invoking Theorem 7.2, we deduce that $t(B) = t(D)$ and so $p^{d-1} < t(D) < p^d$, proving (iv).

Assume that (iii) holds. Then, by Theorem 7.5.2, G has p-length 1. Hence, by Theorem 7.2, $t(B) = t(D) = p^{d-1} + p - 1$, proving (i). ∎

We close this section by quoting the following interesting results, the proof of which is beyond the scope of this book.

7.7. THEOREM. (Koshitani (1977b)). Let F be an algebraically closed field of characteristic $p > 0$ and let B be a block of FG with cyclic defect group D.

(i) $t(B) \leqslant |D|$ and, in particular, $t(G) \leqslant |P|$ if $P \in \mathrm{Syl}_p(G)$ is cyclic

(ii) The following conditions are equivalent

(a) $t(B) = |D|$

(b) B is a uniserial ring

(c) The Cartan matrix of B is of the form

$$\begin{bmatrix} c+1 & c & \cdots & c \\ c & c+1 & \cdots & c \\ \cdot & & \cdot & \cdot \\ \cdot & & \cdot & \cdot \\ \cdot & & \cdot & \cdot \\ c & \cdot & \cdots c & c+1 \end{bmatrix}$$

8. A SURVEY OF SOME FURTHER RESULTS.

Throughout this section, G denotes a finite group, F a field of characteristic $p > 0$, B a block of FG and $t(B)$ the nilpotency index of $J(B)$.

We first examine those blocks B for which $t(B) \leqslant 3$. Let D denote a defect group of B. We know, from Theorem 3.6.4, that $t(B) = 1$ (equivalently, $J(B) = 0$) if and only if $D = 1$. Okuyama (1986) asserts that "it is true that $J(B)^2 = 0$ if and only if $p = 2$ and $|D| = 2$", which should read "it is true that $J(B)^2 = 0$ and $J(B) \neq 0$ if and only if $p = 2$ and $|D| = 2$". However, we were unable to find an explicit reference for this result.

All information pertaining to the Brauer tree of a block B may be extracted from Feit (1982). The following result describes all blocks B with $t(B) = 3$.

8.1. THEOREM. (Okuyama (1986)). Let F be an algebraically closed field of characteristic $p > 0$ and let D be a defect group of the block B of FG. Then $t(B) = 3$ if and only if one of the following conditions hold:

(a) $p = 2$, $D \cong \mathbb{Z}_2 \times \mathbb{Z}_2$ and B is isomorphic to the matrix ring over FD or is Morita-equivalent to FA_4, where A_4 is the alternating group of degree 4.

(b) p is odd, $|D| = p$, the number of irreducible FG-modules in B is p-1 or $(p-1)/2$ and the Brauer tree of B is a straight line segment such that the exceptional vertex is at an end point (if it exists). ∎

The principal block of the following groups satisfies conditions of Theorem 8.1 (see Okuyama (1986)):

(i) $G \cong \mathbb{Z}_2 \times \mathbb{Z}_2$ or $G \cong A_4$ and $p = 2$

(ii) $G \cong S_p$ or $G \cong A_p$, where p is odd.

The next result is of independent interest

8.2. THEOREM. (Okuyama (1986)). Let F be an algebraically closed field of characteristic 2, let 1_G be the trivial FG-module and let $P(1_G)$ be the projective cover of 1_G. If the Loewy length of $P(1_G)$ is 3, then a Sylow 2-subgroup of G is dihedral. ∎

The next result provides a family of groups G for which the Loewy length of $P(1_G)$ is 3.

8.3. THEOREM. (Erdmann (1977, Theorem 4)). Let $G = PSL(2,q)$, where q is a power of a prime with $q \equiv 3(\text{mod} 4)$, and let F be a splitting field of characteristic 2. Then the Loewy length of $P(1_G)$ is 3. ∎

Let G be a p-solvable group and let B be a block of FG of defect d. We know, from Theorem 7.5, that

$$t(B) \geqslant d(p-1) + 1$$

Owing to Theorem 3.16.14, there exists an irreducible FG-module in B a vertex of which is a defect group of B. Hence the inequality $t(B) \geqslant d(p-1) + 1$ can also be proved as a consequence of the following general result.

8.4. THEOREM. (Ninomiya (1984b)). Let G be a finite p-solvable group, let F be a field of characteristic $p > 0$ and let V be an irreducible FG-module. If a vertex of V has order p^n, then the Loewy length of the projective cover of V is greater than or equal to $n(p-1) + 1$. ∎

Consider the following chain

$$J(FG) \supseteq FG \cdot J(Z(FG)) \supseteq J(Z(FG))$$

and assume that F is algebraically closed. Then, by Theorem 3.11.5,

$$J(FG) = FG \cdot J(Z(FG))$$

if and only if G is p-nilpotent with abelian Sylow p-subgroups. The equality

$$FG \cdot J(Z(FG)) = J(Z(FG))$$

holds if and only if $J(Z(FG))$ is an ideal of FG. It is therefore appropriate to ask: What are necessary and sufficient conditions for $J(Z(FG))$ to be an ideal of FG ? This problem was solved by Clarke (1969) under the assumption that G is p-solvable. The following result shows that this assumption is redundant.

8.5. THEOREM. (Koshitani (1978)). Let F be an algebraically closed field of characteristic $p > 0$. If $J(Z(FG))$ is an ideal of FG, then G is p-solvable. ∎

Combining Theorem 8.5 with the results of Clarke (1969, Lemma 8 and Theorem), we obtain the following solution of the mentioned problem.

8.6. THEOREM. (Koshitani (1978), Clarke (1969)). Let G be a finite group and let F be an algebraically closed field of characteristic $p > 0$. Then $J(Z(FG))$ is an ideal of FG if and only if one of the following conditions hold:

(i) $p \nmid |G|$

(ii) G is abelian

(iii) G is a p-nilpotent group with an abelian Sylow p-subgroup P, and $G'P$ is a Frobenius group with complement P and kernel G'

(iv) $p = 2$, G is 2-nilpotent, $P' = Z(P)$ has order 2 and $G'P$ is a Frobenius group with complement P and kernel $G' \cap O_{2'}(G)$, where P is a Sylow 2-subgroup of G.

(v) G has an elementary abelian Sylow p-subgroup P and normal subgroups H, L such that $G \supseteq L \supseteq H$, $L/H \cong P$, $L = G'$ is a Frobenius group with complement P and kernel H, $G/H = \bar{G}$ is a semidirect product of $\bar{P}$ and $\bar{A}$ such that $\bar{P} \cong P$, $\bar{A}$ acts on $\bar{P} - \{1\}$ transitively, and every $x \in \bar{A}$ satisfies $x \in C_{\bar{G}}(\bar{P})$ or $x^{-1}yx \neq y$ for all $y \in \bar{P} - \{1\}$.

(vi) $p = 2$, G has normal subgroups H and L such that $G \supseteq L \supseteq H$ and $L/H \cong P$ where P is a Sylow 2-subgroup of G, $G' = H \cdot Z(P)$, $Z(P)$ has order 2, and L is a Frobenius group with complement P and kernel H. ■

Although Theorem 8.5 is quite adequate for the purpose of characterizing all groups G for which $J(Z(FG))$ is an ideal of FG, it can nevertheless be significantly improved. Namely, the following result holds:

8.7. THEOREM. (Motose (1979)). Let F be an algebraically closed field of characteristic $p > 0$. If $J(Z(FG))$ is an ideal of FG, then G' is either a p-nilpotent group or is a p'-group. ■

The proof of the above theorem does not use Clark's characterization. It is based on the following two results which are of independent interest.

8.8. THEOREM. (Külshammer (1979), Motose (1979)). Let A be a finite-dimensional indecomposable algebra over a field F. If $J(A) = A \cdot J(Z(A))$, then A

is primary. Moreover, if F is a splitting field for A, then A is a full matrix ring over $Z(A)$. ∎

8.9. THEOREM. (Asano (1961), Motose (1979)). Let A be a finite-dimensional indecomposable quasi-Frobenius algebra. Then the following conditions are equivalent:

(i) $J(A)^2 \neq 0$

(ii) $eJ(A)^2 \neq 0$ for every primitive idempotent e of A

(iii) $\ell(J(A)) \subseteq J(A)^2$. ∎

The following lemma due to Motose (1983) is often useful in investigating the nilpotency index of $J(FG)$.

8.10. LEMMA. Let B, I and J be subsets of a ring A which satisfy the following conditions:

(i) $IAI = IBI$

(ii) $IJ = JI$

(iii) $BJ \subseteq JB$

Then $(JIA)^n \subseteq J^nIA$. Moreover, if $J^n = 0$ then $JIA \subseteq J(A)$

Proof. The case $n = 1$ being trivial, we use induction on n. Assume that

$$(JIA)^n \subseteq J^nIA$$

Then we have

$$\begin{aligned}
(JIA)^{n+1} &\subseteq J^nIAJIA = J^nIAIJA && \text{(since } IJ = JI)\\
&= J^nIBIJA && \text{(since } IAI = IBI)\\
&= J^nIBJIA && \text{(since } IJ = JI)\\
&\subseteq J^nIJBIA && \text{(since } BJ \subseteq JB)\\
&= J^{n+1}IBIA && \text{(since } IJ = JI)\\
&\subseteq J^{n+1}IA, && \text{(since } IAI = IBI)
\end{aligned}$$

proving the first assertion. Now assume that $J^n = 0$. Then $(JIA)^n = 0$ and so JIA is a nilpotent right ideal of A. Thus $JIA \subseteq J(A)$ as asserted. ∎

To illustrate an application of Lemma 8.10, we establish the following results.

8.11. THEOREM. (Motose (1983)). Suppose that a group G has subgroups H and U

such that $G = UN_G(H)U$ and $H \subseteq N_G(U)$. Then

$$(J(FH)U^+FG)^{t(H)} = 0$$

and

$$J(FH)U^+ \subseteq J(FG)$$

Proof. Put $A = FG$, $B = FN_G(H)$, $I = \{U^+\}$, and $J = J(FH)$ and apply Lemma 8.10. ■

8.12. THEOREM. (Motose (1983)). Let N be a normal subgroup of G, let e be a block idempotent of FN and let H be the inertia group of e. Then

$$J(FH)e \subseteq J(FG)$$

Proof. Put $A = FG$, $B = FH$, $I = \{e\}$ and $J = J(FH)$. Let $\{a_1, a_2, \ldots, a_s\}$ be a right transversal for H in G. Then

$$\begin{aligned} IAI = eFGe &= \sum_{i=1}^{s} FHea_i ea_i^{-1} a_i \\ &= FHe = eFHe \\ &= IBI \end{aligned}$$

A similar argument shows that $IJ = JI$ and $BJ \subseteq JB$. Invoking Lemma 8.10, we derive $J(FH)e \subseteq J(FG)$ as required. ■

The next two results are easy applications of Theorem 8.11. For the terminology and properties of finite groups of Lie type we refer to Curtis (1970).

8.13. THEOREM. (Motose (1983)). Suppose that G has a split (B,N)-pair of characteristic r such that B is a semidirect product of a normal r-subgroup U and an abelian r'-subgroup $H = B \cap N$. Then

$$J(FH)U^+ \subseteq J(FG)$$

8.14. THEOREM. (Motose (1983)). Let q be a prime power, let $G = SL(2,q)$ and let p be an odd prime divisor of $q-1$. Then $t(G)$ is the p-part of $q-1$. ■

The next result is a useful companion of Theorem 8.14.

8.15. THEOREM. (Alperin (1979)). Let $G = SL(2,2^n)$ and let F be a field of characteristic 2. Then

$$t(G) = \begin{cases} 2n + 1 & \text{if } n > 1 \\ 2 & \text{if } n = 1 \end{cases}$$

To state our next result, we recall the following piece of information (see Broue and Puig (1980)). For any p-subgroup P of G the natural projection $Br_P : FG \longrightarrow FC_G(P)$ defined by $Br_P(g) = g$ for $g \in C_G(P)$ and $Br_P(g) = 0$ otherwise induces a homomorphism of F-algebras $C_{FG}(P) \longrightarrow FC_G(P)$. For a block B of FG, a B-*subpair* is a pair (Q,b) consisting of some p-subgroup Q of G and some block b of $C_G(Q)$ such that

$$Br_Q(1_B)1_b \neq 0$$

Let F be an algebraically closed field of characteristic $p > 0$ and let B be a block of FG. Then B is said to be *nilpotent* if $N_G(Q,b)/C_G(Q)$ is a p-group for all B-subpairs (Q,b), where $N_G(Q,b)$ denotes the stabilizer of (Q,b) in G. The significance of a nilpotent block stems from the fact that it is isomorphic to a full matrix algebra over the group algebra of its defect group (see Puig (1986)). Whether the converse of this is true is not however known. We are now ready to quote the following result.

8.16. THEOREM. (Okuyama (1981)). Let F be an algebraically closed field of characteristic p and let $B = B(e)$ be a block of FG of defect d. Then

$$J(Z(FG))^{p^d-1}e \neq 0$$

if and only if B is nilpotent with a cyclic defect group. ■

Let $t(G)$ be the nilpotency index of $J(FG)$. The groups G for which $t(G) = 2$ have been characterized by Theorem 3.12.4. The case where $t(G) = 3$ and G is p-solvable is settled by Proposition 7.2.6. It is therefore appropriate to investigate the case where $t(G) = 4$ and G is p-solvable. Koshitani (1982b) determined the structure of G with $t(G) = 4$ under either of the following hypotheses:

(i) $O_{2'}(G)$ is abelian

(ii) $O_{2'}(G)$ is metacyclic

(iii) The order of $O_{2'}(G)$ is not divisible by 3.

He also obtained the following result.

8.17. THEOREM. (Koshitani (1982b)). Let G be a p-solvable group and let $P \in Syl_p(G)$. If $t(G) = 4$, then $p = 2$ and one of the following properties hold:

(i) P is cyclic of order 4

(ii) P is elementary abelian of order 8

(iii) $G/O_{2'}(G) \cong S_4$. ∎

We close by quoting the following results.

8.18. THEOREM. (Motose (1983)). Let G be a p-solvable group, let $P \in Syl_p(G)$ be of order p^r and let F be a field of characteristic $p > 0$. If $O_{p',p,p'}(G)/O_{p',p}(G)$ is abelian, P is regular and $t(G) = r(p-1) + 1$, then P is elementary abelian. ∎

8.19. THEOREM. (Ninomiya (1981)). Let F be an algebraically closed field of characteristic p and let G be a p-nilpotent group. For each block idempotent e of $FO_{p'}(G)$, let P_e denote a Sylow p-subgroup of the inertia group of e. Then the following conditions are equivalent:

(i) G is p-radical

(ii) If e is a block idempotent of $FO_{p'}(G)$, then $ex = e$ for all $x \in [O_{p'}(G), P_e]$

(iii) If e is a block idempotent of $FO_{p'}(G)$, then $P_e \subseteq C_{FG}(FO_{p'}(G)e)$.

8.20. COROLLARY. (Ninomiya (1981)). Further to the assumptions and notation of Theorem 8.19, assume that $P \in Syl_p(G)$ and that $P \cap gPg^{-1} = 1$ for all $g \in G - N_G(P)$. Then G is p-radical if and only if one of the following conditions hold:

(i) $P \lhd G$

(ii) G has a subnormal subgroup H which is a Frobenius group with complement P. ∎

8.21. THEOREM. (Tshushima (1986)). Let G be a p-radical group. Then, for any element g of order p in G, $[O_{p'}(G), g]$ is nilpotent. ∎

8.22. THEOREM. (Motose (1977)). Let F be an algebraically closed field of characteristic $p > 0$, let u be the F-dimension of the projective cover of the trivial FG-module and let B be the principal block of FG. If $P \in \mathrm{Syl}_p(G)$ is cyclic, then

$$\dim_F J(B) = \dim_F B(1 - 1/u)$$

if and only if G is p-solvable of p-length 1. ∎

8.23. THEOREM. (Müller (1971)). Let S_n be the symmetric group of degree n. Then

$$\lim_{n \to \infty}(\dim_F J(FS_n)/\dim_F FS_n) = 1$$

8.24. THEOREM. (Koshitani (1982b)). Assume that $p = 2$ and G is a 2-solvable group with a Sylow 2-subgroup P which satisfies one of the following conditions:

(i) P is cyclic of order 4

(ii) P is elementary abelian of order 8

(iii) $G = S_4$

Then $t(G) = 4$. ∎

Bibliography

ALPERIN, J.L

[1979] Projective modules for $SL(2,2^n)$, J. Pure Appl. Algebra 15, 219-234.

[1986] Local Representation Theory, Cambridge Studies in Advanced Mathematics 11, Cambridge Univ. Press.

ALPERIN, J.L., COLLINS, M. J., and SIBLEY, D.A.

[1984] Projective modules, filtrations and Cartan invariants, Bull. London Math. Soc. 16, No4, 416-420.

AMITSUR, S.A

[1951] Nil *PI*-rings, Proc. Amer. Math. Soc. 2, 538-540.

[1953] The identities of *PI*-rings, Proc. Amer. Math. Soc. 4, 27-34.

[1956a] Radicals of polynomial rings, Canad. J. Math. 8, 355-361.

[1956b] Algebras over infinite fields, Proc. Amer. Math. Soc., 35-48

[1957] The radical of field extensions, Bull. Res. Council Israel, 7F, 1-10.

[1959] On the semisimplicity of group algebras, Mich. Math. J. 6, 251-253.

ANDERSON, F.W. and FULLER, K.R.

[1973] Rings and categories of modules, Springer GTM13, New York

ANDERSEN, H.H, JORGENSEN J., and LANDROCK, P.

[1983] The projective indecomposable modules of $SL(2,p^n)$, Proc London Math. Soc. (3), 46, 38-52.

ARTIN, E., NESBITT, C.J. and THRALL, R.M.

[1944] Rings with minimum condition, Univ. of Michigan, Ann Arbor

ASANO, S.

[1961] On the radical of quasi-Frobenius algebras, Kodai Math. Sem. Rep. 13, 135-151.

ASANO, S. and MOTOSE, K.

[1980] On *QF*-2 algebras with commutative radicals, Math. J. Okayama Univ. 22, Nol, 17-20.

AZUMAYA, G.

[1951] On maximally central algebras, Nagoya Math. J. 2, 119-150.

AZUMAYA, G. and NAKAYAMA, T.

[1948] On absolutely uniserial algebras, Japan J. Math. 19,263-273.

BEDI, S.S.

[1979] The Jacobson radical of the group algebra of a finite group, Proc. Amer. Math. Soc., 75, Nol, 13-18.

BOURBAKI, N

[1958] Algébre, Chap. 8. Herman, Paris.

BOVDI, A.A

[1960] Group rings of torsion-free groups, Sibirsk. Math. Z. 1, 251-253.

[1974] Group rings (in Russian), Uzhgorod.

BRAUER, R.

[1935] Uber die Darstellungen von Gruppen in Galoischen Feldern, Act. Sci. Ind. 195, Paris.

[1941a] On the Cartan invariants of groups of finite order, Ann. of Math. (2), 42, 53-61.

[1941b] On the connection between the ordinary and the modular characters of groups of finite order, Ann. of Math. (2)42, 926-935.

[1955] Number theoretical investigations on groups of finite order, Proc. Internat. Sympos. on Algebraic number theory, Tokyo and Nikko, 55-62.

[1956] Zur Darstellungstheorie der Gruppen endlicher Ordnung, Math. Z. 63, 406-444.

[1963] Representations of finite groups, Lectures on modern mathematics, Vol.1, John Wiley & Sons, New York, London.

[1967] On blocks and sections in finite groups, I, Amer. J. Math. 89, 1115-1136.

[1971] Some applications of the theory of blocks of characters of finite groups, IV, J.Algebra 17, 489-521.

BRAUER, R. and NESBITT, C.J.

[1941] On the modular characters of groups, Ann. of Math. 42, 556-590.

BROCKHAUS, P.

[1982] A remark on the radical of a group algebra, Proc. Edinb. Math. Soc. 25, 69-71.

[1985] On the radical of a group algebra, J. Algebra, 95, 454-472.

BROCKHAUS, P. and MICHLER, G. O.

[1985] Finite simple groups of Lie type have non-principal p-blocks, $p \neq 2$, J. Algebra, 94, 113-125.

BROUÉ, M.

[1976] Radical, hauteur, p-sections et blocs, C.R. Acad. Sci. Paris Ser A-B, 283, A563-A565.

[1978a] Radical, hauteurs, p-sections et blocs, Ann. of Math. 107, 89-107.

[1978b] Remarks on blocks and subgroups, J. Algebra 51, 228-232.

BROUÉ, M. and PUIG, L.

[1980] A Frobenius theorem for blocks, Invent. Math. 56, 117-128.

BURNSIDE, W.

[1955] Theory of groups of finite order, 2nd ed., Dover, New York.

CARLSON, J.F.

[1971] Block idempotents and the Brauer correspondence. Bull. Austral. Math. Soc., 5, 337-340.

CARTAN, H. and EILENBERG, S.

[1956] Homological algebra, Princeton University Press, Princeton.

CHALABI, A.

[1973] Modules over group algebras and their application to the study of semisimplicity Math. Ann. 201, 57-63.

CHOUINARD, L.

[1976] Projectivity and relative projectivity over group rings, J. Pure and Applied Algebra 7, 278-302.

CLARKE, R.J.

[1969] On the radical of the centre of a group algebra, J. London Math. Soc.(2), 1, 565-572.

[1972] On the radical of the group algebra of a p-nilpotent group, J. Austral. Math. Soc. 13, 119-123.

CLIFFORD, A.H.

[1937] Representations induced in an invariant subgroup, Ann. of Math. 38, 129-137.

COLEMAN, D.B.

[1964] On the modular group ring of a p-group, Proc. Amer. Math. Soc. 15, 511-514.

CONLON, S.B.

[1964] Twisted group algebras and their representations, J. Austral. Math. Soc. 4, 152-173.

CONNELL, I.G.

[1963] On the group ring, Canad. J. Math. 15, 650-685.

CURTIS, C.W.

[1970] Modular representations of finite groups with split (B,N)-peirs, Lecture Notes in Math. 131, Springer, Berlin-Heidelberg-New York, 57-95.

CURTIS, C.W. and REINER, I.

[1962] Representation theory of finite groups and associative algebras, Interscience, New York.

[1981] Methods in representation theory with applications to finite groups and orders, Vol. 1, Wiley-Interscience, New York, Chichester, Brisbane, Toronto.

DADE, E.C.

[1966] Blocks with cyclic defect groups, Ann. of Math. (2), 84, 20-48.

[1968] Degrees of modular irreducible representations of p-solvable groups, Math. Z. 104, 141-143.

DeMEYER F.R. and JANUSZ G.J.

[1983] Group rings which are Azumaya algebras, Trans. Amer. Math. Soc. 279, 389-395.

DESKINS, W.E.

[1958] On the radical of a group algebra, Pacif. J. Math. 8, 693-697.

DICKSON, L.E.

[1902] On the group defined for any given field by the multiplication table of any given finite group, Trans. Amer. Math. Soc. 3, 285-301.

[1907a] Modular theory of group matrices, Trans. Amer. Math. Soc. 8, 384-398.

[1907b] Modular theory of group characters, Bull. Amer. Math. Soc. 13, 477-488.

DIVINSKY, N.J.

[1965] Rings and radicals, University of Toronto Press, London.

DORNHOFF, L

[1971] Group representation theory, Part A, Marcel Dekker, New York.

[1972] Group representation theory, Part B, Marcel Dekker, New York.

EILENBERG, S. and NAKAYAMA, T

[1955] On the dimensions of modules and algebras, II (Frobenius algebras and quasi-Frobenius rings). Nagoya Math. J., 9, 1-16.

EISENBUD, D. and GRIFFITH, P.

[1971] Serial rings, J. Algebra 17, 389-400

ERDMANN, K.

[1977] Principal blocks of groups with dihedral Sylow 2-subgroups, Comm. Algebra 5, 665-694.

[1979] Blocks whose defect groups are Klein four groups, J. Algebra 59, 452-465.

[1982] Blocks whose defect groups are Klein four groups: a correction, J Algebra 76, 505-518.

FEIT, W.

[1967] Characters of finite groups, Benjamin, New York.

[1982] The representation theory of finite groups, North Holland, Amsterdam.

FONG, P.

[1960] Some properties of characters of finite solvable groups, Bull. Amer. Math. Soc. 66, 116-117.

[1961] On the characters of p-solvable groups, Trans. Amer. Math. Soc. 98, 263-284.

[1962] Solvable groups and modular representation theory, Trans. Amer. Math. Soc. 103, 484-494.

[1963] A note on a conjecture of Brauer, Nagoya Math. J. 22, 1-13.

FONG, P. and GASCHÜTZ, W.

[1961] A note on the modular representations of solvable groups, J. Reine Angew Math. 208, 73-78.

FORMANEK, E

[1972] A problem of Passman on semisimplicity, Bull. Lond. Math. Soc. 4, 375-376.

FORMANEK, E. and SNIDER, R.L.

[1977] The Jacobson radical of the group algebra of a p-solvable group, unpublished.

GASCHUTZ, W.

[1952] Uber den Fundamentalsatz von Maschke zur Darstellungstheorie der endlichen Gruppen, Math. Z, 56, 376-387.

GLAUBERMAN, G.

[1968] A characteristic subgroup of a p-stable group, Canad. J. Math. 20, 1101-1135.

[1971] Global and local properties of finite groups, Finite simple groups (eds M.B. Powell and G. Higman), Academic Press, New York, 1-64.

[1972] A sufficient condition for p-stability, Proc. London Math. Soc.(3)25, 253-287.

GORENSTEIN, D.

[1968] Finite groups, New York, Harper and Row.

GOW, R.

[1978] A note on p-blocks of a finite group, J. London Math. Soc. (2),18,61-64.

GREEN. J. A.

[1959] On the indecomposable representations of a finite group, Math. Z., 70, 430-445.

[1962] Blocks of modular representations, Math. Z. 79, 100-115.

[1964] A transfer theorem for modular representations, J. Algebra 1, 73-84.

[1968] Some remarks on defect groups, Math. Z. 107, 133-150.

GREEN, J.A. and HILL, R.

[1969] On a theorem of Fong and Gaschutz, J. London Math. Soc. (2), 1, 573-576.

GREEN, J.A. and STONEHEWER, S.E.

[1969a] The radicals of some group algebras, J. Algebra 13, 137-142.

[1969b] Note on the paper "The radicals of some group algebras", J. Algebra 13, 297.

GUPTA, R. N.

[1984] Semisimplicity of group rings-an exposition, Algebra and its applications (New Delhi, 1981), 93-97, Lecture Notes in Pure and Applied Math., 91, Marcel Dekker, New York.

HALL, M.

[1938] Group rings and extensions, I, Ann. Math. 39, 220-234.

[1940] The position of the radical in an algebra, Trans. Amer. Math. Soc. 48, 391-404.

[1959] The theory of groups, Macmillan, New York.

HALL, P. and HIGMAN, G.

[1956] On the p-length of p-solvable groups and reduction theorems for Burnside's problem, Proc. London Math. Soc.(3), 1-42.

HAMERNIK, W.

[1975] Group structure and properties of block ideals of the group algebra, Glasgow Math. J. 16, 22-28.

HAMPTON, C.R. and PASSMAN, D.S.

[1972] On the semisimplicity of group rings of solvable groups, Trans. Amer. Math. Soc. 173, 289-301.

HARRIS, M.E.

[1985] On the p-deficiency class of a finite group, J. Algebra 94, 411-424.

HERSTEIN, I.N.

[1968] Noncommutative rings, Carus Math. Monographs, No. 15, Math. Assoc. Amer.

HIGMAN, D.G.

[1954] Indecomposable representations at characteristic p, Duke Math. J.21, 377-381.

[1955] Induced and produced modules, Canad. J. Math. 7, 490-508.

HILL, E.T.

[1970] The annihilator of radical powers in the modular group ring of a p-group, Proc. Amer. Math. Soc. 25, 811-815.

[1971] Ideals in the modular group ring of a p-group, Proc. Amer. Math. Soc. 28, 389-390.

HOCHSCHILD, G.

[1956] Relative homological algebra, Trans. Amer. Math. Soc. 82, 246-269.

HOLVOET, R.

[1968/69] The group algebra of a finite p-group over a field of characteristic p, Simon Stevin 42, 157-170.

HUPPERT, B.

[1967] Endliche Gruppen, I, Springer, Berlin-Heidelberg-New York.

HUPPERT, B.

[1975] Bemerkungen zur modularen Darstellungstheorie I, Absolut unzerlegbare Moduln, Arch Math. (Basel), 26, 242-249.

HUPPERT, B. and BLACKBURN, N.

[1982a] Finite Groups II, Springer-Verlag, Berlin.

[1982b] Finite Groups III, Springer-Verlag, Berlin.

HUPPERT, B. and WILLEMS, W.

[1975] Bemerkungen zur modularen Darstellungstheorie, II, Darsettungen von Normalteirn, Arch. Math. (Basel), 26, 486-496.

IIZUKA, K.

[1956] Note on blocks of group characters, Kumamoto J. Sci. SerA 2, 309-321.

[1960a] On Osima's blocks of group characters, Proc. Japan Acad. 36, 392-396.

[1960b] On Osima's blocks of characters of groups of finite order, Kumamoto J. Sci. SerA4, 275-283.

[1960c] On the blocks and sections of finite groups, Kumamoto J. Sci. SerA5, 53-62.

[1961] On Brauer's theorem on sections in the theory of blocks of group characters, Math. Z. 75, 299-304.

[1972] A note on blocks of characters of a finite group, J. Algebra 20, 196-201.

IIZUKA, K. and WATANABE, A.

[1973] On the number of blocks of irreducible characters of a finite group with a given defect group, Kumamoto J. Sci. (Math) 9, 55-61.

JACOBSON, N.

[1956] Structure of rings, AMS Colloquim Publ. V. 37, AMS, Providence, R.I.

[1964] Lectures in abstract algebra, Vol. III, GTM, Springer-Verlag, New York, Heidelberg, Berlin.

JENNINGS, S.A.

[1941] The structure of the group ring of a p-group over a modular field, Trans. Amer. Math. Soc. 50, 175-185.

JORDAN, C.R.

[1975] The Jacobson radical of the group ring of a generalised free product, J. London Math. Soc.(2), 11, No3, 369-376.

KARPILOVSKY, G.

[1982] The Jacobson radical of commutative group rings, Arch. Math. 39, 428-430.

[1983] Commutative group algebras, Marcel Decker, New York.

KARPILOVSKY, G.

[1985] Projective representations of finite groups, Marcel Dekker, New York.

[1986] Group and Semigroup Rings, Proceedings of the International Conference on Group and Semigroup Rings, University of the Witwatersrand, Johannesburg, 1985, North Holland Mathematics Studies 126, Notas de Matematica 111, North Holland, Amsterdam-New York-Oxford-Tokyo.

[1987] Structure of blocks of group algebras, Longman, to appear.

KASCH, F., KNESER, M. and KUPISCH, H.

[1957] Unzerlegbare modulare Darstellungen endicher Gruppen mit zyklischer p-Sylow gruppe, Arch. Math. (Basel) 8, 320-321.

KAWADA, Y.

[1966] On blocks of group algebras of finite groups, Sci. Rep. Tokyo Kyoiku Daigaku A9, 87-110.

KHATRI, D.C.

[1973] Relative projectivity, the radical and complete reducibility in modular group algebras, Trans. Amer. Math. Soc. 186, 51-63.

[1974] Projectivity-sensitivity in modular group algebras, Math. Japan 19, 71-78.

KHATRI, D.C. and SINHA, I.

[1969] Projective pairings in groups II, Math. Japan 14, 127-135.

KNÖRR, R.

[1976] Blocks, vertices and normal subgroups, Math. Z. 148, 53-60.

[1977] Semisimplicity, induction, and restriction for modular representations of finite groups, J. Algebra 48, 347-367.

[1979] On the vertices of irreducible modules, Ann. of Math. (2), 110, 487-499.

KOSHITANI, S.

[1976] Group rings of metacyclic p-groups, Proc. Japan Acad, 52, No.6, 269-272.

[1977a] On the nilpotency indices of the radicals of group algebras of p-groups which have cyclic subgroups of index p, Tsukuba J. Math. 1, 137-148.

[1977b] On the nilpotency indices of the radicals of group algebras of p-solvable groups, Proc. Japan Acad. 53, SerA, No1, 13-16.

[1978] A note on the radical of the centre of a group algebra, J. London Math. Soc. (2), 18, 243-246.

[1979] Remarks on the commutativity of the radicals of group algebras, Glasgow Math. J. 20, 63-68.

[1980] Some remarks on the modular group algebras, Lecture Notes in Math, No. 903, Springer-Verlag, 186-195.

KOSHITANI, S.

[1981] On the kernels of representations of finite groups, Glasgow Math. J. 22, 151-154.

[1982a] A remark on the nilpotency index of the radical of a group algebra of a p-solvable group, Proc. Edinb. Math. Soc. 25, 31-34.

[1982b] Group algebras of finite p-solvable groups with radicals of the fourth power zero, Proc. Roy. Soc. Edinb., 92A, 205-231.

[1982c] A remark on blocks with dihedral defect groups in solvable groups, Math. Z., 179, 401-406.

[1983] On the Jacobson radical of a block ideal in a finite p-solvable group for $p \geqslant 5$, J. Algebra, 80, No1, 134-144.

[1984] On lower bounds for the radical of a block ideal in a finite p-solvable group, Proc. Eding. Math. Soc., 27, 65-71.

[1985a] A remark on the Jacobson radical of a block ideal in a finite p-solvable group, J. Algebra, 94, No1, 106-112.

[1985b] On the Loewy series of the group algebra of a finite p-solvable group with p-length > 1, Comm. Algebra, 13(10), 2175-2198.

[1986] On group algebras of finite groups, Representation theory II, (Ottawa, Ont. 1984), 109-128, Lecture Notes in Math., 1178, Springer, Berlin-New York.

KÖTHE, G.

[1931] Die Struktur der Ringe, deren Restklassenring nach dem Radikal vollständig reduzibelist, Math. Z 32, 161-186.

KÜLSHAMMER, B.

[1979] Zentral separable gruppen algebren, unpublished

[1980] On the structure of block ideals in group algebras of finite groups, Comm. Algebra, 8(19), 1867-1872.

[1981a] Bemerkungen über die Gruppenalgebra als symmetrische Algebra, J. Algebra, 72, No1, 1-7.

[1981b] On p-blocks of p-solvable groups, Comm. Algebra, 9(17), 1763-1785.

[1982] Bemerkungen über die Gruppenalgebra als symmetrische Algebra, II, J. Algebra, 75, No1, 59-64.

[1983] On blocks with central defect groups, Arch Math. (Basel), 41, 117-120.

[1984a] Symmetric local algebras and small blocks of finite groups, J. Algebra 88, No1, 190-195.

[1984b] Bemerkungen über die Gruppenalgebra als symmetrische Algebra, III, J. Algebra, 88, No1, 279-291.

KÜLSHAMMER, B.

[1984c] Quotients, Cartan matrices and Morita equivalent blocks, J. Algebra, 90, No2, 364-371.

[1985a] Bemerkungen über die Gruppenalgebra als symmetrische Algebra, IV, J. Algebra, 93, No2, 310-323.

[1985b] Crossed products and blocks with normal defect groups, Comm. Algebra, 13(i), 147-168.

[1986] Blocks of modular group algebras, Group and semigroup rings, North Holland, North-Holland Mathematics Studies, Vol. 126, 117-137.

LAMBEK, J.

[1966] Lectures on rings and modules, Blaisdell, Waltham, Mass.

LANDROCK, P.

[1980] Some remarks on Loewy lengths of projective modules, Proc. Internat. Conf. Representations of algebras, Carleton Univ., Ottawa, Lecture Notes in Mathematics No 831, Springer, Berlin, 369-381.

[1983a] Finite group algebras and their modules, London Math. Soc. Lecture Notes Series 84, Cambridge Univ. Press.

[1983b] The Cartan matrix of a group algebra modulo any power of its radical, Proc. Amer. Math. Soc. 88, 205-206.

LAZARD, M.

[1953] Sur les groupes nilpotents et les anneaux de Lie, Ann. Ecole Norm. Sup. (3) LXXI, 101-190.

LOMBARDO- RADICE, L

[1939] Intorno alle algebre legate ai gruppi di ordine finito, II, Rend. Sem. Mat. Roma 3, 239-256.

[1947a] Il difetto di regolaritá di un gruppo finito rispetto a un divisore primo del suo ordine, Atti Accad. Naz. Lincei. Rend, Cl. Sci. Fis. Mat. Nat. (8) 2, 766-767.

[1947b] Sugli elementi eccezionali dell'algebra legata a un gruppo di ordine finito in un corpo a caratteristica p, Atti Accad. Naz. Lincei. Rend. Cl. Sci. Fis. Mat. Nat. (8) 2, 170-174.

[1948] Il difetto di regolaritá di un gruppo finito rispetto a un divisore primo del suo ordine, Univ. Roma, Ist. Naz. Alta Mat. Rend. Mat. *e* Appl.(5)7, 169-183.

LONCORN, C.

[1971] Radical d'une algebra d'un product direct de groupes finis, Bull. Soc. Math. Belg. 23, 423-435.

LORENZ, M.

[1985] On Loewy lengths of projective modules for p-solvable groups, Comm. Algebra 13, No5, 1193-1212.

MARGOT, J.L.

[1974] Idéal fondamental et radical de Jacobson, C.R. Acad. Sci, Paris, t.279, A87-90.

[1975] Radical de Jacobson d'un anneau de groupe sur un groupe de Frobenius fini, C.R. Acad. Sci. Paris, t.280, A551-554.

MELDRUM, J.D.P and WALLACE, D.A.R.

[1972] Closure operations and group algebras, Proc. Edinb. Math. Soc. Vol.18 (Series II), Part 2, 149-158.

MICHLER, G.O.

[1972] Blocks and centers of group algebras, Lectures on rings and modules, Lecture Notes in Math. 246, Springer, Berlin, 429-563.

[1973] The kernel of a block of a group algebra, Proc. Amer. Math. Soc. 37, 47-49.

MORITA, K.

[1951] On group rings over a modular field which possess radicals expressible as principal ideals, Sci. Rep. Tokyo Bunrika Daikagu (A) 4, 177-194.

MOTOSE, K.

[1969] On group rings over semi-primary rings, Math. J. Okayama Univ., 14 (1969), 23-26.

[1971a] On group rings over semi-primary rings II, J. Fac. Sci., Shinshu Univ., 6, No2, 97-99.

[1971b] On the radical of a group ring, Math. J. Okayama Univ. 15, 35-36.

[1972] Note on results of D.A.R. Wallace, J. Fac. Sci. Shinshu Univ., 7, 119-122.

[1974a] On radicals of group rings of Frobenius groups, Hokkaido Math. J. 3, 23-34

[1974b] On C. Loncour's results, Proc. Japan Acad., 50, 570-571.

[1974c] On a theorem of Wallace and Tsushima, Proc. Japan Acad. 50, 572-575.

[1977] On radicals of principal blocks, Hokkaido Math. J. 6(1977), 255-259.

[1978] On a theorem of S. Koshitani, Math. J. Okayama Univ., 20, 59-65.

[1979] Note on the radical of a finite dimensional algebra, J. Fac. Sci, Shinshu Univ., 14, No2, 95-97.

[1980] On the nilpotency index of the radical of a group algebra, II, Math. J. Okayama Univ., 22, No2, 141-143.

MOTOSE, K.

[1982] On the nilpotency index of the radical of a group algebra, III, J. London Math. Soc. (2), 25, 39-42.

[1983] On the nilpotency index of the radical of a group algebra, IV, Math. J. Okayama Univ. 25, No1, 35-42.

[1984a] On the nilpotency index of the radical of a group algebra, V, J. Algebra 90, 251-258.

[1984b] On a theorem of Y. Tsushima, Math. J. Okayama Univ., 25, 11-12.

[1984c] On a result of S. Koshitani, Proc. Edinb. Math. Soc., 27, 57.

[1984d] On the nilpotency index of the radical of a group algebra, VIII, Proc. Amer. Math. Soc., 92, No3, 327-328.

[1985] On the nilpotency index of the radical of a group algebra, VI, J. Algebra, 94, No2, 347-351.

[1986a] On the nilpotency index of the radical of a group algebra, VII, J. Algebra 101, 299-302.

[1986b] On the nilpotency index of the radical of a group algebra IX, Group and Semigroup Rings, G. Karpilovsky (ed.), North-Holland, 193-196.

MOTOSE K., and NINOMIYA, Y.

[1975a] On the nilpotency index of the radical of a group algebra, Hokkaido Math. J., 4, No2, 261-264.

[1975b] On the subgroups H of a group G such that $J(KH)KG \supset J(KG)$, Math. J. Okayama Univ., 17, 171-176.

[1977] On generalized uniserial blocks, Math. J. Okayama Univ. 19, 141-145.

[1980] On the commutativity of the radical of the group algebra of a finite group, Osaka J. Math. 17, 23-26.

MULLER, W.

[1971] Über die grösse des radikals in gruppenringen über der symmetrischen gruppe, Manuscripta Math. 4, 39-60.

[1974a] Symmetriche Algebren mit injective Zentrum, Manuscripta Math., 11, 283-289.

[1974b] Radikal und Sockel in Gruppenalgebren über lokalendlichen Gruppen, Arch. Math(Basel) Vol. XXV, 476-482.

[1974c] Gruppenalgebren über nichtzyklischen p-Gruppen, I, J. Reine Angew Math., 266, 10-48

[1974d] Gruppenalgebren über nichtzyklischen p-Gruppen, II, J. Reine Angew Math., 267, 1-19.

MURAI, M.

[1984] A note on the number of irreducible characters in a p-block of a finite group, Osaka J. Math. 21, No2, 387-398.

NAKAJIMA, A. and TOMINAGA, H.

[1968] A note on group rings of p-groups, Math. J. Okayama Univ. 13, 107-109.

NAKAYAMA, T.

[1938] Some studies on regular representations, induced representations, and modular representations, Ann. of Math. 39, 361-369.

[1939] On Frobenius algebras I, Ann. of Math. 40, 611-633.

[1940] Note on uniserial and generalized uniserial rings, Proc. Imp. Acad. Japan, 16, 285-289.

[1941] On Frobenius algebras II, Ann. of Math. 42, 1-21.

NAKAYAMA, T and SHODA, K.

[1936] Über die Darstellung einer endlichen Gruppen durch halblinear Transformationen, Japan. J. Math. 12, 109-122.

NINOMIYA, Y.

[1974] On the nilpotency index of the radical of a group algebra, Symposium on algebra (in Japanese)

[1980] On the commutativity of the radical of the group algebra of an infinite group, Osaka J. Math. 17, 27-33.

[1981a] On the radical of the group algebra of a p-nilpotent group, Math. J. Okayama Univ. 23, 189-193.

[1981b] On the Jacobson radical of the center of an infinite group algebra, Hokkaido Math. J., 10, 254-263.

[1982] On p-nilpotent groups with extremal p-blocks, Hokkaido Math. J., 11. 229-233.

[1983] On the Cartan invariants of p-solvable groups, Math. J. Okayama Univ. 25, No1, 57-68.

[1984a] On the radical of an infinite group algebra, Math. J. Okayama Univ. 26, 13-18.

[1984b] On a theorem of S. Koshitani, Math. J. Okayama Univ. 26, 19-22.

OKUYAMA, T.

[1978] A note on the Brauer correspondence, Proc. Japan Acad., 54, Ser.A, 27-28.

[1980] Some studies on group algebras, Hokkaido Math. J. 9. 217-221.

[1981] On the radical of the center of a group algebra, Hokkaido Math. J. 10, 406-408.

OKUYAMA, T.

[1983] On finite group algebras with radical cube zero, Proc. 16th Sympos. on Ring Theory, Okayama Math. Lect., Okayama Univ. Okayama, 105-111.

[1986] On blocks of finite groups with radical cube zero, preprint.

OKUYAMA, T and TSUSHIMA Y.

[1983] Local properties of p-block algebras of finite groups, Osaka J. Math. 20, 33-41

[1984] Ext_G^1 for irreducible modules, Comm, Algebra 12(18), 2273-2277.

[1986] On a conjecture of P. Landrock, preprint.

OSIMA, M.

[1942] On primary decomposable group rings, Proc. Phys.-Math. Soc. Japan (3) 24, 1-9.

[1955] Note on blocks of group characters, Math. J. Okayama Univ. 4, 175-188.

[1966] On block idempotents of modular group rings, Nagoya Math. J. 27, 429-433.

PASSMAN, D.S.

[1962] Nil ideals in group rings, Michigan Math. J. 9, 375-384

[1969a] Central idempotents in group rings, Proc. Amer. Math. Soc. 22, 555-556.

[1969b] Blocks and normal subgroups, J. Algebra 12, 569-575.

[1969c] On the semisimplicity of modular group algebras, Proc. Amer. Math. Soc. 20, 515-519.

[1969d] On the semisimplicity of modular group algebras II, Canad. J. Math., 21, 1137-1145.

[1970a] Radicals of twisted group rings, Proc. London Math. Soc. (3), 20, 409-437.

[1970b] On the semisimplicity of twisted group algebras, Proc. Amer. Math. Soc. 25, 161-166.

[1973a] On the semisimplicity of group rings of linear groups, Pac. J. Math. 45, 221-228.

[1973b] On the semisimplicity of group rings of linear groups II, Pac. J. Math., 48, 215-234.

[1974a] The semisimplicity problem for group rings, Symp. Math. 8, 333-342.

[1974b] Advances in group rings, Israel J. Math. 19, 67-107.

[1975] On the semisimplicity of group rings of some locally finite groups, Pac.J. Math. 58, 179-207.

[1977] The algebraic structure of group rings, Wiley, New York.

PASSMAN, D.S.

[1980] The radical of the center of a group algebra, Proc. Amer. Math. Soc. 78, No.3, 323-326.

POTTER, G.

[1977] On the radical of a group algebra over a commutative ring, Glasgow Math.J. 18, Nol, 101-104.

PUIG, L.

[1986] The Source algebra of a nilpotent block, preprint.

PUTTASWAMAIAH, B.H. and DIXON J.D.

[1977] Modular representations of finite groups, Academic Press, New York, London.

RENAULT, G.

[1971] Sur les anneaux de groupes, C.R. Acad. Sci. Paris, 273, 84-87.

REYNOLDS, W.F.

[1963] Blocks and normal subgroups of finite groups, Nagoya Math. J. 22, 15-32.

[1966] Block idempotents of twisted group algebras, Proc. Amer. Math. Soc. 17, 280-282.

[1972] Sections and ideals of centers of group algebras, J. Algebra, 20, 176-181.

RICKART, C.

[1950] The uniqueness of norm problem in Banach algebras, Ann. Math. 51, 615-628.

ROBINSON, G.R.

[1983] The number of blocks with a given defect group, J. Algebra, 84, 493-502.

ROSEN, M.I.

[1966] The Jacobson radical of a group algebra, Michigan Math. J. 13, 477-480.

ROZENBERG, A.

[1961] Blocks and centers of group algebras, Math. Z, 76, 209-216.

SAKSONOV, A.I.

[1971] On the decomposition of a permutation group over a characteristic field, Soviet Math. Dokl. 12, No.3, 786-790.

SCHNEIDER, H. and WEISSGLASS, J.

[1967] Group rings, semigroup rings and their radicals, J. Algebra 5, 1-15.

SCHWARTZ, W.

[1979] Die struktur modularer Gruppenringe Endlicher Gruppen der p-Lange 1, J. Algebra 60, 51-75.

SERRE, J.P.

[1977] Linear representations of finite groups, Springer, New York, Heidelberg, Berlin.

SINHA, I.

[1974] Semisimplicity of group rings with trivial units, Tamkang J. Math. 5, 1, 107-108.

SPIEGEL, H.

[1970] Das Radikal der modularen Gruppenalgebra einer endlichen Gruppe, Diplomarbeit, Tübingen.

[1974] Blockkorrespondenzen und p'-Normalteiler, Arch. Math. (Basel) 25, 483-487.

SRINIVASAN, B.

[1960] On the indecomposable representations of a certain class of groups, Proc. London Math. Soc. (3) 10, 497-513.

STONEHEWER, S.E.

[1969] Group algebras of some torsion-free groups, J. Algebra 13, 143-147.

SWAN, R.G.

[1963] The Grothendieck group of a finite group, Topology 2, 85-110.

TSUSHIMA, Y.

[1967] Radicals of group algebras, Osaka Math. J. 4, 179-182

[1968] A group algebra of a p-solvable group, Osaka Math. J. 5, 89-98.

[1971a] On the annihilator ideals of the radical of a group algebra, Osaka Math. J. 8, 91-97.

[1971b] On the block of defect zero, Nagoya Math. J. 44, 57-59.

[1974] On the existence of characters of defect zero, Osaka Math. J.11, 417-423.

[1977] On the weakly regular p-blocks with respect to $O_{p'}(G)$, Osaka Math. J. 14, 465-470.

[1978a] On the p'-section sum in a finite group ring, Math. J. Okayama Univ. 20, 83-86.

[1978b] Some notes on the radical of a finite group ring, Osaka Math. J. 15, 647-653.

[1978c] On the second reduction theorem of P. Fong, Kumamoto J. Sci. (Math) 13, 6-14.

[1978d] Problems on the radical of a finite group rings, Proc. 10th. Sympos. on Ring theory (Shinhu Univ. Matsumoto, 1977), Dept. of Math. Okayama Univ, Okayama, 116-120.

TSUSHIMA, Y.

[1979] Some notes on the radical of a finite group ring II, Osaka J. Math. 16, 35-38.

[1983] A note on Cartan integers for p-solvable groups, Osaka J. Math. 20, 675-679.

[1986] On p-radical groups, preprint.

VILLAMAYOR, O.E.

[1958] On the semisimplicity of group algebras, Proc. Amer. Math. Soc. 9, 621-627.

[1959] On the semisimplicity of group algebras, II, Proc. Amer. Math. Soc. 9, 621-627.

WADA, T.

[1977] On the existence of p-blocks with given defect groups, Hokkaido Math. J. 6, 243-248.

[1981] Blocks with a normal defect group, Hokkaido Math. J. 10, 319-322.

[1983] On the number of irreducible characters in a finite group, Hokkaido Math. J. 12, 74-82.

[1985] On the number of irreducible characters in a finite group, II, Hokkaido Math. J. 14, No2, 149-154.

WALLACE, D.A.R.

[1958] Note on the radical of a group algebra, Proc. Camb. Phil. Soc. 54, 128-130.

[1961] On the radical of a group algebra, Proc. Amer. Math. Soc. 12, 133-137.

[1962a] Group algebras with radicals of square zero, Proc. Glasgow Math. Assoc. 5, 158-159.

[1962b] Group algebras with central radicals, Proc. Glasgow Math. Assoc. 5, 103-108.

[1965] On the commutativity of the radical of a group algebra, Proc. Glasgow Math. Assoc. 7, 1-8.

[1967] The Jacobson radicals of the group algebras of a group and of certain normal subgroups, Math. Z. 100, 282-294.

[1968] Lower bounds for the radical of the group algebra of a finite p-solvable group, Proc. Edinb Math. Soc(2), 16, 127-134.

[1969] On commutative and central conditions on the Jacobson radical of the group algebra of a group, Proc. London Math. Soc. 19, 385-402.

[1970] The radical of the group algebra of a subgroup, of a polycyclic group and of a restricted SN-group, Proc. Edinb. Math. Soc. Vol, 17 (Series II Part 2, (165-171).

[1971] On commutative and central conditions on the Jacobson radical of the group algebra of a group II, J. London Math. Soc. 4, 91-99.

WALLIS, W.D.

[1968] A reduction of the problem of semisimplicity, J. Algebra 10, 501-502.

WARD, H.N.

[1961] Some results on the group algebra of a group over a prime field, Seminar on Finite Groups and Related Topics, Harvard Univ., 13-19.

[1968] The analysis of representations induced from a normal subgroup, Michig. Math. J. 15, 417-428.

WATANABE, A.

[1979] On Fong's reductions, Kumamoto J. Sci. (Math), 13, 48-54.

[1982a] Relations between blocks of a finite group and its subgroup, J. Algebra 78, 282-291.

[1982b] p-blocks and p-regular classes in a finite group, Kumamoto J. Sci., (Math), 15, 33-38.

WILLEMS, W.

[1976] Bemerkungen zur modularen Darstellungstheorie III. Induzierte und eingeschränkte Moduln, Arch. Math. (Basel), 26, 497-503.

[1978] Über die Existenz von Blöcken, J. Algebra 53, 402-409.

[1980] On the projectives of a group algebra, Math. Z. 171, 163-174.

ZALESSKII, A.E.

[1965] On the semisimplicity of a crossed product, Sibirsk. Math. Zh. 6, 1009-1013.

[1970] On group rings of solvable groups, Izv. Akad. Nauk BSSR, ser, Fiz-Mat., 13-21.

[1973] On the semisimplicity of a modular group algebra of a solvable group, Soviet Math. 14, 101-105.

[1974] The Jacobson radical of the group algebra of a solvable group is locally nilpotent, Izv. Akad. Nauk. SSSR, Ser. Mat. 38, 983-994.

ZASSENHAUS, H.

[1949] The theory of groups, 2nd ed. Chelsea, New York.

Notation

Number Systems

$\mathbb{N}$	the natural numbers
$\mathbb{Z}$	the rational integers
$\mathbb{Q}$	the rational numbers
$\mathbb{R}$	the real numbers
$\mathbb{C}$	the complex numbers
$\mathbb{Z}/m\mathbb{Z}$	the integers mod m

Set Theory

$\subset$	proper inclusion
$\subseteq$	inclusion
$\|X\|$	the cardinality of the set X
$X-Y$	the complement of Y in X

Number Theory

$a\|b$	a divides b
$a\nmid b$	a does not divide b
(a,b)	greatest common divisor of a and b
n_p	the p-part of n

$n_{p'}$	the p'-part of n

Group Theory

F^*	the multiplicative group of a field F
$\langle X \rangle$	the subgroup generated by X
$\mathbb{Z}_n$	the cyclic group of order n
$G_1 \times G_2$	direct product of G_1 and G_2
$H \lhd G$	H is a normal subgroup of G
$D \rtimes S$	the semidirect product of a normal subgroup D and a subgroup S
$C_G(X)$	the centralizer of X in G
$N_G(X)$	the normalizer of X in G
$N_G(X,Y)$	$= N_G(X) \cap N_G(Y)$
$[x,y]$	$= x^{-1}x^{-1}xy$
$[X,Y]$	$= \langle [x,y] \; x \in X, y \in Y \rangle$
$G' = [G,G]$	the commutator subgroup of G
$\prod_{i\in I} G_i$	direct product of G_i, $i \in I$
$\bigoplus_{i\in I} G_i$	restricted direct product (or direct sum) of G_i, $i \in I$
$\exp G$	the exponent of G
$Z(G)$	the centre of G
g_p	the p-part of g
$g_{p'}$	the p'-part of g
$C\ell(G)$	the set of all conjugacy classes of G
$\mathrm{Syl}_p(G)$	the set of all Sylow p-subgroups of G
$G(x)$	the stabilizer of x
G_x	the orbit of x
$(G:H)$	the index of H in G
$O_{p'}(G)$	the maximal normal p'-subgroup of G
$O_p(G)$	the maximal normal p-subgroup of G

$O_{p',p}(G)$	the image of $O_p(G/O_{p'}(G))$ in G
$GL(n,p)$	the general linear group of degree n over the field $\mathbb{F}_p$ of p elements
$SL(n,p)$	the special linear group of degree n over $\mathbb{F}_p$
$C_G(N/M)$	$= \{g \in G \mid gM \in C_{G/M}(N/M)\}$
$\gamma_n(G)$	the n-th term of the lower central series of G
G^n	$= \langle g^n \mid g \in G\rangle$
d_P	the number of double (P,P)-cosets in G
$M_{n,p}(G)$	the n-th term of the Brauer-Jennings-Zassenhaus series for G
$L_{n,p}(G)$	the n-th term of Lazard series for G
$D_n(FG)$	$= G \cap (1 + I(G)^n)$
$w(\lambda)$	the weight of λ
$\Phi(G)$	the Frattini subgroup of G
S_n	symmetric group of degree n
A_n	alternating group of degree n
$GL(V)$	the group of all nonsingular transformations of the vector space V
$GL(n,R)$	the group of all $n \times n$ nonsingular matrices over R
Aut G	the automorphism group of G
$\mathrm{Aut}_R V$	the automorphism group of an R-module V
$C(p)$	$= \{g_p \mid g \in C \in C\ell(G)\}$
$\ell_p(G)$	the p-length of G
$Qd(p)$	the semidirect product of $\mathbb{Z}_p \times \mathbb{Z}_p$ and $SL(2,p)$
$c(P)$	the nilpotency class of P
D_n	the dihedral group of order 2^n, $n \geq 2$
Q_n	the generalized quaternion group of order 2^n, $n \geq 3$
$S(n)$	the semidihedral group of order 2^n, $n > 3$
$M_n(p)$	$= \langle a,b \; a^{p^{n-1}} = 1,\ b^p = 1,\ b^{-1}ab = a^{1+p^{n-2}} \rangle$
$M(p)$	the extra-special p-group of order p^3 and exponent p, p odd.
$H_1 \underset{G}{\subset} H_2$	H_1 is G-conjugate to a subgroup of H_2

$H_1 =_G H_2$	H_1 and H_2 are G-conjugate

Rings, Modules, and Characters

$Z(R)$	the centre of R
$A \otimes B$	tensor product
char$R$	the characteristic of R
$\bigoplus_{i \in I} R_i$	direct sum
$\prod_{i \in I} R_i$	direct product of rings
FG	the group algebra of G over F
$I(G)$	the augmentation ideal of FG
aug(x)	the augmentation of x
$F^{\alpha}G$	the twisted group algebra of G over F
R°	the opposite ring of R
$M_n(R)$	the ring of all $n \times n$-matrices over R
$J(R)$	the Jacobson radical of a ring R
$J(V)$	the radical of an R-module V
Soc$R$	the socle of a ring R
Soc$V$	the socle of a module V
$\mathrm{Hom}_R(V,W)$	the additive group of all R-homomorphisms $V \longrightarrow W$
V^n	the n-th direct power of V
$c(V)$	the composition length of V
ann(V)	the annihilator of V
$P(V)$	the projective cover of V
$U(R)$	the unit group of a ring R
$L(V)$	the Loewy length of V
$\Omega(V)$	the Heller module of V
V^*	$= \mathrm{Hom}_F(V,F)$
$W^{\perp}$	$= \{f \in V^* \mid f(W) = 0\}$

$S_i(R)$	the i-th socle of R
$S_i(V)$	the i-th socle of V
Supp$x$	support of x
$tr(x)$	trace of x
$V \# W$	the outer tensor product of V and W
$U \otimes_F V$	the inner tensor product of FG-modules U and V
V^G	the induced module
V_H	the restriction of V to FH
1_G or 1_{FG}	the trivial FG-module
V_L	$= L \otimes_F V$
A_L	$= L \otimes_F A$
$Id(X)$	the sum of all two-sided ideals of FG contained in X
${}^gV$	G-conjugate of V
V^*	contragradient module
$i(V,W)$	the intertwinning number
$\dim_F V$	dimension of V over F
inf(V)	inflated module
ext(V)	extension of V
χ_V	the character afforded by V
${}^g\chi$	the character afforded by ${}^gV$
χ^G	the induced character
X^+	$= \sum_{x \in X} x$
$[A,A]$	the commutator subspace of A
$\delta(C)$	defect group of $C \in C\ell(G)$
$\delta(B)$	defect group of a block B
$I_D(G)$	the F-linear span of all C^+ with $C \in C\ell(G)$ and $\delta(C) \subseteq_G D$
$\hat{I}_D(G)$	the F-linear span of all C^+ with $C \in C\ell(G)$ such that $D \supset \delta(C)$

$V_D(G)$	the F-linear span of all C^+ where C is a p-regular class of G with $\delta(C) = D$
$t(G)$	the nilpotency index of $J(FG)$
A^*	$= \mathrm{Hom}_F(A,F)$
FX	the F-linear span of X
${}^{\perp}X$	$= \{a \in A \mid \psi(aX) = 0\}$
$X^{\perp}$	$= \{a \in A \mid \psi(Xa) = 0\}$
$\ell(X)$	$= \{a \in A \mid aX = 0\}$
$r(X)$	$= \{a \in A \mid Xa = 0\}$
$T(A)$	$= \{a \in A \mid a^{p^n} \in [A,A]$ for some $n \geqslant 1\}$
$tr: FG \longrightarrow F$	the trace map
$\omega_G(V)$	obstruction to the extension of V to FG
$\rho_1 \otimes \rho_2$	the tensor product of representations
c	the sum of all p-elements of G including 1
$\mathrm{Ker}\chi_V$	$= \{x \in FG \mid \chi_V(gx) = 0$ for all $g \in G\}$
$\mathrm{Ker}T$	$= \{x \in FG \mid T(gx) = 0$ for all $g \in G\}$
$n(V,W)$	the multiplicity of V as a composition factor of W
$R(FG)$	$= \{a \in FG \mid tr(a^{p^n}) = 0$ for all sufficiently large $n \in \mathbb{N}\}$
$Z_{p'}$	$= \sum_{i=1}^{r} FS_i^+$ where $S_1,\ldots,S_r$ are all p-regular sections of G.
trA	the trace of the matrix A
$\mathrm{Ker}V$	the kernel of the representation afforded by V
$\mathrm{Ker}B$	the kernel of the block B
$T_n(FG)$	$= \{x \in FG \mid x^{p^n} \in [FG,FG]\}$
X^{p^n}	$= \{x^{p^n} \mid x \in X\}$
X^{1/p^n}	$= \{g \in G \mid g^{p^n} \in X\}$
$\Delta(A)$	a complement of $J(A)$ in A
$Z^*(FN)$	$= Z(FG) \cap Z(FN)$
$G(b)$	the inertia group of the block b

e^*	the sum of all G-conjugates of e
$T_{H,S}: C_{FG}(S) \longrightarrow C_{FG}(H)$	the relative trace map
$C = (c_{ij})$	the Cartan matrix of FG
$t(B)$	the nilpotency index of $J(B)$
$t(Z(B))$	the nilpotency index of $J(Z(B))$
b^G	the unique block corresponding to λ^G
$vx(V)$	the vertex of V
$\delta(B)$	the defect group of B
$\delta(e)$	the defect group of e

Cohomology Theory

$Z^2(G,A)$	the group of all A-valued 2-cocycles of G
δt	a coboundary
$B^2(G,A)$	the subgroup of all coboundaries
$H^2(G,A)$	$= Z^2(G,A)/B^2(G,A)$
$\bar{f}$	the cohomology class of $f \in Z^2(G,A)$
$\omega_G(V)$	obstruction cocycle
$\inf\alpha$	$= (\inf\alpha)(x,y) = \alpha(xN,yN)$

Index

U.C.W. ABERYSTWYTH
LIBRARY